Animal Husbandry and Biotechnology

Animal Husbandry and Biotechnology

Animal Husbandry and Biotechnology

Editor: Dominic Fasso

RCALLISTO
REFERENCE

www.callistoreference.com

Callisto Reference,
118-35 Queens Blvd., Suite 400,
Forest Hills, NY 11375, USA

Visit us on the World Wide Web at:
www.callistoreference.com

ISBN: 978-1-63239-845-1 (Hardback)

The publisher's policy is to use permanent paper from mills that operate a sustainable forestry policy. Furthermore, the publisher ensures that the text paper and cover boards used have met acceptable environmental accreditation standards.

Trademark Notice: Registered trademark of products or corporate names are used only for explanation and identification without intent to infringe.

Printed in the United States of America.

Cataloging-in-publication Data

Animal husbandry and biotechnology / edited by Dominic Fasso.
    p. cm.
Includes bibliographical references and index.
ISBN 978-1-63239-845-1
1. Animal biotechnology. 2. Animal culture. 3. Biotechnology. I. Fasso, Dominic.
SF140.B54 A55 2017
636--dc23

Table of Contents

Table of Contents

Preface

Animal husbandry is the art of management and care of farm animals by humans, in which the qualities and behavior are considered to be advantageous to humans. The relationship of animal husbandry and biotechnology is also termed as animal biotechnology by scientists and experts working in this field. From theories to research to practical applications, case studies related to all contemporary topics of relevance to this field have been included in this book. It aims to cover a broad spectrum of topics including the good animal husbandry practices which is beneficial both for animals and for humans. The book is appropriate for students seeking detailed information in this area as well as for experts.

This book unites the global concepts and researches in an organized manner for a comprehensive understanding of the subject. It is a ripe text for all researchers, students, scientists or anyone else who is interested in acquiring a better knowledge of this dynamic field.

I extend my sincere thanks to the contributors for such eloquent research chapters. Finally, I thank my family for being a source of support and help.

Editor

Establishment of *HRAS*^{G12V} Transgenic Medaka as a Stable Tumor Model for *In Vivo* Screening of Anticancer Drugs

Yuriko Matsuzaki[1], **Haru Hosokai**[2], **Yukiyo Mizuguchi**[1], **Shoji Fukamachi**[3], **Atsushi Shimizu**[4], **Hideyuki Saya**[1,5]*

1 Division of Gene Regulation, Institute for Advanced Medical Research, School of Medicine, Keio University, Tokyo, Japan, **2** Division of Material and Biological Sciences, Japan Women's University, Tokyo, Japan, **3** Laboratory of Evolutionary Genetics, Department of Chemical and Biological Sciences, Japan Women's University, Tokyo, Japan, **4** Department of Molecular Biology, School of Medicine, Keio University, Tokyo, Japan, **5** Japan Science and Technology Agency, Core Research for Evolutional Science and Technology, Tokyo, Japan

Abstract

Most targeted anticancer drugs have been identified by screening at the molecular or cellular level *in vitro*. However, many compounds selected by such costly and time-consuming screening do not prove effective against tumors *in vivo*. The development of anticancer drugs would thus be facilitated by the availability of an *in vivo* screening system based on a multicellular organism. We have now established a transgenic line of the freshwater fish medaka in which melanophores (melanocytes) proliferate in a manner dependent on heat shock–induced signaling by a human RAS oncoprotein. The human *HRAS*^{G12V} oncogene was expressed under the control of a melanophore-specific gene promoter in order to allow visualization of tumor growth in live fish maintained in a water tank. The expression of *HRAS*^{G12V} was induced as a result of Cre-mediated recombination by exposure of the fish to a temperature of 37°C for 30 min, given that the Cre gene was placed under the control of a medaka heat shock promoter. One of the stable transgenic lines developed abnormal pigment cell proliferation in the eyes and epidermis with 100% penetrance by 6 months postfertilization. Sorafenib, an inhibitor of RAS signaling, was administered to the transgenic fish and was found both to reduce the extent of melanophore proliferation and to improve survival. The transgenic medaka established here thus represents a promising *in vivo* system with which to screen potential anticancer drugs that target RAS signaling, and this system can readily be adapted for the screening of agents that target other oncogenes.

Editor: Gayle E. Woloschak, Northwestern University Feinberg School of Medicine, United States of America

Funding: The study was supported by Japan Science and Technology Agency, Core Research for Evolutional Science and Technology. The funders had no role in study design, data collection and analysis, decision to publish, or preparation of the manuscript.

Competing Interests: The authors have declared that no competing interests exist.

* E-mail: hsaya@a5.keio.jp

Introduction

Further improvements in cancer treatment will require the development of new molecularly targeted drugs that inhibit the growth and spread of tumors. Most such drugs developed to date were first identified by screening at the cell or molecular level and were subsequently tested in animals, usually mice, before being entered into clinical trials. Although progress in imaging technology has allowed evaluation of tumor growth in live experimental animals, the availability of a system that does not require such imaging would be expected to facilitate the testing of anticancer drugs.

The zebrafish is often adopted as a vertebrate model for chemical or genetic screening in part because it is readily raised in large numbers and its embryos are transparent [1,2]. The medaka is a small egg-laying freshwater fish native to East Asia and has several advantages over zebrafish for such studies. The size of the medaka genome (~800 Mb) is thus about half that of the zebrafish genome, with its draft sequence having been recently published [3], and medaka is able to survive in both cold and warm conditions. Medaka has also proved suitable for toxicology and carcinogenesis studies [4,5]. In Organisation for Economic Co-operation and Development (OECD) test guidelines, the medaka fish was recommended as a model acute (from 1992) or prolonged (from 1984) toxicity test (http://www.oecd-ilibrary.org/environment/oecd-guidelines-for-the-testing-of-chemicals-section-2-effects-on-biotic-systems_20745761).

Mutations in the components of the RAS/RAF/MEK/ERK and RAS/PI3K/PTEN/AKT/mTOR signaling cascades have been detected in many types of human cancer [6]. Given that tumor cells become addicted to the activation of these signaling pathways [7], interruption of such oncogenic signaling has the potential to improve the prognosis of affected cancer patients. RAS and RAS-related signal transduction pathways are thus promising targets for inhibition of the growth of cancer cells. Several effective inhibitors of such signaling components have been examined in clinical trials. One such inhibitor, sorafenib, targets BRAF and has completed phase III clinical trials, having been found to be effective for the treatment of various cancer types including hepatocellular carcinoma, renal cell carcinoma, and melanoma [8]. On the other hand, no clinically effective agents that inhibit the GTP-bound active form of RAS, an upstream regulator of RAF, have been identified, although inhibitors have been developed that block the farnesylation of RAS, which allows the protein to bind to cell membranes [6]. The further development of inhibitors of RAS or RAS pathway components may thus yield new and effective anticancer drugs.

We have now developed a transgenic medaka tumor model suitable for the testing of potential inhibitors of oncogenic signaling. To construct this model, we made use of a constitutively active mutant of human HRAS, HRAS^{G12V}, which has been detected in various tumors at a relatively high frequency. Transgenic zebrafish that express HRAS^{G12V} and develop melanoma [9] or that express KRAS^{G12V} and develop liver tumors [10] were recently established and found to be suitable tumor models for further study. In our medaka model, we made use of the Cre-loxP system and a medaka heat shock promoter to ensure that HRAS^{G12V} is expressed specifically in melanophores (melanocytes) and in a manner dependent on water temperature. The resulting transgenic fish developed readily visible melanoma-like tumors with 100% penetrance at 6 months postfertilization (mpf). We treated the fish with sorafenib as a test of their suitability for screening of drugs that target RAS signaling.

Materials and Methods

Medaka maintenance

Medaka (*Oryzias latipes*) (provided by NBRP Medaka, Aichi, Japan) was maintained according to established protocols (http://www.shigen.nig.ac.jp/medaka/medakabook). All animal experiments were performed in accordance with protocols approved by the Animal Care and Use Committee of Keio University (Permit Number: 12038-0).

Plasmid construction

The medaka *tyr* promoter sequence [11](TYRO_ORYLA in http://medakagb.lab.nig.ac.jp/Oryzias_latipes) was retrieved from the bacterial artificial chromosome (BAC) clone Ola1-014A12(provided by NBRP Medaka) by the polymerase chain reaction (PCR) with a specific primer set containing restriction endonuclease (*Xho*I and *Eco*RI) sites (meTyr-p-F, 5'-AAACT-CGAGTCTGACCAATCTCTTCTGGAAGGCCAGCTG-3'; meTyr-p-R, 5'-TATGAATTCAAGATTCACCCACATCTGT-CCAGACCTCAG-3'). The 3.4-kb PCR product was digested with *Xho*I and *Eco*RI and was then inserted into *Xho*I- and *Eco*RI-digested pTIS10-d-loxGFP-OlBMP4 (Figure S1A). A *Not*I site of the resultant recombinant was replaced to *Bam*HI site using *Bam*HI-linker. The human HRAS^{G12V} cDNA sequence was retrieved from pMX-HRASV12-IRES-EGFP [12] by PCR with a specific primer set containing *Bam*HI sites (HRAS-F, 5'-ATATGGATCCATGACGGAATATAAGCTGGTGGTGGT-GG-3'; HRAS-R1, 5'-ATGCGGATCCTCAGGAGAGCACA-CACTTGCAGCTCAT-3'). The 0.6-kb PCR product was digested with *Bam*HI and then inserted into the *Bam*HI-digested plasmid containing the medaka *tyr* promoter region to yield pTIS10-tyr-loxGFP-HRASV12. The inducible Cre expression plasmid pTIS9-hBhsp70creCherry (Figure S1B) contains the DNA sequence for a fusion protein of Cre recombinase and the red fluorescent protein mCherry under the control of the medaka *hsp70* promoter. This plasmid also contains the *TagBFP* gene which encodes blue fluorescent protein (Evrogen, Moscow, Russia) under the control of the promoter of a medaka embryonic globin (hemoglobin a-0) gene [13] as a marker for selection of transgenic lines. This marker allowed the selection

of embryos harboring the *cre* transgene on the basis of the blue fluorescence of their blood without heat treatment. The sequences of all recombinant constructs were verified.

Microinjection of plasmid DNA

We preferred to use a black wild-type medaka line for microinjection of plasmids so as to allow visualization of melanophore proliferation. The black wild-type strain HB11A(-provided by NBRP Medaka) was difficult to maintain in our facility, however, so we crossed this line with the propagative H1bRHi line (orange-red variety) and selected highly pigmented offspring (HBK11). Fertilized eggs were obtained from this hybrid, and one-cell-stage embryos were injected with approximate 5 nl of a mixture containing 50 pg of recombinant DNA, 0.5× I-*Sce*I buffer, 0.025 U of I-*Sce*I, 0.5× Yamamoto's medium, and 0.05% phenol red. Embryos were raised and outcrossed with wild-type fish to identify potential transgenic animals.

Extraction of genomic DNA

Genomic DNA was extracted from the fin of adults or from the whole body of larvae. The tissue samples were incubated for 10 min at room temperature in 200 μl of a lysis buffer containing 20 mM Tris-HCl (pH 8.0), 5 mM EDTA, 400 mM NaCl, 0.3% SDS, and proteinase K (200 μg/ml; Sigma-Aldrich, St. Louis, MO, USA), after which the lysates were heated for 5 min at 95°C and subjected directly to PCR.

Selection of *Tg(tyr:HRAS^{G12V})/Tg(hsp:cre)* double-transgenic medaka

A homozygous *Tg(tyr:HRAS^{G12V})* transgenic strain was crossed with a heterozygous *Tg(hsp:cre)* strain, and resulting Cre-positive embryos were selected on the basis of the blue fluorescence of their blood.

Heat treatment

Tg(tyr:HRAS^{G12V})/Tg(hsp:cre) double-transgenic medaka at 1 or 4 week postfertilization (wpf) was subjected to heat treatment by incubation at 37°C for 30 minutes. The medaka *hsp70* promoter used for construction of the Cre plasmid also responds to stressors other than heat treatment, but this was not a problem for the present study. We studied both double-transgenic medaka showing melanophore hyperplasia without heat treatment as well as fish in which such hyperplasia was induced by heat treatment at 1 or 4 wpf.

Extraction of total RNA and RT

Total RNA was isolated from various medaka tissues with the use of an RNeasy Mini Kit (Qiagen, Hilden, Germany) and with RNase-Free DNase Set (Qiagen) to eliminate genomic DNA from the samples. The extracted RNA was subjected to reverse transcription (RT) with the use of a Transcriptor First Strand cDNA Synthesis Kit (Roche Applied Science, Rotkreuz, Switzerland).

PCR assay conditions

The presence of the recombination-specific sequence in *Tg(tyr:HRAS^{G12V})* medaka was confirmed by PCR with a forward primer targeted to the enhanced green fluorescent protein (EGFP) gene (EGFP-F, 5'-AGCTGGACGGCGACGTAAACGG-3') and a reverse primer targeted to human *HRAS* (HRAS-R2, 5'-ACCACCACCAGCTTATATTCCGTC-3'). The amplification protocol included an initial incubation at 95°C for 10 min; 35 cycles of 94°C for 20 s, 57°C for 30 s, and 72°C for 40 s; and a final incubation at 72°C for 6 min. The reaction mixture contained 10 μl of 2× Ampdirect plus (Shimadzu, Kyoto, Japan), 0.5 μl of genomic DNA, 0.5 μM of each primer, and 0.5 U of BIOTAQ HS DNA Polymerase (Bioline, London, UK) in a total volume of 20 μl. The presence of human *HRAS* cDNA among RT products was determined with a human *HRAS* forward primer (HRAS-F, 5′-AGCTGGACGGCGACGTAAACGG-3′) and a human *HRAS* reverse primer (HRAS-R2). Intact medaka *hras* cDNA was amplified as a positive control with the primers Ol-HRAS-F (5′-ATTGGCCCGGTCCTACGGTA-3′) and Ol-HRAS-R (5′-GCGACAGCTCATGCAGTCCTGA-3′). Medaka cytoplasmic actin mRNA was examined as an RNA quality control (OlCA1-F, 5′-GGGAAATTGTCCGTGACATC-3′; OlCA1-R, 5′-GACTCATCGTACTCCTGCTT-3′). The amplification protocol for human *HRAS* included an initial incubation at 95°C for 2 min; 35 cycles of 95°C for 10 s, 57°C for 20 s, and 72°C for 20 s; and a final incubation at 72°C for 6 min. The amplification protocol for medaka *hras* comprised an initial incubation at 95°C for 2 min; 35 cycles of 95°C for 10 s, 55°C for 20 s, and 72°C for 20 s; and a final incubation at 72°C for 6 min. The amplification protocol for medaka cytoplasmic actin cDNA included an initial incubation at 95°C for 2 min; 30 cycles of 95°C for 10 s, 57°C for 20 s, and 72°C for 30 s; and a final incubation at 72°C for 6 min. These latter three reaction mixtures contained 2 μl of 10×LA buffer (Takara, Kyoto, Japan), 0.5 μl of cDNA, 0.5 μM of each primer, and 0.5 U of LA Taq DNA Polymerase (Takara) in a total volume of 20 μl.

Histochemical staining

The entire fish body was fixed in 4% paraformaldehyde, embedded in paraffin, serially sectioned, and stained with hematoxylin-eosin. Histochemical images were captured with a Biorevo BZ-9000 microscope (Keyence, Osaka, Japan).

Drug administration

For protocol 1 of drug administration, *Tg(tyr:HRAS^{G12V})/Tg(hsp:cre)* double-transgenic medaka manifesting abnormal proliferation of melanophores at 4 mpf were divided into two groups and either sorafenib (BAY-43-9006; Cayman Chemical, Michigan, USA)or dimethyl sulfoxide (DMSO) vehicle was added to the water at a final concentration of 0.1 μM and 0.1%, respectively. The fish were exposed to these agents for four 24-h periods over 2 weeks. For protocol 2, fish at 2 mpf were divided into two groups and either sorafenib or DMSO was added to the water at 0.3 μM or 0.1%, respectively. The fish were exposed to these agents for 24-h periods four times over 4 weeks. The geometric mean of the maximum plasma concentration (C_{max}) of sorafenib is ∼5 mg/l (∼10 μM) in patients taking the drug orally [14], and the median inhibitory concentration of sorafenib for wild-type BRAF is 22 nM [15]. In preliminary experiments, we found that administration of 4 μM sorafenib for 48 h was lethal to medaka, whereas three of five fish died after administration of 2 μM sorafenib five times for 48 h over the course of a month. We therefore selected sorafenib concentrations of 0.1 and 0.3 μM for our experiments. Fish were photographed before and after drug administration.

Photography

Fish were photographed with a Leica DFC300FX camera, and images were processed with Leica Application Suite version 2.7.1.R1 software (Leica Microsystems, Wetzlar, Germany). The black color associated with melanophore hyperplasia in the fish body was quantified with the use of LAS-3000 Mini and Multi Gauge version 3.0 software (Fujifilm, Tokyo, Japan).

Statistical analysis

Differences in mean melanophore hyperplasia and in survival were evaluated with Student's *t* test and the log-rank test, respectively. A *P* value of <0.05 was considered statistically significant.

Results

Transgenic medaka expressing human *HRAS^{G12V}*

We first attempted to establish a transgenic medaka line constitutively expressing human *HRAS^{G12V}* under the control of the promoter for the medaka tyrosinase gene (*tyr*), which is expressed specifically in cells having melanosome such as melanophores and retinal pigment epithelial cells, but the animals died before achieving adulthood. We therefore developed a conditional gene expression system in which *HRAS^{G12V}* is expressed in response to the induction of Cre recombinase by heat shock (Figure 1). The transgene vector pTIS10-tyr-loxGFP-HRASV12 was thus designed to express *HRAS^{G12V}* under the control of the *tyr* promoter in response to Cre-mediated recombination. The *cre* gene was placed under the control of the medaka *hsp70* (*heat shock protein 70*) gene promoter. Cre-mediated recombination removes the EGFP gene from the *HRAS^{G12V}* construct, allowing expression of the human *HRAS^{G12V}* to be induced when the fish are exposed to a temperature of 37°C for 30 min. The Cre protein irreversibly excises nucleotides contain sequences of the EGFP after the heat treatment, and the human *HRAS^{G12V}* is constitutively expressed.

The plasmid pTIS10-tyr-loxGFP-HRASV12 was injected into ∼200 embryos at the one-cell stage, which were then raised to adult fish. The resulting mature animals were outcrossed with wild-type fish to detect DNA fragments derived from pTIS10-tyr-loxGFP-HRASV12. We obtained four stable transgenic lines harboring the *HRAS^{G12V}* construct. These four *Tg(tyr:HRAS^{G12V})* strains were crossed with the *Tg(hsp:cre)* strain (harboring the *cre* gene under the control of the medaka *hsp70* promoter). The resulting *HRAS^{G12V}* heterozygotes harboring *cre* were expected to express *HRAS^{G12V}* in response to heat treatment. However, the heat shock element in medaka is activated not only by heat but also by various other environmental stressors such as ammonia, nitrate, nitrite, a change in pH, or the proliferation of bacteria, mold, or parasites in the water tank. We therefore maintained the *Tg(tyr:HRAS^{G12V})* and *Tg(hsp:cre)* strains independently, and the two strains were crossed only when offspring were required for study.

The double-transgenic offspring resulting from crosses between each of the four *Tg(tyr:HRAS^{G12V})* lines and the *Tg(hsp:cre)* strain manifested abnormal proliferation of melanophores in the eyes and epidermis with different frequencies after heat treatment (Table 1). The prevalence of melanophore proliferative lesions (MPLs) at 6 mpf (with heat treatment applied at 4 wpf) thus ranged from 12.5 to 100%. These differences in penetrance may be attributable to differences in *HRAS^{G12V}* expression level resulting from differences in transgene copy number or insertion sites. All subsequent experiments were performed with offspring of *Tg(tyr:HRAS^{G12V})* line 1 and *Tg(hsp:cre)*, which developed MPLs with 100% penetrance at 6 mpf.

We examined expression of *HRAS^{G12V}* in the double-transgenic medaka at 8 mpf (with heat treatment at 4 wpf) by RT-PCR analysis (Figure 2). Expression of *HRAS^{G12V}* was detected prominently in brain, eye, and liver as well as to a lesser extent in fin and heart of *Tg(tyr:HRAS^{G12V})/Tg(hsp:cre)* medaka but not in wild-type fish.

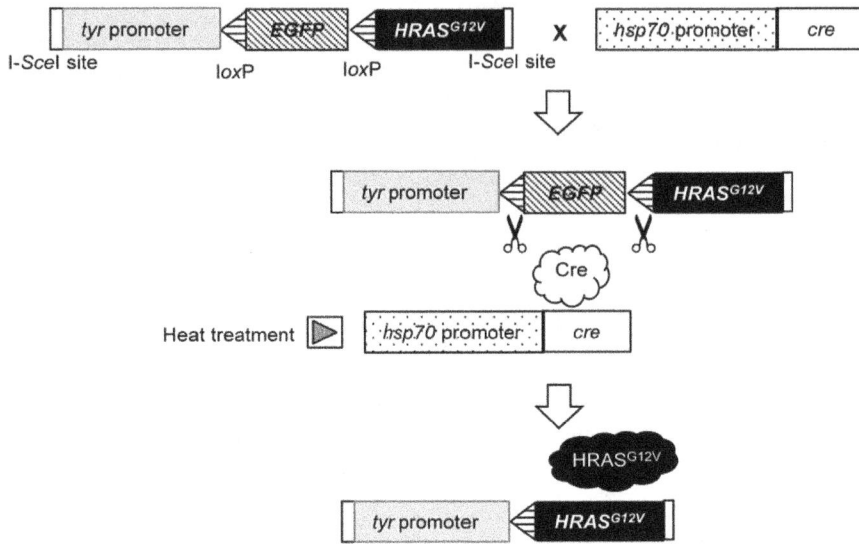

Figure 1. Generation of transgenic medaka expressing human *HRAS^G12V*. A plasmid containing the human *HRAS^G12V* cDNA downstream of the medaka tyrosinase gene (*tyr*) promoter as well as the *EGFP* gene positioned between *loxP* sequences was constructed for the generation of transgenic medaka. These transgenic fish were crossed with another transgenic line harboring the gene for Cre recombinase (*cre*) under the control of the medaka *hsp70* promoter. The resulting double-transgenic animals express Cre recombinase after exposure to heat shock, resulting in excision of the *EGFP* gene and expression of *HRAS^G12V*.

Characteristics of *Tg(tyr:HRAS^G12V)*/*Tg(hsp:cre)* double-transgenic fish

Among eight *Tg(tyr:HRAS^G12V)*/*Tg(hsp:cre)* double-transgenic medaka born on the same day and subjected to heat treatment at 1 wpf, five fish manifested hyperproliferation of pigment cells around the eyes by 8 wpf (Figure 3). All of the fish had developed melanophore-derived infiltrative tumors by 20 wpf, with one animal that developed large MPLs having died. Small ectopic lesions potentially reflecting metastases of the primary lesions were also observed in seven fish (Figure 3). Melanophore infiltration became accelerated after 28 wpf, with all animals having died by 43 wpf (the average additional life span of these transgenic medaka survived more than 28 wpf was 9.2±0.90 wpf). The external surface of the fish manifested large black areas corresponding to melanophores in the skin, eyes, gills, abdomen, and bones. Hematoxylin-eosin staining also revealed abnormal proliferation of melanophores in the transgenic fish (Figure 4). Consistent with the fact that melanin synthesis normally takes place in the eyes (retinal pigment epithelium) and the melanophore layer that covers internal organs, the abnormal proliferation was apparent predominantly around the eyes and in the abdomen. The melanophore layers of the transgenic fish were thicker and manifested abnormal proliferation (Figure 4 C–E) compared with those of wild-type animals (Figure 4 B). The transgenic medaka also variously manifested the presence of an infiltrative mass in the eye (Figure 4 F) as well as melanophore infiltration into vertebrae, heart, muscle, gill, kidney, and the digestive duct (Figure 4 G–M). Cells of an atrium and ventricle of one animal were mostly replaced by melanophores (Figure 4 H). A magnified image of the digestive duct of one fish also showed melanophore infiltration into the stroma around the duct (Figure 4 M). These various pathological findings are similar to those of invasive melanoma in humans.

MPL incidence and survival in *Tg(tyr:HRAS^G12V)*/*Tg(hsp:cre)* medaka

Both *Tg(tyr:HRAS^G12V)*/*Tg(hsp:cre)* and *Tg(hsp:cre)* medaka had a survival rate of <50% between the fertilized egg stage and 4 wpf

Table 1. Penetrance of melanophore proliferative lesions (MPLs) among *Tg(tyr:HRAS^G12V)*/*Tg(hsp:cre)* double-transgenic medaka derived from four different *Tg(tyr:HRAS^G12V)* lines (Tg lines 1 to 4).

	Tg line 1	Tg line 2	Tg line 3	Tg line 4
Total number observed	16	11	12	8
Number with MPLs	16	6	5	1
Penetrance (%)	100	54.5	41.7	12.5

The fish were subjected to heat treatment by incubation at 37°C for 30 minutes at 4 wpf and examined at 6 mpf.

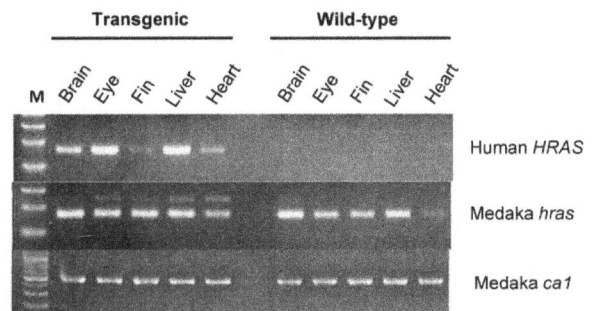

Figure 2. Double-transgenic medaka expresses human *HRAS^G12V*. Total RNA isolated from the indicated tissues both of *Tg(tyr:HRAS^G12V)*/*Tg(hsp:cre)* double-transgenic medaka at 8 mpf (with heat treatment at 4 wpf) and of wild-type medaka was subjected to RT-PCR analysis of human *HRAS* and medaka *hras*. Transcripts of the medaka cytoplasmic actin gene (*ca1*) were examined as an internal control of RNA quality. Lane M, DNA size markers.

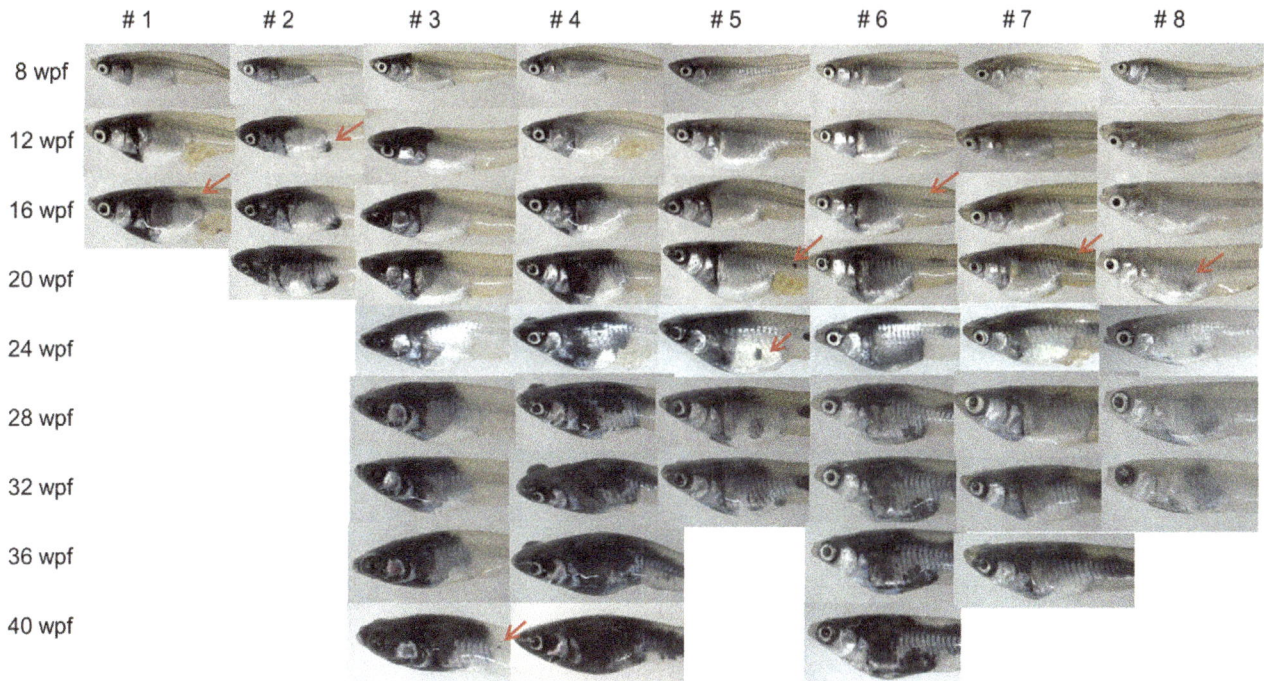

Figure 3. Characteristics of *Tg(tyr:HRAS^{G12V})/Tg(hsp:cre)* **medaka.** Age-dependent changes in eight double-transgenic fish were monitored after heat treatment at 1 wpf. Five of the eight fish manifested hyperproliferation of pigment cells around the eyes by 8 wpf. All of the fish had developed MPLs with characteristics of infiltrative tumors by 20 wpf, and all had died by 43 wpf. Arrows indicate the first appearance of metastasis-like lesions.

when they were subjected to heat treatment at 1 wpf, whereas the corresponding value for wild-type fish was >80%. This finding suggested that Cre expression may have a toxic effect on the fish. We therefore decided to examine double-transgenic fish that survived beyond 4 wpf. Less that 40% of *Tg(tyr:HRAS^{G12V})/Tg(hsp:cre)* medaka manifested MPLs at 4 wpf, whereas >80% of the fish had developed MPLs at 12 wpf (Figure 5). The death rate increased after 16 wpf, and all of the fish had developed MPLs by 28 wpf. These findings suggested that the double-transgenic medaka developed MPLs by 6 mpf were suitable for screening of antitumor drugs that inhibit RAS signaling.

Treatment of double-transgenic fish with sorafenib

Given that *Tg(tyr:HRAS^{G12V})* homozygotes have a low viability, we were not able to use them as parents for the generation of offspring for large-scale screening. Instead, we used the offspring of *Tg(tyr:HRAS^{G12V})/Tg(hsp:cre)* intracrosses and subjected them to heat treatment at 4 wpf. Fish with obvious melanophore hyperplasia were selected at 4 mpf and were divided into two groups, one of which was treated with 0.1 μM sorafenib ($n = 14$) and the other with DMSO as a control ($n = 13$) (Figure 6 A). Sorafenib is a multikinase inhibitor that targets BRAF, c-KIT receptor, VEGFR and PDGFR. Among them, it has substantial activity to inhibit BRAF, a downstream kinase of RAS.

The animals were photographed 7 days before (day −7) and 29 days after (day +29) the first drug administration. Three fish of the control group died before day +29, whereas all of the sorafenib-treated fish survived.

We quantified the black areas caused by melanophore hyperplasia in the fish body by image analysis and calculated the fold change from day −7 to day +29 (Figure 6 B), obtaining average values of 1.04 for the 14 sorafenib-treated fish and 1.26 for the 10 control fish ($P = 0.015$, Student's t test). Furthermore, the

survival rate for the sorafenib-treated group was significantly greater ($P = 0.0267$, log-rank test) than that for the control group (Figure 6 C). At 110 days after the first drug administration, the survival rate for the sorafenib-treated group was thus 35.7% (5/14), whereas that for the control group was only 7.7% (1/13). We performed a second trial with fish at 2 mpf treated with 0.3 μM sorafenib ($n = 15$) or DMSO ($n = 15$) (Figure S2 A, B) and calculated the fold change in melanophore hyperplasia between day −4 and day +31, obtaining values of 1.65 and 1.79, respectively ($P = 0.066$). At 170 days after the first drug administration, the survival rate for the sorafenib-treated group was 33.3% (5/15), whereas that for the control group was 0% (0/15), a statistically significant difference ($P = 0.0484$) (Figure S2 C). These results thus showed that sorafenib inhibited MPL growth and increased survival time in the double-transgenic medaka.

Discussion

We obtained four transgenic medaka lines, *Tg(tyr:HRAS^{G12V})*, that harbor human *HRAS^{G12V}*, which encodes a constitutively active mutant of HRAS. The *Tg(tyr:HRAS^{G12V})/Tg(hsp:cre)* double-transgenic animals derived from a cross between *Tg(tyr:HRAS^{G12V})* and *Tg(hsp:cre)* medaka express *HRAS^{G12V}* in response to heat treatment.

The promoter of zebrafish *mitf* (microphthalmia transcription factor gene) has often been used to control the expression of human oncogenes such as *HRAS* or *BRAF* in melanocytes in transgenic zebrafish models [16,17,18,19]. However, *mitf* gene is expressed in cell types other than melanocytes, the expression of oncogenes under the *mitf* gene promoter may cause diverse toxic effects on the fishes. We therefore chose the medaka *tyr* promoter for our model of melanophore hyperplasia, given that tyrosinase is a key enzyme in the melanin biosynthetic pathway, catalyzing the conversion of tyrosine to dopaquinone. The *tyr* gene is expressed

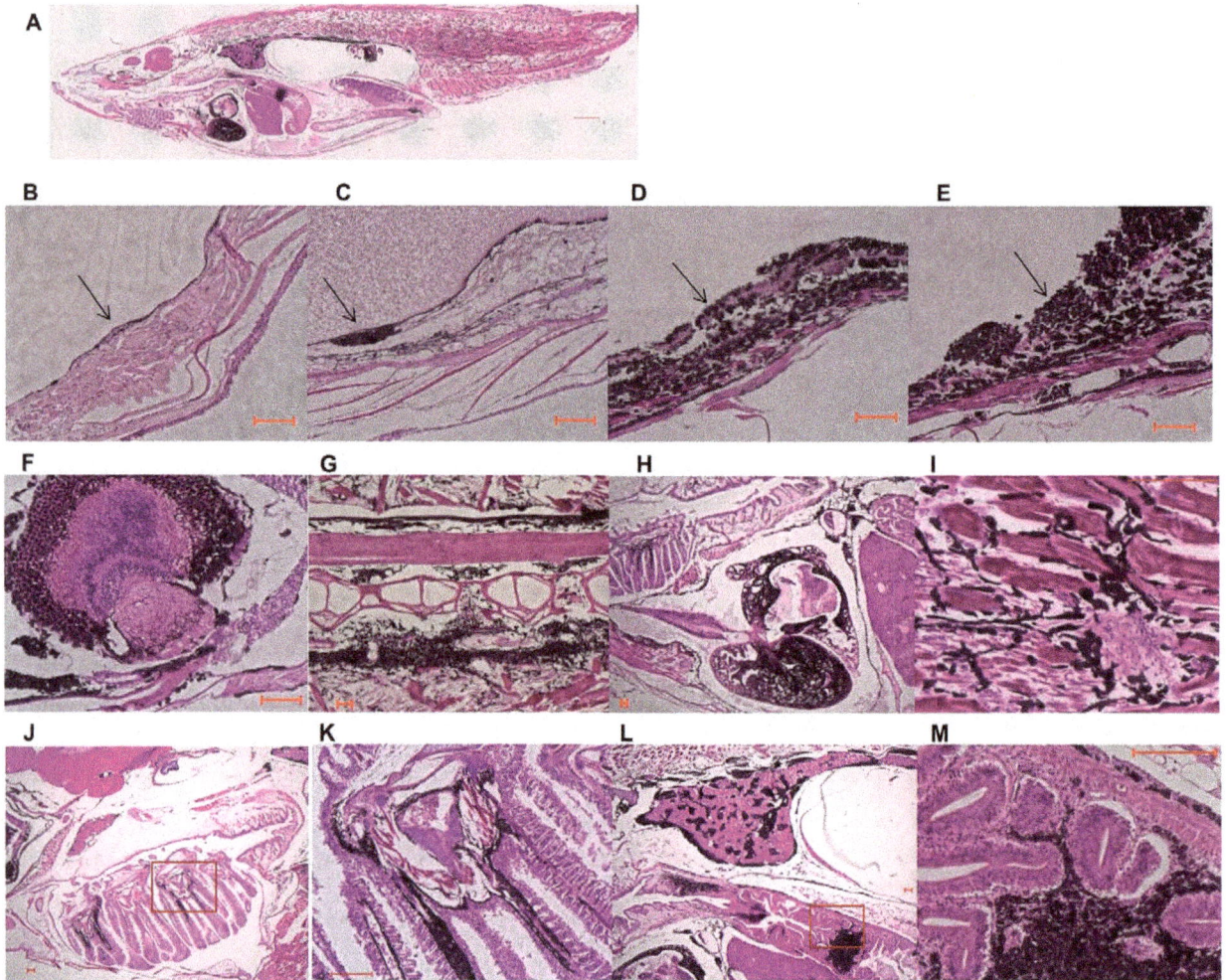

Figure 4. Hematoxylin-eosin staining of *Tg(tyr:HRAS^{G12V})/Tg(hsp:cre)* **medaka.** (A) Overall image of fish #7 in Figure 3. (**B**)– (**E**), Melanophore layers (arrows) of a wild-type adult fish (**B**) as well as of fish #7 (**C**), #6 (**D**), and #4 (**E**) in Figure 3. (**F**)– (**J**), Eye, bone and spinal cord, heart, muscle, and gill, respectively, of fish #7 in Figure 3. (**K**) Enlarged image of the boxed region in (**J**). (**L**) Kidney and digestive duct of fish #7 in Figure 3. (**M**) Enlarged image of the boxed region in (**L**). Scale bars: 1 mm (**A**) and 0.1 mm (**B**–**M**).

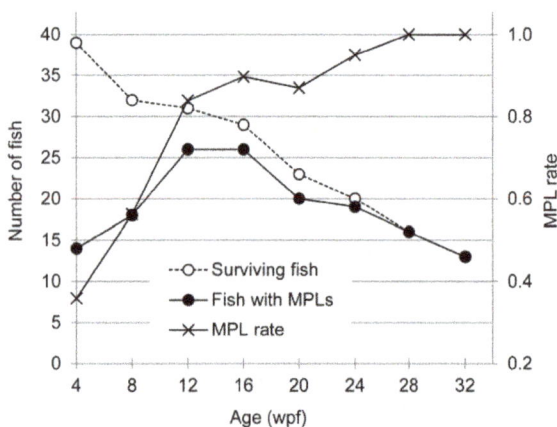

Figure 5. Age dependence of MPL incidence and survival in *Tg(tyr:HRAS^{G12V})/Tg(hsp:cre)* **medaka.** The transgenic fish were subjected to heat treatment at 1 wpf and then monitored for the development of MPLs and death. A single monitoring was performed with 39 *Tg(tyr:HRAS^{G12V})/Tg(hsp:cre)* medaka at 4 wpf.

specifically during melanocyte differentiation, and the promoter of the mouse *tyr* gene has been used to achieve melanocyte-specific expression of human oncogenes in several mouse models [20]. Given that constitutive expression of active HRAS under the control of *tyr* gene promoter resulted in poor viability of the transgenic medaka due to toxicity caused by mutant HRAS, we employed a heat shock inducible system. An inducible *KRAS^{G12V}* transgenic zebrafish model for liver tumorigenesis was previously reported [10]. However, since their system requires a steroid (mifepristone) to induce the *KRAS^{G12V}* gene, the drug screening is done in the presence of mifepristone and, thereby, the effects are based on the combination of a candidate drug and mifepristone. Our system using a heat shock-inducible melanophore-specific expression of *HRAS^{G12V}* has an advantage over those previous model systems for drug screening. Our transgenic medaka may not be sufficient for a preclinical model for cancer therapy but provide a good tool to assess the efficacy of selected candidate drugs.

Larvae of *Tg(hsp:cre)* medaka showed a high mortality rate, with >50% of the animals dying before 4 wpf. Cre toxicity has been described previously [21,22], with Cre expression having been found to inhibit the proliferation of cultured cells as well as to

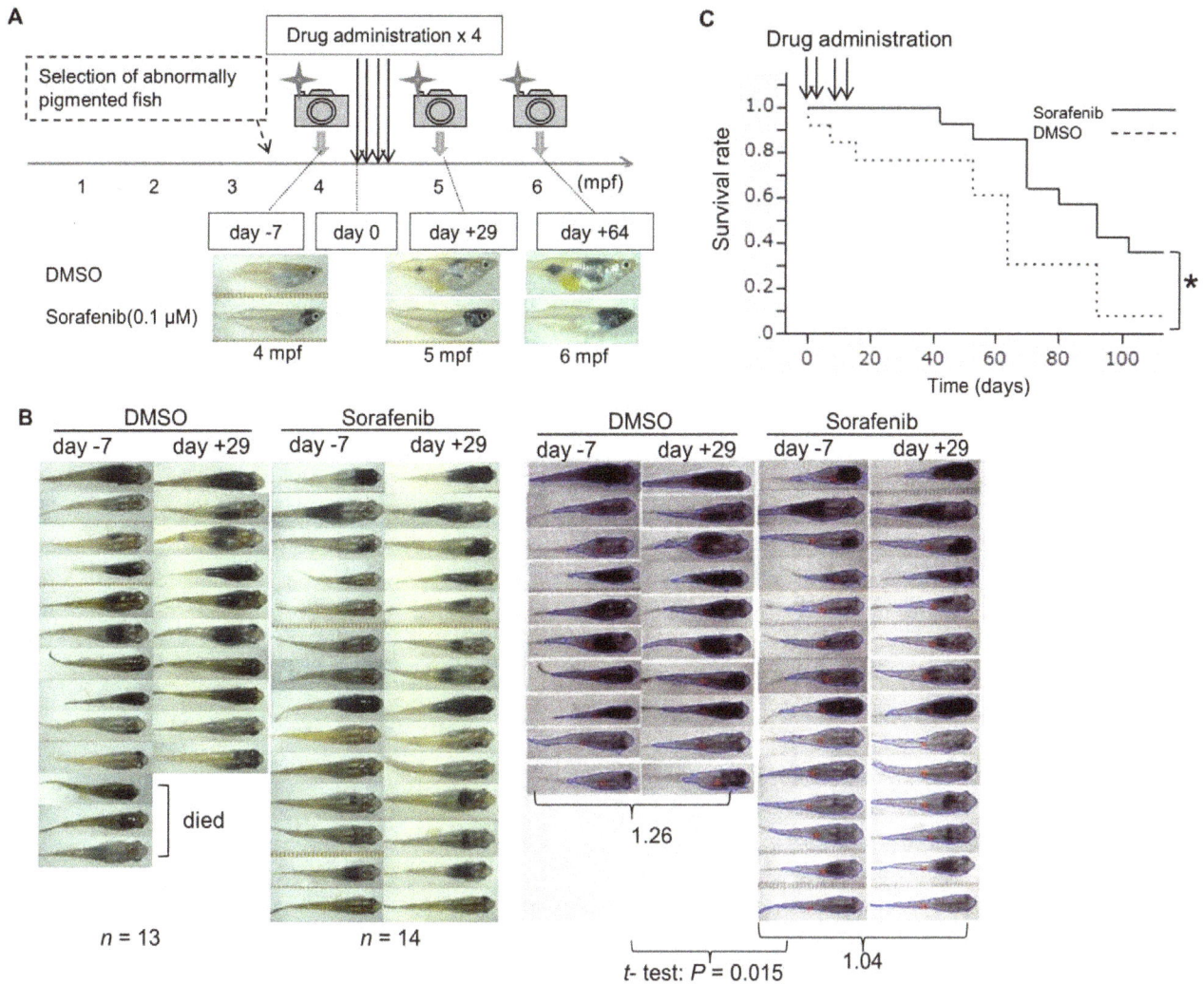

Figure 6. Effect of sorafenib (0.1 μM) treatment on melanophore hyperplasia and overall survival in *Tg(tyr:HRAS^G12V^)/Tg(hsp:cre)* medaka. (**A**) Schedule of drug administration. (**B**) Photos taken from the dorsal side of all fish at 7 days before (day –7) and 29 days after (day +29) the first drug administration. The area of MPLs in the fish body was measured based on the set of captured images on the right, and the average fold change from day –7 to day +29 was calculated for each group. (**C**) Kaplan-Meier survival curves for the sorafenib-treated and control (DMSO-treated) groups were generated from the experiment shown in (**B**). Fish with obvious melanophore hyperplasia were divided into two groups, one of which was treated with 0.1 μM sorafenib (*n* = 14) and the other with DMSO as a control (*n* = 13). *P = 0.0267 (log-rank test).

induce numerous chromosomal aberrations or sister chromatid exchanges. These effects are thought to result from DNA damage caused by the recombinase activity of Cre at cryptic *loxP* sites in the genome. The *Tg(hsp:cre)* medaka established in the present study should thus be preserved in the form of frozen sperm in order to prevent the accumulation of DNA damage.

We performed image analysis to quantify the black areas corresponding to melanophore hyperplasia in the double-transgenic medaka. Such analysis revealed that the extent of MPLs was significantly reduced by sorafenib treatment, and this effect was associated with a significant increase in overall survival. Although we cannot rule out the possibility that the photographs analyzed do not accurately reflect the actual extent of melanophore hyperplasia, we believe that this approach is reliable. The tumor-like black tissue composed of proliferating melanophores appeared in a manner dependent on *HRAS^G12V^* activation and expanded with age. The *Tg(tyr:HRAS^G12V^)/Tg(hsp:cre)* medaka established in the present study thus represents a promising model

system for the screening of inhibitors of RAS signaling *in vivo*. Given that MPL expansion was found to be dependent on RAS activation and that melanophores are readily observed by visual inspection of live fish, our system is potentially applicable to the development of *in vivo* tumor models based on the activation of other oncogenes.

The cause of death for transgenic medaka is not obvious. But, it is possible that the reduction of feeding activity is a main cause because there was marked infiltration of tumor-like cells into digestive tract of the dead medaka. Our data showed that sorafenib inhibited MPL growth and increased survival time in the double-transgenic medaka. However, it should be noted that the tumors are in the medaka background and, thereby, it could impact therapeutic response. In addition, the effect of sorafenib on MPL expansion in our second protocol of drug administration (Figure S2) was not as marked as that in the first protocol (Figure 6). Sorafenib has substantial activity to inhibit BRAF, a downstream kinase of RAS, and it therefore might not be expected to block the RAS/PI3K/PTEN/AKT/mTOR signaling pathway. Tumor growth observed in the presence of sorafenib might thus be due to signaling by this latter pathway in an adaptive response. The combination of an inhibitor of RAS/PI3K/PTEN/AKT/mTOR signaling with sorafenib might thus be expected to be more effective for inhibition of tumor growth than sorafenib alone.

Finally, in addition to the potential of $Tg(tyr:HRAS^{G12V})/Tg(hsp:cre)$ medaka for *in vivo* screening of potential anticancer drugs that target RAS signaling, the establishment of a melanoma-like cell line from these animals may prove useful for studies on the mechanism of *HRAS*-dependent tumorigenesis *in vitro*.

Supporting Information

Figure S1 Map of plasmids used. (A) pTIS10-d-loxGFP-OlBMP4: The promoter of medaka *tyr* was inserted between the *Xho*I and *Eco*RI recognition sites replacing the promoter of the medaka desmin gene, and the human $HRAS^{G12V}$ cDNA sequence was inserted between the *Bam*HI and *Not*I recognition sites replacing the medaka *bmp4* sequence. **(B)** pTIS9-hBhsp70cre-Cherry: The DNA sequence for a fusion protein of Cre recombinase and the red fluorescent protein mCherry under the control of the medaka *hsp70* promoter. This plasmid also contains the TagBFP gene under the control of the promoter of a medaka embryonic globin gene.

Figure S2 Effect of sorafenib (0.3 μM) treatment on melanophore hyperplasia and overall survival in $Tg(tyr:HRAS^{G12V})/Tg(hsp:cre)$ medaka. (A) Schedule of drug administration. **(B)** Photos taken from the dorsal side of all fish at 4 days before (day −4) and 31 days after (day +31) the first drug administration. The area of MPLs in the fish body was measured based on the set of captured images on the right, and the average fold change from day −4 to day +31 was calculated for each group. **(C)** Kaplan-Meier survival curves for the sorafenib-treated and control (DMSO-treated) groups were generated from the experiment shown in **(B)**. Fish with obvious melanophore hyperplasia were divided into two groups, one of which was treated with 0.3 μM sorafenib ($n = 15$) and the other with DMSO as a control ($n = 15$). *$P = 0.0484$ (log-rank test).

Acknowledgments

We thank Dr. Pier P. Pandolfi (Harvard Medical School, Boston, MA) for the $HRAS^{G12V}$ gene; Dr. Shinji Makino (Keio University School of Medicine, Tokyo, Japan) for fish facility; Ikuyo Ishimatsu for technical assistance in the histological analyses; Kuniko Arai for secretarial assistance.

Author Contributions

Conceived and designed the experiments: Y. Matsuzaki HS. Performed the experiments: Y. Matsuzaki HH Y. Mizuguchi AS. Analyzed the data: Y. Matsuzaki HH HS SF HS. Contributed reagents/materials/analysis tools: Y. Matsuzaki HH Y. Mizuguchi AS SF. Wrote the paper: Y. Matsuzaki HS.

References

1. Kaufman CK, White RM, Zon L (2009) Chemical genetic screening in the zebrafish embryo. Nat Protoc 4: 1422–1432.
2. White RM, Cech J, Ratanasirintrawoot S, Lin CY, Rahl PB, et al. (2011) DHODH modulates transcriptional elongation in the neural crest and melanoma. Nature 471: 518–522.
3. Kasahara M, Naruse K, Sasaki S, Nakatani Y, Qu W, et al. (2007) The medaka draft genome and insights into vertebrate genome evolution. Nature 447: 714–719.
4. Klaunig JE, Barut BA, Goldblatt PJ (1984) Preliminary studies on the usefulness of medaka, Oryzias latipes, embryos in carcinogenicity testing. Natl Cancer Inst Monogr 65: 155–161.
5. Hawkins WE, Walker WW, Fournie JW, Manning CS, Krol RM (2003) Use of the Japanese medaka (Oryzias latipes) and guppy (Poecilia reticulata) in carcinogenesis testing under national toxicology program protocols. Toxicol Pathol 31 Suppl: 88–91.
6. Chappell WH, Steelman LS, Long JM, Kempf RC, Abrams SL, et al. (2011) Ras/Raf/MEK/ERK and PI3K/PTEN/Akt/mTOR inhibitors: rationale and importance to inhibiting these pathways in human health. Oncotarget 2: 135–164.
7. Weinstein IB, Joe AK (2006) Mechanisms of disease: Oncogene addiction–a rationale for molecular targeting in cancer therapy. Nat Clin Pract Oncol 3: 448–457.
8. Flaherty KT, Hodi FS, Fisher DE (2012) From genes to drugs: targeted strategies for melanoma. Nat Rev Cancer 12: 349–361.
9. Santoriello C, Gennaro E, Anelli V, Distel M, Kelly A, et al. (2010) Kita driven expression of oncogenic HRAS leads to early onset and highly penetrant melanoma in zebrafish. PLoS One 5: e15170.
10. Nguyen AT, Emelyanov A, Koh CH, Spitsbergen JM, Parinov S, et al. (2012) An inducible kras(V12) transgenic zebrafish model for liver tumorigenesis and chemical drug screening. Dis Model Mech 5: 63–72.
11. Inagaki H, Bessho Y, Koga A, Hori H (1994) Expression of the tyrosinase-encoding gene in a colorless melanophore mutant of the medaka fish, Oryzias latipes. Gene 150: 319–324.
12. Kobayashi Y, Shimizu T, Naoe H, Ueki A, Ishizawa J, et al. (2011) Establishment of a choriocarcinoma model from immortalized normal extravillous trophoblast cells transduced with HRASV12. Am J Pathol 179: 1471–1482.
13. Maruyama K, Yasumasu S, Iuchi I (2002) Characterization and expression of embryonic and adult globins of the teleost Oryzias latipes (medaka). J Biochem 132: 581–589.
14. Abou-Alfa GK, Schwartz L, Ricci S, Amadori D, Santoro A, et al. (2006) Phase II study of sorafenib in patients with advanced hepatocellular carcinoma. J Clin Oncol 24: 4293–4300.
15. Wilhelm SM, Carter C, Tang L, Wilkie D, McNabola A, et al. (2004) BAY 43-9006 exhibits broad spectrum oral antitumor activity and targets the RAF/MEK/ERK pathway and receptor tyrosine kinases involved in tumor progression and angiogenesis. Cancer Res 64: 7099–7109.
16. Dovey M, White RM, Zon LI (2009) Oncogenic NRAS cooperates with p53 loss to generate melanoma in zebrafish. Zebrafish 6: 397–404.
17. Michailidou C, Jones M, Walker P, Kamarashev J, Kelly A, et al. (2009) Dissecting the roles of Raf- and PI3K-signalling pathways in melanoma formation and progression in a zebrafish model. Dis Model Mech 2: 399–411.
18. Schartl M, Wilde B, Laisney JA, Taniguchi Y, Takeda S, et al. (2010) A mutated EGFR is sufficient to induce malignant melanoma with genetic background-dependent histopathologies. J Invest Dermatol 130: 249–258.
19. Patton EE, Widlund HR, Kutok JL, Kopani KR, Amatruda JF, et al. (2005) BRAF mutations are sufficient to promote nevi formation and cooperate with p53 in the genesis of melanoma. Curr Biol 15: 249–254.
20. Larue L, Beermann F (2007) Cutaneous melanoma in genetically modified animals. Pigment Cell Res 20: 485–497.
21. Loonstra A, Vooijs M, Beverloo HB, Allak BA, van Drunen E, et al. (2001) Growth inhibition and DNA damage induced by Cre recombinase in mammalian cells. Proc Natl Acad Sci U S A 98: 9209–9214.
22. Schmidt-Supprian M, Rajewsky K (2007) Vagaries of conditional gene targeting. Nat Immunol 8: 665–668.

2

Knockdown of Myostatin Expression by RNAi Enhances Muscle Growth in Transgenic Sheep

Shengwei Hu[1,2], Wei Ni[1,2], Wujiafu Sai[1], Ha Zi[1], Jun Qiao[1], Pengyang Wang[1], Jinliang Sheng[1], Chuangfu Chen[1,2]*

1 College of Animal Science and Technology, Shihezi University, Shihezi, China, **2** Key Laboratory of Agrobiotechnology, Shihezi University, Shihezi, China

Abstract

Myostatin (MSTN) has been shown to be a negative regulator of skeletal muscle development and growth. MSTN dysfunction therefore offers a strategy for promoting animal growth performance in livestock production. In this study, we investigated the possibility of using RNAi-based technology to generate transgenic sheep with a double-muscle phenotype. A shRNA expression cassette targeting sheep MSTN was used to generate stable shRNA-expressing fibroblast clones. Transgenic sheep were further produced by somatic cell nuclear transfer (SCNT) technology. Five lambs developed to term and three live lambs were obtained. Integration of shRNA expression cassette in three live lambs was confirmed by PCR. RNase protection assay showed that the shRNAs targeting MSTN were expressed in muscle tissues of three transgenic sheep. MSTN expression was significantly inhibited in muscle tissues of transgenic sheep when compared with control sheep. Moreover, transgenic sheep showed a tendency to faster increase in body weight than control sheep. Histological analysis showed that myofiber diameter of transgenic sheep M17 were bigger than that of control sheep. Our findings demonstrate a promising approach to promoting muscle growth in livestock production.

Editor: Se-Jin Lee, Johns Hopkins University School of Medicine, United States of America

Funding: This work was supported by grants from the 973 Program (2010CB530200), The National Natural Science Foundation of China (NSFC) (31201800, 31260534) and the Bingtuan Twelve Five-year Special Projects (2012BB051). The funders had no role in study design, data collection and analysis, decision to publish, or preparation of the manuscript.

Competing Interests: The authors have declared that no competing interests exist.

* E-mail: chencf1962@yahoo.com

Introduction

Myostatin (MSTN), a member of the transforming growth factor beta (TGF-β) superfamily, functions as a negative regulator of skeletal muscle development and growth. MSTN gene knockout mice have about a doubling of skeletal muscle weights throughout the body as a result of a combination of muscle fiber hyperplasia and hypertrophy [1]. Natural gene mutations of MSTN have also been reported in some cattle breeds [2–4], sheep [5], dogs [6] and human [7]. These animals show a double-muscled phenotype of dramatically increased muscle mass, and still viable and fertile [2–7]. These findings have suggested that strategies capable of disrupting MSTN function may be applied to enhance animal growth performance.

RNA interference (RNAi) is a process of sequence-specific, post transcriptional gene silencing, which has been used to analyse gene function and develop novel animal models [8]. Several groups, including us, produced transgenic RNAi mice which showed a gene knockdown phenotype that was functionally similar to gene knockout [9,10]. The ability to generate RNAi transgenics is especially significant for livestock animal for which stem cells have yet to be derived. Recently, transgenic RNAi zebrafish with MSTN knockdown were successfully produced, which resulted in giant- or double-muscle in transgenic zebrafish [11,12]. These findings suggest that animal growth performance could be improved by knocking down MSTN using RNAi technology.

In this study, we investigated the possibility of using RNAi technology to generate transgenic sheep with a double-muscle phenotype. Our results showed that shRNA targeting MSTN effectively inhibited endogenous MSTN expression in transgenic sheep. Moreover, transgenic sheep showed a tendency to faster increase in body weight than normal controls. Our study provide a promising approach for the production of transgenic double-muscle animals.

Materials and Methods

Ethics Statement

All experiments involving animals were conducted under the protocol (SU-ACUC-08032) approved by the Animal Care and Use Committee of Shihezi University. All sheep involved in this research were raised and breed followed the guideline of Animal Husbandry Department of Xinjiang, P.R.China.

Plasmid Construction

shRNAs targeting sheep MSTN were reported in our previous reports [13]. The shMSTN3 (5′-CAAAGATGCTATAAGACAA-3′) targeted the first exon of sheep MSTN gene. The shMSTN3 expression cassette was amplified using primers U6-F (5′-AGT AGT TGC CAG GAT CAC CGT GC-3′) and U6-R (5′-CCT AAT GAG TG A GCT AAC TCA CA-3′), and then cloned into BamH I and SwaI site of ploxP vector to generate ploxP-shMSTN3 (Figure 1).

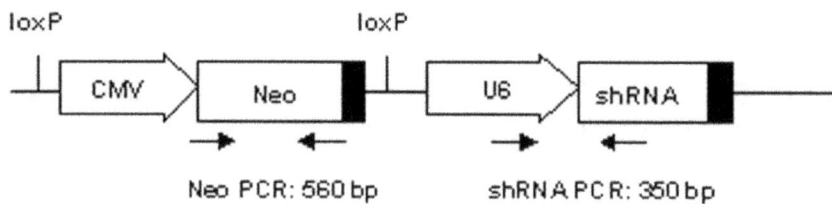

Figure 1. Schematic illustration representing ploxP-shMSTN3 vector used in this study. Loxp: recombination site of Cre recombinase for bacteriophage P1; CMV: CMV promoter; Neo: neomycin gene; U6: polymerase III U6-RNA gene promoter, shRNA: short hairpin RNA. Arrowhead indicated localization of the primers specific for shRNA expression cassette and Neo gene. The size of the PCR amplicons is indicated.

Cell Culture, Transfection and Selection

China Merino sheep fibroblast cells (SF) were isolated and cultured as previously described [14]. 2×10^5 cells per well were seeded in 12-well plate and cultured in fresh DMEM without antibiotics to achieve 80–90% confluency on the day of transfection. The cells were then transfected with 1.8 μg/well of ploxP-shMSTN3 vectors using Lipofectamine 2000 (Invitrogen) according to the manufacturer's protocol. After 48 h transfection, cells were split into 100 mm dish at an appropriate dilution for G418 selection (500 μg/ml). Single G418-resistant colonies were obtained after 14 days of selection.

Construction of Transgenic Sheep by Somatic Nuclear Transfer

Transgenic fibroblast colonies (TF-s2 and TF-s19) were used to construct transgenic sheep. Sheep nuclear transfer (NT) was performed as described reports [15,16]. Briefly, ovaries were collected from a local abattoir and transported to our laboratory within 4 h after slaughter. Cumulus-oocyte complexes (COCs) were aspirated from 2 to 5 mm follicles with PBS (containing 5% FCS) by using a 5 ml syringe fitted with a 20-gauge needle. The COCs were cultured in maturation medium at 38.5°C in a humidified atmosphere for 22 h. Cumulus cells were removed by exposure to 1 mg/mL hyaluronidase. Oocytes with a first polar body were enucleated manually in the presence of 7.5 μg/ml of cytochalasin B. A single intact donor cell was injected into the perivitelline space and placed adjacent to the recipient cytoplasm. After injection, reconstructed embryos were transferred into an electrical fusion chamber overlaid with Zimmermann's fusion medium. Cell fusion was induced with two direct current pulses (1.0 kV/cm, 60 μs, 1 s apart). All fused reconstructed embryos were further activated in 5 μM ionomycin for 4 min, followed by exposure to 1.9 mM 6-dimethylaminopurine in synthetic oviduct fluid with amino acids (SOFaa) for 4 h. Following activation, reconstructed embryos were transferred and cultured in SOFaa. 429 embryos at the 2- to 4-cell stages were surgically transferred into 45 synchronized recipient ewes (7–12 embryos per recipient). Pregnancies were monitored by ultrasound scanning using a transabdominal linear probe every two weeks until days 90. Normal control sheep were produced by normal sexual reproduction.

PCR Analysis

Genomic DNA was isolated from ear biopsy of each lamb using TIANamp Genomic DNA kit (Tiangen Biotech, China). Transgene integration was identified by two independent PCR assays. PCR was performed on 20 ng of genomic DNA using specific primers (Figure 1) for Neo (Neo-F: 5'-ATT CGG CTA TGA CTG GGC ACAC-3'; Neo-R: 5'-CCA GAA AAG CGG CCA TTT TCCA-3') and for shRNA expression cassette (U6-F: 5'-AGT AGT TGC CAG GAT CAC CGT GC-3'; U6-R: 5'-CCT

AAT GAG TGA GCT AAC TCACA-3'). PCR reaction consisted of 95°C for 4 min; 30 cycles at 95°C for 35 s, 58°C for 30 s and 72°C for 40 s; an extension at 72°C for 10 min. PCR products were analyzed by gel electrophoresis.

shRNA Expression Analysis

Biceps brachii muscle tissues were obtained by surgical biopsy from transgenic and control sheep. shRNA expression was identified by RNase protection assay. Small RNAs were isolated from muscle tissues by using the mirVana miRNA isolation Kit (Ambion). The ^{32}P-labeled RNA probes (29 nt) were generated by in vitro transcription using the mirVana miRNA Probe Construction Kit (Ambion). A DNA oligonucleotide under T7 promoter control, which was reverse complement with the target RNA, was used for in vitro transcription. Protection assay carried out using the mirVana miRNA Detection Kit (Ambion) according to kit manufacturer's instructions. The protected RNA probe was detected by autoradiography.

Western Blot Analysis

Protein extracts were taken from transgenic and control biceps brachii muscles, and then were analyzed by western blot analysis as previously described [13]. A primary rabbit anti-MSTN antibody (1:1000 dilution) (Sigma-Aldrich) and anti-GAPDH (Sigma-Aldrich) were used in the western blotting. Band intensities were estimated by densitometry and corrected by the respective GAPDH band intensities.

Real-time RT-PCR Analysis

Total RNAs were isolated from biceps brachii muscle tissues using Trizol (Invitrogen) according to the manufacturer's instructions. The primer sets were used for amplifying MHCII (MCF: 5'-AAC GAT ACC GTG GTT GGG-3'; MCR: 5'-CAG CAC GCC GTT ACA CCT-3'), MyoD (MDF: 5'-GCG GAT GAC TTC TAT GAT GACC-3'; MDR: 5'-GTG CAG CGT TTG AGC GTCT-3'), Myogenin (MF: 5'-AAG CGG AAG TCG GTG TCTG-3'; MR: 5'-ATT GTG GGC ATC TGT AGG GT-3') and Smad2 (sMF: 5'-GGG ATG GAA GAA GTC AGC-3'; sMR: 5'-ATG GGA CAC CTG AAG ACG-3'). Real-Time PCR (Stratagene MX3000P) was carried out using SYBR Green (TaKaRa Biotech, Dalian) following the manufacturer's protocol. The PCR thermal cycle reactions consisted of denaturation at 95°C for 4 min followed by 45 cycles at 95°C for 15 s, 60°C for 1 min. Cycle threshold (Ct) values were normalized to GAPDH, and comparative quantification of mRNA was done by the ΔΔCt method.

Growth Evaluation and Muscle Histological Analysis

Control sheep (n = 3, male) and three transgenic sheep (male) were weighted at 1, 20, 40, 60 and 90 days after birth. The transgenic M17 showed a fastest increase in body weight and was used for further muscle histological analysis. Biceps brachii muscles were isolated by surgical biopsy from transgenic M17 and control sheep (n = 3). The muscle tissues were fixed in formol for 10 h followed by routine paraffin sectioning and Haematoxilin/Eosin staining. Myofiber diameter and number were determined as previously described [17]. Five representative images (one central and four peripheral) were captured from muscle sections, compounding to 0.7 mm^2. Myofiber diameter and number were measured with Scion Image software at three independent pictures. Fiber size-distribution diagram was generated and showed as percentage of 500 fibers analyzed. Mean diameter of sheep myofibers represented average value of 500 fibers.

Statistical Analysis

Expression levels of MHCII, MyoD, Myogenin and Smad2 were analysed statistically by one-way ANOVA and Tukey's test. Data were representative of three independent experiments performed in triplicate. Differences were considered significant at $P<0.05$.

Results

Production of Cloned Sheep

The positive cells were selected as donor cells for constructing cloned embryos by SCNT. The cleavage and blastocyst development rates were 74.4% (131/176) and 14.2% (17/120), respectively. In total, 429 embryos at the 2- to 4-cell stages were transferred to the oviducts of 45 recipient sheep (Table 1). Pregnancy rates were 24.4% (11/45) at forty-five days after embryo transfer. Six recipients were spontaneously aborted and five recipients developed to term. Five lambs were named M17, M18, M21, M23 and M24, respectively. M21 and M24 died 1 h and 22 days after birth, respectively. M17, M18 and M23 have survived for more than six months.

Evaluation of MSTN Knockdown in Transgenic Sheep

Genomic DNAs were prepared from the ear tissues of transgenic sheep M17, M18 and M23. PCR was used to detect the presence of the transgene. All the three cloned sheep were positive for the shRNA expression cassette and for Neo gene, respectively (Figure 2A, 2B).

To determine whether shRNAs targeting MSTN were expressed in muscles of transgenic sheep, shRNA expression was analyzed by RNase protection assay. As shown in the Figure 2C, shRNA expression was detected in the muscles of three transgenic sheep, whereas not in the muscles of control sheep. shRNA expression was also observed in transgenic fibroblast cells TF-s2 and TF-s19 as positive controls.

To further confirm whether the shRNA inhibited MSTN expression in vivo, biceps brachii muscles were obtained by surgical biopsy from M17, M18, M23 and three control sheep, and MSTN precursor expression was detected by western blot analysis (Figure 2D). Compared with average expression levels

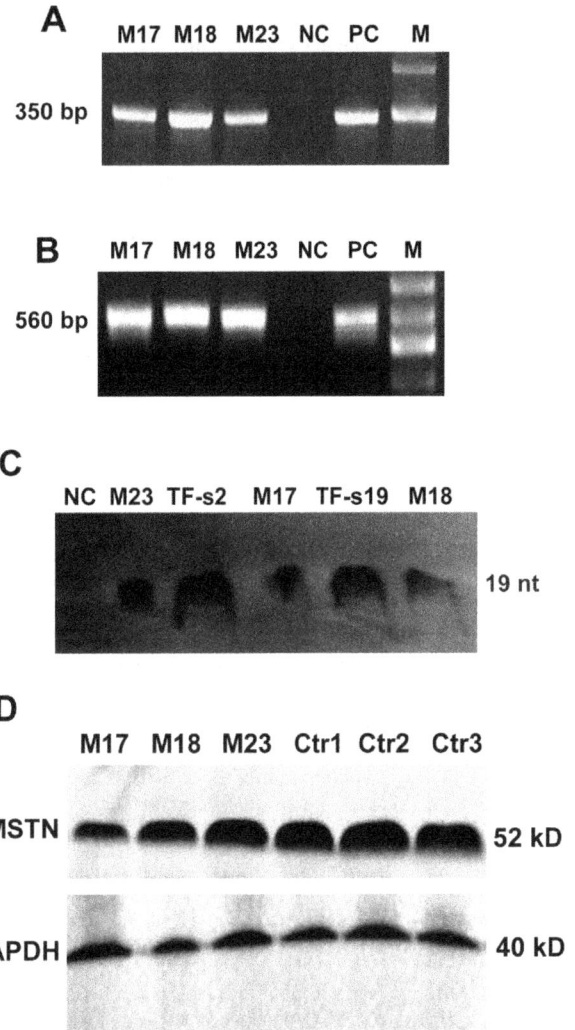

Figure 2. Transgene integration and expression analysis in three cloned lambs. PCR analysis using primers specific for the shRNA expression cassette (A) and Neo (B). (C) Expression of the shRNA targeting MSTN in muscle tissues of transgenic sheep M17, M18 and M23. Negative control (NC): control sheep; positive control (PC): ploxP-shMSTN3 vector. M: Maker; TF-s2: Transgenic cell clone TF-s2; TF-s19: Transgenic cell clone TF-s19. (D) Western blot analysis of MSTN protein expression in muscle tissues of transgenic sheep M17, M18 and M23 and control sheep (Ctr1, Ctr2 and Ctr3).

of the controls, MSTN expression was reduced by about 61.12%, 31.32% and 10.11% in the transgenic M17, M18 and M23, respectively. This results suggested that MSTN expression was silenced by shRNA in transgenic sheep.

Table 1. Summary of SCNT results.

No. embryos transferred	No. recipient sheep	Pregnancy rates	Liveborn lamb	No. lamb alive over six month	Cloning efficiency (%)
429	45	24.4% (11/45)	5	3	1.2%

Expression of MSTN-related Genes in Transgenic Sheep

We further determined whether shRNA-mediated MSTN knockdown affected expression of MSTN-related genes in transgenic sheep. Expression levels of MHCII, MyoD, Myogenin and Smad2 were analysed by Real-time RT-PCR. As indicated in Figure 3, expression levels of MHCII, MyoD and Myogenin in transgenic M17 and M18 were significantly higher than that of the controls. Expression levels of Smad2 were decreased in transgenic sheep, but without significant difference compared with the controls.

Body Weight and Muscle Growth of Transgenic Sheep

Based on body weight records, transgenic sheep showed a faster increase tendency in body weight than control sheep during the 90 days after birth (Figure 4). The 3-month-old M17 (25.42 kg), M18 (23.52 kg) and M23 (22.93 kg) were heavier than control sheep (21.22 kg, n = 3), respectively. Daily weight gain of M17 was 0.23 kg/d, when compared with 0.18 kg/d of control sheep.

Muscle tissues of M17 were isolated by surgical biopsy and sectioned for morphometric analysis. H&E staining showed larger myofiber sizes in the transgenic sheep (Figure 5A). Digital morphometric analysis of 500 myofibers from M17 muscle tissues revealed an increase in myofiber diameter compared with control sheep (Figure 5B). More myofibers were in the large myofiber diameter range (35–65 μm) compared with controls (Figure 5B). Mean diameter of M17 myofibers was 58 μm, whereas that of control myofibers was 47 μm (Figure 5C). No significant difference was found in the myofiber number per an unbiased 0.7 mm^2 frame between M17 (912±13) and controls (869±16; n = 3). These results suggested that MSTN suppression may caused hypertrophy in transgenic sheep.

Discussion

MSTN dysfunction resulted in dramatic increase of animal muscle mass due to hypertrophy and hyperplasia of muscle fibers [18–20]. Inhibition of MSTN expression by gene knockout or RNAi could promote the muscle growth and meat production of livestock animals. Due to the low efficiency of gene target in livestock animal somatic cells, RNAi is an ideal alternative for production of MSTN-knockdown transgenic livestock. In addition, there are some disadvantages to double-muscled cattle, including the reduction in female fertility, lower viability of offsprings, and delay in sexual maturation [21,22]. The partial silencing of MSTN by RNAi in livestock may weaken some negative effect of null mutations, while at the same time increasing meat performance [23]. In the present study, we generated MSTN-knockdown transgenic sheep with increased muscle phenotype by RNAi and SCNT technology. To our knowledge, this is the first report in which MSTN-knockdown transgenic sheep were generated and showed increased muscle phenotype.

In our previous study, a shRNA expression cassette targeting MSTN had been constructed and induced significant decrease of MSTN expression by 90% in sheep fibroblasts [13]. This shRNA expression cassette included a sheep U6 and shRNA construct. Here this shRNA expression cassette was used to generate stable shRNA-expressing fibroblasts, and subsequently generate transgenic sheep by SCNT technology. shRNA expression was confirmed in muscles of transgenic sheep by RNase protection assay, indicating that the sheep U6 promoter could efficiently drive shRNA expression for gene silencing in vivo. However, in fact sheep M17 and M18 were derived from the same transgenic cell clones (TF-s19), but some variation in shRNA and MSTN expression levels were observed in muscles of M17 and M18

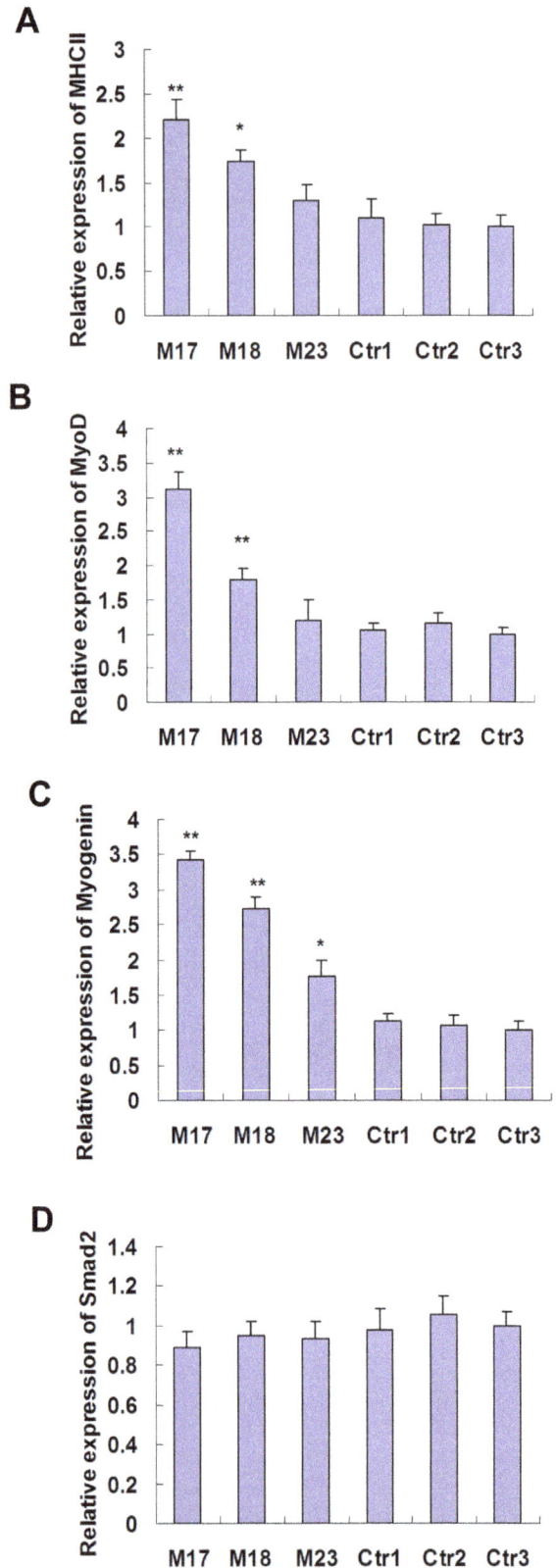

Figure 3. Expression levels of MHCII, MyoD, Myogenin and Smad2 in the muscle tissues of transgenic sheep. mRNA expression of MHCII (A), MyoD (B), Myogenin (C) and Smad2 (D) were determined using Real-time RT-PCR and normalized to GAPDH expression. * P<0.05, ** P<0.01.

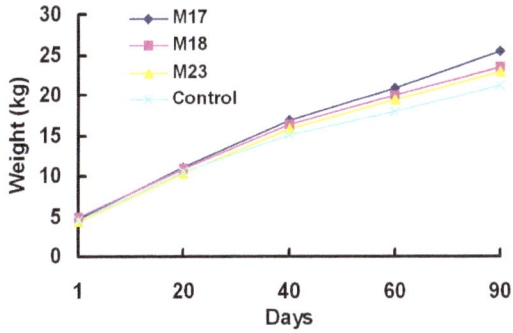

Figure 4. Body weight of transgenic sheep and controls. M17, M18, M23 and three control sheep were weighted at 1, 20, 40, 60 and 90 days after birth. Control values are average weights of three control sheep.

(Figure 2C, 2D). The possible reason may include: (1) Sheep U6 promoter may be influenced by epigenetic modifications as well as CMV and SV40 promoter hypermethylated in transgenic animal [24–26]. The hypermethylated CMV promoter probably resulted in silencing of transgene in animal cells [27,28]; (2) Transgenic cell clones used for nuclear transfer may be mixed cell clones with different copies of shRNA gene integration. Different copies of shRNA resulted in variation of shRNA expression levels in transgenic sheep.

We transferred 429 embryos to the oviducts of 45 recipient sheep. Pregnancy rates were 24.4% at forty-five days after embryo transfer. Five lambs were born from five recipients. The clone efficiency was approximately 1.2%, consistent with other reports [29]. Transgenic sheep M21 and M24 died after birth, which may

be as a result of developmental abnormalities of cloned fetus and placenta [30]. In addition, we did not see any abnormal development or behaviour in three live transgenic sheep up to at least six months of age. These results suggested that MSTN-knockdown may not affect development of cloned embryo and lamb.

Transgenic sheep M17, with largest reduction in MSTN expression levels (Figure 2D), showed a tendency to faster increase in body weight.than control sheep. Myofiber mean diameter of M17 was bigger than that of the non-transgenic controls, suggesting that MSTN-knockdown possibly caused hypertrophy of M17 myofiber. However, owing to our morphometric analysis of only one muscle biopsy from M17, further research is required to clarify whether MSTN-knockdown in sheep causes myofiber hyperplasia or hypertrophy and the effect on fiber types. The increased muscle mass in MSTN null mice and transgenic mice expressing high levels of the propeptide, follistatin, or a dominant negative form of activin receptor type IIB (ActRIIB) resulted from both hyperplasia and hypertrophy [17–19]. In contrast, missense mutant MSTN caused hyperplasia but not hypertrophy in mouse muscles, whereas dominant negative MSTN produced muscle hypertrophy without hyperplasia [31,32]. These results indicated that this hypertrophic response and lack of hyperplasia may be due to the incomplete inhibition of MSTN gene expression.

We acknowledge that the present study only contained analysis of three transgenic sheep and growth performance of transgenic sheep need to be further observed. Work is currently underway to generate more MSTN-knockdown sheep by reclone transgenic sheep M17. In addition, application of transgenic animals for meat production is forbidden in some countries (e.g. European Union) and meat from cloned animals must be carefully assessed before entering the food chain. In summary, we successfully generated MSTN-knockdown transgenic sheep by RNAi and SCNT. Our

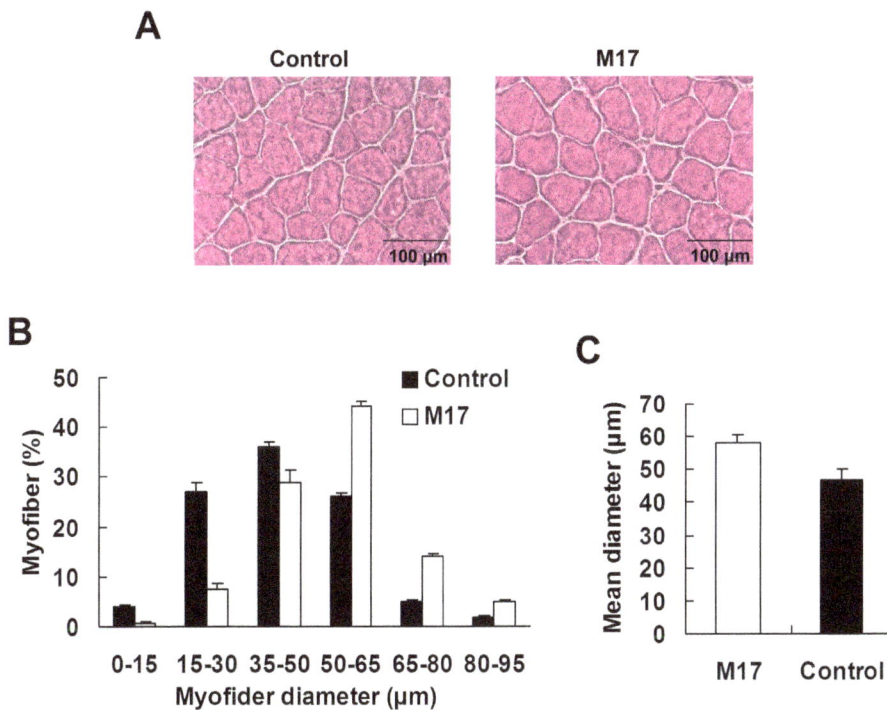

Figure 5. Morphometric analysis of muscles. (A) H&E staining displayed myofiber hypertrophy in muscles of M17 compared with control sheep. (B) Distribution of muscle fiber sizes in M17 and control sheep. A total of 500 fibers from each sheep were measured. (C) Mean myofiber diameter for M17 and controls.

findings demonstrate a promising approach to promoting muscle growth in livestock production.

Acknowledgments

We express our thanks to all members in Chuangfu chen's laboratory who contributed to the MSTN Knockdown project. We also acknowledge Ren Meng, Zhirui He and Jixing Guo for technical help, and Aide Lijiang for collection of sheep tissue samples.

Author Contributions

Conceived and designed the experiments: SH CC JQ. Performed the experiments: SH WN WS HZ PW JS. Analyzed the data: SH CC JQ. Contributed reagents/materials/analysis tools: PW JS. Wrote the paper: SH CC.

References

1. McPherron AC, Lawler AM, Lee SJ (1997) Regulation of skeletal muscle mass in mice by a new TGF-ß superfamily member. Nature 387: 83–90.
2. McPherron AC, Lee SJ (1997) Double muscling in cattle due to mutations in the myostatin gene. Proc Natl Acad Sci USA 94: 12457–12461.
3. Kambadur R, Sharma M, Smith TPL, Bass JJ (1997) Mutations in myostatin (GDF8) in double-muscled Belgian Blue and Piedmontese cattle. Genome Res 7: 910–915.
4. Grobet L, Martin LJR, Poncelet D, Pirottin D, Brouwers B, et al. (1997) A deletion in the bovine myostatin gene causes the double-muscled phenotype in cattle. Nat Genet 17: 71–74.
5. Clop A, Marcq F, Takeda H, Pirottin D, Tordoir X, et al. (2006) A mutation creating a potential illegitimate microRNA target site in the myostatin gene affects muscularity in sheep. Nat Genet 38: 813–818.
6. Mosher DS, Quignon P, Bustamante CD, Sutter NB, Mellersh CS, et al. (2007) A mutation in the myostatin gene increases muscle mass and enhances racing performance in heterozygote dogs. PLoS Genet 3: 779–786.
7. Schuelke M, Wagner KR, Stolz LE, Hübner C, Riebel T, et al. (2004) Myostatin mutation associated with gross muscle hypertrophy in a child. N Engl J Med 350: 2682–2688.
8. Meister G, Tuschl T (2004) Mechanisms of gene silencing by double-stranded RNA. Nature 431: 343–349.
9. Hemann MT, Fridman JS, Zilfou JT, Hernando E, Paddison PJ, et al. (2003) An epi-allelic series of p53 hypomorphs created by stable RNAi produces distinct tumor phenotypes in vivo. Nat Genet 33: 396–400.
10. Wang PY, Jiang JJ, Li N, Sheng JL, Ren Y, et al. (2010) Transgenic mouse model integrating siRNA targeting the foot and mouth disease virus. Antivir Res 87: 265–268.
11. Lee CY, Hu SY, Gong HY, Chen MHC, Lu JK, et al. (2009) Suppression of myostatin with vector-based RNA interference causes a double-muscle effect in transgenic zebrafish. Biochem Biophys Res Commun 387: 766–771.
12. Acosta J, Carpio Y, Borroto I, Gonzalez O, Estrada MP (2005) Myostatin gene silenced by RNAi show a zebrafish giant phenotype. J Biotechnol 119: 324–331.
13. Hu S, Ni W, Sai W, Zhang H, Cao X, et al. (2011) Sleeping Beauty-mediated knockdown of sheep myostatin by RNA interference. Biotechnol Lett 33: 1949–1953.
14. Hu S, Ni W, Hazi W, Zhang H, Zhang N, et al. (2011) Cloning and functional analysis of sheep U6 promoters. Anim Biotechnol 22: 170–174.
15. Sai WJF, Peng XR, Li XC, An ZX, Zheng YM, et al. (2008) Effects of ovine sperm extract on the development competence of homogeneous and heterogeneous embryos reconstructed by nuclear transfer. Acta Vet Zootech Sinica 39: 1343–1348.
16. Hu S, Ni W, Chen C, Sai W, Hazi W, et al. (2012) Comparison between the effects of valproic acid and trichostatin a on in vitro development of sheep somatic cell nuclear transfer embryos. Journal of Animal and Veterinary Advances 11: 1868–1872.
17. Haidet AM, Rizo L, Handy C, Umapathi P, Eagle A, et al. (2008) Long-term enhancement of skeletal muscle mass and strength by single gene administration of myostatin inhibitors. Proc Natl Acad Sci USA 105: 4318–4322.
18. Lee SJ, McPherron AC (2001) Regulation of myostatin activity and muscle growth. Proc Natl Acad Sci USA 98: 9306–9311.
19. Lee SJ (2007) Quadrupling muscle mass in mice by targeting TGF-ß signaling pathways. PLoS ONE 2: e789.
20. Lee SJ (2008) Genetic analysis of the role of proteolysis in the activation of latent myostatin. PLoS ONE 3: e1628.
21. Bellinge RH, Liberles DA, Iaschi SP, O'Brien PA, Tay GK (2005) Myostatin and its implications on animal breeding: A review. Anim Genet 36: 1–6.
22. Arthur PF (1995) Double muscling in cattle: A review. Aust J Agric Res 46: 1493–1515.
23. Tessanne K, Golding MC, Long CR, Peoples MD, Hannon G, et al. (2012) Production of transgenic calves expressing an shRNA targeting myostatin. Mol Reprod Dev 79: 176–185.
24. Mehta AK, Majumdar SS, Alam P, Gulati N, Brahmachari V (2009) Epigenetic regulation of cytomegalovirus major immediateearly promoter activity in transgenic mice. Gene 428: 20–24.
25. Kong Q, Wu M, Huan Y, Zhang L, Liu H, et al. (2009) Transgene expression is associated with copy number and cytomegalovirus promoter methylation in transgenic pigs. PloS ONE 4: e6679.
26. Duan B, Cheng L, Gao Y, Yin FX, Su GH, et al. (2012) Silencing of fat-1 transgene expression in sheep may result from hypermethylation of its driven cytomegalovirus (CMV) promoter. Theriogenology 78: 793–802.
27. Guo Z, Wang L, Eisensmith R, Woo S (1996) Evaluation of promoter strength for hepatic gene expression in vivo following adenovirus-mediated gene transfer. Gene Ther 3: 802–810.
28. Toth M, Lichtenberg U, Doerfler W (1989) Genomic sequencing reveals a 5-methylcytosine-free domain in active promoters and the spreading of preimposed methylation patterns. Proc Natl Acad Sci USA 86: 3728–32.
29. Schnieke AE, Kind AJ, Ritchie WA, Mycock K, Scott AR, et al. (1997) Human factor IX transgenic sheep produced by transfer of nuclei from transfected fetal fibroblasts. Science 278: 2130–2133.
30. Fletcher CJ, Roberts CT, Hartwich KM, Walker SK, McMillen IC (2007) Somatic cell nuclear transfer in the sheep induces placental defects that likely precede fetal demise. Reproduction 133: 243–255.
31. Nishi M, Yasue A, Nishimatu S, Nohno T, Yamaoka T, et al. (2002) A missense mutant myostatin causes hyperplasia without hypertrophy in the mouse muscle. Biochem Biophys Res Commun 293: 247–251.
32. Zhu X, Hadhazy M, Wehling M, Tidball JG, McNally EM (2000) Dominant negative myostatin produces hypertrophy without hyperplasia in muscle. FEBS Lett 474: 71–75.

3

A Site-Specific Recombinase-Based Method to Produce Antibiotic Selectable Marker Free Transgenic Cattle

Yuan Yu[⑨], Yongsheng Wang[⑨], Qi Tong, Xu Liu, Feng Su, Fusheng Quan, Zekun Guo, Yong Zhang*

Key Laboratory of Animal Biotechnology of the Ministry of Agriculture, College of Veterinary Medicine, Northwest A&F University, Yangling, Shaanxi, People's Republic of China

Abstract

Antibiotic selectable marker genes have been widely used to generate transgenic animals. Once transgenic animals have been obtained, the selectable marker is no longer necessary but raises public concerns regarding biological safety. The aim of this study was to prepare competent antibiotic selectable marker free transgenic cells for somatic cell nuclear transfer (SCNT). PhiC31 intergrase was used to insert a transgene cassette into a "safe harbor" in the bovine genome. Then, Cre recombinase was employed to excise the selectable marker under the monitoring of a fluorescent double reporter. By visually tracking the phenotypic switch from red to green fluorescence, antibiotic selectable marker free cells were easily detected and sorted by fluorescence-activated cell sorting. For safety, we used phiC31 mRNA and cell-permeant Cre protein in this study. When used as donor nuclei for SCNT, these safe harbor integrated marker-free transgenic cells supported a similar developmental competence of SCNT embryos compared with that of non-transgenic cells. After embryo transfer, antibiotic selectable marker free transgenic cattle were generated and anti-bacterial recombinant human β-defensin-3 in milk was detected during their lactation period. Thus, this approach offers a rapid and safe alternative to produce antibiotic selectable marker free transgenic farm animals, thereby making it a valuable tool to promote the healthy development and welfare of transgenic farm animals.

Editor: Xiuhcun (Cindy) Tian, University of Connecticut, United States of America

Funding: This work was supported by the National Major Project for Production of Transgenic Breeding (No. 2011ZX08007-004) and the National High Technology Research and Development Program of China (863 Program) (No. 2011AA100303). The funders had no role in study design, data collection and analysis, decision to publish, or preparation of the manuscript.

Competing Interests: The authors have declared that no competing interests exist.

* E-mail: zhangynwsuaf@gmail.com

⑨ These authors contributed equally to this work.

Introduction

Transgenic farm animals are important materials for biomedical and agricultural research [1,2]. However, the present approaches to generate transgenic animals are still hampered by low efficiency [3], variable expression levels of the transgene [4], and the residual antibiotic-resistance gene that is required to select transgenic cells but provokes public concerns regarding biological safety. Thus, efficient and safe methods are urgently needed to improve the current situation.

Currently, somatic cell nuclear transfer (SCNT) has been proven to be the most effective protocol for the production of transgenic animals [5,6]. Therefore, preparation of competent transgenic donor cells is a key step for successful SCNT. Many methods are available to produce transgenic donor cells, and the traditional method relies on random integration of the transgene of interest. However, random integration into chromosomes suffers from low stable integration [7], and variable expression levels of the genes due to positional effects and the number of inserted copies [8–11]. Homologous recombination provides site specificity, but at a very low efficiency [12]. Furthermore, many virus-based gene transfer approaches are limited by their preference for integration into the gene-coding region [13,14], which is a safety risk of transgenic animal production. Thus, an efficient and safe gene delivery approach is important for transgenic cell preparation.

PhiC31 integrase, the *Streptomyces* phage-derived recombinase, has been developed as a non-viral gene therapy tool, because it has the ability to integrate a transgene-containing plasmid carrying an *attB* site into pseudo *attP* sites in mammalian genomes [15,16]. This enzyme has been previously shown to integrate genes effectively and prolong transgene expression in several mammalian cell culture systems including those for human keratinocytes [17], muscle-derived stem cells and myoblasts [18], and a human T cell line [19]. In addition, it has been recently reported that phiC31-mediated integration events usually occur in "genomic safe harbors" in mammalian cells [20], the regions of the genome where the integrated material is adequately expressed without perturbing endogenous gene structure or function, following a process that is amenable to precise mapping and minimizing occult genotoxicity [21], which makes phiC31 integrase an ideal tool for gene delivery and transgenic animal production.

Furthermore, the integration of antibiotic-resistance genes into transgenic animals may cause many problems such as disturbing the expression of neighboring genes [22], confounding the evaluation of food safety of these transgenic animals, and increasing worldwide public concern regarding the release of such antibiotics resistance genes into the environment. Recent studies have shown that selective marker genes can be successfully knocked out from transgenic cells using the Cre/loxP system [23–26]. However, because of the reversibility of Cre recombinase, labor-intensive procedures must be performed to identify complete excision events among randomly picked colonies. Although a lot of effort is being invested into solving this issue, an efficient and reliable method has yet to be developed.

Here, we demonstrate an efficient and safe approach to produce transgenic cattle, which consists of single-copy integration of *human β-defensin-3* (*HBD3*) gene into a genomic safe harbor and visual removal of the antibiotic-resistance marker. In addition, this procedure can be prospectively applied for breeding other antibiotic marker-free disease-resistant transgenic animals as well as production of human recombinant pharmaceuticals in transgenic cattle.

Results

Generation of Transgenic Cells using phiC31 Integrase mRNA

Delivery of the *human β-defensin-3* transgene into bovine fetal fibroblasts was performed by co-electroporation of phiC31 integrase mRNA produced by *in vitro* transcription, and a transgenic plasmid, pARNG-HBD3. The pARNG-HBD3 plasmid was an *attB*-containing human β-defensin-3 mammary gland expression vector (Fig. 1). A fluorescence double reporter was constructed to monitor Cre-mediated recombination in living cells before and after Cre recombination by expression of two different fluorescent proteins. The fluorescence double reporter construct contained the ubiquitous active CMV IE promoter that drove transcription of the fluorescent gene. Downstream of the CMV IE promoter was a *loxP*-flanked *DsRed* gene and P_{SV40e}-driven neomycin-resistance (*neoR*) expression cassette coupled to an *enhanced green fluorescent protein* (*EGFP*) reporter.

To test the effect of phiC31 integrase on site-specific integration, 5 μg pARNG-HBD3 was electroporated into 1 × 10^6 bovine fetal fibroblasts in the presence of 1 μg phiC31 integrase mRNA (integrase group) or 1 μg inactive mutant phiC31 integrase mRNA (mutant integrase group). At 8–12 days after electroporation, individual cell colonies were obtained by G418 screening. A total of 46 colonies were obtained by integrase mediated electroporation while 30 colonies were obtained by the mutant integrase mediated electroporation. Both G418-resistant and red fluorescent protein (RFP)-positive (RFP+) colonies were picked and expanded in culture. A total of 38 out of 46 G418-resistant colonies were RFP+ in the integrase group, whereas 18 out of 30 colonies were RFP+ in the mutant integrase group. These RFP+ colonies were further analyzed by PCR. A band of 426 bp for an un-cleaved *attB* site was found in 15 RFP+ colonies of the integrase group (Fig. S1a) and all 18 RFP+ colonies of the mutant integrase group (Fig. S1b), indicating that these colonies were involved with random integration events. The remaining 23 RFP+ colonies of the integrase group, which lacked the 426 bp band, were preliminarily considered to be the result of *attB* site cleavage caused by site-specific integration. These *attB*-cleaved RFP+ colonies were evaluated for the number of integration events by absolute quantity PCR. Among the 23 integrants, 11 exhibited single-copy integration, 12 exhibited double integration, and three or more integrants were not found in this study (Fig. S1c).

Of the 11 single-copy site-specific integrants we evaluated by half-nested inverse PCR, seven pseudo *attP* sites were identified. Among the seven sites, three sites were found in intergenic regions, the other four were located within an intron, and none of the sites were located in an exon. Notably, of the 11 single-copy integrants, five were found to be integrated in the intergenic regions on chromosome 2, whereas the others were identified only once as shown in Table 1. We found that repetitive sequences were expanded in DNA sequences surrounding these pseudo *attP* sites. Short interspersed elements (SINEs) were major repetitive elements flanking all integration sites, which included both intergenic and intronic insertions. The mammalian-wide interspersed repeat (MIR) and bovine tRNA pseudo gene coupled to A element (BovA) subfamilies were the most frequently identified types of SINEs. We also evaluated the integration sites obtained in our study according to the criteria articulated in a recently published study by Papapetrou et al. [21], which defined so-called "genomic safe harbors". Of the seven integration sites, only the intergenic integration site on chromosome 2 met the criteria proposed by the previous study (Table S1). Using junction PCR with primers specific for the phiC31 integrase *attB* site and the genomic site, 13 out of 46 colonies of the integrase group were found to be integrated in the genomic safe harbor on chromosome 2, and demonstrated a total integration frequency of 28% (Fig. S1d). However, none of the 18 colonies of the mutant integrase group was detected to be integrated in the safe harbor on chromosome 2 (data not shown).

Excision of the Antibiotic Selectable Marker from Transgenic Cells using Cell-permeant Cre Recombinase

Cell-permeant Cre protein was used to remove the selectable *neoR* marker gene. His-NLS-TAT-Cre protein (Fig. 2a) used in this study is a recombinant fusion protein consisting of a basic protein translocation peptide derived from HIV-TAT (TAT), a nuclear localization sequence (NLS) derived from SV40 large T antigen, the Cre protein and an N-terminal histidine tag for efficient purification from *E. coli* (Fig. 2b).

For simplicity and as a proof of concept, we focused on three clones, SC5, SC6 and SC27, each with a single integration site in the safe harbor. At day 5 after His-NLS-TAT-Cre protein transduction, these colony cells were prepared for flow cytometry by trypsinization and resuspension in PBS containing 10% fetal bovine serum. By visually tracking the phenotypic switch from red to green fluorescence, more than 70% GFP+ cells were easily detected (Fig. S2a) and isolated for clonal expansion by fluorescence-activated cell sorting (FACS) using a BD FACSAria. Excision of the *neoR* gene was verified by fluorescence visual detection (Fig. 3a) and Southern blotting (Fig. 3b). Moreover, the absence of either phiC31 integrase-encoding DNA or Cre recombinase-encoding DNA was demonstrated by PCR (Fig. S3a).

The integration sites of these excised transgenic cells were accurately confirmed by junction PCR (Fig. S3b). As expected, the integration sites of these excised cells remained the same as those of the un-excised ones. Thus, we tested the effect of the insertion and excision on the expression of two neighboring genes, *KIAA1486* and *IRS-1* genes, using qRT-PCR and compared the expression of these two genes with that in untransfected bovine fetal fibroblasts as the control (n = 3). We found no significant difference in the gene expression levels of *KIAA1486* and *IRS-1* in un-excised and excised transgenic cells, and untransfected bovine fetal fibroblasts (Fig. S3c), indicating that the insertion and excision did not alter the expression of flanking genes. In addition, chromosome spreads of metaphase cells of the selected clones were analyzed and revealed the correct chromosome number (2n = 58+ XX) and no major differences among un-excised and excised transgenic cells, and untransfected bovine fetal fibroblasts (Fig. S4). However, more

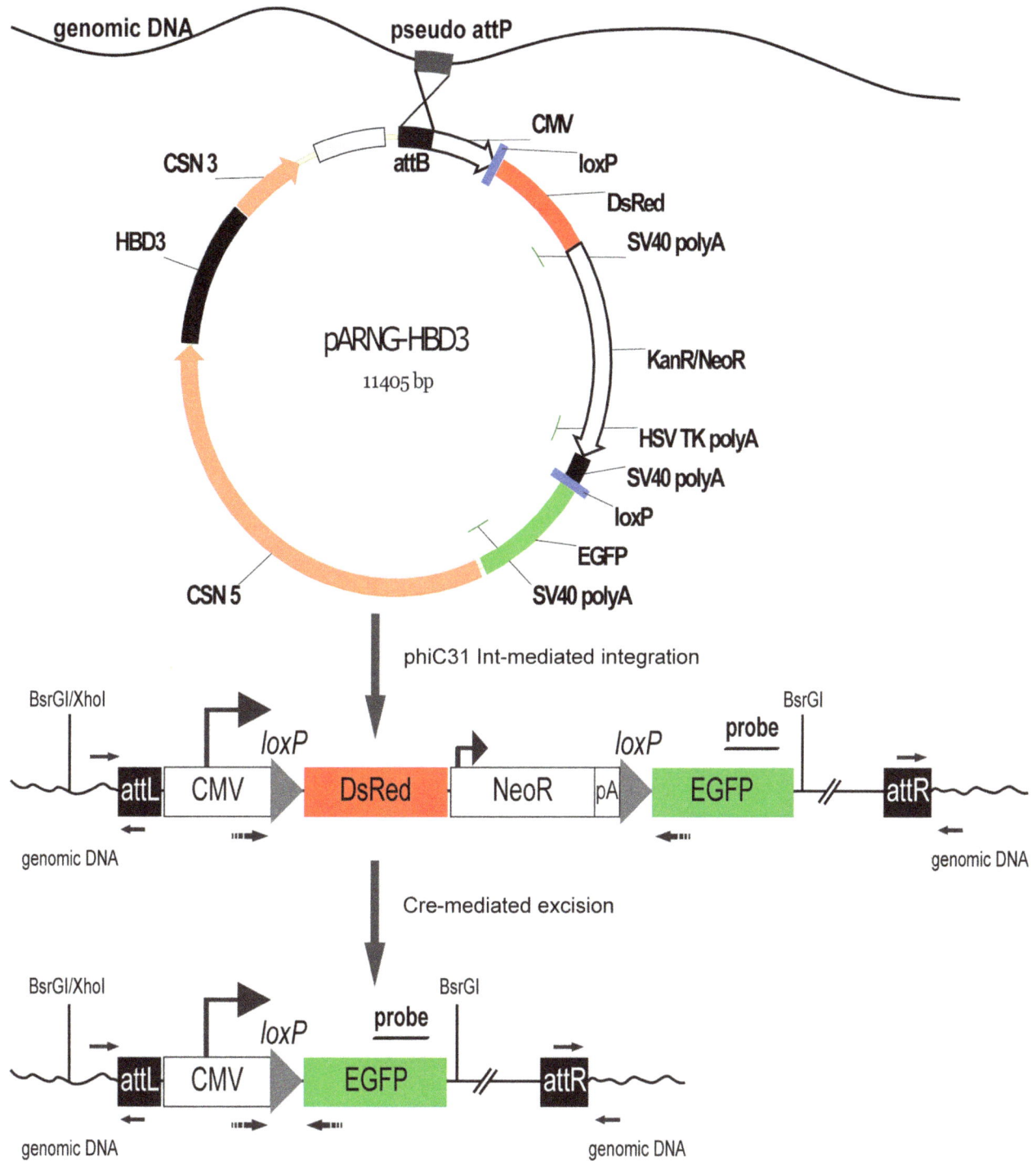

Figure 1. Schematic overview of the fluorescence double reporter construct pARNG-HBD3. The *attB* site is used for site-specific integration mediated by phiC31 integrase. *DsRed* is ubiquitously expressed under the control of the CMV promoter. A neomycin resistance gene (*neoR*) under the control of the SV40 early promoter allows selection of stable-transfected colonies. Cre-mediated recombination causes excision of *DsRed* and *neoR* resulting in expression of the second reporter, *EGFP*.

refined cytogenetic techniques would be required to reveal more subtle chromosomal rearrangements that may occur in transgenic cells derived by this protocol. Such analyses will be carried out in the next study.

Production of Antibiotic Selectable Marker Free Transgenic Embryos by SCNT

Antibiotic selectable marker free cells from different clones with different integration sites were used as nuclei donors for production of transgenic SCNT embryos. In addition, normal bovine fetal fibroblasts derived from the same fetus and at a similar

Table 1. Overview of the seven integration events analyzed by half-nested inverse PCR and sequencing.

Pseudo site	GenBank accession No.	Chromosome	Within Intron	Intergenic Distance to upstream gene gene	Distance to downstream gene	Repetitive elements SINEs	DNA elements	Total length of repeats	Identity with WT attP
BFF2	NW_003103850	2		121 kb to IRS-1	1068 kb to KIAA1486	SINE/MIR, SINE/BovA		341 bp (63%)*	33%
BFF4a	NW_001499941	4	PODXL					–	39%
BFF4b	NW_003103903	4		3.4 kb to NAA38	1909 kb to KCND2	SINE/MIR		111 bp (21%)	33%
BFF13	NW_001493120	13	KIF5B				low complexity	30 bp (6%)	33%
BFF19	NW_003104490	19		5.9 kb to UNC45B	86 kb to SLFN11	SINE/MIR		111 bp (21%)	28%
BFF22	NW_003104541	22	LOC100852122			SINE/tRNA-Glu, SINE/MIR		256 bp (47%)	41%
BFF25	NW_001494275	25	CLCN7			SINE/BovA		33 bp (6%)	21%

*Total length of interspersed repeats/540 bp of genomic sequences flanking phiC31 integration sites.

passage number (passage 5–10) were used as a control to investigate the effect of transgenic procedures on the developmental potential of SCNT embryos. As shown in Table 2, no significant differences were observed in the cleavage and blastocyst formation rates among these antibiotic marker free clones and control groups.

PCR analysis of single blastocyst showed that a 3323 bp fragment could be amplified from the un-excised transgenic cell derived blastocyst but only a 313 bp fragment could be amplified from the excised transgenic cell derived blastocyst (Fig. 4a). The 3323 and 313 bp PCR fragments were sequenced. In addition, the fluorescence phenotype of excised transgenic cell derived blastocyst (Fig. 4b) confirmed that the selectable *neoR* marker gene had been successfully removed prior to SCNT.

Generation of Cloned Transgenic Cattle that Express Human β-defensin-3

After SCNT, 1012 antibiotic selectable marker free transgenic blastocysts were transferred into 506 recipient cattle. The BFF2 site integrated clones showed similar pregnancy rate (17.1%– 17.9%) and calf birth rate (7.3%–7.7%) with nontransgenic control (18.7% and 8.7%, respectively), while the pregnancy rate and birth rate of the other site integrated clones were lower than those of nontransgenic control (Table 2). A total of 18 calves were born, two calves died within a few hours after birth, and four calves died within 6 months after birth. Twelve calves survived and were healthy after weaning. As shown by PCR analysis, *human β-defensin-3* transgene was integrated into the genome of the 12 transgenic cattle (Fig. S5), but only five transgenic cattle, mTG1, mTG2, mTG3, mTG4 and mTG5, were lactating naturally when the observation started. Thus the data from subsequent experiments to be report just focus on these five transgenic cattle. Of the five cattle, mTG1 and mTG4 were generated from different clones with the same safe harbor integration, mTG2 was generated from BFF19 site integrated colony, mTG3 was derived from BFF22 site integrant, and mTG5 was from BFF4b site integrant. In addition, five unrelated non-transgenic, nonclone Holstein cow, matched by age and lactation, were chosen as control group. The breeding conditions were identical for the two groups.

Milk samples from transgenic and non-transgenic cattle were collected each month for 6 months during their natural lactation period. There were no significant differences in milk yield (P = 0.239) and percentage of fat, protein, lactose, and milk solids in the milk of transgenic and non-transgenic cattle (P = 0.793, 0.569, 0.696, and 0.976, respectively), as shown in Table 3. Milk proteins of the transgenic animals, as visualized on a polyacrylamide gel, appeared essentially identical to those from a nontransgenic Holstein cow (Fig. 5a). A single protein of the predicted size was immunologically reactive to antibodies against human β-defensin-3. The protein was observed in the milk of five transgenic cows, but not in that of non-transgenic cows (Fig. 5b). GFP was not detected in the milk of transgenic or non-transgenic cows (Fig. 5b). Human β-defensin-3 concentrations, as measured by ELISA, ranged from 3.9 to 10.4 μg/mL in milk of the five transgenic cows and no significant decline of human β-defensin-3 expression was observed during the natural lactation period of 6 months (Fig. S6). The lytic activity of milk samples from various transgenic cows was pre-estimated by an agar diffusion test (Fig. S7). The transgenic cows' ability to resist infection by *S. aureus* and *E. coli* was tested by intramammary infusion of viable bacterial cultures. Of the mammary glands infused with *S. aureus*, 14 of 15 glands became infected in nontransgenic animals compared to 5 of 15 glands in transgenic animals (P = 0.001, Table 4); While of the mammary glands infused with *E. coli*, 13 of 15 glands became infected in

A

T7Lac

| H6 | NLS | TAT | Cre |

MGSSHHHHHHSSGLVPRGSHM PKKKRKV YGRKKRRQRRR

B

M   E   W   FT   S   I   CL

100kDa
62 kDa

40 kDa

30 kDa

24 kDa

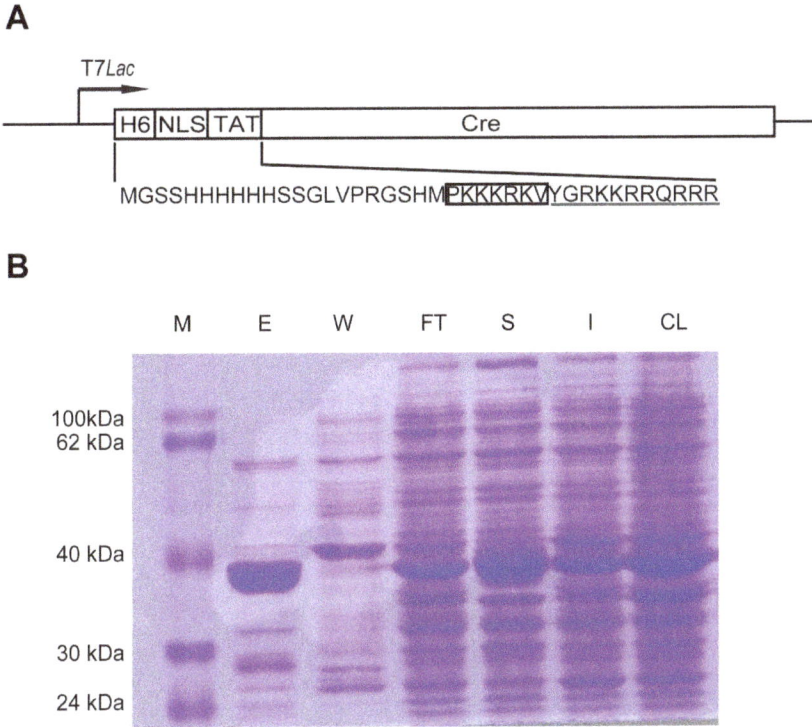

Figure 2. Recombinant modified Cre protein-mediated recombination. (a) Schematic of cell-permeant His-NLS-TAT-Cre fusion protein. H6, 6×His tag. The amino acid sequence of NLS is boxed and TAT is underlined. (b) Purification of recombinant His-NLS-TAT-Cre protein from bacteria as analyzed by Coomassie blue staining of an SDS-PAGE gel. CL: Cleared lysate; I: Insoluble; S: Supernatant; FT: Flow-through; W: Washing; E: Eluate; M: Marker.

nontransgenic animals compared to only 1 of 15 glands in transgenic animals (P = 0.001, Table 4). With the higher human β-defensin-3 expression level, transgenic cow mTG1 and mTG4, which derived from the evaluated safe harbor integrated transgenic cells, never became infected after *S. aureus* or *E. coli*

infusion. Transgenic cow mTG2, mTG3 and mTG5, expressing lower level of human β-defensin-3, were infected twice, twice and once out of three infusions of *S. aureus*, respectively. Transgenic cow mTG2 was infected once out of three infusions of *E. coli*.

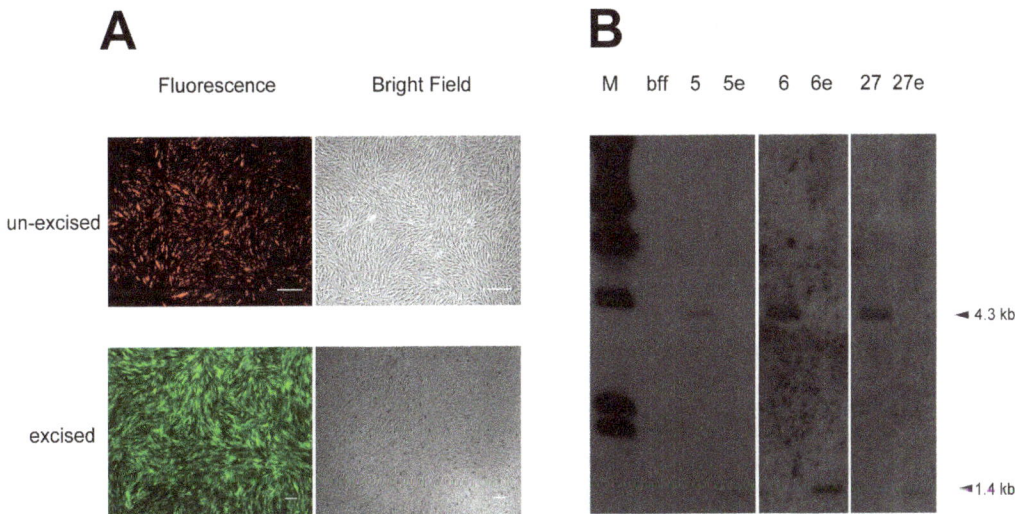

A

Fluorescence   Bright Field

un-excised

excised

B

M   bff   5   5e   6   6e   27   27e

◄ 4.3 kb

◄ 1.4 kb

Figure 3. Identification of excision of the antibiotic selectable neoR marker in the transgenic cells. (a) Fluorescence phenotype of transgenic cells before and after Cre-mediated excision. The scale bar indicates 20 μm. (b) Southern blot analysis of three representative transgenic cell linesbefore and after Cre-mediated excision of the selectable marker gene using an *EGFP* probe. The un-excised clones (SC5, SC6 and SC27) carried a single integration of the transgene cassette (4.3 kb), whereas the excised clones (SC5E, SC6E and SC27E) also presented a single lower band (1.4 kb).

Table 2. Summary of *in vitro* and *in vivo* development following SCNT with antibiotic selectable marker-free cells.

Donor cells*	Insertion site	No. cultured	No. cleaved (%)[‡]	No. blastocysts (%)[‡]	No. recipients[§]	No. pregnancies (%)[¶]	No. calve born (%)[‖]	No. calve survived
SC5E	BFF2[†]	475	372 (78.3±1.4)	129 (27.2±0.8)	41	7 (17.1)	3 (7.3)	1
SC6E	BFF2	781	615 (78.7±2.0)	210 (26.9±0.5)	67	12 (17.9)	5 (7.5)	4
SC27E	BFF2	602	473 (78.6±2.7)	163 (27.1±1.4)	52	9 (17.3)	4 (7.7)	3
SC10E	BFF4a	705	545 (77.3±2.4)	182 (26.4±0.4)	72	4 (5.6)	1 (1.4)	1
SC11E	BFF4b	603	472 (78.3±2.1)	160 (26.5±0.9)	51	5 (9.8)	2 (3.9)	1
SC2E	BFF13	457	353 (77.3±2.3)	117 (26.4±1.1)	46	2 (4.3)	–	–
SC4E	BFF19	812	627 (77.2±1.3)	215 (26.5±2.0)	68	4 (5.9)	2 (2.9)	1
SC1E	BFF22	786	615 (78.4±5.6)	209 (26.6±1.6)	67	5 (7.5)	1 (1.5)	1
SC16E	BFF25	411	324 (78.6±3.2)	104 (26.9±1.2)	42	2 (4.8)	–	–
Control	–	1699	1304 (76.9±5.9)	481 (28.3±2.1)	150	28 (18.7)	13 (8.7)	9

*Donor nuclei from antibiotic selectable marker-excised cells with single-copy integration of transgene into different locus, or from untransfected cells (control), were used for nuclear transfer experiments.
[†]BFF2 was evaluated as a genomic safe harbor according to the criteria articulated in a recently published study by Papapetrou et al.
[‡]Values within one column are not significantly different (P>0.05).
[§]Day-7 blastocysts were nonsurgically transferred to synchronized recipient cows with two embryos per recipient.
[¶]No. pregnancies: number of pregnant recipients on day 90 d after embryo transfer/number of recipient cows.
[‖]Percentage of transferred recipients.

Discussion

In this study, we attempted to establish a rapid and safe procedure to generate antibiotic-resistant marker-free transgenic cattle. We generated safe harbor integrated bovine fetal fibroblast colonies in the presence of phiC31-NLS mRNA and employed a fluorescence double reporter system to sort antibiotic-resistant marker-free transgenic cells after cell-permeant Cre protein treatment. Our success of generating live cloned cattle using the antibiotic selectable marker free fibroblasts as nuclear donors in SCNT provided strong evidence that no severe damage was caused to the host cells using this method.

Because selectable marker elimination in plants was first reported in 1991 [27], marker-free technology has been already applied to many plant species such as rice [28], maize [29,30], and tomato [31,32]. With the development of transgenic animals, especially farm animals, removal of antibiotic selectable markers in transgenic animals has been recently brought to the forefront of research. Several antibiotic selectable marker free transgenic

animals or donor cells prepared for SCNT has been established in various studies. Xu et al. [25] obtained two live cloned goats with removal of the selectable gene cassette. Wang et al. [26] successfully excised the *neoR* gene in transgenic cloned cattle. Generally, two steps are required to prepare antibiotic selectable marker free transgenic donor cells for SCNT: generation of transgenic animals with a *loxP*-flanked selectable marker and removal of the selectable marker gene from fibroblasts isolated from the generated transgenic animals by Cre recombinase. The two-step manipulation does not affect the developmental competence of recloned preimplantation embryos [25,26], however, it is costly and time consuming to prepare *loxP*-flanked transgenic animals (almost a year, take cattle for example) and requires labor-intensive work to identify complete excision events among randomly picked colonies.

In our study, the fluorescence double reporter was used to monitor Cre-mediated recombination in living cells before and after Cre recombination by expression of two different fluorescent proteins. Before Cre-mediated excision, transgenic cell colonies

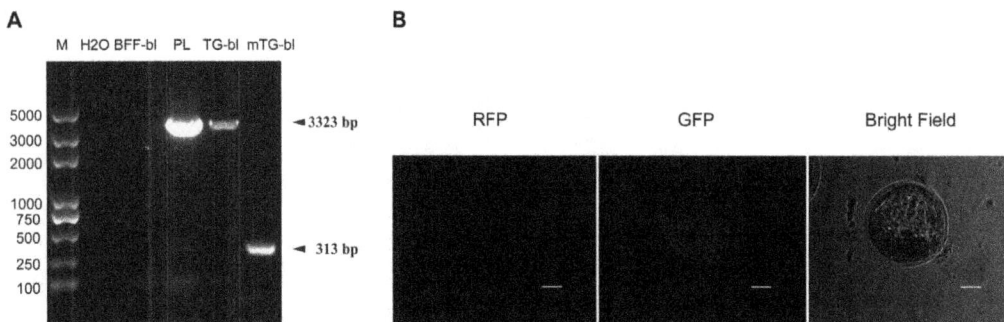

Figure 4. Identification of excision of the antibiotic selectable neoR marker in the transgenic embryo. (a) PCR analysis of single blastocyst to verify excision of the selectable marker gene in the excised transgenic cell derived blastocyst. BFF-bl represented untransfected bovine fetal fibroblast-derived blastocyst, was used as a negative control; PL (pARNG-HBD3 plasmid) was used as a positive control. TG-bl and mTG-bl were un-excised and excised colony derived blastocysts, respectively. (b) Fluorescence phenotype of the excised transgenic cell derived blastocyst. The scale bar indicates 200 μm. Only green fluorescence was observed.

Table 3. Raw components of transgenic milk compared with conventional milk.

	Transgenic (n = 5)	Non-transgenic (n = 5)
Fat (g/100 mL)	4.07±0.32	4.01±0.35
Protein (g/100 mL)	3.45±0.24	3.60±0.52
Lactose (g/100 mL)	4.65±0.25	4.57±0.36
Solids (g/100 mL)	13.66±0.49	13.64±0.70

No significant differences were detected between transgenic group and non-transgenic group (P>0.05).

were easily generated by the design of both a selectable marker and fluorescent reporter. The *DsRed* reporter was employed to exclude either few false positive cells that were insensitive to G418 or stably-transfected colonies that integrated in a transcriptionally

silent region on chromosomes. Subsequently, Cre recombination resulted in replacement of *DsRed* expression with *EGFP* expression. Cells showing both GFP+ and RFP− were considered as cells with the selectable marker completely removed and were sorted by FACS for SCNT. Finally, the absence of the selectable marker gene in both sorted GFP+ cells and blastocysts derived from such cells was verified by Southern blot and PCR, thereby providing strong evidence that the Cre/loxP system combined with fluorescence double reporter offers a highly efficient and reliable tool to remove selectable maker genes completely and visually.

To avoid unwanted translocation events caused by multiple *loxP* sites on different chromosomes [33], the phiC31 integrase system was employed to generate a single copy of *loxP*-flanked *neoR*. PhiC31-mediated integration events were analyzed by absolute quantity PCR to identify single integrants that were used for subsequent steps. In this study, at least 24% (11/46) of all G418-resistant colonies showed a single integration of the pARNG-HBD3 plasmid. This fraction is superior to that of single-copy integrants obtained with other non-viral methods such as PB

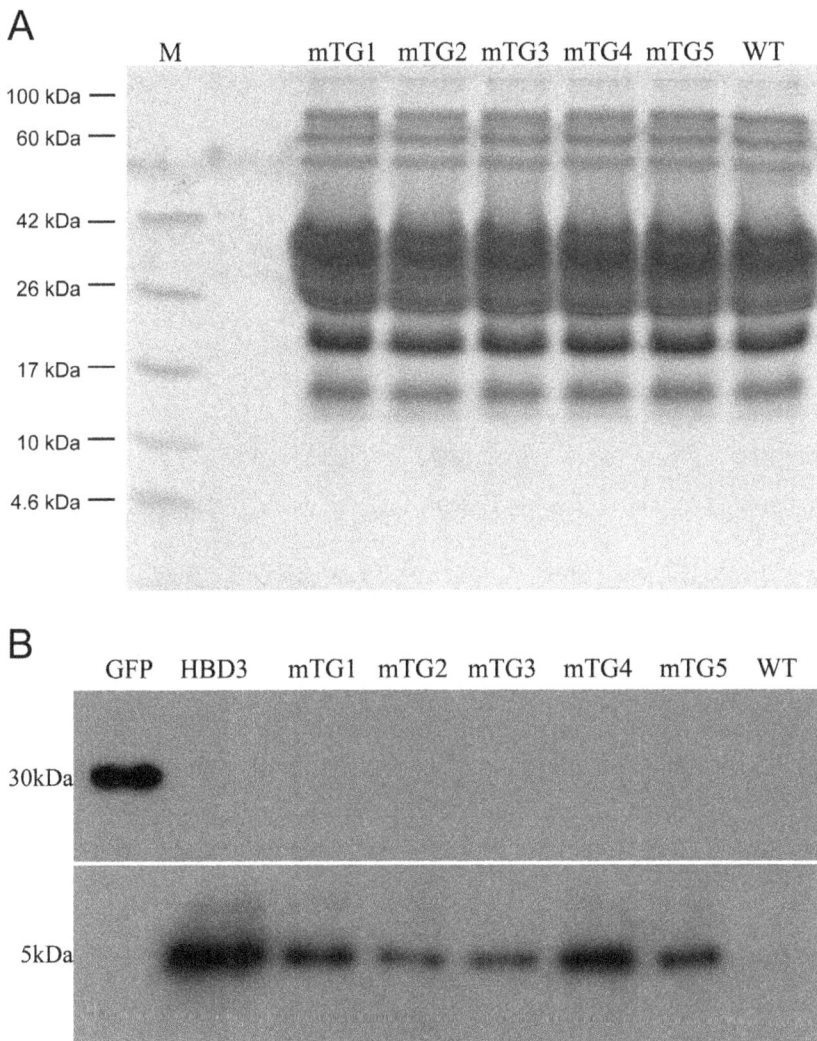

Figure 5. Protein analysis of milk from antibiotic selectable marker free transgenic cattle and non-transgenic cow. (a) Total proteins from transgenic cows mTG1–5 and non-transgenic animals were separated on a 10% Tricine-SDS-PAGE gel and visualized by Coomassie blue staining. Equal amounts of total milk protein were loaded onto each lane of the gel. (b) Western blot analysis of milk probed with an anti-human β-defensin-3 antibody or an anti-EGFP antibody. HBD3, commercial human β-defensin-3 (positive control); GFP, cell lysate of excised transgenic colony (positive control); WT, non-transgenic milk; and the other lanes, milk from transgenic cattle mTG1-mTG5.

Table 4. Infection rate of *S. aureus* and *E. coli* infused into mammary glands of five transgenic and five non-transgenic lactating cows.

Cow*	*S. aureus*	*E. coli*	Overall
mTG1	0/3[†]	0/3	0/6
mTG2	2/3	1/3	2/6
mTG3	2/3	0/3	1/6
mTG4	0/3	0/3	0/6
mTG5	1/3	0/3	0/6
Non-tg (n = 5)	14/15	13/15	27/30

During each challenge experiment each gland was infused with *S. aureus*, *E. coli* or PBS, the fourth gland remained untreated. Infection was defined by the presence of viable *S. aureus* or *E. coli* in two consecutive milk samples. None of transgenic or nontransgenic glands infused with PBS became infected, so were the untreated transgenic and nontransgenic gland.
*Transgenic cows or non-transgenic cows were challenged three times at interval of two months during their natural lactation period.
[†](Number of glands infected)/(number of glands infused).

transposon-based approaches (15%) [34,35]. Very recently, a profile of bovine native integration sites for phiC31 integrase has been well described [36]. To evaluate the feasibility of our approach for the bovine genome, we re-analyzed previous DNA sequence data of phiC31 integration sites in bovine cells [36–38], according to all of the safe harbor criteria proposed by Papapetrou et al. [21]. This analysis revealed that of the 33 sequenced native pseudo *attP* sites, 17 were intergenic, representing 52%, and 5 of the intergenic sites met the criteria of genomic safe sites, representing 15% of the total integration sites (Table S1). Notably, in our study, 13 out of 46 colonies of the integrase group showed integration at BFF2 site in intergenic region between *KIAA1486* and *IRS-1*, which met all of the criteria of a genomic safe harbor [21]. The BFF2 site has 33% sequence identity with wild-type *attP*, while other integration sites that have the similar or even better homology did not show such integration frequency. It is possible that some integration sites that have a lower level of sequence identity have a more highly preferred genomic context. Thus, contextual factors, such as repetitive sequences surrounding the integration sites, may play a strong role in site selection as well as primary sequence.

Another advantage of the integrase system is that the distance of the single insertion from genes of known and unknown function can be determined precisely [39]. Hence, it is possible to determine whether the intergenic insertion affects the expression of flanking genes. We tested three colonies in which the integration occurred in the intergenic region on chromosome 2, both before and after Cre-mediated excision. qRT-PCR showed that the expression of neighboring genes was unaffected even after removal of the selectable marker by Cre treatment. At the very least, using the integrase system, transgenic donor cells can be prepared in a rather safe manner.Recently, a similar site-specific recombinase strategy was reported to create reprogramming factor-free induced pluripotent stem (iPS) cells [20]. However, delivery of phiC31 and Cre expression plasmids may result in unwanted insertion, and constitutive expression of phiC31 integrase [40,41] or Cre recombinase [42–44] can induce cytotoxicity and genotoxicity. In our study, we used phiC31 mRNA for site-specific integration and cell-permeant Cre protein for removal of the selectable marker in a highly efficient and safe manner. Furthermore, no abnormal chromosome numbers were found in transgenic cells

before or after cell-permeant Cre treatment. There is evidence indicating that the extracellular concentration of cell-permeable Cre protein can be precisely optimized to induce recombination, while causing no apparent toxicity [45,46]. Therefore, this controllable and mild method provides the possibility for a second genomic manipulation immediately after the stably transfected cells are generated, which saves time and minimizes expenses compared with that using the traditional two-step method as discussed above.

The development ability of the transgenic embryos or non-transgenic controls in this study was somewhat lower than that in others' work [47–49]. Numerous factors, such as the state and source of donor cells [49–51], cytoplast source and quality [52], methods of manipulation and activation [53,54], and embryo culture conditions [55], could have an impact on the efficiency of SCNT embryo development and caused the difference. In this study, no significant difference was found in cleavage rate, blastocyst rate, pregnancy rate and birth rate between the experimental (BFF2) and non-transgenic group, which indicated that the developmental ability of the antibiotic selectable maker free transgenic embryos we produced was not compromised by using this procedure. However, other transgenic embryos derived from the non-safe harbor integrated transgenic clones we produced showed relatively lower developmental competence compared with the control group, which might suggest that integration site of transgene is crucial in this procedure, some undetermined integration sites might be detrimental to the nuclear donor cell.

Human β-defensin-3, a kind of antimicrobial peptide, is widely expressed in many tissues [56,57]. Human β-defensin-3 has broad-spectrum antimicrobial activity against bacteria, fungi, and enveloped viruses, and has an important role in immunity [58]. In previous studies, transgenic livestock expressing some kinds of antimicrobial peptides in milk inhibited bacterial pathogens causing mastitis [59–61]. However, both lysostaphin [61] and lysozyme [62] were effective against *S. aureus* but not *E. coli*, and other human β-defensins exhibited less activity against gram-positive bacteria, including *S. aureus*, than against gram-negative bacteria [63]. While human β-defensin-3 showed antibacterial activity against both *S. aureus* and *E. coli* at very low concentrations [56,62] and exhibited less salt sensitive than other human β-defensins [64]. Therefore, *human β-defensin-3* might be a candidate gene for enhancing mastitis resistance. By simple evaluation of transgenic milk samples from five transgenic cows, which were lactating normally in the observation period, the bacteriolytic activity of the milk was confirmed. Transgene expression levels in the animals reported here were similar to those found in some livestock bioreactors [61], but less than those found in other livestock bioreactors [47,60,65]. Some reasons might cause this difference. Firstly, more regulatory elements in transgenic vector may lead to high expression level, for example using bacterial artificial chromosome (BAC) [65] or insulators [60]. Secondly, the different levels of expression among transgenic cattle may relate to the copy numbers of the transgene [66]. Thirdly, the signal peptide of human β-defensin-3 we used in this study may have lower secretory capacity than that of others, such as human growth hormone [67]. In addition, many factors of the SCNT procedure, such as the quality of the donor cells, the nuclear reprogramming, could also lead to different levels of expression [68–70]. The exact mechanisms remain to be determined, however.

However, the human β-defensin-3 expression levels remained stable during the natural lactation period of 6 months, and milk from these transgenic cows repressed the growth of both *S. aureus* and *E. coli*, the most prevalent species of gram-positive and gram-negative bacteria that induce clinical mastitis [71]. The transgenic cows' ability to resist infection by *S. aureus* and *E. coli* was confirmed by intramammary infusion of viable bacterial cultures. The infection rate between nontransgenic animals and those expressing human β-defensin-3 provided strong evidence that the transgene product conferred a protective effect. Because of the small transgenic sample size it is not possible to make any statistical inferences regarding to the infection frequency and human β-defensin-3 milk concentration. However, the data are suggestive. The three cows with the lower human β-defensin-3 expression level had the higher *S. aureus* infection rate (2/3, 2/3 and 1/3, respectively), and impressively, only one out of all fifteen *E. coli* infused transgenic glands became infected. Notably, the higher expressing transgenic cows, which derived from the evaluated safe harbor integrated transgenic cells, never became infected after *S. aureus* or *E. coli* infusion.

Interestingly, only human β-defensin-3, but not GFP, was detected in the transgenic milk, which might be caused by the fluorescent protein lacking the signal peptide. GFP, as a fluorescent reporter to monitor the excision of the selectable marker, remained constitutively expressed in transgenic donor cells generated by this protocol. Multiple studies have used the green fluorescent protein gene transfected donor cells and subsequently used for NT to create bovine embryos and calves [72–75]. The positive GFP embryos demonstrate easily detectable bright fluorescence and no genetic screening (PCR) is required prior to transfer. Recently, it has been reported that GFP expression does not appear to have unexpected or undesirable effects on transgenic animals, and it is unlikely that the welfare of transgenic animals is compromised [76]. Thus far, in our study, it has not been found that the expression of the transgene influences the welfare and behavioral phenotype of the generated antibiotic selectable marker free transgenic cattle. Thus, we considered that *EGFP* is a useful marker for screening preimplantation embryos to increase the overall efficiency of transgenic animal production, and its non-secreting character makes it an acceptable tool for generation of an antibiotic selectable marker free transgenic mammary bioreactor.

In conclusion, we performed safe harbor integration in the bovine genome using phiC31 integrase and the combination of the Cre/loxP system with a fluorescent double reporter to remove the selectable marker efficiently. The non-viral protocol (Fig. S8) offers a rapid and safe alternative to produce antibiotic selectable marker free transgenic large animals, thereby making it a valuable tool for promoting the healthy development of transgenic large animals, and their recombinant products may be more easily accepted by the public.

Materials and Methods

Ethics Statement

The experimental procedure was approved by the Animal Care Commission of the College of Veterinary Medicine, Northwest A&F University. Bovine ovaries from slaughtered mature cows were collected from Tumen abattoir, a local abattoir in Xi'An, China. A newborn female Holstein calf was obtained for nuclear donor cell cultures and Beef-breed Angus cows were obtained for recipient animals from Yangling Keyuan Cloning Co., Ltd., PR China.

Chemicals

All chemicals and reagents were purchased from Sigma-Aldrich (St. Louis, USA) unless specifically stated otherwise. Disposable,

sterile plasticware was purchased from Nunclon (Roskilde, Denmark).

Plasmid Construction

Primer sequences, restriction sites and templates are depicted in Table S2.

For construction of the universal fluorescent double reporter plasmid, pARNG, the pEGFP-C1 plasmid (Clontech, Mountain View, CA) was digested with BamHI/BglII and self-ligated to remove the original multiple cloning sites. To construct pAttB-eGFP, this vector (pEGFP-B2) was digested with AseI and ligated to a 287 bp *attB* fragment generated using splicing by overlap extension (SOE) PCR with attB1_F~attB4_R primers listed in Table S2. A *loxP*-flanked MCS was obtained by SOE-PCR with LoxPMCS F1~LoxPMCS R2 primers and subsequently cloned into pAttB-eGFP. The pAttB-LoxP2-eGFP plasmid was digested with SpeI/AflII and ligated to a new KanR/neoR fragment generated with Kneo_F and Kneo_R primers. Also, an additional SV40 poly(A) was obtained and cloned downstream of the KanR/neoR expression cassette. This vector (pAL2G-neopA) was digested with MluI/EcoO109I to replace the original KanR/neoR expression cassette with another synthetic MCS for subcloning. To construct pDsRed1-B2, the pDsRed1-C1 plasmid (Clontech) was digested with BamHI/BglII and self-ligated to remove the multiple cloning sites. Then the final pARNG plasmid was generated by cloning a 900 bp *DsRed* ORF and an SV40 poly(A) sequence derived from pDsRed1-B2 into the pANG-MCS plasmid. For subcloning of the *human β-defensin-3* expression cassette into pARNG, the *human β-defensin-3* genomic gene (1 kb) cloned from human placenta, as well as the CSN2 promoter (3.7 kb) and 3′ flanking region (0.6 kb) cloned from the Holstein female bovine genome were ligated to pARNG in correct order as shown in Figure 1.

To transcribe phiC31 mRNA *in vitro*, pcDNA3.1-Int plasmid was constructed by cloning the phiC31 integrase ORF into pcDNA3.1(+) (Invitrogen, Carlsbad, California) with an addition of a Kozak site and an NLS sequence using IVTInt_F and IVTInt_R primers. The pcDNA3.1-mtInt plasmid was constructed by deleting a 537 bp in-frame fragment in the integrase gene using PmlI (New England Biolabs, Beijing, China). The deletion reduced the polypeptide length from 621 amino acids (68 kDa) to 442 amino acids (48 kDa) and encompassed the entire DNA binding region of the protein to render it inactive [41].

The pET-28a(+)-His-NLS-TAT-Cre plasmid was constructed for prokaryotic expression of recombinant Cre protein. A NLS-TAT-encoding fragment was generated by annealing the NT_F and NT_R primers. This fragment was cloned into pET-28a(+) (Novagen, Madison, WI) via the NdeI and SacI restriction sites, resulting in pET-28a(+)-His-NLS-TAT. To obtain pET-28a(+)-His-NLS-TAT-Cre, a Cre-encoding PCR product was generated from the pCAG-Cre-IP template [77] using CreTAA_F and CreTAA_R primers. This fragment was subsequently cloned into pET-28a(+)-His-NLS-TAT via SacI and XhoI sites.

Preparation of phiC31 mRNA

pcDNA3.1-Int or pcDNA3.1-mtInt plasmid DNA was linearized by PmcI (New England Biolabs) downstream of the coding region as the template (1 µg). PhiC31 integrase or mutant integrase mRNA was transcribed *in vitro* using a mMESSAGE mMACHINE® T7 mRNA transcription kit (Ambion, Austin, TX), and the polyadenylation reaction was performed using a Poly(A) Tailing Kit (Ambion). Purification of the tailed mRNA was performed using a MEGAClear™ kit (Ambion). All procedures followed the kit protocols. mRNA concentrations were determined and diluted to about 500 ng/μL for electroporation, and stored at −80°C.

Cell Culture and DNA/RNA Co-transfection

Bovine fetal fibroblasts were isolated from a 50–60-day-old Holstein female fetus (Yangling Keyuan Biotechnology Inc) by disaggregation of the body without the head and viscera, followed by culturing in Dulbecco's modified Eagle's medium (DMEM; GIBCO, Carlsbad, CA) supplemented with 10% fetal bovine serum (FBS; GIBCO) at 38.5°C in a humidified atmosphere with 5% CO_2. At confluency, bovine fetal fibroblasts were collected by trypsinization for passaging or cryopreservation.

To co-electroporate the *human β-defensin-3* donor plasmid and phiC31 mRNA, 1×10^6 passage 3 cells were re-suspended in 0.2 mL diluted electroporation working buffer [electroporation buffer (120 mM KCl, 0.15 mM $CaCl_2$, 10 mM K_2HPO_4, and 5 mM $MgCl_2$) : opti-MEM (GIBCO) = 3:1 (v/v)] after being washed twice in opti-MEM and once in ice-cold PBS. Then, 5 μg pARNG-HBD3 plasmid DNA and 1 μg phiC31 integrase or mutant integrase mRNA were added to the cells. Electroporation was performed with a BTX ECM2001 set to a single pulse of 2 ms at 510 V. At 24 h after electroporation, ~30,000 cells were diluted in a 10-cm dish containing fresh medium supplemented with G418 (600 mg/mL) for selection of stably transfected cells. Individual cell colonies were isolated at 8–12 days after dilution culture.

Quantitative RT-PCR Analyses

Real-time PCR was performed using SYBR Premix Ex TaqTM (TaKaRa, Dalian, China) and a StepOne Plus thermocycler (Applied Biosystems) with the following parameters: 95°C for 10 sec, followed by 40 cycles at 95°C for 5 sec and 60°C for 30 sec. For RT-PCR, total RNA was extracted from each sample using TRIZOL reagent (Invitrogen), and reverse transcription was performed to generate cDNA using a PrimeScript^{TM} RT Reagent Kit (TaKaRa). The RFP gene primers used in absolute quantitative PCR were 5′-GCCACAACACCGTGAAGCT-GAA-3′ (forward) and 5′-ACACCTTGGAGCCGTACTG-GAA-3′ (reverse). The GAPDH gene, as a reference gene in relative quantitative RT-PCR, was amplified using the following primers: 5′-TCAACGGGAAGCTCACTGG-3′ (forward) and 5′-CCCCAGCATCGAAGGTAGA-3′ (reverse). For each DNA and cDNA sample, target and reference genes were amplified independently on the same plate and in the same experimental run in triplicate. PCR specificity was confirmed by gel electrophoresis on a 2.5% agarose gel and by a single peak in the melting curve. For relative quantitative RT-PCR, the amount of target normalized to the reference was calculated by the $2^{-\Delta\Delta Ct}$ method.

Establishment of the Absolute Quantitative Standard Curve

To examine the copy number, generation of an absolute quantitative standard curve was necessary. A series of standard samples containing 0.5, 1, 2, 4, 8, and 16 copies of the RFP gene were prepared as described previously [66] by mixing the wild-type genome of an Holstein cow with the pARNG-HBD3 plasmid. To prepare a standard sample containing one copy of the RFP gene, the quantity of the plasmid mixed with genomic DNA was calculated by the following formula: $\frac{a \times b \times 0.5}{2.45 \times 10^9}$ ("a" represents the size of plasmid). The absolute quantitative standard curve was drawn by plotting Ct values against the log of RFP gene copies of the corresponding standard samples. The parameters of the standard curve were: $\log 2N = -0.9568$ Ct$+25.335$ ($R^2 = 0.9887$, $P < 0.001$).

Half-nested Inverse PCR and Integration Site Analysis

Genomic DNA was isolated from stably transfected cell colonies by phenol/chloroform extraction and ethanol precipitation. Half-nested inverse PCR was performed to identify the pseudo *attP* sites as described previously [38]. Briefly, 5 μg genomic DNA was digested with the isocaudamer enzymes NheI and SpeI (New England Biolabs). After extraction with phenol/chloroform and ethanol precipitation, digests were ligated with a DNA Ligation Kit (TaKaRa) at 4°C overnight in a total reaction volume of 50 μL. Ligated DNA was extracted with phenol/chloroform and ethanol precipitation, and suspended in 25 μL nuclease-free water. Primary PCR was performed with HNI_F1 and HNI_R primers that were designed against *attL* sites (Table S2) as follows: 94°C for 5 min, followed by 30 cycles of 94°C for 30 s, an annealing temperature gradient of 51–61°C for 45 s and 72°C for 5 min, and then 72°C for 10 min. One microliter of the PCR product was used as a template in the secondary reaction. The secondary PCR was performed with the HNI_F2 and HNI_R primers (Table S2). The PCR conditions were the same as those for the primary PCR except for an annealing temperature gradient of 50–60°C. PCR products were run on agarose gels and the resulting bands were excised. The purified product was cloned into a pMD19-T vector (TaKaRa) and sequenced with M13 and reverse M13 primers. Sequences were examined by BLAST searching of the bovine genome databases (http://www.ncbi.nlm.nih.gov/genome/seq/BlastGen/BlastGen.cgi?taxid = 9913).

To analyze repetitive sequences surrounding pseudo *attP* sites, genomic sequences flanking phiC31 integration sites (540 bp) were aligned to bovine repeat sequences using the Repeat Masker Web Server (http://www.repeatmasker.org, Institute for Systems Biology), microRNA around the pseudo *attP* sites was searched using miRBase 18.0 (http://www.mirbase.org/), a cancer related gene and UCRs were obtained from statistics reported previously [21], and the data were downloaded from http://users.soe.ucsc.edu/jill/ultra.html.

Preparation of Cre Recombinant Protein and Cre Protein Transduction

pET-28a(+)-His-NLS-TAT-Cre plasmids were used to transform *E. coli* strain BL21 (DE3) for IPTG-inducible expression of His-tagged Cre protein. Bacteria were inoculated at 1:50 into 500 mL Luria Bertani (LB) medium containing 50 μg/mL kanamycin and grown at 37°C until an OD_{600} of 0.7. Overexpression was induced by addition of IPTG at a final concentration of 0.7 mM, followed by incubation for 8 h at 28°C. Cells were harvested by centrifugation and stored at –20°C. Cell pellets were thawed by resuspension in 20 mL binding buffer (50 mM NaH_2PO_4, pH 8.0, 300 mM NaCl, and 10 mM imidazole). Cleared lysate was obtained after incubation with 1 mg/mL lysozyme and 5 U/mL benzonase (Novagen) and centrifugation for 20 min at $12,000 \times g$ and 4°C. The supernatant was filtered through a 0.45 mm filter (Millipore, Billerica, MA), subsequently added to binding buffer and passed through a His60 Ni Gravity Flow Column (Clontech) to bind the His-tagged Cre protein to the Nickel column. The column was washed with 10 column volumes of His Ni wash buffer (50 mM NaH_2PO_4, pH 8.0, 300 mM NaCl, and 20 mM imidazole). The His-tagged Cre protein was eluted with 10 column volumes of elution buffer (50 mM NaH_2PO_4, pH 8.0, 300 mM NaCl, and 250 mM imidazole) and then dialyzed against opti-MEM for immediate use or storage at −

80°C. Protein concentrations were measured by a BCA Protein Assay Kit (Beyotime, Jiangsu, China).

For Cre transduction experiments, 2×10^6 cells were seeded on a 24-well plate and grown for 24 h or until 90% confluency. Cre protein was dialyzed against opti-MEM and sterilized by filtration using a 0.22 µm filter (Millipore, Billerica, MA). Cells were incubated for 24 h in medium containing 100 µg/mL His-NLS-TAT-Cre protein. After transduction, cells were washed with PBS and cultured for an additional 72 h in normal medium before by flow cytometric analysis and FACS.

Chromosome Analysis

The number of chromosomes in untransfected bovine fetal fibroblasts, and un-excised and excised transgenic cell colonies was determined by Giemsa staining. Slides were prepared by standard techniques. The appropriate spreads were photographed and used to count the number of chromosomes according to the size and position of the centromere.

Southern Blot Analysis

Ten micrograms of genomic DNA from untransfected cells or stably transfected cell colonies was digested overnight with BsrGI and XhoI (New England Biolabs), and resolved by agarose gel electrophoresis. After transfer and UV crosslinking onto a Hybond N+ nylon membrane (Roche, South San Francisco, CA), the DNA was hybridized with an *EGFP* probe generated by a DIG High Prime Labeling and Detection Starter Kit II (Roche). The primers sequences were 5′-AAGTTCATCTGCACCACCG-3′ (forward) and 5′-TGCTCAGGTAGTGGTTGTCG-3′ (reverse).

SCNT, Activation, and Culture of SCNT Embryos

SCNT, activation of reconstructed embryos, and culture of SCNT embryos were performed as described previously [78]. Briefly, matured oocytes were enucleated using a 20 mm inner diameter glass pipette to remove the first polar body and a small amount of surrounding cytoplasm. Successful enucleation was confirmed by Hoechst 33342 staining. A single disaggregated donor cell was injected into the pre-vitelline space of an enucleated oocyte. The oocyte-cell fusion was performed using a pair of platinum electrodes connected to a micromanipulator in microdrops of Zimmermann's fusion medium, and a double electrical pulse of 35 V for 10 ms was used for fusion. Reconstructed SCNT embryos were kept in synthetic oviductal fluid (SOFaa) containing 5 mg/mL cytochalasin B for 2 h until activation. The mSOF medium was prepared according to a formula described previously [79] and supplemented with 8 mg/mL bovine serum albumin, 1% MEM non-essential amino acids and 2% BME essential amino acids. Activation of reconstructed embryos was performed in 5 mM ionomycin for 4 min followed by 4 h exposure to 1.9 mM dimethynopyridine in SOFaa. After activation, embryos were cultured in G1.3/G2.3 sequential media (Vitrolife AB, Gothenburg, Sweden). Droplets of 150 µL G1.3 were prepared in a 35-mm cell culture dish under mineral oil and equilibrated for 2 h before loading the embryos (20 embryos/microdrop). Embryos were transferred to G2.3 droplets on day 3 of culture (day 0 being the day of SCNT).

Western Blot Analysis

Daily milk weights were recorded and milk sample was collected once a month for 6 months. The percentages (w/vol) of fat, protein, lactose, and solids were determined using a MilkoScan 6000 (Foss, Hillerod, Denmark). Milk samples and standards were separated by 10% Tricine-SDS-PAGE and electrotransferred onto polyvinylidene fluoride membranes (Millipore, Billerica, MA) for western blot analysis using a standard protocol. A rabbit polyclonal antibody against human β-defensin-3 (1:1000, Sigma) and horseradish peroxidase-conjugated goat anti-rabbit IgG (1:1000, Beyotime) were used to detect human β-defensin-3, and a mouse monoclonal antibody against EGFP (1:5000, Clontech) and horseradish peroxidase-conjugated goat anti-mouse IgG (1:1000, Beyotime) were used to detect GFP. Positive controls were commercial human β-defensin-3 (Sigma) for detection of human β-defensin-3, and a cell lysate of excised colonies for detection of GFP. The negative control was non-transgenic milk. Human β-defensin-3 milk concentration was determined by ELISA as per manufacturer's directions (USCNK Life Sciences, Wuhan, China).

Agar Diffusion Test

The disc diffusion method was performed with LB agar (Sigma). *S. aureus* (ATCC 25923) and *E. coli* (ATCC 25922) at mid-log phase ($A_{600}<0.6$–0.7) was mixed with 20 mL solid culture medium containing 1.5% agar. Each sample was placed on a sterile, quantitative filter paper disc (6 mm in diameter) on a plate, and incubated for 24 h at 37°C. The results were assessed by inhibition zones around disc paper. Milk sample from a non-transgenic cow were placed on the filter paper disc as a negative control. The experiment was repeated three times.

Challenge Study

Forty-eight hours before initiating the bacterial challenges, health of the animals was assessed by differential leukocyte and milk somatic cell counts to verify that they were within normal ranges. Milk from each quarter, cultured overnight also had to be free of bacterial growth for the animal to be included in the study. Since the udder contains four glands that function independently of one another, it is possible to expose a cow to multiple treatments, as was done in these experiments. After the morning milking, an aseptic milk sample was collected from each of the four glands before infusing 2 mL of different bacterial cultures of *S. aureus* (40 ± 2 c.f.u/mL, diluted with PBS) or *E. coli* (40 ± 3 c.f.u./mL, diluted with PBS) into one of two contralateral quarters via the streak canal. A third quarter received a similar infusion of sterile PBS only, and the fourth gland remained no infusion. Treatments were assigned such that no gland received the same strain in subsequent trials. The milk samples were taken at 12-h intervals or more frequently throughout the study. Milk samples (20 µL) were plated on blood agar and MacConkey agar incubated at 37°C for 18 to 24 h. An infection was confirmed by the presence of viable *S. aureus* growing on blood agar plates or *E. coli* growing on MacConkey agar in two consecutive milk samples.

Statistical Analysis

Each experiment was performed at least three times. All data were analyzed using SPSS 20.0 statistical software (IBM Corporation, Somers, NY, USA). Data were tested by one-way ANOVA and least-significant difference tests, and reported as the mean ± SEM. For all analyses, $P<0.05$ was considered significant.

Supporting Information

Figure S1 Identification of transgenic colonies in the integrase group and the mutant integrase group. *AttB* cleavage assay of the phiC31 integrase (a) and the mutant integrase (b) groups. All RFP+ colonies were subjected to the PCR test for the full-length *attB* site. A band of 426 bp for the un-cleaved *attB* site indicated non-site-specific integration. Genomic DNA from untransfected bovine fetal fibroblasts was used as a negative control, and PL (pARNG-HBD3) was used as a positive control. (c) Copy number assay of the 23 *attB*-cleaved colonies by absolute quantity PCR. Black columns represent a transgenic cell colony with a single copy integration, and gray columns represent double copy integration. Three or more copy integration events were not detected. Error bars denote SEM. (d) All 46 G418-resistant colonies were subjected to junction PCR with primers specific for the safe harbor and adjacent *attB* site. Genomic DNA from untransfected bovine fetal fibroblasts was used as a negative control.

Figure S2 FACS analysis after His-NLS-TAT-Cre protein transduction into three RFP+ cells lines. At Day 5 after protein transduction, cells were trypsinized and resuspended in PBS containing 10% FBS, and then analyzed for RFP and GFP expression by flow cytometry. More than 70% of the transduced cells were GFP+ as shown by flow cytometry. BFF cells were untransfected control.

Figure S3 Further analysis of antibiotic selectable marker free transgenic cells. (a) PCR to test for pCMVInt and pCAG-Cre-IP showing the absence of either phiC31 integrase or Cre recombinase encoding DNA in established excised transgenic colonies. Genomic DNA from untransfected bovine fetal fibroblasts was used as a negative control, and PL1 (pCMVInt plasmid) and PL2 (pCAG-Cre-IP plasmid) were used as positive controls. (b) Verification of the genomic integration sites of the established excised cells by junction PCR using pairs of the respective genomic and plasmid-binding primers. (c) Relative real-time RT-PCR analysis of *KIAA1486* and *IRS-1* gene expression in un-excised and excised transgenic cells compared with that in untransfected bovine fetal fibroblasts normalized to 1. Error bars denote SEM.

Figure S4 Chromosome counts. Metaphase spreads of un-excised and excised transgenic cells were counted and compared to untransfected bovine fetal fibroblasts.

Figure S5 PCR analysis of transgenic cattle. M, DNA ladder; WT, non-transgenic cattle; and PL, positive control from the constructed vector; line 1–12, marker-free transgenic cattle mTG1-12.

Figure S6 Human β-defensin-3 concentration during the first lactation period of transgenic cows. Milk was collected once a month for 6 months. Error bars denote SEM.

Figure S7 Lytic activity of human β-defensin-3 and milk samples from transgenic cows against *S. aureus* and *E. coli*. The small circles (6 mm in diameter) consist of quantitative filter paper with 10 μL of test sample or skim milk from non-transgenic cows (control). The larger circles are the inhibition zones. A, commercial human β-defensin-3; B–F, milk samples from transgenic cows mTG1–5; and G, milk samples from non-transgenic cows (control). The bar indicates 10 mm.

Figure S8 Generation of antibiotic selectable marker free transgenic cells using site-specific recombinase. To generate competent transgenic donor cells for SCNT efficiently, five main steps were used in our study as follows. Step 1: Co-electroporation with the transgene incoming plasmid and phiC31 integrase mRNA; Step 2: Generation of G418-resistant and RFP+ colonies; Step 3: Identification (*attB* cleavage assay → copy number analysis of colonies with cleaved *attB* site → integration site analysis of single-copy integrants) and proliferation; Step 4: Cre protein transduction into "safe harbor" integrated colonies; Step 5: Sorting cells showing both GFP+ and RFP− by FACS for SCNT.

Table S1 PhiC31-mediated integration into the bovine genome. Pseudo *attP* sites in Italian were from statistics reported in several other papers, the rest ones were found in our study. The pseudo sites presented in bold met all the safe harbor criteria in Papapetrou et al.

Table S2 Primers used for SOE-PCR, gene amplification, qRT-PCR and the detection of pCMVInt and pCAG-Cre-IP.

Acknowledgments

We thank Dr. M. P. Calos (Stanford University School of Medicine, USA) for the pCMV-Int plasmid and Dr. Yongyan Wu for his technical assistance. We also thank Mr. Younan Wang for transportation of the Holstein cow ovaries used in this study.

Author Contributions

Conceived and designed the experiments: YY YZ. Performed the experiments: YY YSW QT XL FS FSQ ZKG. Analyzed the data: YY YSW YZ. Contributed reagents/materials/analysis tools: YY YSW. Wrote the paper: YY YSW YZ.

References

1. Kind A, Schnieke A (2008) Animal pharming, two decades on. Transgenic Res 17: 1025–1033.
2. Wheeler MB (2007) Agricultural applications for transgenic livestock. Trends Biotechnol 25: 204–210.
3. Jura J, Smorag Z, Slomski R, Lipinski D, Gajda B (2007) Factors affecting the production of potential transgenic pigs by DNA microinjection; a six-year retrospective study. Journal of Animal and Feed Sciences 16: 636–645.
4. Hofmann A, Kessler B, Ewerling S, Kabermann A, Brem G, et al. (2006) Epigenetic regulation of lentiviral transgene vectors in a large animal model. Mol Ther 13: 59–66.
5. Baguisi A, Behboodi E, Melican DT, Pollock JS, Destrempes MM, et al. (1999) Production of goats by somatic cell nuclear transfer. Nat Biotechnol 17: 456–461.
6. Hyun S, Lee G, Kim D, Kim H, Lee S, et al. (2003) Production of nuclear transfer-derived piglets using porcine fetal fibroblasts transfected with the enhanced green fluorescent protein. Biol Reprod 69: 1060–1068.
7. Gorman C, Bullock C (2000) Site-specific gene targeting for gene expression in eukaryotes. Curr Opin Biotechnol 11: 455–460.

8. Dobie KW, Lee M, Fantes JA, Graham E, Clark AJ, et al. (1996) Variegated transgene expression in mouse mammary gland is determined by the transgene integration locus. PROCEEDINGS OF THE NATIONAL ACADEMY OF SCIENCES OF THE UNITED STATES OF AMERICA 93: 6659–6664.
9. Clark AJ, Harold G, Yull FE (1997) Mammalian cDNA and prokaryotic reporter sequences silence adjacent transgenes in transgenic mice. Nucleic Acids Res 25: 1009–1014.
10. Ramirez A, Milot E, Ponsa I, Marcos-Gutierrez C, Page A, et al. (2001) Sequence and chromosomal context effects on variegated expression of keratin 5/lacZ constructs in stratified epithelia of transgenic mice. Genetics 158: 341–350.
11. Pilbrough W, Munro TP, Gray P (2009) Intraclonal protein expression heterogeneity in recombinant CHO cells. Plos One 4: e8432.
12. Costa M, Dottori M, Sourris K, Jamshidi P, Hatzistavrou T, et al. (2007) A method for genetic modification of human embryonic stem cells using electroporation. Nat Protoc 2: 792–796.
13. Schroder AR, Shinn P, Chen H, Berry C, Ecker JR, et al. (2002) HIV-1 integration in the human genome favors active genes and local hotspots. Cell 110: 521–529.

14. Mitchell RS, Beitzel BF, Schroder AR, Shinn P, Chen H, et al. (2004) Retroviral DNA integration: ASLV, HIV, and MLV show distinct target site preferences. PLoS Biol 2: E234.

15. Thyagarajan B, Olivares EC, Hollis RP, Ginsburg DS, Calos MP (2001) Site-specific genomic integration in mammalian cells mediated by phage phiC31 integrase. Mol Cell Biol 21: 3926–3934.

16. Chalberg TW, Portlock JL, Olivares EC, Thyagarajan B, Kirby PJ, et al. (2006) Integration specificity of phage phiC31 integrase in the human genome. J Mol Biol 357: 28–48.

17. Ortiz-Urda S, Thyagarajan B, Keene DR, Lin Q, Fang M, et al. (2002) Stable nonviral genetic correction of inherited human skin disease. Nat Med 8: 1166–1170.

18. Quenneville SP, Chapdelaine P, Rousseau J, Beaulieu J, Caron NJ, et al. (2004) Nucleofection of muscle-derived stem cells and myoblasts with phiC31 integrase: stable expression of a full-length-dystrophin fusion gene by human myoblasts. Mol Ther 10: 679–687.

19. Ishikawa Y, Tanaka N, Murakami K, Uchiyama T, Kumaki S, et al. (2006) Phage phiC31 integrase-mediated genomic integration of the common cytokine receptor gamma chain in human T-cell lines. J Gene Med 8: 646–653.

20. Karow M, Chavez CL, Farruggio AP, Geisinger JM, Keravala A, et al. (2011) Site-specific recombinase strategy to create induced pluripotent stem cells efficiently with plasmid DNA. Stem Cells 29: 1696–1704.

21. Papapetrou EP, Lee G, Malani N, Setty M, Riviere I, et al. (2011) Genomic safe harbors permit high beta-globin transgene expression in thalassemia induced pluripotent stem cells. Nat Biotechnol 29: 73–78.

22. Pham CT, MacIvor DM, Hug BA, Heusel JW, Ley TJ (1996) Long-range disruption of gene expression by a selectable marker cassette. Proc Natl Acad Sci U S A 93: 13090–13095.

23. Kuroiwa Y, Kasinathan P, Matsushita H, Sathiyaselan J, Sullivan EJ, et al. (2004) Sequential targeting of the genes encoding immunoglobulin-mu and prion protein in cattle. Nat Genet 36: 775–780.

24. Sendai Y, Sawada T, Urakawa M, Shinkai Y, Kubota K, et al. (2006) alpha1,3-Galactosyltransferase-gene knockout in cattle using a single targeting vector with loxP sequences and cre-expressing adenovirus. Transplantation 81: 760–766.

25. Xu Y, Liu S, Yu G, Chen J, Xu X, et al. (2008) Excision of selectable genes from transgenic goat cells by a protein transducible TAT-Cre recombinase. Gene 419: 70–74.

26. Wang S, Sun X, Ding F, Zhang K, Zhao R, et al. (2009) Removal of selectable marker gene from fibroblast cells in transgenic cloned cattle by transient expression of Cre recombinase and subsequent effects on recloned embryo development. Theriogenology 72: 535–541.

27. Dale EC, Ow DW (1991) Gene transfer with subsequent removal of the selection gene from the host genome. Proc Natl Acad Sci U S A 88: 10558–10562.

28. Endo S, Sugita K, Sakai M, Tanaka H, Ebinuma H (2002) Single-step transformation for generating marker-free transgenic rice using the ipt-type MAT vector system. Plant J 30: 115–122.

29. Huang S, Gilbertson LA, Adams TH, Malloy KP, Reisenbigler EK, et al. (2004) Generation of marker-free transgenic maize by regular two-border Agrobacterium transformation vectors. Transgenic Res 13: 451–461.

30. Li B, Li N, Duan X, Wei A, Yang A, et al. (2010) Generation of marker-free transgenic maize with improved salt tolerance using the FLP/FRT recombination system. J Biotechnol 145: 206–213.

31. Iamtham S, Day A (2000) Removal of antibiotic resistance genes from transgenic tobacco plastids. Nat Biotechnol 18: 1172–1176.

32. Zhang Y, Liu H, Li B, Zhang JT, Li Y, et al. (2009) Generation of selectable marker-free transgenic tomato resistant to drought, cold and oxidative stress using the Cre/loxP DNA excision system. Transgenic Res 18: 607–619.

33. Van Deursen J, Fornerod M, Van Rees B, Grosveld G (1995) Cre-mediated site-specific translocation between nonhomologous mouse chromosomes. Proc Natl Acad Sci U S A 92: 7376–7380.

34. Woltjen K, Michael IP, Mohseni P, Desai R, Mileikovsky M, et al. (2009) piggyBac transposition reprograms fibroblasts to induced pluripotent stem cells. Nature 458: 766–770.

35. Yusa K, Rad R, Takeda J, Bradley A (2009) Generation of transgene-free induced pluripotent mouse stem cells by the piggyBac transposon. Nat Methods 6: 363–369.

36. Qu L, Ma Q, Zhou Z, Ma H, Huang Y, et al. (2012) A Profile of Native Integration Sites Used by phiC31 Integrase in the Bovine Genome. J Genet Genomics 39: 217–224.

37. Ma QW, Sheng HQ, Yan JB, Cheng S, Huang Y, et al. (2006) Identification of pseudo attP sites for phage phiC31 integrase in bovine genome. Biochem Biophys Res Commun 345: 984–988.

38. Ou HL, Huang Y, Qu LJ, Xu M, Yan JB, et al. (2009) A phiC31 integrase-mediated integration hotspot in favor of transgene expression exists in the bovine genome. FEBS J 276: 155–163.

39. Ye L, Chang JC, Lin C, Qi Z, Yu J, et al. (2010) Generation of induced pluripotent stem cells using site-specific integration with phage integrase. Proc Natl Acad Sci U S A 107: 19467–19472.

40. Liu J, Jeppesen I, Nielsen K, Jensen TG (2006) Phi c31 integrase induces chromosomal aberrations in primary human fibroblasts. Gene Ther 13: 1188–1190.

41. Liu J, Skjorringe T, Gjetting T, Jensen TG (2009) PhiC31 integrase induces a DNA damage response and chromosomal rearrangements in human adult fibroblasts. BMC Biotechnol 9: 31.

42. Loonstra A, Vooijs M, Beverloo HB, Allak BA, van Drunen E, et al. (2001) Growth inhibition and DNA damage induced by Cre recombinase in mammalian cells. Proc Natl Acad Sci U S A 98: 9209–9214.

43. Silver DP, Livingston DM (2001) Self-excising retroviral vectors encoding the Cre recombinase overcome Cre-mediated cellular toxicity. Mol Cell 8: 233–243.

44. Baba Y, Nakano M, Yamada Y, Saito I, Kanegae Y (2005) Practical range of effective dose for Cre recombinase-expressing recombinant adenovirus without cell toxicity in mammalian cells. Microbiol Immunol 49: 559–570.

45. Nolden L, Edenhofer F, Haupt S, Koch P, Wunderlich FT, et al. (2006) Site-specific recombination in human embryonic stem cells induced by cell-permeant Cre recombinase. Nat Methods 3: 461–467.

46. Peitz M, Pfannkuche K, Rajewsky K, Edenhofer F (2002) Ability of the hydrophobic FGF and basic TAT peptides to promote cellular uptake of recombinant Cre recombinase: a tool for efficient genetic engineering of mammalian genomes. Proc Natl Acad Sci U S A 99: 4489–4494.

47. Brophy B, Smolenski G, Wheeler T, Wells D, L'Huillier P, et al. (2003) Cloned transgenic cattle produce milk with higher levels of beta-casein and kappa-casein. Nat Biotechnol 21: 157–162.

48. Zakhartchenko V, Mueller S, Alberio R, Schernthaner W, Stojkovic M, et al. (2001) Nuclear transfer in cattle with non-transfected and transfected fetal or cloned transgenic fetal and postnatal fibroblasts. Mol Reprod Dev 60: 362–369.

49. Heyman Y, Zhou Q, Lebourhis D, Chavatte-Palmer P, Renard JP, et al. (2002) Novel approaches and hurdles to somatic cloning in cattle. Cloning Stem Cells 4: 47–55.

50. Batchelder CA, Hoffert KA, Bertolini M, Moyer AL, Mason JB, et al. (2005) Effect of the nuclear-donor cell lineage, type, and cell donor on development of somatic cell nuclear transfer embryos in cattle. Cloning Stem Cells 7: 238–254.

51. Oback B, Wells D (2002) Donor cells for nuclear cloning: many are called, but few are chosen. Cloning Stem Cells 4: 147–168.

52. Rizos D, Burke L, Duffy P, Wade M, Mee JF, et al. (2005) Comparisons between nulliparous heifers and cows as oocyte donors for embryo production in vitro. Theriogenology 63: 939–949.

53. Bhak JS, Lee SL, Ock SA, Mohana Kumar B, Choe SY, et al. (2006) Developmental rate and ploidy of embryos produced by nuclear transfer with different activation treatments in cattle. Anim Reprod Sci 92: 37–49.

54. Shin MR, Park SW, Shim H, Kim NH (2002) Nuclear and microtubule reorganization in nuclear-transferred bovine embryos. Mol Reprod Dev 62: 74–82.

55. Lim KT, Jang G, Ko KH, Lee WW, Park HJ, et al. (2007) Improved in vitro bovine embryo development and increased efficiency in producing viable calves using defined media. Theriogenology 67: 293–302.

56. Harder J, Bartels J, Christophers E, Schroder JM (2001) Isolation and characterization of human beta -defensin-3, a novel human inducible peptide antibiotic. J Biol Chem 276: 5707–5713.

57. Dunsche A, Acil Y, Dommisch H, Siebert R, Schroder JM, et al. (2002) The novel human beta-defensin-3 is widely expressed in oral tissues. Eur J Oral Sci 110: 121–124.

58. Dhople V, Krukemeyer A, Ramamoorthy A (2006) The human beta-defensin-3, an antibacterial peptide with multiple biological functions. Biochim Biophys Acta 1758: 1499–1512.

59. Maga EA, Cullor JS, Smith W, Anderson GB, Murray JD (2006) Human lysozyme expressed in the mammary gland of transgenic dairy goats can inhibit the growth of bacteria that cause mastitis and the cold-spoilage of milk. Foodborne Pathog Dis 3: 384–392.

60. Yang B, Wang J, Tang B, Liu Y, Guo C, et al. (2011) Characterization of bioactive recombinant human lysozyme expressed in milk of cloned transgenic cattle. Plos One 6: e17593.

61. Wall RJ, Powell AM, Paape MJ, Kerr DE, Bannerman DD, et al. (2005) Genetically enhanced cows resist intramammary Staphylococcus aureus infection. Nat Biotechnol 23: 445–451.

62. Chen X, Niyonsaba F, Ushio H, Okuda D, Nagaoka I, et al. (2005) Synergistic effect of antibacterial agents human beta-defensins, cathelicidin LL-37 and lysozyme against Staphylococcus aureus and Escherichia coli. J Dermatol Sci 40: 123–132.

63. Midorikawa K, Ouhara K, Komatsuzawa H, Kawai T, Yamada S, et al. (2003) Staphylococcus aureus susceptibility to innate antimicrobial peptides, beta-defensins and CAP18, expressed by human keratinocytes. Infect Immun 71: 3730–3739.

64. Joly S, Maze C, McCray PB Jr, Guthmiller JM (2004) Human beta-defensins 2 and 3 demonstrate strain-selective activity against oral microorganisms. J Clin Microbiol 42: 1024–1029.

65. Yang P, Wang J, Gong G, Sun X, Zhang R, et al. (2008) Cattle mammary bioreactor generated by a novel procedure of transgenic cloning for large-scale production of functional human lactoferrin. Plos One 3: e3453.

66. Kong Q, Wu M, Huan Y, Zhang L, Liu H, et al. (2009) Transgene expression is associated with copy number and cytomegalovirus promoter methylation in transgenic pigs. Plos One 4: e6679.

67. Kerr DE, Plaut K, Bramley AJ, Williamson CM, Lax AJ, et al. (2001) Lysostaphin expression in mammary glands confers protection against staphylococcal infection in transgenic mice. Nat Biotechnol 19: 66–70.

68. Edwards JL, Schrick FN, McCracken MD, van Amstel SR, Hopkins FM, et al. (2003) Cloning adult farm animals: a review of the possibilities and problems associated with somatic cell nuclear transfer. Am J Reprod Immunol 50: 113–123.

69. Wuensch A, Habermann FA, Kurosaka S, Klose R, Zakhartchenko V, et al. (2007) Quantitative monitoring of pluripotency gene activation after somatic cloning in cattle. Biol Reprod 76: 983–991.

70. Krepulat F, Lohler J, Heinlein C, Hermannstadter A, Tolstonog GV, et al. (2005) Epigenetic mechanisms affect mutant p53 transgene expression in WAP-mutp53 transgenic mice. Oncogene 24: 4645–4659.

71. Bannerman DD, Paape MJ, Lee JW, Zhao X, Hope JC, et al. (2004) Escherichia coli and Staphylococcus aureus elicit differential innate immune responses following intramammary infection. Clin Diagn Lab Immunol 11: 463–472.

72. Arat S, Gibbons J, Rzucidlo SJ, Respess DS, Tumlin M, et al. (2002) In vitro development of bovine nuclear transfer embryos from transgenic clonal lines of adult and fetal fibroblast cells of the same genotype. Biol Reprod 66: 1768–1774.

73. Bordignon V, Keyston R, Lazaris A, Bilodeau AS, Pontes JH, et al. (2003) Transgene expression of green fluorescent protein and germ line transmission in cloned calves derived from in vitro-transfected somatic cells. Biol Reprod 68: 2013–2023.

74. Chen SH, Vaught TD, Monahan JA, Boone J, Emslie E, et al. (2002) Efficient production of transgenic cloned calves using preimplantation screening. Biol Reprod 67: 1488–1492.

75. Roh S, Shim H, Hwang WS, Yoon JT (2000) In vitro development of green fluorescent protein (GFP) transgenic bovine embryos after nuclear transfer using different cell cycles and passages of fetal fibroblasts. Reprod Fertil Dev 12: 1–6.

76. Huber RC, Remuge L, Carlisle A, Lillico S, Sandoe P, et al. (2011) Welfare assessment in transgenic pigs expressing green fluorescent protein (GFP). Transgenic Res.

77. Li P, Tong C, Mehrian-Shai R, Jia L, Wu N, et al. (2008) Germline competent embryonic stem cells derived from rat blastocysts. Cell 135: 1299–1310.

78. Wang Y, Su J, Wang L, Xu W, Quan F, et al. (2011) The effects of 5-aza-2′-deoxycytidine and trichostatin A on gene expression and DNA methylation status in cloned bovine blastocysts. Cell Reprogram 13: 297–306.

79. Takahashi Y, First NL (1992) In vitro development of bovine one-cell embryos: Influence of glucose, lactate, pyruvate, amino acids and vitamins. Theriogenology 37: 963–978.

4

RNAi-Mediated Gene Suppression in a GCAP1(L151F) Cone-Rod Dystrophy Mouse Model

Li Jiang[1]*, Tansy Z. Li[1], Shannon E. Boye[2], William W. Hauswirth[2], Jeanne M. Frederick[1], Wolfgang Baehr[1,3,4]*

1 Department of Ophthalmology and Visual Sciences, University of Utah Health Science Center, Salt Lake City, Utah, United States of America, 2 Department of Ophthalmology, University of Florida College of Medicine, Gainesville, Florida, United States of America, 3 Department of Biology, University of Utah, Salt Lake City, Utah, United States of America, 4 Department of Neurobiology and Anatomy, University of Utah Health Science Center, Salt Lake City Utah, United States of America

Abstract

Dominant mutations occurring in the high-affinity Ca^{2+}-binding sites (EF-hands) of the *GUCA1A* gene encoding guanylate cyclase-activating protein 1 (GCAP1) cause slowly progressing cone-rod dystrophy (CORD) in a dozen families worldwide. We developed a nonallele-specific adeno-associated virus (AAV)-based RNAi knockdown strategy to rescue the retina degeneration caused by GCAP1 mutations. We generated three genomic transgenic mouse lines expressing wildtype (WT) and L151F mutant mouse GCAP1 with or without a C-terminal GFP fusion. Under control of endogenous regulatory elements, the transgenes were expressed specifically in mouse photoreceptors. GCAP1(L151F) and GCAP1(L151F)-GFP transgenic mice presented with a late onset and slowly progressive photoreceptor degeneration, similar to that observed in human GCAP1-CORD patients. Transgenic expression of WT GCAP1-EGFP in photoreceptors had no adverse effect. Toward therapy development, a highly effective anti-mGCAP1 shRNA, mG1hp4, was selected from four candidate shRNAs using an *in-vitro* screening assay. Subsequently a self-complementary (sc) AAV serotype 2/8 expressing mG1hp4 was delivered subretinally to GCAP1(L151F)-GFP transgenic mice. Knockdown of the GCAP1(L151F)-GFP transgene product was visualized by fluorescence live imaging in the scAAV2/8-mG1hp4-treated retinas. Concomitant with the mutant GCAP1-GFP fusion protein, endogenous GCAP1 decreased as well in treated retinas. We propose nonallele-specific RNAi knockdown of GCAP1 as a general therapeutic strategy to rescue any GCAP1-based dominant cone-rod dystrophy in human patients.

Editor: Anand Swaroop, National Eye Institute, United States of America

Funding: Supported by National Institutes of Health Grants EY08123 and EY019298 (WB), Grant EY014800-039003 (National Eye Institute core grant), EY021721 (to WWH); a Center Grant from the Foundation Fighting Blindness, Inc. (http://www.blindness.org/), to the University of Utah; and unrestricted grants to the Department of Ophthalmology at the University of Utah from Research to Prevent Blindness. WB is a recipient of a Research to Prevent Blindness (http://www.rpbusa.org/rpb/) Senior Investigator Award. WWH was supported by grants from the Foundation Fighting Blindness (http://www.blindness.org/), Macula Vision Research Foundation (http://www.mvrf.org/index.php), and Research to Prevent Blindness, Inc.(http://www.rpbusa.org/rpb/). The funders had no role in study design, data collection and analysis, decision to publish, or preparation of the manuscript.

Competing Interests: WWH and the University of Florida at Gainesville have a financial interest in the use of AAV therapies, and own equity in a company (AGTC, Inc.) that may commercialize portions of this work.

* E-mail: li.jiang@hsc.utah.edu (LJ); wbaehr@hsc.utah.edu (WB)

Introduction

Cone-rod dystrophy (CORD, with a prevalence of 1/40,000) is a rare, highly heterogeneous class of hereditary retinal disease inherited in a dominant, recessive or X-linked fashion [1]. The disease manifests with photoaversion, reduced central visual acuity, achromatopsia at early stages, and eventually loss of peripheral vision attributed to progressive loss of first cone and then rod photoreceptors. Thus far, 27 genes have been linked to cone-rod dystrophy; of these, ten genes are associated with dominant CORD, 15 with recessive CORD, two are X-linked (RetNet, https://sph.uth.tmc.edu/retnet/). The protein products of these genes are involved in multiple aspects of photoreceptor structure and function [2]. One of the best characterized dominant CORD genes is *GUCA1A* encoding guanylate cyclase-activating protein 1 (GCAP1 [3]. Worldwide, about one dozen families with more than 100 affected members have been identified to date [4].

GCAP1 plays a key role in accelerating guanylate cyclase activity in retinal photoreceptors. Rod phototransduction is regulated by two guanylate cyclases (GC1 and GC2) and two

GCAPs (GCAP1 and GCAP2) [5]. The two GCAP genes (*Guca1a* and *Guca1b*) are arranged in a tail-to-tail array on mouse chromosome 17 [6] while the GC genes (*Gucy2e* and *Gucy2f*) are located on different chromosomes (chromosomes 11 and X, [7]). The two GCAPs overlap partially in regulating the GCs of rods, and both contribute to rod recovery after photolysis [8–11]. In cones, only GCAP1 is involved in regulating GC1, the predominant GC of cone phototransduction. Germline deletion of both GCAPs renders GC activity Ca^{2+} insensitive; flash responses from dark-adapted rods were larger and slower, and recovery to the dark state was delayed [8]. By contrast, cone ERGs recorded from GCAPs$^{-/-}$ mice had normally saturated a-wave and b-wave amplitudes and increased sensitivity of both M- and S-cone systems [12]. Both the cone driven b-wave and a-wave were delayed similarly as observed in rods. Transgenic GCAP1 could restore normal rod and cone response recovery [9,12].

As members of the calmodulin superfamily, GCAPs feature four EF-hand motifs (EF1-4) for Ca^{2+}-binding [3,13]. Three of these (EF2-4) have been established as high-affinity Ca^{2+}-binding sites, whereas motif EF1 is incompatible with Ca^{2+}-binding as key residues essential for Ca^{2+} coordination are absent [14]. Mutations in GCAP1, but not GCAP2, have been associated with autosomal dominant CORD3 [15]. Missense mutations in GCAP1 include EF3 mutations (E89K, Y99C, D100E, N104K) and EF4 mutations (I143NT, **L151F**, E155G, G159V) (**Fig. 1**) [16–23]. These dominant GCAP1 mutations alter Ca^{2+}-association, decrease Ca^{2+} sensitivity and produce constitutive activity of photoreceptor guanylate cyclase 1 at normal dark Ca^{2+} levels, and persistent stimulation of GC1 in the dark [21,22]. Elevated intracellular cGMP and Ca^{2+} trigger cell death which can be ameliorated by increased cGMP hydrolysis [24].

The L151F mutation was identified in two independent families with cone and cone-rod dystrophy, respectively [17,23]. In the first family [17], dyschromatopsia, hemeralopia and reduced visual acuity were evident by the second-to-third decades of life and photopic electroretinographic (ERG) responses were non-recordable. In the second, five-generation family [23], 11 of 24 individuals were affected by photoaversion, color vision defects, central acuity loss and legal blindness in decades 2–3 of life. The causative GCAP1(L151F) mutation, a conservative substitution, disrupted Ca^{2+} coordination at EF-hand 4 and changed Ca^{2+} sensitivity. Molecular dynamics suggested a significant decrease in Ca^{2+}-binding to EF-hand motifs 2 and 4, with an overall shape change of the L151F-GCAP1 molecule relative to wildtype [17].

Two mouse models of retinal dystrophy associated with GCAP1 mutations have been generated previously. The first, a transgenic mouse model expressing bovine GCAP1(Y99C) under control of the opsin promoter, produced severe and rapid rod-cone degeneration resembling retinitis pigmentosa [25]. The second model, a GCAP1(E155G) knock-in mouse, displayed late-onset and slowly progressive cone-rod photoreceptor degeneration, mimicking human CORD [26]. In this study, we generated three transgenic mouse lines expressing wildtype and L151F mutant GCAP1 with or without a C-terminal EGFP fusion. The EGFP

tag was instrumental toward developing an effective knockdown RNA interference and served to distinguish transgenic GCAPs from endogenous GCAPs. All transgenes contained the complete mouse *Guca1a* gene, including native regulatory elements. Mutant transgenic mice develop retina pathology slowly and recapitulate features of human CORD. To develop a vector suitable for knockdown gene therapy, we generated scAAV2/8 virus that expresses a nonallele-specific shRNA that targets both mutant and native *Guca1a* mRNAs. By immunoblot and *in-vivo* fundoscopy, we show that both transgenic and endogenous GCAP1 were down-regulated effectively in AAV-treated retinas. These data establish shRNA-mediated RNAi as a potential therapeutic strategy for adCORD patients carrying any EF-hand *GUCA1A* mutation.

Materials and Methods

Mice (Ethics Statement)

Procedures for the animal experiments of this study were IACUC-approved by the University of Utah and conformed to recommendations of the Association of Research for Vision and Ophthalmology (ARVO). Transgenic and wildtype (WT) mice were maintained under 12-hour cyclic dark/light conditions.

Cloning of mGCAP1 Genomic Constructs

A 14,832 bp mouse GCAP1 genomic sequence (mG1) was modified to generate three transgenes which expressed either wildtype GCAP1 fused to EGFP (G1-GFP), or one of two mutant proteins, G1(L151F) and G1(L151F)-GFP. To generate a GFP fusion mGCAP1 transgenic construct, G1-GFP, a chloramphenicol-resistant cassette flanked by AsisI and AscI sites was first inserted into the wildtype mGCAP1 genomic construct right before the stop codon using a homologous recombination method, termed ET cloning [27]. We amplified by PCR the AsisI-AscI chloramphenicol-resistant cassette with primer pair, G1_CmRasisIF: 5′-GCGAACACGAGGAGGCAGGCACCGGC-

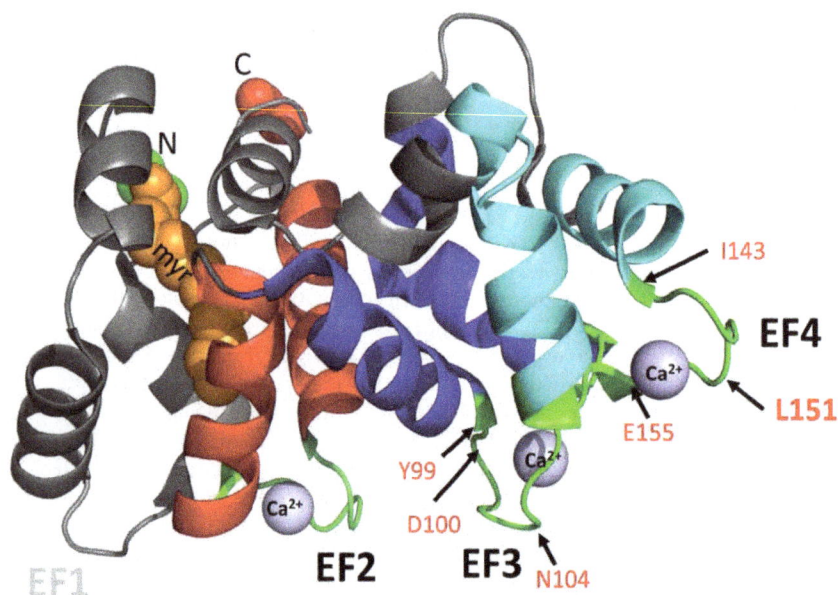

Figure 1. GCAP1 mutations causative for CORD. Ribbon structure of N-myristoylated GCAP1. N, N-terminus; C, C-terminus; myr, myristoyl side chain (orange) attached to Gly-2; Ca^{2+} ions (blue-gray). EF-hands consist of helix-loop-helix domains. Two helices flanking EF-hand 2 (EF2, red), two helices flanking EF3 (dark blue) and the two helices flanking EF4 (light blue) are shown. Amino acid residues mutated in dominant CORD cluster around EF3 and EF4. Positions (black arrows) of mutant residues (red) are indicated: E89K, Y99C, D100E and N104K are located in EF-hand 3 (EF3), while I143NT, L151F and E155G are located in EF-hand 4 (EF4).

GACCTGGCAGCGGAGGCTGCGGGTGCGATCGCagcat-tacacgtcttgagcgattgt, and G1_CmRascIR: 5'-AC-CGCACGGGGCCAGCCCTCAGCAGGCAGAAGCCA-CAGGGTGAATGCTCAGGCGCGCCCacttaacggctgacatgg-gaatta. Relevant to primer design are regions homologous to 5' and 3' flanking sequences (50-bp, shown in black) of the mGCAP1 stop codon (bold-faced) in the transgene, AsisI and AscI restriction sites (underlined), and regions complementary to the 5' and 3' sequences of a chloramphenicol-resistant cassette (lower case), respectively. The purified PCR product of the chloramphenicol-resistant cassette was co-electroporated with the mGCAP1 transgenic construct (containing an ampicillin-resistant cassette) into competent cells containing an inducible Red recombinase. Ampicillin and chloramphenicol 'double'-resistant colonies were selected, in which a Red homologous recombination occurred and indicated that the chloramphenicol-resistant cassette with flanking AsisI and AscI sites was inserted immediately before the mGCAP1 stop codon. Subsequently, we replaced the chloramphenicol-resistant cassette in the recombinant mGCAP1 genomic construct with an AsisI-AscI EGFP cassette.

To generate the G1 (L151F) mutant transgene, a C to T point mutation in codon 151 was introduced by site-directed mutagenesis into the AatII-AatII mGCAP1 fragment (3.3-kb) containing codon L151 (Stratagene, La Jolla, CA). The pair of DNA oligonucleotides used for mutagenesis was: G1(L151F)_F: 5'-TTTCTCTCCATCCCAGGGGAATTGTCCCTGGAG-GAG, and G1(L151F)_R: 5'- CATGAACTCCTCCAGGGA-CAATTCCCCTGGGATGGA. The mutated AatII-AatII fragment was then substituted for the wild-type counterpart in the mG1 genomic construct. The GFP fusion mutant transgene, G1(L151F)-GFP, was generated by replacing a 2-kb SgrAI-ClaI fragment in G1 (L151F) with a 2.8-kb SgrAI-ClaI fragment containing GFP from G1-GFP. All transgenes were confirmed by DNA sequencing within the pBSKS (+) vector.

Generation of Genomic Transgenic Mice

The three transgenes were released from pBSKS (+) vector by NotI digestion and microinjected into the pronuclei of fertilized mouse FVB/N oocytes to produce transgenic mice at the University of Utah core facility. Transgenic mice were identified

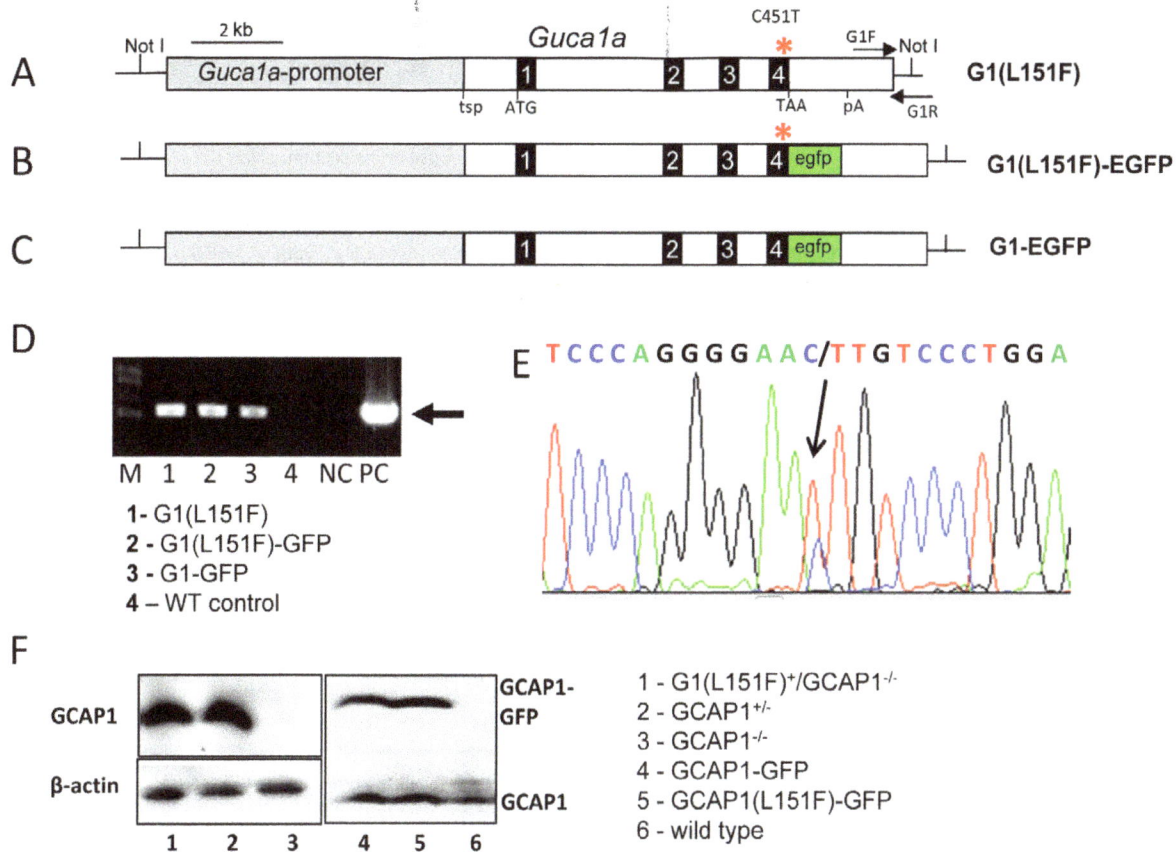

Figure 2. Generation of transgenic mouse lines, GCAP1-EGFP and GCAP1(L151F). Schematic of three mouse GCAP1 genomic transgene constructs: G1(L151F) (**A**), G1(L151F)-GFP (**B**) and G1-GFP (**C**). Black boxes depict exons 1–4. Tsp, transcription start point; ATG and TAA, translation start and stop codons; pA, polyadenylation signal. All transgenes contain the entire GCAP1 genomic sequence, including 6.2 kb promoter and 2.5 kb 3'-UTR sequence. Oligonucleotide pair (horizontal arrows, **A**), G1F and G1R, were used for genotyping. The C451T point mutation was introduced at exon 4 (red stars) resulting in a L151F mutation of G1(L151F) and G1(L151F)-GFP transgenes. In transgenes G1(L151F)-GFP and G1-GFP, EGFP was inserted immediately upstream of the stop codon. **D.** Genotyping of wildtype and three GCAP1 transgenic mice by PCR-amplification. The 540-bp DNA fragment amplified from tail DNA of mouse lines A–C, but not from a wildtype control (mouse 4). Wildtype mouse DNA and transgene plasmids were used as negative (NC) and positive controls (PC), respectively. M, 1 Kb plus DNA ladder; **E.** Representative DNA sequence showing the point mutation C451T (black arrow) in transgenic mice lines A and B. **F.** Immunoblot analysis of transgenic GCAP1 expression using anti-GCAP1 antibody, UW 101. The source of each retina lysate and its lane (1–6) on the blot is indicated (right). β-actin (lanes 1–3) and endogenous GCAP1 (lanes 4–6) are loading controls.

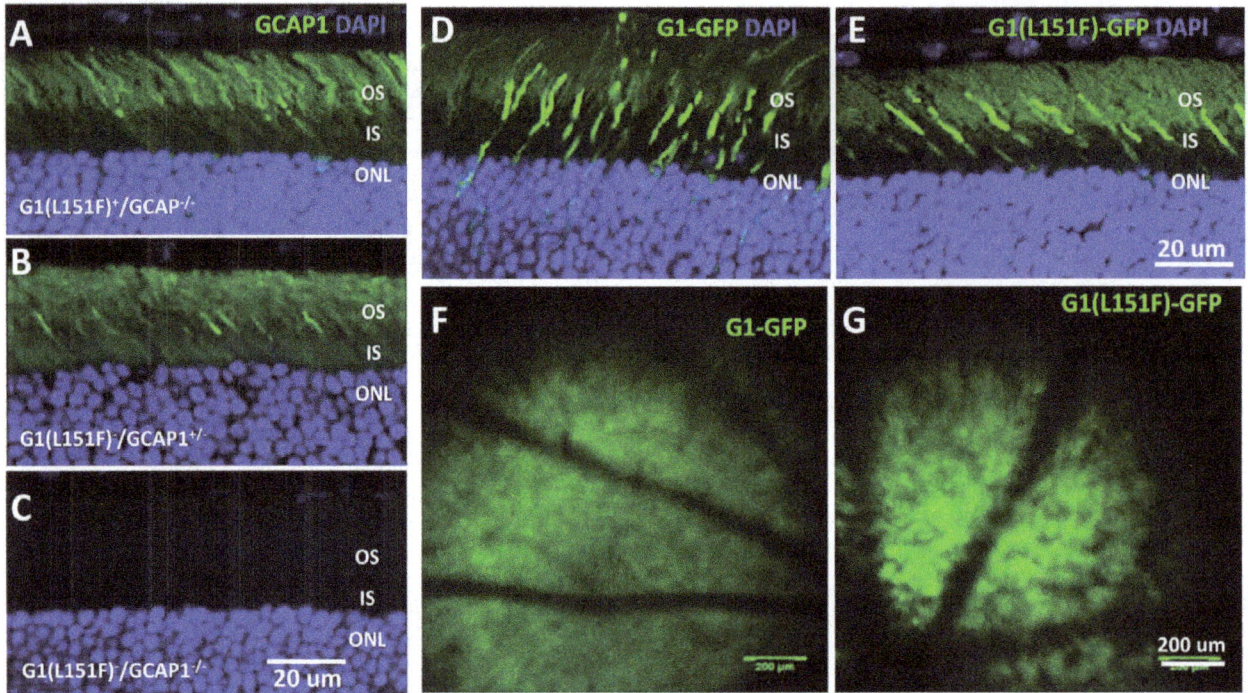

Figure 3. Confocal immunolocalization of transgene products (A-C), direct fluorescence (D,E) and fundoscopy (F,G) of transgenic mouse retinas. Immunohistochemical distribution of G1(L151F)-GCAP1 protein in transgenic mouse retina sections using the anti-GCAP1 antibody, UW101. When the mutant transgene is bred onto a GCAP1/2 double-knockout background, label attributed to mutant GCAP1 (green) is observed in rods and cones of the G1(L151F)$^{+}$/GCAP$^{-/-}$ transgenic retina (A). Comparison to GCAP$^{+/-}$ (heterozygous knockout of endogenous GCAP1, B) retina reveals similar immunolabel intensity, with slight enrichment in cones whereas label is absent in GCAP$^{-/-}$ retina (C). Direct fluorescence microscopy to investigate subcellular localization of G1-GFP and G1(L151F)-GFP transgenes expressed in the mouse retinas. Both G1-GFP (D) and G1(L151F)-GFP (E) transgenes are expressed in photoreceptors, particularly in cones, as shown by GFP fluorescence. F,G. *In-vivo* fluorescence fundus photography visualizing retinal expression of G1-GFP and G1(L151F)-GFP in the transgenic mice. Abbreviations: OS, outer segment; IS, inner segment; ONL, outer nuclear layer stained by DAPI, 4',6-diamidino-2-phenylindole, a nuclear marker (blue).

by PCR genotyping with primers **G1T_F:** 5'-ATAGGGCGTC-GACTCGATCACGCAGC, and **G1T_R:** 5'- TAAGGGCG-GAAGATCACGGAGGTAGC). The primers are located across the 3' boundary of the transgene and the pBSKS (+) backbone. A diagnostic fragment of 540-bp discriminates the transgenes from the endogenous GCAP1 gene. The C to T point mutation (L151F) was confirmed by DNA sequencing in the G1(L151F) and G1(L151F)-GFP transgenic founder mice. Transgenic mice were outbred to C57BL/J mice.

Western Blotting

Cultured cells and mouse retinas were lysed by sonication in RIPA buffer (150 mmol/l NaCl, 1% NP-40, 0.5% sodium deoxycholate, 1% SDS, 50 mmol/l Tris pH 8.0). The supernatant of each lysate was separated on a 10% SDS-PAGE($\sim$15 µg protein/well), and then transferred to a nitrocellulose membrane (Biorad, Hercules, CA). Subsequently, the membrane was probed with primary antibodies (anti-GCAP1 polyclonal antibody, UW101, 1:5,000 or anti-β actin monoclonal antibody, 1:3,000) followed by HRP-conjugated secondary antibody. Phosphorescence (ECL system, NEN Life Science, Boston, MA) was used to visualize the signal on X-ray film.

Immunocytochemistry

Mouse eyes were dissected and immediately immersion-fixed with 4% paraformaldehyde in 0.1 M phosphate buffer, pH 7.4, for 2 hours on ice. After removal of the anterior segment, the eyecups

were equilibrated sequentially with 15% and 30% sucrose in phosphate buffer for cryoprotection, and then embedded in OCT. Subsequently, retina cryosections of 12 µm thickness were cut and used for direct fluorescence confocal microscopy or immunohistochemical analysis. For immunohistochemistry, retina sections were incubated with primary antibody (UW101, 1:2,000, 4°C) overnight, followed by fluorescence-conjugated secondary antibody (1:300, 21°C) for 1 hour in a humidified chamber. Sections were washed with phosphate buffer for 5 minutes (X3) between incubations. To label nuclei, 4',6-diamidino-2-phenylindole (DAPI, 1:5,000) was added to the solution containing secondary antibody; retina sections were incubated with DAPI for one hour if imaging by direct fluorescence microscopy. After applying a drop of anti-fade agent, the sections were imaged with an Olympus Fluoview (model FV 1000) inverted confocal microscope.

Histology

Mouse eyeballs were fixed by immersion in 2% glutaraldehyde-1% paraformaldehyde in 0.1 M cacodylate, pH 7.4, overnight at 4°C. Following removal of the anterior segment, the eyecups were postfixed in 1% osmium oxide in the same buffer for 1 hour, dehydrated with an ascending series of ethanols, infiltrated with ethanol:plastic mixtures and embedded in Spurr's resin (Ted Pella, Inc., Redding, CA). Retina sections (1 µm) passing through the optic nerve were cut with an ultramicrotome and contrasted with Richardson's stain. Bright-field images of retina histology were acquired using a Zeiss Axiovert 200 (Carl Zeiss Inc., Thornwood,

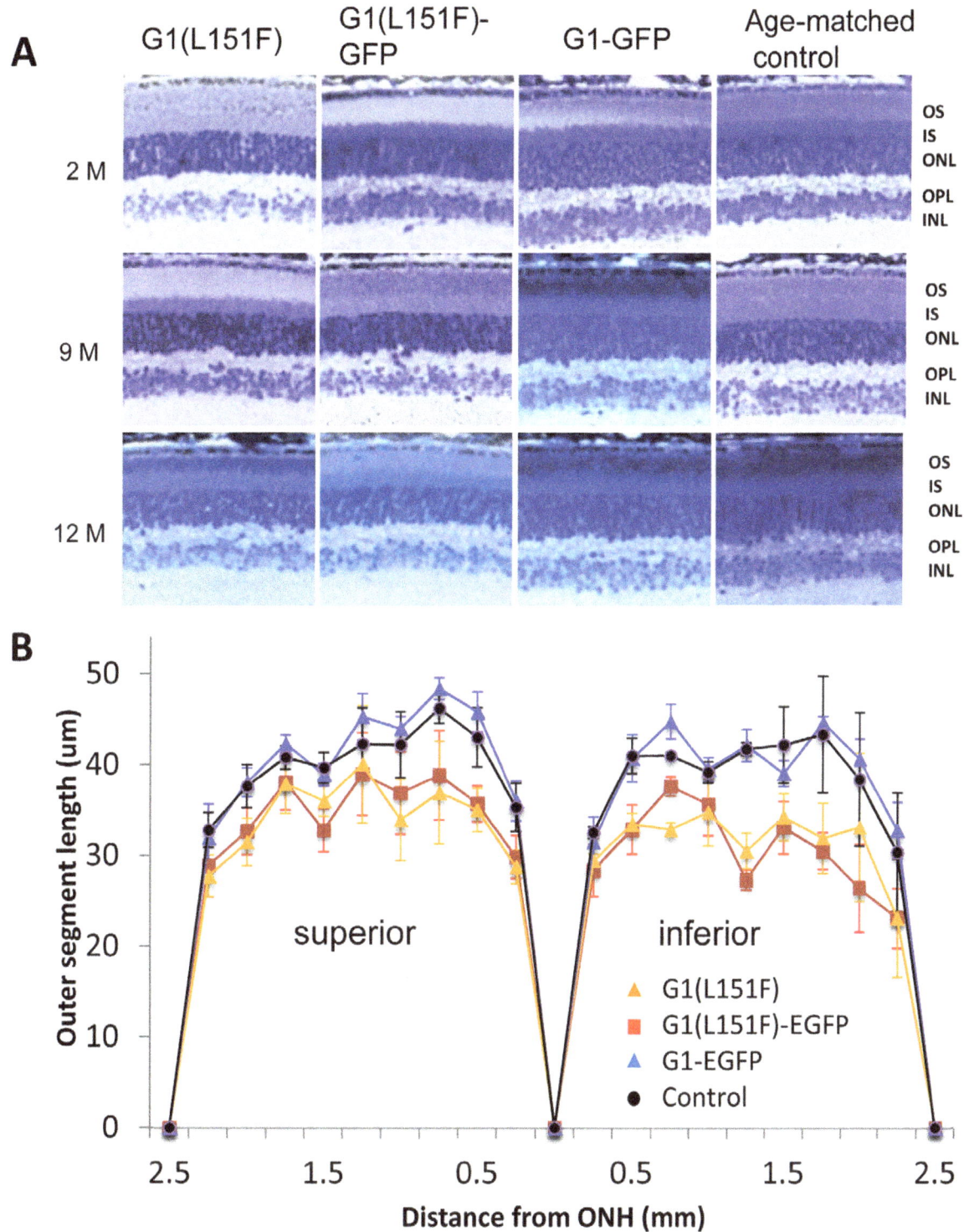

Figure 4. Expression of mutant GCAP1(L151F) results in slowly progressive photoreceptor degeneration. Retinal morphology of the three GCAP1 transgenic lines and age-matched controls were analyzed with plastic retinal sections at ages 2, 9 and 12 months (**A**). Retinas from G1 (L151F) and G1(L151F)-GFP mutant transgenic mice show no morphological change at age 2 months (top, left two panels) compared to control retinas (top, right two panels). At 9 months of age, a thinned photoreceptor layer (including OS, IS and ONL) becomes evident with ~10% reduction (middle, left two panels). Photoreceptor layer thinning became more obvious at 12 months with ~20–30% reduction (bottom, left two panels). In contrast, G1-GFP and age-matched wildtype photoreceptor layers are normal at all three ages (bottom, right two panels). Abbreviations: OS, outer segment; IS, inner segment; ONL, outer nuclear layer; OPL, outer plexiform layer; INL, the inner nuclear layer. **B.** Quantitation of the photoreceptor outer nuclear layer (ONL) thickness in three GCAP1 transgenic mice and age-matched controls at 12 months of age. ONL thickness was measured every 250 μm moving from the optic nerve head (ONH) to the periphery, both inferiorly and superiorly. ONL thickness was reduced in retinas of G1(L151F) and G1(L151F)-GFP transgenic mice, but not in G1-GFP transgenic or wildtype mice. (n = 3 each group).

Table 1. Outer nuclear layer (ONL) thickness of GCAP1 transgenic mice and age-matched controls.

	Distance from ONH (mm)	0.25	0.5	0.75	1	1.25	1.5	1.75	2	2.25	2.5
1	G1(L151F)	29.4±0.8	33.4±1.2	32.8±0.8	34.7±3.7	30.5±2.0	34.2±2.6	31.9±3.9	33.1±8.2	23.1±6.5	0
2	G1(L151F)-GFP	28.3±2.9	32.8±2.7	37.6±1.1	35.6±3.5	27.3±1.1	33.1±2.9	30.5±2.0	26.4±4.8	23.1±3.3	0
3	G1-GFP	31.5±2.7	40.7±2.6	44.7±1.9	39.5±1.3	42.1±1.8	39.0±1.4	44.6±0.7	40.6±2.2	32.8±3.1	0
4	Age-matched Control	32.5±0.6	41.0±1.9	41.0±0.4	39.2±1.2	41.7±0.7	42.2±4.2	43.3±6.4	38.4±7.3	30.4±6.5	0
	Distance from ONH (mm)	0.25	0.5	0.75	1	1.25	1.5	1.75	2	2.25	2.5
5	G1(L151F)	28.8±1.8	35.0±2.4	36.9±5.7	33.9±4.5	40.0±6.5	36.0±1.7	37.9±3.3	31.4±2.6	27.7±2.3	0
6	G1(L151F)-GFP	29.8±2.4	35.7±2.0	38.8±4.9	36.9±4.6	39.0±4.6	32.8±2.4	38.0±3.0	32.7±2.6	28.9±3.5	0
7	G1-GFP	36.0±2.2	45.8±2.2	48.4±1.2	44.0±1.3	45.3±2.6	38.9±1.2	42.3±1.0	38.1±1.6	32.0±3.7	0
8	Age-matched Ctr	35.3±2.7	43.0±3.3	46.2±1.7	42.2±3.6	42.7±4.0	39.7±1.7	40.8±1.3	37.6±2.3	32.8±1.9	0
9	G1(L151F) vs Ctr					p<0.001*					
10	G1(L151F)-GFP vs Ctr					p<0.001*					
11	G1(L151F)-GFP vs G1-GFP					p<0.001*					
12	G1-GFP vs Ctr					p=0.09					

Rows 1–4, distances from the optic nerve head (ONH) in inferior hemispheres; rows 5–8, distances from ONH in inferior hemispheres at 0.25 mm intervals. Rows 9–10, p values of distance comparison between different groups.
*p<0.05.

Table 2. ERG a- and b- wave amplitudes recorded from wildtype and GCAP1 transgenic mice.

	Light intensity (dB)	−10	0	10	20	25
		a-wave amplitude of scotopic ERG response (µV)				
1	G1(L151F)	95±13	141±11	226±9	285±28	296±34
2	G1(L151F)_GFP	80±11	117±21	240±40	294±35	311±43
3	G1_GFP	147±15	209±17	347±27	403±30	428±31
4	Age-matched Ctr	141±16	196±9	338±17	401±20	436±39
5	G1(L151F) vs Ctr		p<0.001*			
6	G1(L151F)_GFP vs Ctr		p<0.001*			
7	G1(L151F)_GFP vs G1_GFP		p<0.001*			
8	G1_GFP vs Ctr		p=0.61			
		b-wave amplitude of photopic ERG response (µV)				
9	G1(L151F)	13±4	46±7	95±22	108±17	113±19
10	G1(L151F)_GFP	10±2	45±11	90±11	115±14	121±23
11	G1_GFP	20±5	90±7	154±8	183±14	184±10
12	Age-matched Ctr	19±2	83±3	163±9	185±13	199±12
13	G1(L151F) vs Ctr		p<0.001*			
14	G1(L151F)_GFP vs Ctr		p<0.001*			
15	G1(L151F)_GFP vs G1_GFP		p<0.001*			
16	G1_GFP vs Ctr		p=0.25			

Rows 1–4, a-wave amplitudes of scotopic ERG responses (µV). Rows 5–8, p values of a-wave amplitude comparison between different groups, Rows 9–12, b-wave amplitudes of photopic ERG responses (µV), Rows 13–16, p values of b-wave amplitude comparison between different groups.
*p<0.05.

NY) microscope and 62x objective. Images across entire retina sections were acquired with a 20x objective and imported into image J software with scaling of 3.8 pixels/µm. Outer nuclear layer (ONL) thicknesses were measured at 250 µm intervals from the optic nerve head.

Electroretinography (ERG)

Scotopic and photopic ERG responses of control and transgenic mice were recorded as described [4]. Mice were dark-adapted overnight, anesthetized by intraperitoneal injection of 10 mg/ml ketamine-1 mg/ml xylazine (10 µl/g body weight) in PBS under dim red illumination, and kept comfortable on a heating pad. After dilating pupils by applying a drop of 2.5% phenylephrine (Akorn, Inc., Decatur, IL), ERG responses were recorded from 3–5 mice of each genotype per time point using a UTAS E-3000 system (LKC Technologies,Inc., Gaithersburg, MD). For scotopic ERG, mice were tested at intensities ranging from −40 decibels (db) (−3.4 logcds m^{-2}) to 25 db (2.9 log cds m^{-2}) without initial rod saturation. For photopic ERG, a rod saturating background light of 10 db (1.48 log cds m^{-2}) was applied for 20 minutes before and during recording. Single flash responses were usually recorded at stimulus intensities of −10 db (−0.6 log cds m^{-2}) to 25 db (2.9 log cds m^{-2}). Five or fewer flashes were averaged per intensity level, with longer flash intervals with increasing intensity. Triple-antibiotic ointment was routinely placed on the eye to prevent infection after ERG testing.

OptoMotry

Transgenic and control mice were tested for spatial visual acuity using an Optomotry apparatus which facilitates rapid screening of functional vision [28]. Briefly, mice were placed singly on a platform centered within a quad-square formed by four inward-facing computer screens, and observed by an overhead video camera. Vertical sine-wave gratings were presented at

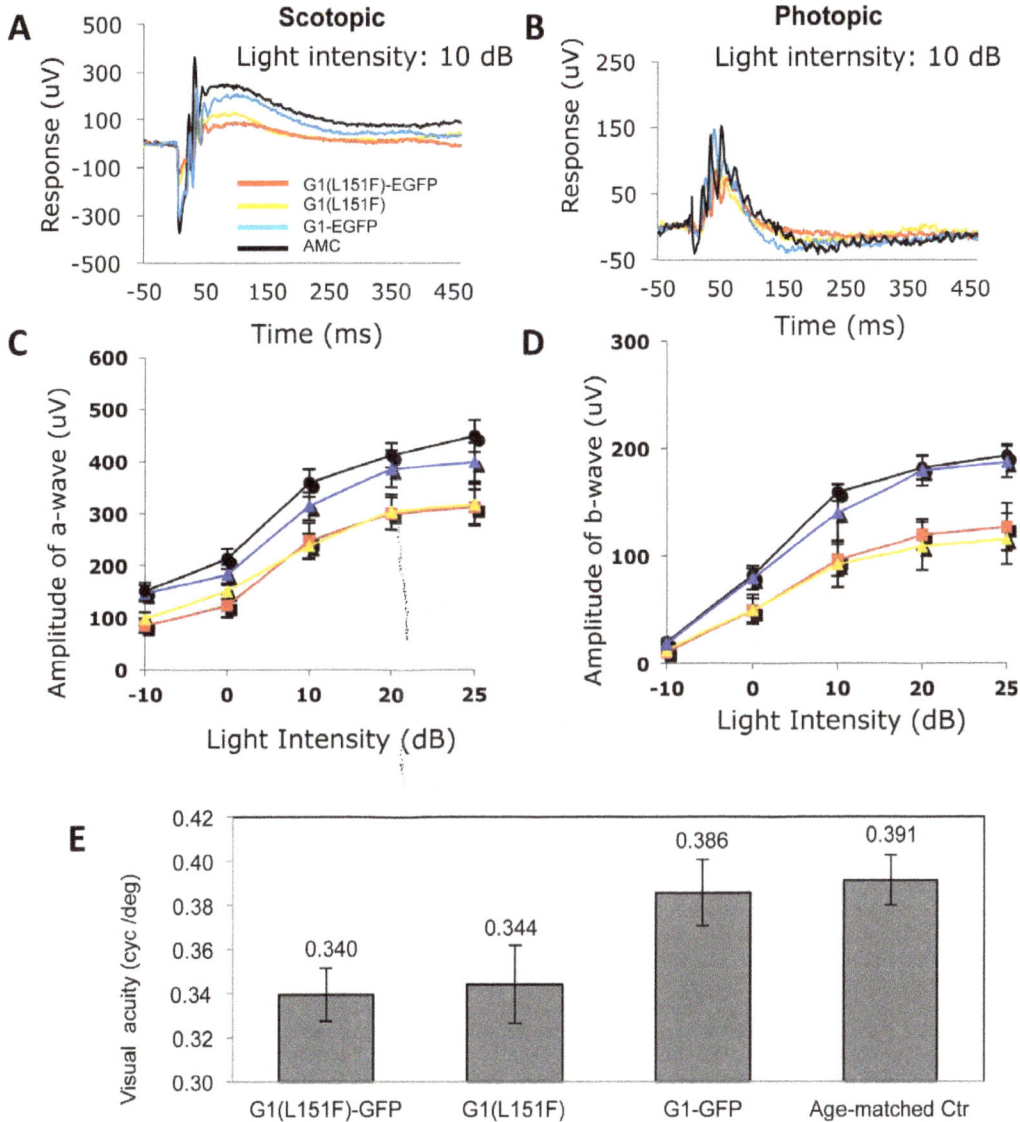

Figure 5. Retinal function of three GCAP1 transgenic mouse lines evaluated by ERG and Optomotry. Scotopic (**A**) and photopic (**B**) ERG analysis at age 12 months (n = 5 mice per genotype). Average a-wave amplitudes of the scotopic responses (**C**) and b-wave amplitudes of the photopic responses (**D**) as a function light intensity. Error bars are standard deviations. **E.** Visual acuities of GCAP1 transgenic mice versus age-matched controls (n = 4) were tested by Optomotry at 12 months of age. Error bars are standard deviations, *p < 0.05.

photopic luminance levels on screens rotating 12 deg/sec (speed) either clockwise or counterclockwise, as determined randomly by OptoMotry© software. The mouse tracks the grating with reflexive head movements when the rotating grating is perceptible. By observing mouse tracking, spatial frequency thresholds were quantified by increasing the spatial frequency of the rotating grating (usually between 0.03–0.35 grating cycles/degree) until finding a maximum frequency at which the mice track.

Constructs Expressing Mouse GCAP1-EGFP and GCAP1_shRNA *in vitro*

Using shRNA prediction tools provided *online* by Whitehead, Ambion, Invitrogen and Dharmacon, we designed four putative shRNAs (mG1hp1-4) specifically targeting mouse GCAP1 and a mismatch control (mG1hp4m2) carrying the central two-nucleotide mutation in the mG1hp4 guide strand (GG to CC).

To avoid possible off-target effects, an NCBI BLAST homology search was performed for the candidate shRNAs and no significant homology with other mouse genes or sequences was identified. A short-hairpin RNA (shRNA) expression vector, pmC_hH1, containing a shRNA expression promoter (human H1 promoter, hH1) and a CMV promoter-driven mCherry reporter, was used to generate anti-mGCAP1 shRNA constructs expressing mG1hp1-4 and mG1hp4m2 [4]. The construct mG1-EGFP, expressing EGFP fusion mouse GCAP1, was generated by cloning the mouse GCAP1 cDNA into pEGFP-N1 vector (Clontech, Mountain View, CA) downstream of the CMV promoter.

Knockdown Efficiency Assay in Cell Culture

HEK293 cells (ATCC, CRL-1573) were cultured in DMEM supplemented with 10% fetal calf serum and 100 u/ml of penicillin-streptomycin (Invitrogen, Carlsbad, CA). In 12-well

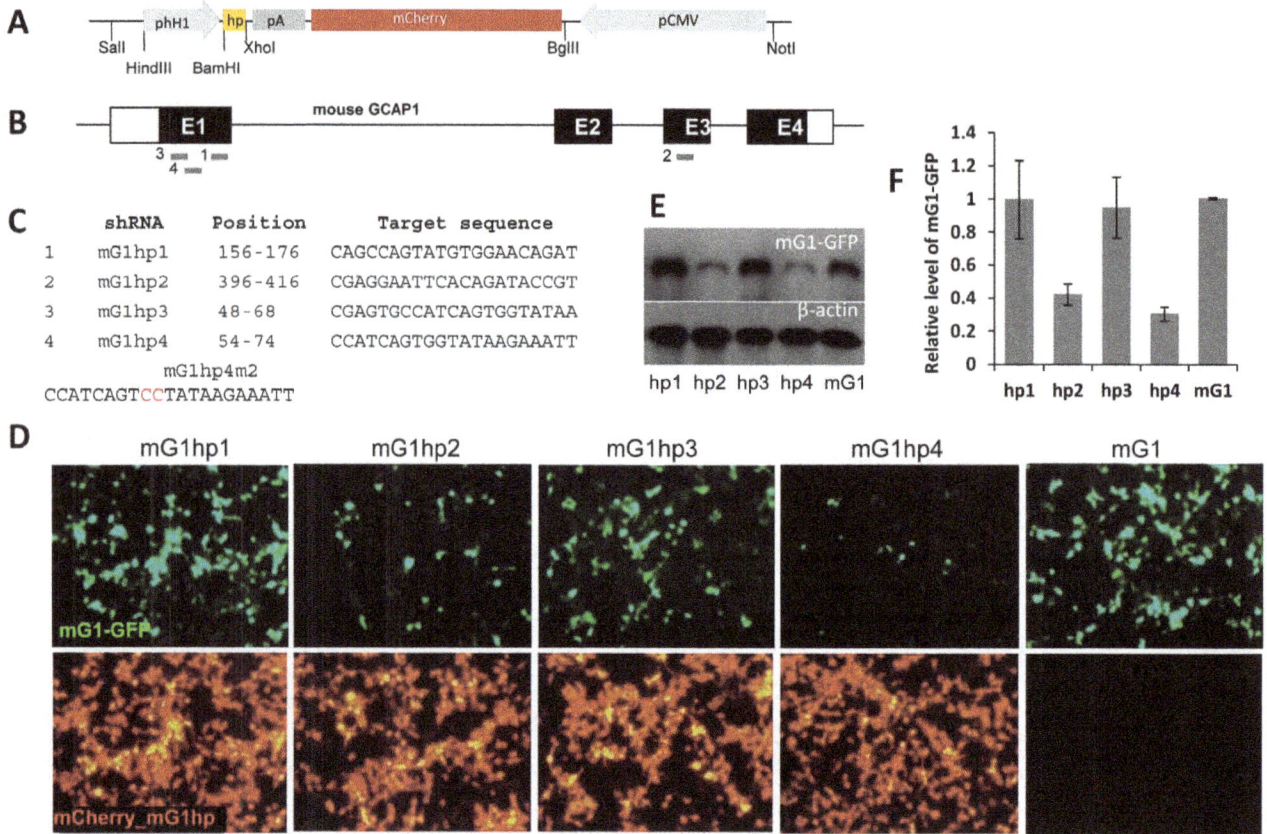

Figure 6. Knockdown of mouse GCAP1 *in vitro* by shRNAs. Schematic (**A**) shows shRNA expression construct that contains a 68-bp shRNA cassette driven by the human H1 promoter, and a CMV promoter-driven mCherry reporter gene. **B.** Schematic of the mouse GCAP1 gene and four candidate anti-mGCAP1 shRNAs, mG1hp1-4, each targeting a different exomic location. **C.** Targeting sequences of mG1hp1-4 and their positions within the GCAP1 gene. mG1hp4m2 is a mismatched control of mG1hp4 in which the central GG was replaced by CC (red). **D.** Knockdown efficiencies of anti-mGCAP1 shRNAs as determined by fluorescence microscopy of HEK293 cells 48 hr after cotransfection with mGCAP1-GFP and each anti-mGCAP1 shRNA, mG1hp1-4. Top row, GFP signal is diminished in cultures transfected with mG1hp2 and mG1hp4, indicating knockdown of mGCAP1 protein. Bottom row, mCherry reporter co-expression of mG1hp1-4 reveals comparable transfection efficiencies among the cultures. In the bottom far right panel, culture "mG1" was transfected with mGCAP1-GFP plasmids only as a non-knockdown control. **E.** Western blot analysis of transfected cell lysates at 50 hr post-transfection. mGCAP1-GFP was detected by anti-GCAP1 antibody, UW101, with β-actin immunostaining as the endogenous loading control. **F.** Relative levels of mGCAP1-GFP expression in the 50 hr transfected cultures were quantified by immunoblot band intensities and normalized against β-actin. The relative level of mG1-GFP in each cotransfected cell sample was averaged from three tests, representing the knockdown efficiency of each shRNA at different concentrations. Knockdown was most pronounced by transfection with mG1hp4, showing ~70% decrease compared to mG1 non-knockdown control.

plates, HEK293 cells were cotransfected with mG1-EGFP plasmids and anti-mGCAP1 shRNA expression plasmids, mG1hp1-4 or mG1hp4m2, at different mass ratios (1:3–3:1) with a total amount of 1.5 μg/well (normalized by pBS vector DNA). At 48 hours post-transfection, expression of mG1-GFP and mCherry reporter in the transfected HEK293 cells was directly analyzed by live cell fluorescence microscopy using a Zeiss Axiovert 200 microscope. At 50 hours post-transfection, the cotransfected cells were harvested to analyze protein levels of mG1-GFP by immunoblotting; β-actin was used as an endogenous control.

Preparation of scAAV2/8 Virus Expressing Anti-mGCAP1 shRNA

Self-complimentary AAV2/8 (scAAV2/8) was used to express mG1hp4 and its mismatched control, mG1hp4m2, in the mouse retina. The scAAV2/8 packaging constructs were generated by cloning hH1 promoter-driven mG1hp4 and mG1hp4m2 cassettes into a shuttle vector pscAAV-CAG-mCherry, which was modified

from pscAAV-CAG-hGFP (provided by W.W. Hauswirth). The two plasmid cotransfection method was used to produce scAAV2/8 viral particles expressing mG1hp4/4m2 in collaboration with the University of Florida [29]. Vector genome-containing viral particles were titered by real-time PCR and resuspended in a balanced salt solution (Alcon Laboratories, Fort Worth, TX) containing 0.014% Tween-20 at a concentration of ~1.0 ×10^{12}vector genomes per milliliter (vg/ml).

Mouse Subretinal Injection

G1(L151F)-GFP mice were injected subretinally with scAAV2/8 on postnatal day 30 (P30) as described [4]. Briefly, after anesthesia with a ketamine-xylazine solution (see ERG protocol), 0.5% proparacaine solution (Alcon Laboratories Inc., Fort Worth, TX) was applied to the cornea as a topical anesthetic. A small puncture hole was made through the cornea with a 30½ -gauge beveled needle (Becton Dickinson & Company, Franklin Lakes, NJ). A 33-gauge blunt needle attached to a microliter syringe (Hamilton Bonaduz, Switzerland) was introduced tangentially

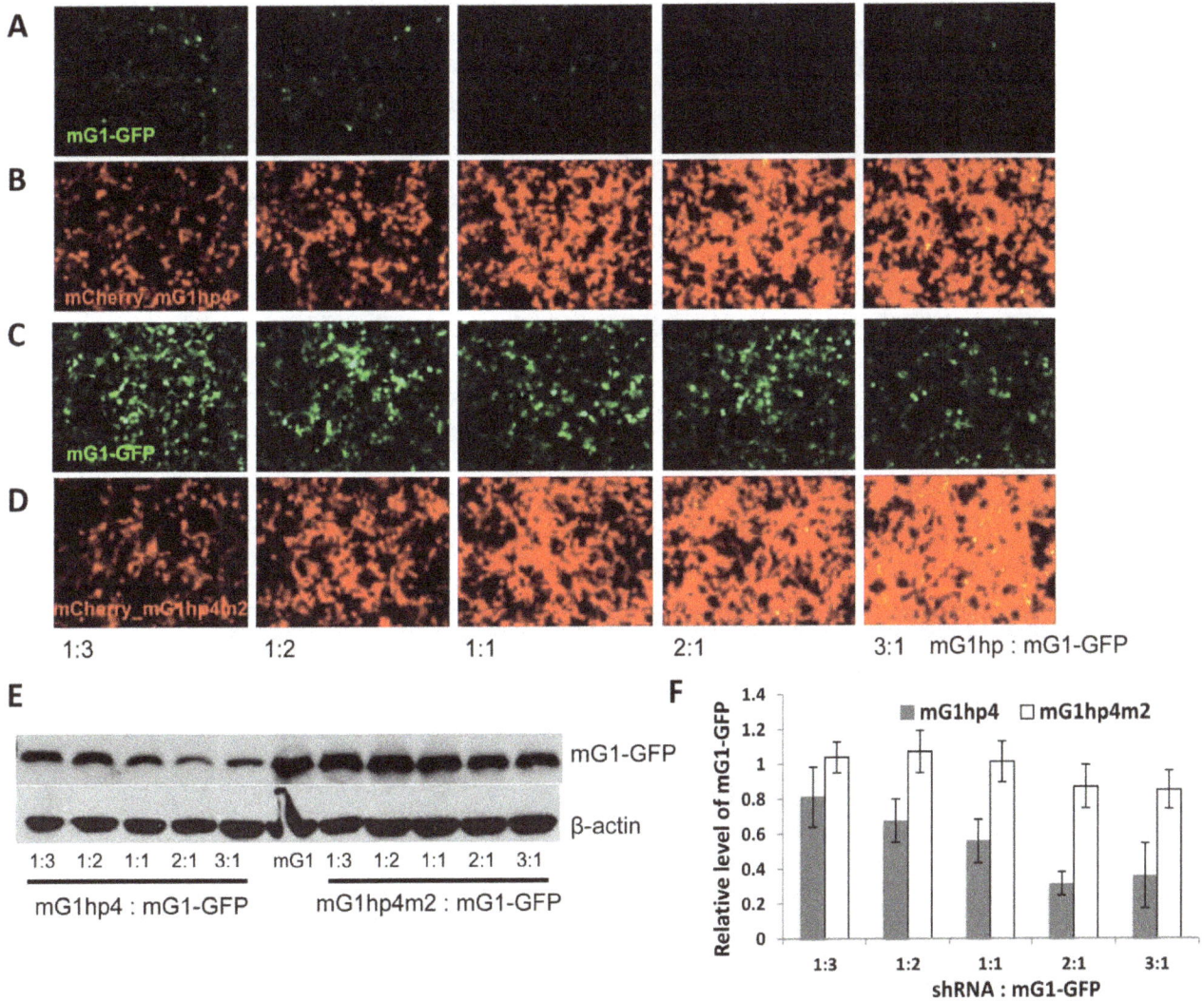

Figure 7. Dose-dependent and sequence-specific knockdown of mGCAP1 by mG1hp4. A–D. Fluorescence microscopy of HEK293 cells 48 hrs after cotransfection with mGCAP1-EGFP and mCherry_mG1hp4 (**A, B**) or its mismatched control mCherry_mG1hp4m2 (**C, D**) at five different mass ratios. GFP signal in row A is gradually diminished as mG1hp4 increases to a saturation level, indicating knockdown dose-dependency from 1:3 to 2:1. In row C there is no obvious change of GFP signal in cultures from 1:3 to 2:1 since the mutant mCherry_mG1hp4m2 is inactive for suppression of mGCAP1. Panels of rows B and D show increase in mCherry signal corresponding to increased mass ratios, left to right. **E.** Representative western blot evaluating mGCAP1-GFP protein levels in the cotransfected cells (50-hr post-transfection); mGCAP1-GFP was detected by anti-GCAP1 antibody, UW101, with β-actin immunostaining as the loading control. **F.** Quantification of relative mGCAP1-GFP levels shown in (**E**). The intensities of the mG1-GFP specific bands in western blot were measured by using ImageJ, and normalized against β-actin. The relative level of mG1-GFP in each cotransfected cell sample was averaged from three tests, representing the knockdown efficiency of each shRNA at different concentrations.

through the hole, and 1 μl of scAAV2/8 virus was delivered subretinally. Following injection, a triple-antibiotic ointment containing bacitracin, neomycin sulfate and polymyxin B (Taro Pharmaceuticals, Inc., Hawthorne, NY) was placed on the eye to prevent infection. Injections were performed in left eyes only, leaving the right eyes as untreated controls.

In-vivo Fluorescence Imaging of the Mouse Retina

Fluorescence live imaging of mouse retinas was used to detect the expression of GFP fusion GCAP1 transgenes, G1-GFP and G1(L151F)-GFP, in mice 1–2 months of age, as well as in the knockdown of mG1(L151F)-GFP in transgenic mice one-month posttreatment with scAAV2/8. Mice were anesthetized with isoflurane gas (2%–3%) and their pupils were dilated with 2.5% phenylephrine (Akorn, Inc., Decatur, IL). After applying 0.5%

proparacaine solution (Alcon Laboratories Inc., Fort Worth, TX) to the cornea as topical anesthetic, a glass coverslip was placed on the mouse eye to maintain corneal moisture and reduce refractive error. The mouse was stabilized during imaging on a custom-designed heated aluminum stage specifically fit for the microscope (Zeiss LSM 700; Carl Zeiss, Inc., Thornwood, NY). Fluorescent fundus images were acquired with a 5x air objective and Zeiss image acquisition software. Triple-antibiotic ointment was placed on the eye to prevent infection after imaging.

Statistical Data Analysis

ONL thicknesses and the ERG response amplitudes were analyzed by two-way ANOVA, while the Optomotry data was analyzed by single-way ANOVA. The level of statistical significance was set at $p = 0.05$.

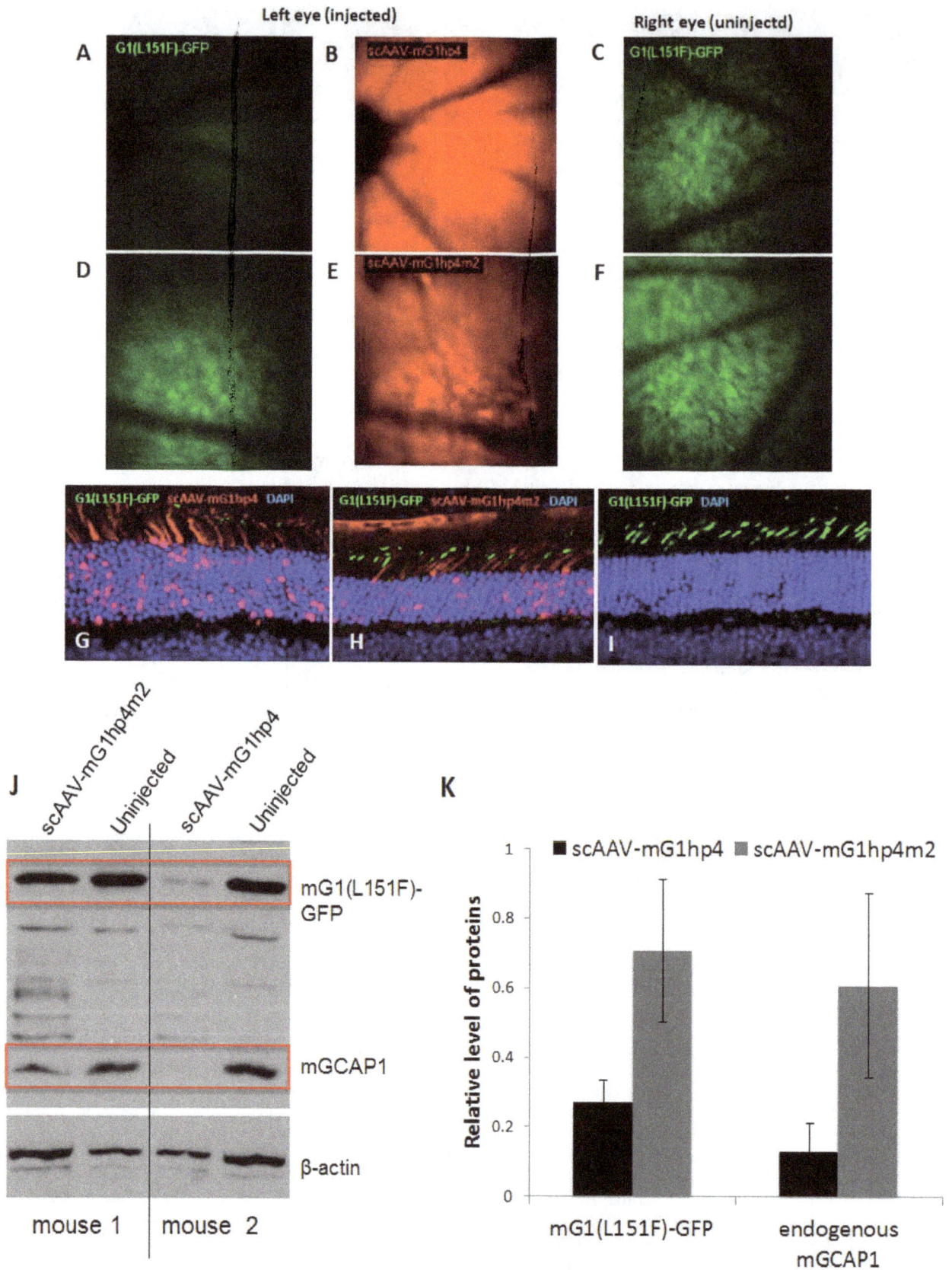

Figure 8. *In-vivo* **knockdown of transgenic mGCAP1 by therapeutic AAV2/8-mG1hp4. A–C.** Representative fluorescence fundus images of G1(L151F)-GFP transgenic mice with subretinal injection of scAAV2/8-mG1hp4 to the left eyes. GFP signal representing the G1(L151F)-GFP level in photoreceptors is diminished in the left eye at 30 days post-injection (**A**), compared to that in the uninjected right eye (**C**), indicating significant

knockdown of G1(L151F)-GFP by mG1hp4 whose expression is shown by mCherry reporter signal (**B**). **D–F**. Representative fluorescence fundus images of G1(L151F)-GFP transgenic mice with subretinal injection of scAAV2/8-mG1hp4m2 (2-bp mismatch control) in their left eyes. In contrast to scAAV2/8-mG1hp4-treated eyes (**A**), there is no obvious knockdown of G1(L151F)-GFP by scAAV2/8-mG1hp4m2 (**D**), compared to the uninjected right eye (**F**). Even mG1hp4m2 is expressed at levels comparable to mG1hp4 in the treated eye, as shown by mCherry reporter (**B, E**). **G–I**. Fluorescence microscopy of retinal sections of G1(L151F)-GFP transgenic mice receiving scAAV2/8-mG1hp4 or -hp4m2 in their left eyes. Compared to uninjected right eyes (**I**), G1(L151F)-GFP signal in the outer nuclear layer decreased significantly with mGhp4 expression, but less with mG1hp4m2 expression. **J**. Immunoblot analysis of G1(L151F)-GFP transgenic mouse retinas 30 days after subretinal injection. Transgenic G1(L151F)-GFP and endogenous GCAP1 proteins were detected by anti-GCAP1 antibody, UW101. β-actin served as endogenous loading control. Compared to the uninjected right retinas (lanes 2 and 4), both G1(L151F)-GFP (~50 kD) and endogenous GCAP1 (~25 kD) are decreased in the left retina treated with AAV-mG1hp4 (lane 3), but not when treated with AAV-mG1hp4m2 (lane 1). **K**. Quantification of relative G1(L151F)-GFP and mGCAP1 levels in retinas 30 days post-injection with AAV8-mG1hp4 and AAV8-mG1hp4m2. G1(L151F)-GFP and mGCAP1 levels were averaged from 3 retinas of each vector treatment, and normalized against β-actin. Relative to uninjected eyes, G1(L151F)-GFP and mGCAP1 were knocked down ~75% and ~90% by AAV-mG1hp4, and 30% and 35% by AAV-mG1hp4m2.

Results

Generation of Transgenic Mice and Expression of Mutant GCAP1

We used the mouse *Guca1a* gene as a foundation to construct three transgenes. The L151F mutation was introduced in exon 4 to establish two GCAP1 transgenic mouse lines, G1(L151F) (**Fig. 2A**) and G1(L151F)-GFP (**Fig. 2B**). The GCAP1-EGFP fusion allowed detection of transgene expression by live fluorescence microscopy, and distinction of transgenic GCAP1 (50 kDa) from native GCAP1 (23 kDa). A third line expressed a GCAP1-EGFP fusion protein containing L151 (no L151 mutation) as a WT control (**Fig. 2C**). The C-terminal in-frame fusion with EGFP was achieved using the homologous recombination method, Red/ET recombination cloning, which is a DNA manipulation technique that is independent of the presence of restriction sites and the size of the DNA molecule to be modified [27] (see Methods). Primers located near the 3′-end of the transgene and in the multiple cloning site of the pBSKS(+) backbone allowed precise genotyping of transgenic mice (**Fig. 2D**). The L151F mutation in G1(L151F) and G1(L151F)-GFP transgenic mice was confirmed by sequencing (**Fig. 2E**).

Relative expression of the GCAP1 transgenes in mouse retinas was examined by immunoblotting retinal lysates at 1–2 months of age (**Fig. 2F**). Antibody directed against mouse GCAP1 (UW101) recognized both endogenous GCAP1 and transgenic GCAP1 fusion proteins. To identify expression of G1(L151F) which co-migrates with native GCAP1, we crossed the G1(L151F) transgenic line with GCAP$^{-/-}$ mice. UW101 detected a 23 kDa polypeptide in G1(L151F)$^{+}$/GCAP$^{-/-}$ retinal lysates (**Fig. 2F**, lane 1) comparable to the expression level of GCAP1$^{+/-}$ retinal lysate expressing only one allele (**Fig. 2F**, lane 2). As expected, no GCAP1 was detected in GCAP$^{-/-}$ retinal lysates (**Lane 3**). In G1(L151F)-GFP and G1-GFP transgenic retinas, GCAP1-GFP fusion proteins (~50 kDa) and native GCAP1 (23 kDa) were co-expressed (**Fig. 2F**, lanes 4 and 5). Immunoblot intensities of the 50 kDa fusion proteins suggest that transgenes G1(L151F)-GFP and G1-GFP are expressed at similar levels.

Subcellular Distribution of Transgenic GCAP1 Proteins

We next analyzed the subcellular distribution of G1(L151F) by immunohistochemistry and confocal microscopy. When expressed on the GCAP1/GCAP2 null background, GCAP1(L151F) localized to the photoreceptor inner and outer segments and exhibited a stronger signal in cones than in rods (**Fig. 3A**), comparable to endogenous GCAP1 (**Fig. 3B**) [5]. As predicted, the GCAP double knockout mouse retina was negative for GCAP1 immunolabeling (**Fig. 3C**). Direct fluorescence microscopy of retinal cryosections of G1-GFP and G1(L151F)-GFP transgenic mice, contrasted with diamidino-2-

phenylindole (DAPI) nuclear stain, revealed that both G1(L151F)-EGFP and G1-EGFP transgene products were expressed in the photoreceptor inner and outer segments (**Fig. 3D, E**). Fluorescence attributable to transgenic wildtype and mutant GCAP1s was observed in photoreceptors residing in the most sclerad three layers of the ONL, i.e., cones, where the perinuclear accumulation is most likely caused by 'overexpression' when the transgene product exists on a background of endogenous (two alleles) GCAP1. Expression of G1(L151F)-GFP and G1-EGFP in transgenic mouse retinas could be discerned easily by *in-vivo* fluorescence (**Fig. 3F, G**).

Slowly Progressive Photoreceptor Degeneration in GCAP1 Mutant Transgenic Mice

The onset of retinal degeneration was determined by histology, electroretinography and behavioral testing of live animals. Transgenic G1(L151F), G1-GFP and age-matched wildtype mouse retinas were indistinguishable early (**Fig. 4A**, top panels at 2 months). Further, all three GCAP1 transgenic lines showed normal scotopic and photopic ERG responses at two months (data not shown). However, slight reduction of scotopic (a-wave) and photopic (b-wave) ERG amplitudes in 9 month-old mutant transgenic mice signaled the onset of photoreceptor degeneration, and this was reflected by a modest thinning of the ONL (**Fig. 4A**, middle panels). Degeneration in both mutant mouse models progressed slowly, revealing ~10–20% reduction of ONL thickness compared to wildtype controls and was particularly obvious in the inferior retina (**Fig. 4A**, bottom panels at 12 months; **Fig. 4B, Table 1**).

Attenuated scotopic and photopic ERG responses became significant at 12 months, especially under high input light intensities. For example, under a 10 dB flash intensity, the mean scotopic a-wave amplitudes measured in G1(L151F) and G1(L151F)-GFP mice were 238 μV(±23 μV, n = 5) and 247 μV(±35 μV, n = 5). These values are reduced by 20–30% relative to age-matched wildtype control (358±26 μV, n = 5) and G1-GFP transgenic mice (314±27 μV, n = 5) (**Fig. 5, Table 2**). The mean photopic b-wave amplitudes measured in G1(L151F) and G1(L151F)-GFP transgenic mice, 92 μV (±22 μV, n = 5) and 96 μV (±15 μV, n = 5) respectively, were reduced by 30–40% relative to age-matched wildtype (159±8 μV, n = 5) and G1-GFP (140±26 μV, n = 5) transgenic mice.

Correspondingly, Optomtry at 12 months of age showed marked visual impairment in mice expressing the mutant GCAP1 transgenes. Cone-mediated behavior of G1(L151F) and G1(L151F)-GFP transgenic mice showed threshold reduction of ~10–15% compared to that of age-matched controls and G1-GFP transgenic mice (**Fig. 5E**). Thus, no significant difference between G1-GFP transgenic and age-matched wildtype mice were observed with regard to morphology (**Fig. 4**), scotopic a-wave (p = 0.085) and photopic b-wave amplitudes (p = 0.074) (**Fig. 5A–D, Table 1**) or visually guided behavior (**Fig. 5E**). These tests reveal that the GCAP1(L151F) mutation causes a slowly progressing cone-rod dystrophy, first recognizable at around 9 months, progressing to a 20% reduction in ONL thickness, 40% reduction in cone b-wave amplitude, and 15% reduction in visual acuity by one year of age.

Identification of Anti-mGCAP1 shRNAs *in vitro*

Germline deletion of GCAPs in mouse display neither gross functional deficiencies nor retinal degeneration [8,9]. A nonallele-specific knockdown of *Guca1a* mRNA in the G1(L151F) transgenic mice may therefore be a promising tool for RNAi- based gene therapy to cure dominant CORD caused by GCAP1 mutations. Using online shRNA prediction tools, we designed four anti-mGCAP1 candidate shRNAs which have no significant homology with other mouse genes. To identify an effective anti-mGCAP1 shRNA, we developed an *in-vitro* shRNA screening system in which disappearance of mGCAP1-EGFP fluorescence would be proportional to the knockdown efficiencies of shRNA candidates [4]. We used a shRNA expression construct with a mCherry reporter (**Fig. 6A**) to test four anti-mGCAP1 shRNA candidates, mG1hp1-4 (**Fig. 6B, C**), in which shRNA transcription is controlled by the human H1 promoter. Target mG1-EGFP plasmids were co-transfected into HEK293 cells with each anti-mGCAP1shRNA plasmid at a 1:3 mass ratio. Live cell imaging revealed that mG1hp2 and 4 significantly suppressed mG1-EGFP expression in the co-transfected cells whereas mG1hp1 or 3 were largely ineffective (**Fig. 6D**). Western blotting of the cell lysates verified that mG1hp2 and 4 knocked down mG1-EGFP with 60% and 70% efficiency, respectively (**Fig. 6E, F**).

To determine knockdown specificity of mG1hp4, a 2-nucleotide mismatch control, mG1hp4m2was generated by mutating the central two nucleotides in the mG1hp4 guide strand (GG to CC) (**Fig. 6C**). We then cotransfected mG1-GFP and mG1hp4/mG1hp4m2 plasmids into HEK293 cells at five different mass ratios (1:3, 1:2, 1:1, 2:1, and 3:1). Fluorescence live cell imaging of cotransfected cells 48 hr post-transfection showed that mG1-GFP levels were incrementally reduced with increasing amounts of mG1hp4 from 1:3 to 2:1, as evidenced by mCherry fluorescence (**Fig. 7A, B**). No significant suppression of mG1-EGFP was observed when mG1hp4 was replaced by mG1hp4m2 at the mass ratio of mG1hp4m2/mG1-GFPfrom 1:3 to 2:1 (**Fig. 7C, D**). However, when shRNA increased to the mass ratio of 3:1, knockdown efficiency of mG1hp4 did not rise further, and mG1hp4m2 presented a nonspecific suppression effect on mG1-EGFP. Western blot analysis of the cell lysates at 50 hr post-cotransfection confirmed that mG1-GFP was reduced from 70% to 30% as shRNA increased from 1:3 to 2:1, whereas mutant shRNA showed ~15% reduction at 2:1 and 3:1 (**Fig. 7E, F**). These results demonstrate that mG1hp4 could efficiently and specifically suppress mG1-GFP expression. A nonspecific effect was observed in mG1hp4m2 at very high shRNA/mGCAP1-EGFP ratios.

Visualization of shRNA-mediated Gene Silencing *in-vivo*

We packaged mG1hp4 and mG1hp4m2 into scAAV2/8 vectors that express shRNA rapidly and persistently in mouse photoreceptors [4]. The scAAV8-mG1hp4 virus was injected subretinally into the left eyes of G1(L151F)-GFP mice at 1 month of age, before appreciable photoreceptor degeneration has occurred. At one week post-injection, mice exhibiting mCherry expression in >50% of the retinal area (visualized by live fluorescence imaging) were retained for further analysis. Fluorescence fundus photo-graphs of both treated and untreated eyes were taken from the G1(L151F)-GFP transgenic mice one month after the injection. In treated left eyes, the extent of mCherry expression covered nearly the whole retina suggesting successful and widespread vector-mediated expression (**Fig. 8B**). In those retinas expression of transgenic G1(L151F)-GFP was significantly suppressed (**Fig. 8A**) compared to that seen in untreated right eyes (**Fig. 7C**). In contrast, scAAV8-mG1hp4m2 treatment in the left eyes had no detectable effect (**Fig. 8D–F**).

To verify the fundoscopy, retinal cryosections of both left and right retinas from scAAV8-mG1hp4 or scAAV8-mG1hp4m2-treated mice were analyzed by direct fluorescence microscopy (**Fig. 8G–I**). G1(L151F)-GFP expression was nearly completely suppressed in the scAAV8-mG1hp4- treated retinas (**Fig. 8G**), but to a much lesser extent in retinas treated with scAAV8-mG1hp4m2 (**Fig. 8H**) consistent with live fundoscopy. Further, expression levels of endogenous GCAP1 in the treated mice were analyzed by immunoblot. In the scAAV-mG1hp4 treated retinas, both transgenic G1(L151F)-GFP and endogenous GCAP1 were significantly suppressed while the mutant shRNA vector had no significant effect (**Fig. 8J**). Quantitative analysis of GCAP1 levels (n = 3) revealed ~70% knockdown of G1(L151F)-GFP transgene, and 90% knockdown of endogenous GCAP1 in the scAAV-mG1hp4-treated G1(L151F)-GFP retinas, but only 30–35% knockdown in the scAAV-mG1hp4m2-treated retinas, which probably was caused by nonspecific effects at a high shRNA concentration (**Fig. 8K**).

Discussion

The purpose of this study was to generate mouse models of dominant CORD based on a GCAP1(L151F) mutation associated with human disease. The mouse models serve as tools to explore the efficiency of a nonallele-specific knockdown strategy suitable for human gene therapy. The transgenes consisted of a 15 kb *Guca1a* genomic fragment containing the native regulatory elements including endogenous promoter, all introns and endogenous polyadenylation sites. In the GCAP1-EGFP and GCAP1(L151)-EGFP fusion models, the coding region of exon 4 of the GCAP1 gene was fused to EGFP cDNA (**Fig. 2B, C**). Fusion of EGFP to GCAP1 at its C-terminus provided means to directly visualize GCAP1 localization by live fluorescence retina imaging (**Fig. 3**) and to monitor the suppression of GCAP1 by RNA interference (**Fig. 8**). The transgenes expressed wildtype and mutant GCAP1s in mouse photoreceptors at a level similar to endogenous GCAP1 in heterozygous mice (GCAP$^{+/-}$). No ectopic transgene expression was observed apart from accumulation of transgenic GCAP1 in perinuclear locales of a few individual cones (**Fig. 3D**).

In human patients carrying GCAP1 mutations, phenotypes are noticed typically within the second or third decade and progress slowly [2]. In transgenic mice expressing GCAP1(L151F), we observed reduced thickness of the photoreceptor layer as early as 9 months of age (**Fig. 4**). At 12 months of age, reduction of photopic ERG b-wave amplitudes was more pronounced than that of scotopic a-waves suggesting that cones degenerate somewhat faster than rods in these mouse models (**Fig. 5**). As mouse cones express GCAP1 only, a more severe L151F dominant negative effect is expected in these cells. Another, more severe, EF4-hand mutation associated with CORD (E155G) causes reductions in cone function in patients starting at an average age of 16 years, about 10 years earlier than for L151F patients [16,17,23]. Consistent with this more aggressive degenerative human phenotype, far greater functional deficits in cones than rods were observed in the

E155G knock-in mouse model [26]. Starting at 3 months postnatally, photopic ERG responses of E155G knock-in mice were reduced and the flicker response was severely depressed. At 12 months, the photopic ERG b-wave was reduced to 42% of normal, compared to 60% in our L151F transgenic mice. The presence of two normal *Guca1a* alleles in our L151F model versus just one in the E155G knock-in model may account for the phenotype differences. Further, the E155G mutation in which an essential charged residue is replaced by a neutral one represents a much more severe EF-hand structural change than seen in L151F. Nevertheless, our 'genomic transgenic' mouse yields phenotypes comparable to the 'knock-in' mouse suggesting that the 15 kb genomic fragment harboring Guca1a contains all the essential regulatory elements.

Functional differences in the transgenic mice at one year of age relative to age-matched wildtype mice are relatively mild (**Figs. 4, 5**), similar to those of second and third decade GCAP1(L151F) patients. The GCAP1(L151F) mutation was identified in two unrelated, five-generation families affected with cone and cone-rod dystrophy [17,23]. In young patients, rod and cone ERGs were near normal. Loss of cone function occurs first and rod function generally persists much longer in CORD families. Although the disease phenotypes were variable among the affected members in both families, most patients experienced late-onset cone dysfunction by the second and third decade of life. Another important CORD mutation is N104K in EF-hand 3 which we investigated previously [20]. A proband carrying a GCAP1(N104K) mutation and diagnosed by fundoscopy with dominant CORD at 39 years, experienced a slow loss of acuity over the course of 12 years [20]. Considering differences in human and mouse lifetimes, our late-onset and slowly progressive transgenic mouse models reveal very similar features and compare well with the human disease. The subtle deviations may be explained by the presence of two normal *Guca1a* alleles in our transgenic mice and by differences in human and mouse retinal anatomy, as the mouse has neither macula nor cone-rich central fovea.

The principal goal of this research is to identify gene therapy vectors that may be used successfully to delay the onset of cone degeneration, and/or cure CORD disease. In an earlier study, we demonstrated the feasibility of shRNA knockdown using an allele-specific approach in a retinitis pigmentosa mouse model carrying the GCAP1(Y99C) mutation [4]. In that mouse line, a GCAP1(Y99C) cDNA was expressed under the control of the mouse opsin promoter producing a retinitis pigmentosa like phenotype [25]. A therapeutic recombinant AAV robustly and persistently expressed shRNAs in photoreceptors at one week post-injection and silenced the disease-causing bovine GCAP1(Y99C) transgene with ~80% efficiency; the gene silencing was effective for nearly one year without apparent off-target interference and significantly improved rod photoreceptor survival, delayed disease onset and increased visual function [4].

Germline deletion of GCAPs in mouse delays recovery to the dark state by seconds as guanylate cyclase activity is not accelerated in low Ca^{2+} [8]. However, there are no detrimental morphological consequences, nor have gross functional deficiencies or retinal degeneration been reported. Patients with GCAP1 null or GCAP1/GCAP2 null mutations have not been identified to date, perhaps because a clear degeneration phenotype is absent [30]. We therefore reasoned that a nonallele-specific shRNA knockdown of both wildtype and mutant GCAP1s could be a general therapeutic strategy to rescue the dominant degeneration caused by any of the eight known EF-hand GCAP1 mutations. The most effective anti-mGCAP1 shRNA (mG1hp4) which degraded more than 70% of mGCAP1-EGFP was identified by an *in-vitro* fluorescent screening assay (**Fig. 6**). The "negative control" shRNA that carried a GG to CC mutation in the center was inactive (**Fig. 7**). After packaging into scAAV2/8vector, the mG1hp4 proved to be a potent GCAP1 knockdown agent *in-vivo* as well, as shown by live fluorescence imaging (**Fig. 8**). One month post-injection with AAV8-mG1hp4, we observed significant suppression of GCAP1(L151F)-GFP by 70% in the treated eyes. As expected, knockdown of mutant GCAP1 occurred concomitantly with knockdown of endogenous GCAP1 (~90%) (**Fig. 8J, K**). As mG1hp4 was designed to specifically target mouse GCAP1 without a C-terminal GFP, AAV-mG1hp4 has shown more efficient suppression of endogenous GCAP1 gene than the GCAP1(L151F)-GFP transgene. These experiments provide proof of principle that RNAi suppression of both wildtype and mutant GCAP1 may be a potent therapeutic strategy, applicable to all GCAP1 mutations of EF-hands 3 and 4, as long as the guide strand of shRNA is located outside the disease-causing mutations. Notably, when shRNA reaches its highest suppression of the target gene, overexpression does not increase efficiency but generates off-target silencing effects on other genes due to saturation of the intracellular RNAi machinery [**31**]. Thus, identification of a highly efficient shRNA may require its titration to find the lowest level efficacious for RNAi-based gene therapy in humans.

An advantage of dominant GCAP1 mutations is that a nonallelic approach promises to be successful, while mutations in other CORD genes require gene replacement for rescue. Dominant CORD is also associated with missense mutations in ***GUCY2D*** (guanylate cyclase 1) (reviewed in [32], ***CRX*** (cone-rod otx-like photoreceptor homeobox transcription factor) [33–35], ***AIPL1*** (arylhydrocarbon-interacting receptor protein-like 1) [36–38], and ***PROM1*** (Prominin 1) [39,40]. Proteins encoded by these genes have very diverse functions, yet null mutations of these genes are associated with recessive RP or LCA suggesting expression is vital for photoreceptor survival.

Author Contributions

Conceived and designed the experiments: LJ WB. Performed the experiments: LJ TZL JMF. Analyzed the data: JL JMF WB. Contributed reagents/materials/analysis tools: SEB WWH. Wrote the paper: LJ WB.

References

1. Hamel CP (2007) Cone rod dystrophies. Orphanet J Rare Dis 2: 7.
2. Jiang L, Baehr W (2010) GCAP1 Mutations Associated with Autosomal Dominant Cone Dystrophy. Adv Exp Med Biol 664: 273–282.
3. Baehr W, Palczewski K (2009) Focus on molecules: guanylate cyclase-activating proteins (GCAPs). Exp Eye Res 89: 2–3.
4. Jiang L, Zhang H, Dizhoor AM, Boye SE, Hauswirth WW et al. (2011) Long-term RNA interference gene therapy in a dominant retinitis pigmentosa mouse model. Proc Natl Acad Sci U S A 108: 18476–18481.
5. Baehr W, Karan S, Maeda T, Luo DG, Li S et al. (2007) The function of guanylate cyclase 1 and guanylate cyclase 2 in rod and cone photoreceptors. J Biol Chem 282: 8837–8847.
6. Howes KA, Bronson JD, Dang YL, Li N, Zhang K et al. (1998) Gene array and expression of mouse retina guanylate cyclase activating proteins 1 and 2. Invest Ophthalmol & Vis Sci 39: 867–875.
7. Yang RB, Fulle HJ, Garbers DL (1996) Chromosomal localization and genomic organization of genes encoding guanylyl cyclase receptors expressed in olfactory sensory neurons and retina. Genomics 31: 367–372.
8. Mendez A, Burns ME, Sokal I, Dizhoor AM, Baehr W et al. (2001) Role of guanylate cyclase-activating proteins (GCAPs) in setting the flash sensitivity of rod photoreceptors. Proc Natl Acad Sci U S A 98: 9948–9953.

9. Howes KA, Pennesi ME, Sokal I, Church-Kopish J, Schmidt B et al. (2002) GCAP1 rescues rod photoreceptor response in GCAP1/GCAP2 knockout mice. EMBO J 21: 1545–1554.

10. Makino CL, Peshenko IV, Wen XH, Olshevskaya EV, Barrett R et al. (2008) A role for GCAP2 in regulating the photoresponse. Guanylyl cyclase activation and rod electrophysiology in GUCA1B knock-out mice. J Biol Chem 283: 29135–29143.

11. Makino CL, Wen XH, Olshevskaya EV, Peshenko IV, Savchenko AB et al. (2012) Enzymatic Relay Mechanism Stimulates Cyclic GMP Synthesis in Rod Photoresponse: Biochemical and Physiological Study in Guanylyl Cyclase Activating Protein 1 Knockout Mice. PLoS ONE 7: e47637.

12. Pennesi ME, Howes KA, Baehr W, Wu SM (2003) Guanylate cyclase-activating protein (GCAP) 1 rescues cone recovery kinetics in GCAP1/GCAP2 knockout mice. Proc Natl Acad Sci U S A 100: 6783–6788.

13. Palczewski K, Sokal I, Baehr W (2004) Guanylate cyclase-activating proteins: structure, function, and diversity. Biochem Biophys Res Commun 322: 1123–1130.

14. Rudnicka-Nawrot M, Surgucheva I, Hulmes JD, Haeseleer F, Sokal I et al. (1998) Changes in biological activity and folding of guanylate cyclase-activating protein 1 as a function of calcium. Biochemistry 37: 248–257.

15. Baehr W, Palczewski K (2007) Guanylate cyclase-activating proteins and retina disease. Subcell Biochem 45: 71–91.

16. Wilkie SE, Li Y, Deery EC, Newbold RJ, Garibaldi D et al. (2001) Identification and functional consequences of a new mutation (E155G) in the gene for GCAP1 that causes autosomal dominant cone dystrophy. Am J Hum Genet 69: 471–480.

17. Sokal I, Dupps WJ, Grassi MA, Brown J, Jr., Affatigato LM et al. (2005) A GCAP1 missense mutation (L151F) in a large family with autosomal dominant cone-rod dystrophy (adCORD). Invest Ophthalmol Vis Sci 46: 1124–1132.

18. Nishiguchi KM, Sokal I, Yang L, Roychowdhury N, Palczewski K et al. (2004) A novel mutation (I143NT) in guanylate cyclase-activating protein 1 (GCAP1) associated with autosomal dominant cone degeneration. Invest Ophthalmol Vis Sci 45: 3863–3870.

19. Kitiratschky VB, Behnen P, Kellner U, Heckenlively JR, Zrenner E et al. (2009) Mutations in the GUCA1A gene involved in hereditary cone dystrophies impair calcium-mediated regulation of guanylate cyclase. Hum Mutat 30: E782–E796.

20. Jiang L, Wheaton D, Bereta G, Zhang K, Palczewski K et al. (2008) A novel GCAP1(N104K) mutation in EF-hand 3 (EF3) linked to autosomal dominant cone dystrophy. Vision Res 48: 2425–2432.

21. Sokal I, Li N, Surgucheva I, Warren MJ, Payne AM et al. (1998) GCAP1(Y99C) mutant is constitutively active in autosomal dominant cone dystrophy. Mol Cell 2: 129–133.

22. Dizhoor AM, Boikov SG, Olshevskaya EV (1998) Constitutive activation of photoreceptor guanylate cyclase by Y99C mutant of GCAP-1. Possible role in causing human autosomal dominant cone degeneration. J Biol Chem 273: 17311–17314.

23. Jiang L, Katz BJ, Yang Z, Zhao Y, Faulkner N et al. (2005) Autosomal dominant cone dystrophy caused by a novel mutation in the GCAP1 gene (GUCA1A). Mol Vis 11: 143–151.

24. Woodruff ML, Olshevskaya EV, Savchenko AB, Peshenko IV, Barrett R, et al. (2007) Constitutive excitation by Gly90Asp rhodopsin rescues rods from

degeneration caused by elevated production of cGMP in the dark. J Neurosci 27: 8805–8815.

25. Olshevskaya EV, Calvert PD, Woodruff ML, Peshenko IV, Savchenko AB, et al. (2004) The Y99C mutation in guanylyl cyclase-activating protein 1 increases intracellular Ca2+ and causes photoreceptor degeneration in transgenic mice. J Neurosci 24: 6078–6085.

26. Buch PK, Mihelec M, Cottrill P, Wilkie SE, Pearson RA et al. (2011) Dominant cone-rod dystrophy: a mouse model generated by gene targeting of the GCAP1/Guca1a gene. PLoS ONE 6: e18089.

27. Zhang Y, Buchholz F, Muyrers JP, Stewart AF (1998) A new logic for DNA engineering using recombination in Escherichia coli. Nat Genet 20: 123–128.

28. Prusky GT, Alam NM, Beekman S, Douglas RM (2004) Rapid quantification of adult and developing mouse spatial vision using a virtual optomotor system. Invest Ophthalmol Vis Sci 45: 4611–4616.

29. Zolotukhin S, Byrne BJ, Mason E, Zolotukhin I, Potter M et al. (1999) Recombinant adeno-associated virus purification using novel methods improves infectious titer and yield. Gene Ther 6: 973–985.

30. Hunt DM, Buch P, Michaelides M (2010) Guanylate cyclases and associated activator proteins in retinal disease. Mol Cell Biochem 334: 157–168.

31. Cullen BR (2006) Enhancing and confirming the specificity of RNAi experiments. Nat Methods 3: 677–681.

32. Karan S, Frederick JM, Baehr W (2010) Novel functions of photoreceptor guanylate cyclases revealed by targeted deletion. Mol Cell Biochem 334: 141–155.

33. Chen S, Wang QL, Xu S, Liu I, Li LY et al. (2002) Functional analysis of cone-rod homeobox (CRX) mutations associated with retinal dystrophy. Hum Mol Genet 11: 873–884.

34. Freund CL, Gregory EC, Furukawa T, Papaioannou M, Looser J et al. (1997) Cone-rod dystrophy due to mutations in a novel photoreceptor-specific homeobox gene (CRX) essential for maintenance of the photoreceptor. Cell 91: 543–553.

35. Furukawa T, Morrow EM, Cepko CL (1997) Crx, a novel otx-like homeobox gene, shows photoreceptor-specific expression and regulates photoreceptor differentiation. Cell 91: 531–541.

36. Jacobson SG, Cideciyan AV, Aleman TS, Sumaroka A, Roman AJ et al. (2011) Human retinal disease from AIPL1 gene mutations: foveal cone loss with minimal macular photoreceptors and rod function remaining. Invest Ophthalmol Vis Sci 52: 70–79.

37. Kirschman LT, Kolandaivelu S, Frederick JM, Dang L, Goldberg AF et al. (2010) The Leber congenital amaurosis protein, AIPL1, is needed for the viability and functioning of cone photoreceptor cells. Hum Mol Genet 19: 1076–1087.

38. Tan MH, Smith AJ, Pawlyk B, Xu X, Liu X et al. (2009) Gene therapy for retinitis pigmentosa and Leber congenital amaurosis caused by defects in AIPL1: effective rescue of mouse models of partial and complete Aipl1 deficiency using AAV2/2 and AAV2/8 vectors. Hum Mol Genet 18: 2099–2114.

39. Arrigoni FI, Matarin M, Thompson PJ, Michaelides M, McClements ME et al. (2011) Extended extraocular phenotype of PROM1 mutation in kindreds with known autosomal dominant macular dystrophy. Eur J Hum Genet 19: 131–137.

40. Pras E, Abu A, Rotenstreich Y, Avni I, Reish O et al. (2009) Cone-rod dystrophy and a frameshift mutation in the PROM1 gene. Mol Vis 15: 1709–1716.

5

Hoxb4 Overexpression in CD4 Memory Phenotype T Cells Increases the Central Memory Population upon Homeostatic Proliferation

Héloïse Frison[1,2,9], Gloria Giono[1,2,9], Paméla Thébault[1,9], Marilaine Fournier[1,2], Nathalie Labrecque[1,2,3], Janet J. Bijl[1,3]*

1 Hospital Maisonneuve-Rosemont Research Center, Montreal, Quebec, Canada, 2 Department of Microbiology, Infectiology and Immunology, University of Montreal, Montreal, Quebec, Canada, 3 Department of Medicine, University of Montreal, Montreal, Quebec, Canada

Abstract

Memory T cell populations allow a rapid immune response to pathogens that have been previously encountered and thus form the basis of success in vaccinations. However, the molecular pathways underlying the development and maintenance of these cells are only starting to be unveiled. Memory T cells have the capacity to self renew as do hematopoietic stem cells, and overlapping gene expression profiles suggested that these cells might use the same self-renewal pathways. The transcription factor Hoxb4 has been shown to promote self-renewal divisions of hematopoietic stem cells resulting in an expansion of these cells. In this study we investigated whether overexpression of Hoxb4 could provide an advantage to CD4 memory phenotype T cells in engrafting the niche of T cell deficient mice following adoptive transfer. Competitive transplantation experiments demonstrated that CD4 memory phenotype T cells derived from mice transgenic for Hoxb4 contributed overall less to the repopulation of the lymphoid organs than wild type CD4 memory phenotype T cells after two months. These proportions were relatively maintained following serial transplantation in secondary and tertiary mice. Interestingly, a significantly higher percentage of the Hoxb4 CD4 memory phenotype T cell population expressed the CD62L and Ly6C surface markers, characteristic for central memory T cells, after homeostatic proliferation. Thus Hoxb4 favours the maintenance and increase of the CD4 central memory phenotype T cell population. These cells are more stem cell like and might eventually lead to an advantage of Hoxb4 T cells after subjecting the cells to additional rounds of proliferation.

Editor: Troy A. Baldwin, University of Alberta, Canada

Funding: This study was supported by a discovery grant from the NSERC (342301-2008). GG is a recipient of a recruitment award from the Faculty of Medicine of the University of Montreal. MF is a recipient of a Scholarship from the Graduate School Faculty of University of Montréal. Furthermore this work was also supported by an NSERC grant of Dr. Nathalie Labrecque: 262146-2009. The funders had no role in study design, data collection and analysis, decision to publish, or preparation of the manuscript.

* E-mail: janettabijl@yahoo.ca

9 These authors contributed equally to this work.

Introduction

Memory T cells develop from a small subset of effector T cells following a primary immune response. While effector T cells undergo apoptosis, memory T cells survive and provide the host an immunological memory allowing a faster and more effective immune response against previously encountered pathogens. Memory T cells are long-lived cells and their survival after antigen clearance depends on the homeostatic cytokines interleukin (IL)-7 and IL-15 [1–5]. Memory T cells persist by undergoing a slow turn-over, also referred to as basal homeostatic proliferation, with a frequency of one division in 2–3 weeks [3]. However, upon transfer into a lymphopenic host, memory T cells divide rapidly due to an increased availability of IL-7 and IL-15 [1–6], a phenomenon indicated as acute homeostatic proliferation. Knock-out mouse models for IL-15, IL-7 and IL-7Rα demonstrated that CD4 and CD8 memory T cells have a differential dependence for these cytokines. In the absence of IL-15 the basal homeostatic proliferation of CD8, but not CD4 memory T cells was severely reduced [1,7,8], while CD4 memory T cells fail to persist upon transfer into IL-7 deficient hosts [9]. However, acute homeostatic proliferation of both CD4 and CD8 memory T cells can be induced by either IL-15 or IL-7Rα signalling [1,3,10]. In addition to IL-7 and IL-15, which are the key factors for the survival and homeostatic proliferation of memory T cells, other cytokines have been shown to boost their homeostatic proliferation, such as IL-2 and interferon-1 (IFN-I) [11–13]. Despite their independence for T cell receptor (TCR) signalling to survive, experiments using knock-out mice showed that antigen specific CD4 memory T cells had reduced responses to antigen re-encounter in the absence of major histocompatibility complex (MHC) II [14]. Moreover, the presence of MHC II signals influenced the homeostatic expansion capacity of memory T cells under lymphopenic conditions, but this appeared to be independent on the avidity for MHC II, in contrast to naïve T cells [15]. This suggests that regulatory mechanisms governing memory homeostasis are different from naïve T cell homeostasis, which is important to maintain optimal diversity of the memory pool.

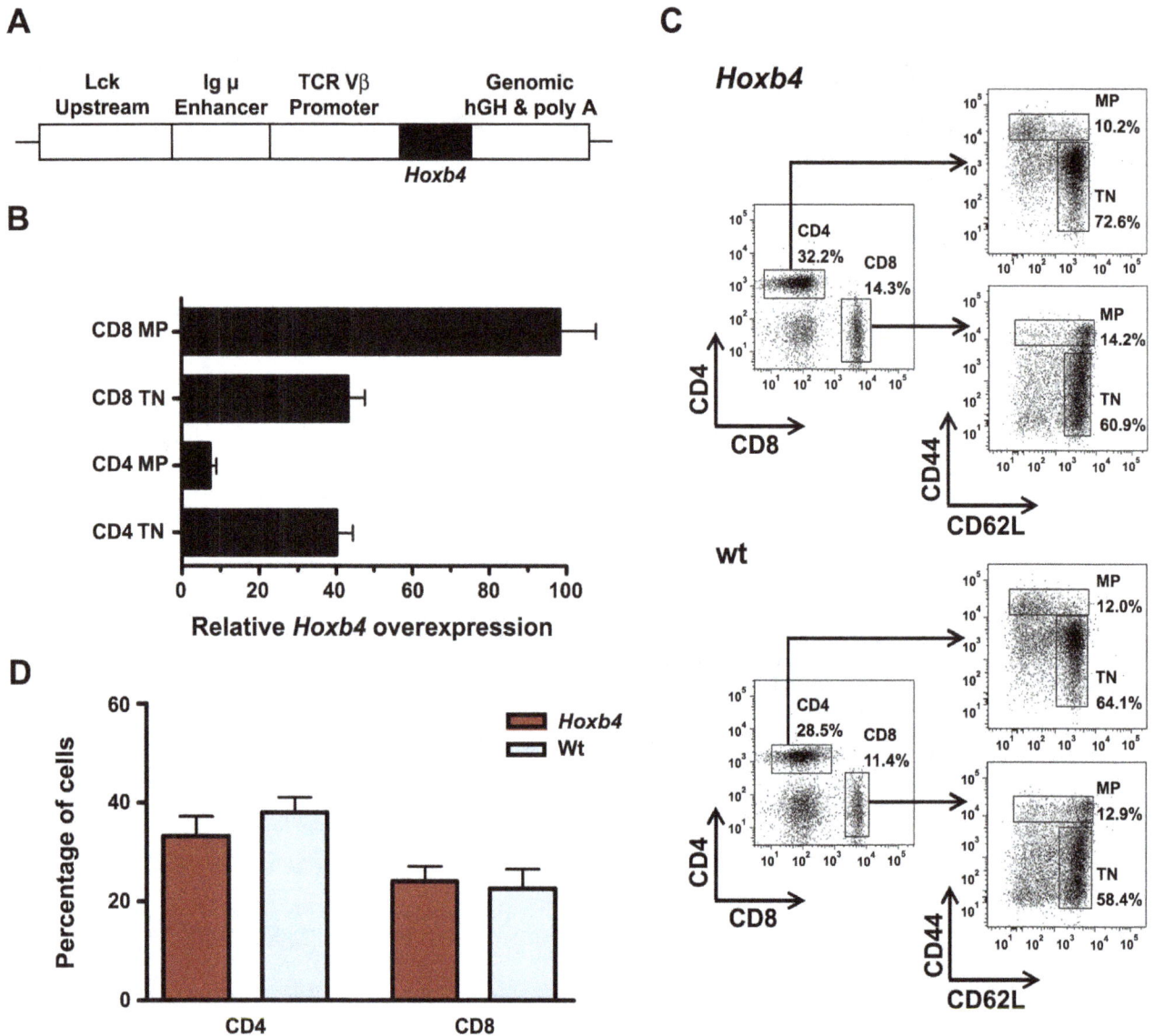

Figure 1. Analysis of T cell populations in *Hoxb4* transgenic mice. (**A**) Scheme of the transgenic construct with lymphoid promoter and enhancer elements from the TCR Vβ, lck and immunoglobulin-μ genes. (**B**) Fold overexpression of *Hoxb4* transgene mRNA in naïve and MP T cells compared to wt. (**C**) Representative FACS profiles of CD4 and CD8 naïve (CD44^{lo}/CD62L^{hi}) and MP (CD44^{hi}) T cell populations in spleen of three months old *Hoxb4* transgenic and wt mice. (**D**) Average percentage of CD4 and CD8 T cells in lymph nodes of *Hoxb4* transgenic (n = 7) and wt mice (n = 6). No significant differences were observed. P>0.05, 2-tailed Student ttest. TN = naïve T cells, MP = memory phenotype T cells, wt = wild type.

In addition to antigen-experienced memory T cells (true memory) a population of immunophenotypically identical memory cells exists that arise from interactions of the T cell receptor with endogenously expressed antigens [16] and are also referred to as memory phenotype (MP) T cells. Similarly to antigen-experienced memory cells, MP T cells are proliferating in response to lymphopenia and at least for CD8 it has been shown that they provide protection against antigen [3,17,18]. The requirements for homeostatic proliferation of MP T cells are slightly different than for true memory T cells. In addition to IL-15 and IL-7 they are dependent on MHCII [3,19,20], likely to avoid competition for signals provided in the niche.

Despite our increasing knowledge on the required signals, the molecular pathways behind homeostatic proliferation are still elusive. Some transcription factors have been shown to induce the expression of IL-7R or CD122 and thus allowing their permissive state to homeostatic survival and proliferation signals. For example, Foxo1 and GABPα promote IL-7R expression in T cells [21,22]. In contrast, transcription factor Gfi-1 downregulates IL-7R expression by inhibiting GABPα following TCR signalling or cytokine stimulation [23]. On the other hand, transcription factors T-bet and Eomes are found to maintain high levels of CD122 on CD8 memory T cells [24]. In addition, genes encoding for epigenetic regulators of transcriptional programs have been attributed important, but distinct functions in Th2 memory cells. First, using a knock-out model it was demonstrated that the polycomb gene *Bmi1* is critical for the survival of CD4 memory T cells through repression of the pro-apoptotic gene *noxa* [25]. Furthermore, the trithorax gene *MLL* was shown to provide activating histone modifications on the *GATA3* and Th2 cytokine loci, which are required for Th2 memory function [26,27]. Interestingly, both *MLL* and *Bmi1* are also critical for the maintenance of hematopoietic stem cells (HSC) [28–30]. Actually, memory T cells and HSCs have several features in common, such as longevity, the ability to self-renew at a very low rate normally followed by re-entering quiescence and the potential to proliferate and differentiate upon cytokine or antigen receptor signalling. Moreover, an overlap was observed in gene expression patterns between memory T, B cells and long-term HSCs [31], supporting the fact that identical molecular pathways might be involved in self-renewal of HSCs and memory T cells.

Hoxb4 is another well known critical regulator of HSCs and belongs to the family of homeobox (*Hox*) genes, which are transcription factors initially found to determine cell fate in the embryo [32]. The expression of *Hox* genes is epigenetically regulated by the antagonistic actions of polycomb and thrithorax genes [33]. *Hoxb4* is expressed in HSCs, but knock-out mouse models for *Hoxb4* showed that *Hoxb4* is not essential for their generation [34,35]. However, retroviral mediated overexpression of *Hoxb4* in bone marrow (BM) cells resulted in the expansion of HSCs through promotion of self-renewal divisions without development of overt leukemia in BM chimeric mice [36,37]. Interestingly, like memory T cells HSCs do not persist in the absence of *Bmi1* [28,29]. However, *Hoxb4* overexpression could not rescue the long-term maintenance of HSCs deficient for *Bmi1*, despite the triggering of self-renewal of *Bmi1*−/− HSCs [38]. Thus while *Hoxb4* has a function in the execution of the self-renewal division, *Bmi1* provides the HSCs their sustainability. With respect to the resemblances between HSCs and memory T cells, it is likely that these distinct functions might also apply to

memory T cells. In this study we set out to evaluate whether *Hoxb4* overexpression would lead to an enhanced self-renewal activity of CD4 MP T cells. Using *Hoxb4* lymphoid specific transgenic mice, we investigated the acute homeostatic proliferation of *Hoxb4* CD4 MP T cells and wild type (wt) MP T cells following transfer in competition into lymphopenic mice. Surprisingly, *Hoxb4* did not provide CD4 MP T cells with an advantage in repopulating the empty T cell niche. In fact the overall contribution of *Hoxb4* cells was significantly lower, but remained rather stable after two additional rounds of homeostatic proliferation. Intriguingly, *Hoxb4* MP T cells consistently comprised a significant larger population of cells expressing surface markers CD62L and Ly6C, indicating a central memory T cell phenotype.

Materials and Methods

Ethics Statement

All animal experiments have been performed in accordance with the guidelines of the Canadian Council on Animal Care and have been approved by the Hospital Maisonneuve-Rosemont animal protection committee (protocol number 2012–20).

Mice

Hoxb4 transgenic mice were generated using the pLIT3 vector (Fig. 1A) and have been described by us before [39]. PCR for the *hGH* gene was performed on genomic tail DNA to identify transgenic mice. Lymphopenic CD3ε$^{-/-}$ recipient mice for transplantation assays have been originally generated by Malissen et al. [40]. Wt mice C57BL/6 (CD45.2) and B6.SJL (CD45.1) were purchased from Jackson Laboratories (Bar Harbor, ME,

Table 1. Percentage of T cell populations in hematopoietic organs of young adult mice.

Hoxb4			Gated on CD4 or CD8 cells					
	n	Total	CD44^{lo}CD62L^{hi}	CD44^{hi}	CD44^{hi}CD62L^{hi}	CD44^{hi}CD62L^{lo}	CD44^{hi}Ly6C$^+$
CD4 T cells								
LN	5	36.99±4.47	71.49±13.87	9.73±4.69	2.18±0.80	6.06±2.83	0.14±0.05
Spleen	7	21.68±8.33	59.22±14.66	24.56±10.32	4.29±0.67	20.59±11.18	0.80±0.37
BM	7	1.86±1.05	11.51±9.97	57.19±8.37	6.65±4.71	50.54±6.90	11.37±7.92
CD8 T cells								
LN	5	24.10±8.05	52.73±25.58	14.83±4.75	9.64±4.56	4.55±3.87	10.47±3.07
Spleen	7	10.56±4.93	56.91±9.64	23.24±4.18	16.38±2.88	7.46±4.56	15.06±1.98
BM	7	1.99±1.34	25.94±19.95	41.68±13.84	14.23±9.19	27.44±18.25	22.16±6.35
Wild Type			Gated on CD4 or CD8 cells					
	n	Total	CD44^{lo}CD62L^{hi}	CD44^{hi}	CD44^{hi}CD62L^{hi}	CD44^{hi}CD62L^{lo}	CD44^{hi}Ly6C$^+$
CD4 T cells								
LN	6	38.05±7.45	55.80±19.07	10.93±3.90	2.23±0.57	8.69±3.98	0.17±0.09
Spleen	7	20.50±5.57	56.07±12.47	22.17±6.24	4.56±1.44	18.99±5.19	1.02±0.48
BM	7	1.68±0.80	15.30±10.15	56.68±17.49	6.56±4.62	50.13±14.97	17.89±11.75
CD8 T cells								
LN	6	22.60±9.60	38.45±20.26	12.36±4.58	6.03±2.95	6.33±5.14	9.17±3.89
Spleen	7	9.36±4.31	52.06±10.42	19.90±4.83	12.72±2.89	7.52±3.23	13.53±2.64
BM	7	1.75±1.65	30.01±21.76	34.78±15.10	12.41±6.61	22.37±12.93	20.60±6.32

Note that no significant differences were observed between T cell populations of *Hoxb4* and wild type mice. 1-tailed Student ttest, comparing *Hoxb4* vs. wild type mice.
LN = Lymph node; BM = bone marrow.

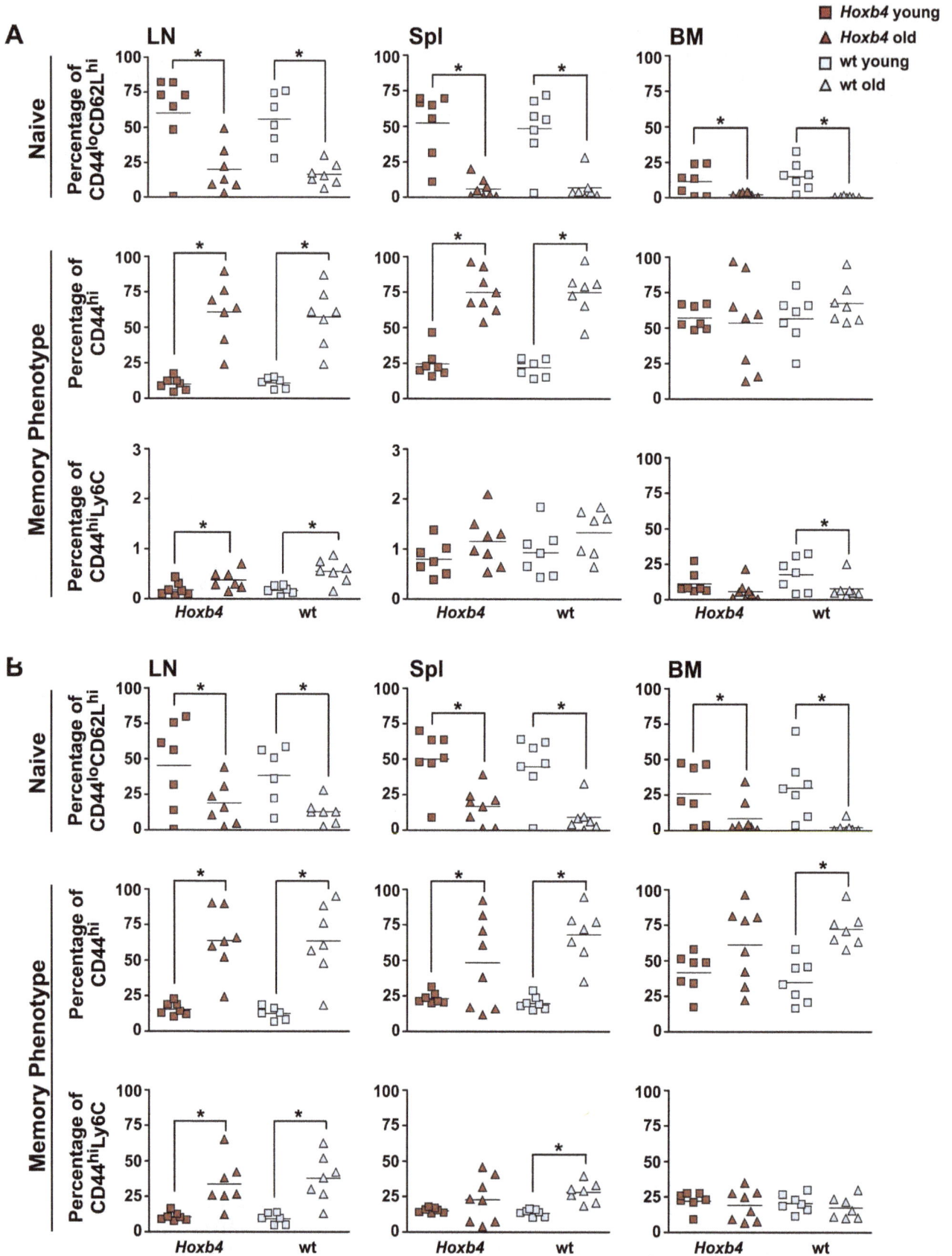

Figure 2. Change of naïve and MP T cell populations in *Hoxb4* transgenic and wt mice with age. Scatter plots showing the percentage of (**A**) CD4 and (**B**) CD8 T cells that are naïve (CD44^lo/CD62L^hi), MP (CD44^hi) or are a subpopulation of MP T cells (CD44^hi/Ly6C^hi) for LN, Spl and BM

derived from individual *Hoxb4* transgenic and wt (n = 6–8) age matched mice. Young mice are between 3–4 months and old mice are all older than 28 months. Naïve and MP populations change significantly with age, but not between *Hoxb4* and wt mice. *P<0.05, 2-tailed Student ttest. MP = memory phenotype, Wt = wild type, LN = lymph node, Spl = spleen and BM = bone marrow.

USA). *Hoxb4* transgenic mice (CD45.2) were bred to B6.SJL mice to generate compound CD45.1/2 *Hoxb4* transgenics. Mice were housed at the animal facility of the Maisonneuve-Rosemont Hospital Research Center under specific pathogen free conditions.

FACS Analysis and Sorting

To analyse memory T cell populations the following antibodies with conjugated fluorochromes were used: CD8α-PerCP, CD62L-Pacific Blue, Ly6C-Alexa647, Sca-1-PE/Cy7 or Pacific Blue (all obtained from BioLegend, San Diego, CA, USA); CD4-APC/Cy7, CD44-PE/Cy7 or -APC, Ly6C-FITC, CD62L-PE-Cy7 (BD Pharmingen, Mississauga, ON, Canada); CD127-biotin and CD62L-eFluor® 605NC (eBioscience, San Diego, CA, USA). To determine the transgenic or wt origin of MP T cell populations in competitive transplantation assays following antibodies against the CD45 alleles were used: CD45.1-Pacific Blue or -FITC (BioLegend) and CD45.2-APC or -V500 (BD Bioscience). Labelled cells were analyzed on a LSR II with an UV laser (BD Bioscience, Missisauga, ON, Canada), using the Diva software. FlowJo software (TreeStar Inc., Ashland, OR, USA) was used to further analyze specific cell populations.

MP T cell populations for transplantations assays or expression analysis were sorted on a FACSAria III sorter (BD Bioscience) using antibodies for CD44-APC (BioLegend) or -PE-Cy7 (BD

Bioscience), CD4 -PerCP, CD8α-PerCP, CD25-biotin and NK1.1-biotin, all from BD Bioscience. Biotin conjugated antibodies were detected with Streptavidin-PE (BioLegend). MP T cells were gated on either CD4+ or CD8+ and further defined as CD44^{hi}/CD25$^-$/NK1.1$^-$.

Antibodies used to determine cytokine production include: IFNγ-FITC (clone B27), TNF-α-PE (clone MP6-XT22), IL-2-APC (JES6-5H4) all purchased from BD Bioscience and their isotype controls: FITC Rat IgG1 (Life Technologies, Burlington, ON, Canada), -PE l5 and -APC l28 (Biolegend).

Competitive Transplantation Assays

CD4 MP T cells were sorted from spleen and lymph nodes (LNs) of 3 to 4 months old *Hoxb4* transgenic (CD45.1/2) and congenic wt (CD45.1) mice. A cell dose of 2×10^5 cells composed of equal numbers of *Hoxb4* and wt MP T cells were transplanted in sex matched CD3ε$^{-/-}$ mice by injection in the tail vein. For serial transplantation, mice were sacrificed two months post-transplantation and 10^7 cells derived from the LNs of each donor were transplanted into a secondary CD3ε$^{-/-}$ host. A third transplantation was repeated again after two months.

For evaluation of short-term proliferation under competitive conditions, sorted *Hoxb4* and wt MP T cells were labelled with CellTrace^{TM} Violet (Life Technologies Inc.) according to the

Table 2. Percentage of T cell populations in hematopoietic organs of aged adult mice.

Hoxb4	n	Total	Gated on CD4 or CD8 cells				
			CD44^{lo}CD62L^{hi}	CD44^{hi}	CD44^{hi}CD62L^{hi}	CD44^{hi}CD62L^{lo}	CD44^{hi}Ly6C$^+$
CD4 T cells							
LN	5	20.44±5.41	19.98±16.36	60.76±21.87	4.39±1.85	56.37±21.20	0.38±0.19
Spleen	7	*11.43±3.09	6.08±7.47	73.77±15.76	3.05±1.85	71.76±15.40	1.16±0.35
BM	7	*2.91±1.60	*2.28±1.61	53.03±34.94	1.32±0.86	52.22±31.56	6.06±6.97
CD8 T cells							
LN	5	13.78±7.82	19.07±15.09	63.75±22.69	30.64±14.72	33.11±16.19	33.71±16.88
Spleen	7	*4.68±1.96	16.02±16.12	56.93±32.64	17.09±8.75	42.17±24.90	28.65±14.06
BM	7	*1.51±0.98	8.45±12.25	63.91±27.87	10.23±7.45	50.98±21.90	19.32±10.93

Wild Type	n	Total	Gated on CD4 or CD8 cells				
			CD44^{lo}CD62L^{hi}	CD44^{hi}	CD44^{hi}CD62L^{hi}	CD44^{hi}CD62L^{lo}	CD44^{hi}Ly6C$^+$
CD4 T cells							
LN	6	22.49±7.55	16.74±7.90	57.35±20.85	5.64±3.04	51.72±17.96	0.56±0.24
Spleen	7	18.12±5.57	6.94±9.65	74.77±16.12	2.98±1.07	71.79±15.60	1.33±0.47
BM	7	5.55±1.50	0.76±0.44	67.47±14.67	2.01±1.22	65.47±14.44	8.14±7.67
CD8 T cells							
LN	6	17.38±10.71	12.88±8.31	63.47±26.18	37.33±26.09	26.15±8.15	37.87±16.45
Spleen	7	9.26±5.48	9.25±10.95	60.09±32.16	25.22±11.54	42.91±10.96	28.36±7.18
BM	7	3.60±2.06	2.41±3.74	72.30±12.48	11.76±9.39	60.54±12.22	17.50±7.77

Note the decrease in total T cells in bone marrow and spleen of *Hoxb4* mice;
*P<0.05;
1-tailed student ttest, comparing *Hoxb4* vs. wild type mice. LN = Lymph node; BM = bone marrow.

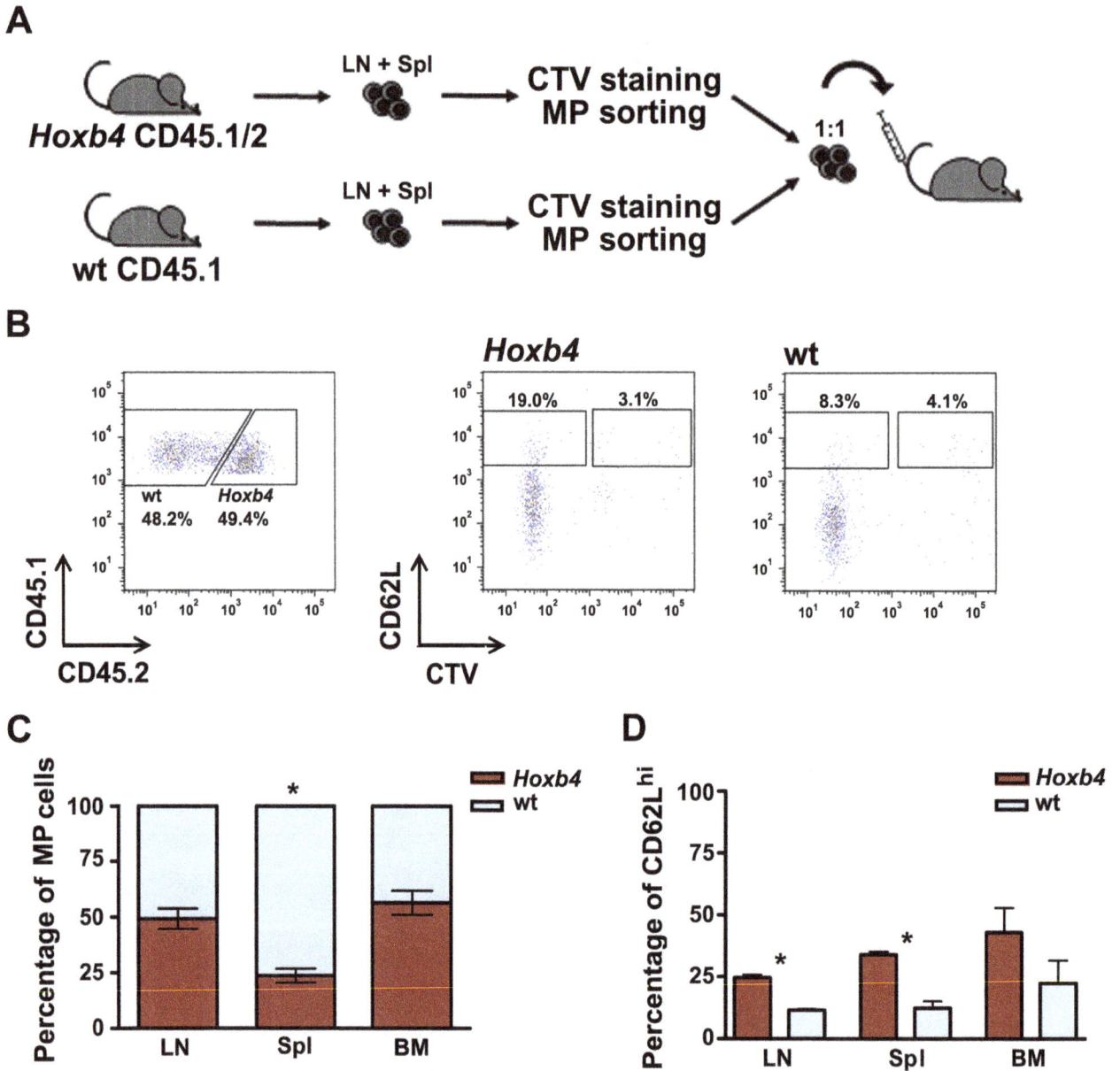

Figure 3. Competitive short-term homeostatic proliferations (7 days) of *Hoxb4* transgenic and wt CD4 MP T cells. (A) Scheme of the experimental approach. CD4 MP T cells are sorted from CellTrace^{TM} Violet (CTV) labelled cells isolated from LN and Spl of *Hoxb4* (CD45.1/2) and congenic wt (CD45.1) mice. Cells of both genotypes are transplanted in a 1:1 ratio in CD3ε$^{-/-}$ (CD45.2) mice. (B) FACS profiles showing *Hoxb4* and wt fractions to donor derived CD4 MP T population (CD45.1) in LN (left panel). Representative FACS profiles for CD62L and CTV on *Hoxb4* and wt populations. Loss of CTV tracer indicates that most cells are dividing rapidly (right panels). (C) Average contribution (%) of *Hoxb4* and wt cells to donor derived MP T cells in LN, Spl and BM (n=3). (D) Percentage of CD62L^{hi} MP T cells in *Hoxb4* and wt population found in lymphoid organs. *P<0.05; paired 2-tailed Student ttest. Wt=wild type, MP=memory phenotype, LN=lymph node, Spl=spleen and BM=bone marrow.

manufacturer's protocol prior to transplantation. Each recipient received a total dose of 7×10^5 cells. Mice were sacrificed after one week for analysis.

Cytokine Analysis

The production of cytokines was measured as described previously [41]. Splenocytes were stimulated with PMA/ionomycin (5 µg/ml) for 2 hours at 37°C, followed by 2 hours incubation with 100 µg/ml Brefeldin A (Sigma-Aldrich Co., St. Louis, MO) to block cytokine secretion. After fixation with formaldehyde (2%)

followed by permabilization with 0.5% saponine (Sigma-Aldrich) cells were stained with antibodies against cytokines.

Quantitative Reverse Transcriptase (Q-RT)-PCR

Total RNA was isolated by Trizol®, DNase-I-treated and cDNA was prepared using MMLV-RT according to the manufacturer's instructions (Invitrogen, Paisley U.K.). Q-RT-PCR was carried out using SYBRGreen® Power mix (Applied Biosystems, Toronto, ON, Canada), using oligonucleotides for *Hoxb4* and Gapdh as designed before [42,43]. Reactions were carried out in triplicate. CT-values were corrected for Gapdh

Table 3. Total donor cells contribution to hematopoietic organs and the proportion of *Hoxb4* versus wild type cells.

	LN	Spleen	BM
primary hosts, n = 9	(%)	(%)	(%)
Total	6.6±3.1	4.5±1.7	0.7±0.8
Hoxb4	38.2±25.5	29.6±18.3	24.2±18.3
wt	61.6±25.5	70.2±18.4	75.5±19.2
secondary hosts, n = 6			
Total	8.7±3.8	3.8±1.6	0.2±0.1
Hoxb4	32.4±15.6	33.4±8.4	34.4±17.2
wt	68.7±17.2	66.4±8.1	65.6±17.2
tertiary hosts, n = 3			
Total	4.4±2.0	1.0±0.7	0.2±0.1
Hoxb4	29.4±4.1	22.5±17.5	24.9±11.0
wt	39.9±24.6	35.0±31.9	40.5±28.7

LN = Lymph node; BM = bone marrow, wt = wild type.

expression and to the expression in a calibrator comprised of BM, spleen and LN cells ($\Delta\Delta CT$). The average fold difference over the expression in the calibrator sample was calculated as $2^{(-\Delta\Delta CT)}$. To compare the expression of *Hoxb4* in transgenic and wt mice, the average fold difference of *Hoxb4* transgenic was divided by those for the wt mice.

Results

To evaluate whether overexpression of *Hoxb4* modulates the size of the MP T cell population, a lymphoid specific transgenic mouse for *Hoxb4* was analysed. No major abnormalities in the thymic and splenic lymphoid populations of these mice have been reported [39], but the MP T cell populations have not been analysed in these mice before. To validate the expression of the transgene in MP T cell populations, Q-RT-PCR for *Hoxb4* was performed on RNA purified from CD4 and CD8 T cell subpopulations sorted from LN and spleen of adult *Hoxb4* transgenic and wt mice. Both naïve and MP T cell populations of wt mice expressed *Hoxb4*. Expression levels for *Hoxb4* were markedly increased in transgenic CD4 (7-fold) and CD8 (90-fold) MP T cells (Fig. 1B).

Memory Phenotype T cell Populations in Young and Aged *Hoxb4* Transgenic Mice

FACS analysis showed that CD4 and CD8 T cell populations in BM, spleen and LNs were comparable in adult 2–3 months old *Hoxb4* transgenic and wt mice (Fig. 1C and D; Table 1). Furthermore, naïve T cell (CD44^{lo}/CD62L^{hi}) and MP T cell populations of the CD4 (CD44^{hi}) or CD8 (CD44^{hi}/Ly6C^{hi}) fractions in these lymphoid organs were not significantly different between *Hoxb4* and wt mice (Fig. 2; Table 1), indicating that intrinsic and extrinsic regulatory mechanisms determining the number of MP T cells are intact in *Hoxb4* transgenic mice.

It has been well documented that T cell populations decline with age as result of an involuted thymus decreasing production of new naïve T cells [44]. The memory T cell population is relatively stable and might even increase due to accumulation of newly generated memory T cells as result of lifelong exposure to antigens. To evaluate whether changes in proportions of naïve and MP cells with age also occur in T cells overexpressing *Hoxb4*, lymphoid populations of old mice (>15 months) were analyzed

and compared to wt mice. First, analysis of wt mice showed that in aged mice the overall percentage of CD4 cells was significantly reduced in LN compared to young mice (Fig. S1A; Table 1 and 2). On the contrary, an increase in CD4 and CD8 T cells was found in the BM of wt mice. As expected naïve CD4 and CD8 T cell subpopulations in aged wt mice were dramatically reduced in LN, spleen and BM (Fig. 2 and Table 1 and 2), while the CD44^{hi} memory populations and their subpopulations CD44^{hi}/Ly6C^{hi} were increased, except in the BM for CD4. In *Hoxb4* transgenic mice the decrease in both CD4 and CD8 T cells with age was more pronounced than in wt mice, reaching significance in both LN and spleen (Fig. S1). Similarly as in wt mice, the proportions of naïve CD4 and CD8 subpopulations in old *Hoxb4* transgenic mice were decreased compared to young mice and those of CD44^{hi} memory phenotype T cells increased (Fig. 2; Table 1 and 2). However, the cellularity in hematopoietic organs tended to be reduced in old *Hoxb4* mice (Table S1). This had as consequence that the numbers of CD44^{hi} MP cells were not significantly increased in these mice. Together, these data show that *Hoxb4* does not affect T cell homeostasis in young mice that are in steady-state hematopoiesis.

Short-term Competitive Homeostatic Proliferation of CD4 Memory T cells

Adoptive transfer of T cells into a lymphopenic host results in a rapid proliferation of these cells to fill the empty niche. To test whether *Hoxb4* CD4 MP T cells have an advantage over wt cells in the initial proliferation phase following transfer, CD4 MP cells (CD44^{hi}/CD25$^-$NK1.1$^-$) were sorted from both *Hoxb4* transgenic (CD45.1/2) and wt congenic mice (CD45.1), stained with CellTrace^{TM} Violet (CTV) and transplanted in a 1:1 ratio into CD3ε$^{-/-}$ mice lacking T cells (Fig. 3A). One week post transplantation, the mice were sacrificed and analysed for the presence of donor cells by fluorescence activated cell sorting (FACS). Distinct populations of donor cells were detected in all organs representing 0.20% of the total cell population in both LNs and spleen and 0.03% in BM (data not shown). *Hoxb4* and wt CD4 MP T cells contributed equally to LNs and BM of the recipient mice, but in the spleen the wt cells dominated over *Hoxb4* MP T cells (Fig. 3B left panel and 3C). Both *Hoxb4* and wt MP T cells underwent several divisions during seven days, because the majority of cells were negative for CTV (Fig. 3B).

Interestingly, the proportions of cells expressing the surface marker CD62L, which is characteristic for central memory T cells, were higher in *Hoxb4* CD4 MP T cells (Fig. 3D).

Together these data show that both *Hoxb4* and wt CD4 MP T cells have divided rapidly to occupy the niche in T cell deficient mice with fluctuations in the different organs. Moreover, overexpression of *Hoxb4* promoted the enrichment of CD62L positive cells following lymphopenia induced proliferation.

Medium-term Competitive Repopulation in a Lymphopenic Host

To evaluate whether *Hoxb4* CD4 MP T cells would dominate over the wt MP T cells with time, equal numbers of *Hoxb4* (CD45.1/.2) and wt (CD45.1) CD4 MP T cells were transplanted in CD3ε$^{-/-}$ mice and sacrificed after two months. No signs of disease were observed in these mice. At this time point the transferred MP T cells occupied at average 6.6±3.1% of the cells in LN, 4.5±1.7% in the spleen and 0.7±0.8% in the BM of recipient mice (Table 3). This is equivalent with a total of ~6.4×10^6 donor cells in these organs, which implies that MP T cells have expanded at least 31-fold over the number of injected

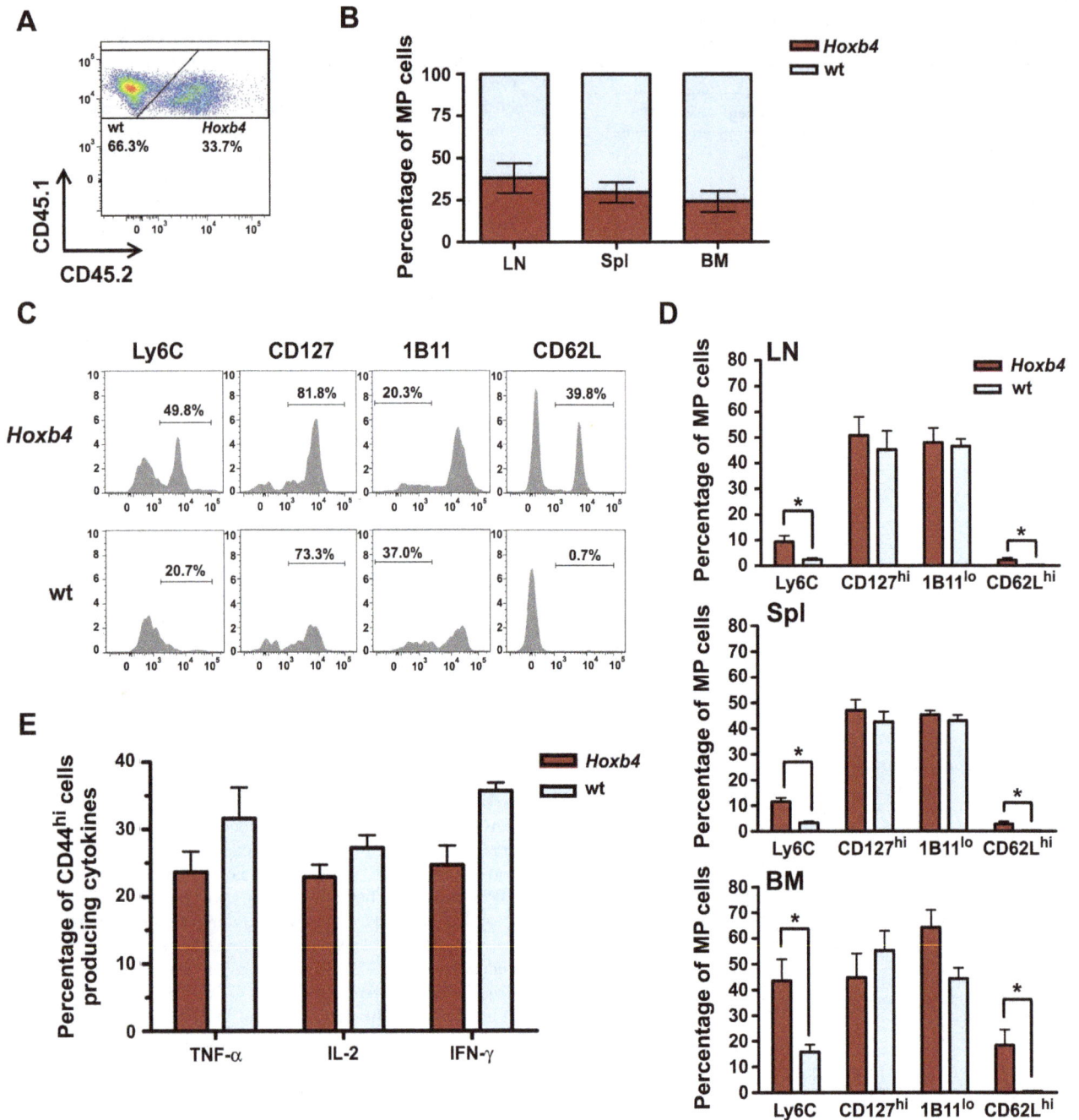

Figure 4. Medium-term competitive homeostatic proliferations (60 days) of *Hoxb4* transgenic and wt CD4 MP T cells. (A) FACS profile showing fractions of *Hoxb4* and wt cells to donor derived MP T population in lymph node (LN). **(B)** Stacked bar graphs indicating the average contributions of *Hoxb4* and wt cells in LN, Spl and BM measured in three independent experiments; n = 9. **(C)** FACS profiles for the expression of typical memory T cell surface markers on *Hoxb4* and wt MP T cells in the BM. **(D)** Average subpopulations of *Hoxb4* and wt MP T cells expressing the indicated surface markers in LN (upper panel), Spl (middle panel) and BM (lower panel). **(E)** Percentage of *Hoxb4* and wt MP T cells (gated on CD44^{hi}) positive for indicated cytokines (n = 3–6). *P<0.05, 2-tailed Student ttest. MP = memory phenotype, wt = wild type, LN = lymph node, Spl = spleen and BM = bone marrow, TNF = tumor necrosis factor; IL-2 = interleukine-2; IFN = interferon.

cells (data not shown). The average contribution of wt CD4 MP T cells to the repopulation of the BM and spleen in nine individual recipients transplanted in three independent experiments was 2- to 3-fold higher in the spleen and BM than that of *Hoxb4* CD4 MP cells (Table 3 and Fig. 4A and B). However, the percentage of the *Hoxb4* and wt CD4 MP T cells in the LNs of host mice was very variable and not significantly different (Fig. 4B and Table 3). Total

Hoxb4 and wt CD4 MP T cells in the combined organs were calculated and showed a net expansion of 20-fold and 43-fold for *Hoxb4* and wt cells, respectively (Fig. S2A). This resulted in a net 2-fold higher contribution of wt CD4 MP T cells over *Hoxb4*. Both *Hoxb4* and wt CD4 MP T cell populations were further characterized for the presence of specific surface markers such as CD62L, Ly6C, CD127 and 1B11. FACS analysis showed that the

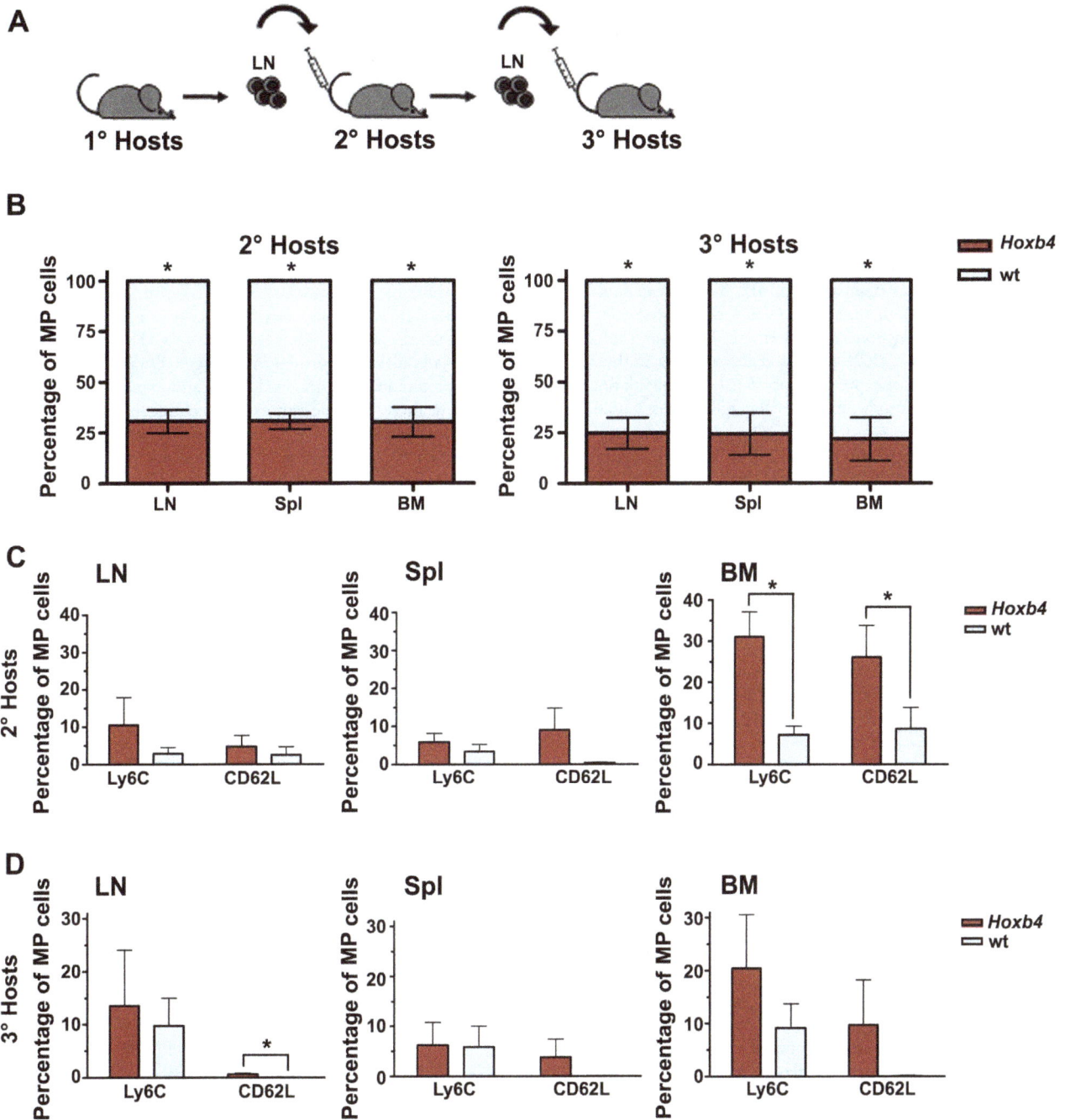

Figure 5. Long-term competitive homeostatic proliferations (180 days) of *Hoxb4* **transgenic and wt CD4 cells. (A)** Scheme of serial transplantations. 10×10^6 cells of the LNs of primary hosts that received a transplant composed of equal doses of *Hoxb4* and wt MP T cells were serially transplanted into secondary and tertiary hosts with a 60 days interval. **(B)** Compilation of *Hoxb4* and wt fractions of donor derived cells in LN, Spl and BM of secondary (n = 6) and tertiary hosts (n = 4) from two independent experiments. **(C)** Bar graphs showing the average percentage of cells positive for CD62L and Ly6C within the *Hoxb4* or wt memory T populations. *P<0.05, 2-tailed Student ttest. Wt = wild type, MP = memory phenotype, LN = lymph node, Spl = spleen and BM = bone marrow.

percentage of Ly6C$^+$ cells was 3- to 4-fold higher in the *Hoxb4* population compared to wt in all organs analysed (Fig. 4C and D). Moreover, in contrast to a negligible proportion of wt CD4 memory T cells that express CD62L (<1.0% in all organs), a distinct proportion of *Hoxb4* cells expressed this marker in BM (13.9±7.1%), spleen (3.4±1.3%) and LN (3.3±1.2%), indicating that *Hoxb4* favours central memory characteristics. All CD62L$^+$

cells carried also the Ly6C marker (data not shown). The percentage of cells expressing CD127, CD43 and CD44 was not different for both genotypes. Since central memory T cells are considered the long-term memory T cells, we analysed if *Hoxb4* overexpression act more specifically on the expansion of CD4 central MP T cells. The total numbers of CD62L$^+$ cells in BM, spleen and LNs of the *Hoxb4* population ranged from 7000 to

41000 (mean 22821 ± 13083) and of the wt population from 160 to 5000 (mean 2030 ± 1912; Fig. S2B). The percentage of CD62L$^+$ CD4 MP T cells in the initial graft were considered equal based on data in Table 1 and estimated at 20000 cells. After competitive homeostatic proliferation the number CD62L$^+$ cells had increased in the *Hoxb4* population in four out of six mice. In contrast the CD62L$^+$ MP T cells were decreased in the wt population for all six mice. In addition to the molecular make-up of the MP T cells, the functional response after stimulation with PMA was measured. The proportion of *Hoxb4* MP T cells that produced TNF-α and IL-2 was somewhat, but not significantly, reduced. However, a modest, but significant decrease in the number of *Hoxb4* cells producing IFN-γ was observed (Fig. 4E). Thus these data show that in the majority of mice *Hoxb4* CD4 MP T cells were less competitive, but functional. Furthermore, the subset of more central memory like CD62L positive cells was larger in the *Hoxb4* MP T cells and in the presence of *Hoxb4* this population was actually expanded in some mice over the initial transplanted cells.

Evaluation of Long-term Competitive Repopulation in Lymphopenic Hosts by Serial Transplantation

Central memory T cells have a more extended live span than effector memory T cells and are considered the long-term memory cells. The *Hoxb4* CD4 MP T population comprised a higher number of CD62L$^+$ memory T cells than the wt after one round of homeostatic proliferation. To evaluate whether the contributions of *Hoxb4* and wt MP T cells changed in favour of *Hoxb4* after several rounds of homeostatic proliferation, 10^7 cells of total LN isolated from six primary hosts were serially transplanted (Fig. 5A), which comprises between 0.5 and 1.0×10^6 donor CD4 MP T cells. The percentage of transferred MP T cells detected in the hematopoietic organs of the secondary hosts after two months of expansion was not significantly different than in primary hosts (Table 3). Evaluation of the proportions of *Hoxb4* and wt MP T cells by FACS showed that the average contribution of *Hoxb4* MP T cells was significantly lower than that of wt in all organs (Fig. 5B and Table 3). However, only in one out of six mice the contribution of *Hoxb4* was higher than that of wt MP T cells in LN and BM. These ratios of *Hoxb4* vs. wt MP T cells were maintained upon transfer and expansion in a tertiary host. Interestingly, the proportion of CD62L$^+$ and Ly6C$^+$ cells remained consistently higher in the *Hoxb4* CD4 MP T cell population following serial transfer (Fig. 5C and D). This translated in an actual higher number of *Hoxb4* than wt CD62L$^+$ cells in five of six secondary hosts (data not shown). Cytokine analysis following stimulation *in vitro* showed that even after several rounds of expansion *Hoxb4* MP T cells were functionally intact (Fig. S3). Thus despite an increased population of CD62L$^+$ MP T cells, the contribution of *Hoxb4* CD4 MP T cells did not change after three rounds of expansion.

Discussion

Memory T cells are triggered to proliferate following the transplantation into a lymphopenic host through increased availability of IL-7 and likely IL-15 [3,6]. In this study we investigated whether *Hoxb4*, known for its potential to expand HSCs, also could increase the pool of CD4 MP T cells. Competitive transplantation of *Hoxb4* transgenic and wt CD4 MP T cells showed that *Hoxb4* overexpression did not provide the total MP T cell population with a proliferative advantage. On the contrary, the overall population of *Hoxb4* MP T cells were less competitive in expansion to occupy the empty niche of CD3$\epsilon^{-/-}$ mice than wt CD4 MP T cells. The disadvantage became already

apparent in the spleen after 7 days of homeostatic proliferation, despite the fact that most *Hoxb4* CD4 MP T cells had undergone a rapid proliferation (Fig. 3B), and was even more pronounced after 2 months. Two additional rounds of homeostatic proliferation did not further change the established ratios between *Hoxb4* and wt MP T cells. We cannot completely exclude that *Hoxb4* CD4 MP T cells were lost by apoptosis, but we think that this is not very likely, because no differences in apoptosis within the MP T cell populations were observed in *Hoxb4* transgenic and wt mice (data not shown).

A major observation is the enrichment of CD62L and Ly6C subpopulations in the presence of *Hoxb4* overexpression. The CD62L molecule is a classical indicator for central memory T cells [45]. Although Ly6C has been recognized as a marker for CD8 central memory T cells, this is less clear for CD4 memory T cells. Tokoyoda et al. demonstrated that antigen specific CD4 memory T cells preferentially reside in the BM and express high levels of Ly6C, but these cells did not express central memory molecules CD62L or CCR7 [46]. The Ly6C expression on CD4 memory T cells still remains controversial as a more recent study by Marshall et al. demonstrates that Ly6C^{lo} effector cells persisted better during the contraction phase than Ly6C^{hi} cells and thus are more prone to develop into memory T cells [47]. In the same study they show that the transcriptional profile of Ly6C^{lo} effector T cells resembled those of memory T cells. Interestingly, high expression of Ly6C was observed once the effector T cells were converted into memory T cells. It is thus not clear whether Ly6C$^+$ cells could be considered central memory cells. We found that CD62L expressing cells expressed also Ly6C, but in addition Ly6C$^+$/CD62L$^-$ cells were present. The expression of Ly6C was mostly low to intermediate, while only a small fraction of cells were Ly6C^{hi} and thus does not contradicts observations mentioned by Marshall et al. It is of interest to note that we observed highest proportions of Ly6C and CD62L cells in the BM, which has been allocated as the principal niche for long-term CD4 memory T cells [46]. The actual increase of the CD62L population by *Hoxb4* found in several mice could be achieved by promotion of self-renewal divisions as has been suggested for *Hoxb4* [36]. Thus our data suggest that *Hoxb4* might indeed favour self-renewal of CD4 MP T cells, but only the stem cell like CD62L$^+$ central memory T cells. Alternatively, *Hoxb4* could activate CD62L or Ly6C expression, however, no binding sites for Hoxb4 or its cofactor Pbx are predicted on the promoter sequences of either gene according to the DECipherment Of DNA Elements (DECODE) database, which compiles predicted binding sites for over 200 transcription factors, suggesting no direct activation. Indirect activation of CD62L by Hoxb4 or through binding to more distant enhancer regions cannot be excluded, but is unlikely as populations of CD44$^+$/CD62L$^+$ in transgenic *Hoxb4* mice are not enhanced compared to wild type.

In our experimental design, memory T cells were purified from non-immunized mice and are considered MP CD4 T cells that have been generated in the absence of antigen during homeostatic proliferation. It is well known that MP CD4 T cells are a heterogenous population of cells [3]. A subset of these cells has been shown to divide more rapidly than antigen specific memory T cells. This proliferation appeared to be independent of homeostatic cytokines, but these MP CD4 cells do require contact with MHCII, possibly loaded with foreign antigens, for their homeostatic proliferation. It has been reported that these fast dividing CD4 MP T cells have some properties of effector cells [3] and it is thus possible that *Hoxb4* is not favouring the expansion of this subpopulation. A potential reason for the reduced proliferation of the *Hoxb4* CD4 MP T cells might be an effect of *Hoxb4* on thymic differentiation and the TCR repertoire. However, we did not observe any anomalies of thymic T cell differentiation as thymic FACS profiles showed a normal distribution of thymic cell subsets [39], indicating that *Hoxb4* does not interfere with T cell development and thus making this possibility less likely. Furthermore, analysis of the TCR Vβ usage by peripheral T cells did not show any difference in repertoire between wt and *Hoxb4* transgenic mice (data not shown), indicating that a change in TCR repertoire is probably not the reason for the reduced homeostatic expansion of *Hoxb4* CD4 MP T cells in our experiments.

Based on the enrichment of CD4 central MP T cells in the *Hoxb4* population it was expected that *Hoxb4* MP T cell would dominate the repopulation lymphopenic hosts. It is not entirely clear why after three rounds of homeostatic proliferation the contribution of *Hoxb4* CD4 MP T cells did not increase. Recently, the gut has been identified as an important reservoir for CD4 memory T cells [48]. One possibility is that *Hoxb4* CD4 MP T cells preferentially migrated to the gut site and were not included in our analysis. This requires the expression of adhesion molecule, integrin α4β7. Interestingly, the DECODE database mentions binding sites for Hoxa9 and Meis1 in the promoter of the *itga4* and *itgb7* genes coding for integrin α4 and β7, respectively. Although Hoxb4 is not included in their list it is possible that Hoxb4 might target this gene as well, either directly or through complex with Pbx and Meis1. Of note is that *Hoxa9* also has the potential to expand HSCs [49], and another integrin sharing the same alpha chain, α4β1 (VLA-4) is expressed on HSCs, allowing for adhesion to the BM stroma [50]. Thus an increase in α4β7 expression on *Hoxb4* CD4 MP T cells resulting in enhanced homing to the gut cannot be excluded, and could explain a lower contribution of *Hoxb4* CD4 memory T cells to the lymphoid organs.

In addition, it is still plausible that CD4 MP T cells might respond differently to *Hoxb4* than HSCs. It might be that homeostatic proliferation of MP T cells is predominantly governed by cytokine signalling, while intrinsic signalling pathways play a larger role in HSCs.

In conclusion we show that *Hoxb4* favours the maintenance and expansion of CD4 central MP cells following acute homeostatic proliferation, which suggest a more robust preservation of CD4 MP T cells in the long-term.

Supporting Information

Figure S1   Analysis of T cell populations in young and old *Hoxb4* transgenic mice. Graphs showing the average size of CD4 (**A**) and CD8 (**B**) populations in lymphoid organs of young (2–3 months of age) and old (>15 months of age) *Hoxb4* transgenic (n = 7) and wt (n = 7) age matched mice. *P<0.05, 2-tailed Student ttest. Wt = wild type, LN = Lymph node; BM = bone marrow.

Figure S2   Absolute CD4 MP T cell numbers following homeostatic proliferation. (**A**) Absolute number of *Hoxb4* and wt CD4 MP T cells in lymphoid organs of primary hosts after 2 months of competitive proliferation. The calculations of the absolute numbers are based on 8 LNs, Spl and BM derived from 2 legs. Data are obtained from 9 mice in 3 independent experiments. *P = 0.03; 2-tailed Student ttest. (**B**) Absolute number of *Hoxb4* and wt CD62L positive CD4 MP T cells in primary hosts (n = 6). The numbers of CD62L MP T cells in the initial graft are calculated based on percentage of CD44^{hi}/CD62L$^+$ population as given in Table 1. Note the expansion of the CD62L population in several mice. *P = 0.01; 2-tailed Student ttest. Wt = wild type, MP = memory phenotype, LN = Lymph node, Spl = spleen, BM = bone marrow.

Figure S3   Production of cytokines after stimulation with PMA/ionomycin. Percentage of *Hoxb4* and wt MP T cells (gated on CD44^{hi}) in secondary and tertiary hosts positive for indicated cytokines (n = 3–6). Wt = wild type, MP = memory phenotype, TNF = tumor necrosis factor; IL-2 = interleukine-2; IFN = interferon.

Table S1   Average cell numbers ($\times 10^6$) in hematopoietic organs of *Hoxb4* and wt mice.

Acknowledgments

The authors wish to thank Martine Dupuis from the flow cytometry service for her expertise and sorting of cells. The staff of the animal facility is thanked for the care of the animals.

Author Contributions

Conceived and designed the experiments: NL JB. Performed the experiments: HF GG PT. Analyzed the data: HF GG PT NL JB. Contributed reagents/materials/analysis tools: NL JB. Wrote the paper: NL JB. Mounting of figures: MF HF GG JB.

References

1. Goldrath AW, Sivakumar PV, Glaccum M, Kennedy MK, Bevan MJ, et al. (2002) Cytokine requirements for acute and Basal homeostatic proliferation of naive and memory CD8+ T cells. J Exp Med 195: 1515–1522.

2. Boyman O, Purton JF, Surh CD, Sprent J (2007) Cytokines and T-cell homeostasis. Curr Opin Immunol 19: 320–326.

3. Purton JF, Tan JT, Rubinstein MP, Kim DM, Sprent J, et al. (2007) Antiviral CD4+ memory T cells are IL-15 dependent. J Exp Med 204: 951–961.

4. Boyman O, Krieg C, Homann D, Sprent J (2012) Homeostatic maintenance of T cells and natural killer cells. Cell Mol Life Sci 69: 1597–1608.

5. Swain SL, Hu H, Huston G (1999) Class II-independent generation of CD4 memory T cells from effectors. Science 286: 1381–1383.

6. Tan JT, Ernst B, Kieper WC, LeRoy E, Sprent J, et al. (2002) Interleukin (IL)-15 and IL-7 jointly regulate homeostatic proliferation of memory phenotype CD8+ cells but are not required for memory phenotype CD4+ cells. J Exp Med 195: 1523–1532.

7. Becker TC, Wherry EJ, Boone D, Murali-Krishna K, Antia R, et al. (2002) Interleukin 15 is required for proliferative renewal of virus-specific memory CD8 T cells. J Exp Med 195: 1541–1548.

8. Lodolce JP, Boone DL, Chai S, Swain RE, Dassopoulos T, et al. (1998) IL-15 receptor maintains lymphoid homeostasis by supporting lymphocyte homing and proliferation. Immunity 9: 669–676.

9. Kondrack RM, Harbertson J, Tan JT, McBreen ME, Surh CD, et al. (2003) Interleukin 7 regulates the survival and generation of memory CD4 cells. J Exp Med 198: 1797–1806.

10. Surh CD, Sprent J (2008) Homeostasis of naive and memory T cells. Immunity 29: 848–862.

11. Blattman JN, Grayson JM, Wherry EJ, Kaech SM, Smith KA, et al. (2003) Therapeutic use of IL-2 to enhance antiviral T-cell responses in vivo. Nat Med 9: 540–547.

12. Boyman O, Kovar M, Rubinstein MP, Surh CD, Sprent J (2006) Selective stimulation of T cell subsets with antibody-cytokine immune complexes. Science 311: 1924–1927.

13. Whitmire JK, Tan JT, Whitton JL (2005) Interferon-gamma acts directly on CD8+ T cells to increase their abundance during virus infection. J Exp Med 201: 1053–1059.

14. Kassiotis G, Garcia S, Simpson E, Stockinger B (2002) Impairment of immunological memory in the absence of MHC despite survival of memory T cells. Nat Immunol 3: 244–250.

15. Kassiotis G, Zamoyska R, Stockinger B (2003) Involvement of avidity for major histocompatibility complex in homeostasis of naive and memory T cells. J Exp Med 197: 1007–1016.

16. Vos Q, Jones LA, Kruisbeek AM (1992) Mice deprived of exogenous antigenic stimulation develop a normal repertoire of functional T cells. J Immunol 149: 1204–1210.

17. Cheung KP, Yang E, Goldrath AW (2009) Memory-like CD8+ T cells generated during homeostatic proliferation defer to antigen-experienced memory cells. J Immunol 183: 3364–3372.

18. Hamilton SE, Wolkers MC, Schoenberger SP, Jameson SC (2006) The generation of protective memory-like CD8+ T cells during homeostatic proliferation requires CD4+ T cells. Nat Immunol 7: 475–481.

19. Hamilton SE, Jameson SC (2008) The nature of the lymphopenic environment dictates protective function of homeostatic-memory CD8+ T cells. Proc Natl Acad Sci U S A 105: 18484–18489.

20. Leignadier J, Hardy MP, Cloutier M, Rooney J, Labrecque N (2008) Memory T-lymphocyte survival does not require T-cell receptor expression. Proc Natl Acad Sci U S A 105: 20440–20445.

21. Kerdiles YM, Beisner DR, Tinoco R, Dejean AS, Castrillon DH, et al. (2009) Foxo1 links homing and survival of naive T cells by regulating L-selectin, CCR7 and interleukin 7 receptor. Nat Immunol 10: 176–184.

22. Xue HH, Bollenbacher J, Rovella V, Tripuraneni R, Du YB, et al. (2004) GA binding protein regulates interleukin 7 receptor alpha-chain gene expression in T cells. Nat Immunol 5: 1036–1044.

23. Chandele A, Joshi NS, Zhu J, Paul WE, Leonard WJ, et al. (2008) Formation of IL-7Ralphahigh and IL-7Ralphalow CD8 T cells during infection is regulated by the opposing functions of GABPalpha and Gfi-1. J Immunol 180: 5309–5319.

24. Intlekofer AM, Takemoto N, Wherry EJ, Longworth SA, Northrup JT, et al. (2005) Effector and memory CD8+ T cell fate coupled by T-bet and eomesodermin. Nat Immunol 6: 1236–1244.

25. Yamashita M, Kuwahara M, Suzuki A, Hirahara K, Shinnakasu R, et al. (2008) Bmi1 regulates memory CD4 T cell survival via repression of the Noxa gene. J Exp Med 205: 1109–1120.

26. Onodera A, Yamashita M, Endo Y, Kuwahara M, Tofukuji S, et al. (2010) STAT6-mediated displacement of polycomb by trithorax complex establishes long-term maintenance of GATA3 expression in T helper type 2 cells. J Exp Med.

27. Yamashita M, Hirahara K, Shinnakasu R, Hosokawa H, Norikane S, et al. (2006) Crucial role of MLL for the maintenance of memory T helper type 2 cell responses. Immunity 24: 611–622.

28. Park IK, Morrison SJ, Clarke MF (2004) Bmi1, stem cells, and senescence regulation. J Clin Invest 113: 175–179.

29. Lessard J, Sauvageau G (2003) Bmi-1 determines the proliferative capacity of normal and leukaemic stem cells. Nature 423: 255–260.

30. Jude CD, Climer L, Xu D, Artinger E, Fisher JK, et al. (2007) Unique and independent roles for MLL in adult hematopoietic stem cells and progenitors. Cell Stem Cell 1: 324–337.

31. Luckey CJ, Bhattacharya D, Goldrath AW, Weissman IL, Benoist C, et al. (2006) Memory T and memory B cells share a transcriptional program of self-renewal with long-term hematopoietic stem cells. Proc Natl Acad Sci U S A 103: 3304–3309.

32. McGinnis W, Krumlauf R (1992) Homeobox genes and axial patterning. Cell 68: 283–302.

33. Hanson RD, Hess JL, Yu BD, Ernst P, van Lohuizen M, et al. (1999) Mammalian Trithorax and polycomb-group homologues are antagonistic regulators of homeotic development. Proc Natl Acad Sci U S A 96: 14372–14377.

34. Bijl J, Thompson A, Ramirez-Solis R, Krosl J, Grier DG, et al. (2006) Analysis of HSC activity and compensatory Hox gene expression profile in Hoxb cluster mutant fetal liver cells. Blood 108: 116–122.

35. Brun AC, Bjornsson JM, Magnusson M, Larsson N, Leveen P, et al. (2004) Hoxb4-deficient mice undergo normal hematopoietic development but exhibit a mild proliferation defect in hematopoietic stem cells. Blood 103: 4126–4133.

36. Cellot S, Krosl J, Chagraoui J, Meloche S, Humphries RK, et al. (2007) Sustained in vitro trigger of self-renewal divisions in Hoxb4hiPbx1(10) hematopoietic stem cells. Exp Hematol 35: 802–816.

37. Sauvageau G, Thorsteinsdottir U, Eaves CJ, Lawrence HJ, Largman C, et al. (1995) Overexpression of HOXB4 in hematopoietic cells causes the selective expansion of more primitive populations in vitro and in vivo. Genes Dev 9: 1753–1765.

38. Faubert A, Chagraoui J, Mayotte N, Frechette M, Iscove NN, et al. (2008) Complementary and independent function for Hoxb4 and Bmi1 in HSC activity. Cold Spring Harb Symp Quant Biol 73: 555–564.

39. Bijl J, Krosl J, Lebert-Ghali CE, Vacher J, Mayotte N, et al. (2008) Evidence for Hox and E2A-PBX1 collaboration in mouse T-cell leukemia. Oncogene 27: 6356–6364.

40. Malissen M, Gillet A, Ardouin L, Bouvier G, Trucy J, et al. (1995) Altered T cell development in mice with a targeted mutation of the CD3-epsilon gene. EMBO J 14: 4641–4653.

41. Lacombe MH, Hardy MP, Rooney J, Labrecque N (2005) IL-7 receptor expression levels do not identify CD8+ memory T lymphocyte precursors following peptide immunization. J Immunol 175: 4400–4407.

42. Thompson A, Quinn MF, Grimwade D, O'Neill CM, Ahmed MR, et al. (2003) Global down-regulation of HOX gene expression in PML-RARalpha+acute promyelocytic leukemia identified by small-array real-time PCR. Blood 101: 1558–1565.

43. Lebert-Ghali CE, Fournier M, Dickson GJ, Thompson A, Sauvageau G, et al. (2010) HoxA cluster is haploinsufficient for activity of hematopoietic stem and progenitor cells. Exp Hematol 38: 1074–1086.

44. Bourgeois C, Kassiotis G, Stockinger B (2005) A major role for memory CD4 T cells in the control of lymphopenia-induced proliferation of naive CD4 T cells. J Immunol 174: 5316–5323.

45. Sallusto F, Lenig D, Forster R, Lipp M, Lanzavecchia A (1999) Two subsets of memory T lymphocytes with distinct homing potentials and effector functions. Nature 401: 708–712.

46. Tokoyoda K, Zehentmeier S, Hegazy AN, Albrecht I, Grun JR, et al. (2009) Professional memory CD4+ T lymphocytes preferentially reside and rest in the bone marrow. Immunity 30: 721–730.

47. Marshall HD, Chandele A, Jung YW, Meng H, Poholek AC, et al. (2011) Differential expression of Ly6C and T-bet distinguish effector and memory Th1 CD4(+) cell properties during viral infection. Immunity 35: 633–646.

48. Yang L, Yu Y, Kalwani M, Tseng TW, Baltimore D (2011) Homeostatic cytokines orchestrate the segregation of CD4 and CD8 memory T-cell reservoirs in mice. Blood 118: 3039–3050.

49. Thorsteinsdottir U, Mamo A, Kroon E, Jerome L, Bijl J, et al. (2002) Overexpression of the myeloid leukemia-associated Hoxa9 gene in bone marrow cells induces stem cell expansion. Blood 99: 121–129.

50. Mazo IB, Gutierrez-Ramos JC, Frenette PS, Hynes RO, Wagner DD, et al. (1998) Hematopoietic progenitor cell rolling in bone marrow microvessels: parallel contributions by endothelial selectins and vascular cell adhesion molecule 1. J Exp Med 188: 465–474.

6

Efficient and Rapid *C. elegans* Transgenesis by Bombardment and Hygromycin B Selection

Inja Radman, Sebastian Greiss*, Jason W. Chin*

Medical Research Council Laboratory of Molecular Biology, Cambridge, United Kingdom

Abstract

We report a simple, cost-effective, scalable and efficient method for creating transgenic *Caenorhabditis elegans* that requires minimal hands-on time. The method combines biolistic bombardment with selection for transgenics that bear a hygromycin B resistance gene on agar plates supplemented with hygromycin B, taking advantage of our observation that hygromycin B is sufficient to kill wild-type *C. elegans* at very low concentrations. Crucially, the method provides substantial improvements in the success of bombardments for isolating transmitting strains, the isolation of multiple independent strains, and the isolation of integrated strains: 100% of bombardments in a large data set yielded transgenics; 10 or more independent strains were isolated from 84% of bombardments, and up to 28 independent strains were isolated from a single bombardment; 82% of bombardments yielded stably transmitting integrated lines with most yielding multiple integrated lines. We anticipate that the selection will be widely adopted for *C. elegans* transgenesis via bombardment, and that hygromycin B resistance will be adopted as a marker in other approaches for manipulating, introducing or deleting DNA in *C. elegans*.

Editor: Bob Goldstein, University of North Carolina at Chapel Hill, United States of America

Funding: This work was supported by the UK Medical Research Council (MRC; http://www.mrc.ac.uk/) (grant numbers U105181009, UD99999908) to J.W.C.; the Louis-Jeantet Foundation Young Investigator Career Award to J.W.C. (http://www.jeantet.ch/); Herchel Smith Foundation studentship to I.R. The funders had no role in study design, data collection and analysis, decision to publish, or preparation of the manuscript.

Competing Interests: The authors have declared that no competing interests exist.

* E-mail: sgreiss@mrc-lmb.cam.ac.uk (SG); chin@mrc-lmb.cam.ac.uk (JWC)

Introduction

To create transgenic *C. elegans* DNA may be introduced by microinjection [1,2] or microparticle bombardment. While simple microinjection leads to extrachromosomal arrays that are lost over time, making it impossible to grow large populations, microparticle bombardment can lead to the creation of stable integrated strains [3,4].

Although the microparticle bombardment method is widely used to generate transgenic *C. elegans*, it has several limitations including: i) the need to use specialized strains with a mutant background - to allow phenotypic isolation of transformants, ii) slow and labor intensive procedures for mutant isolation, iii) the requirement to handle large numbers of worms, and iv) loss of the transgenic array for non-integrated lines, v) the isolation of few, if any, transgenic lines from a given bombardment, and vi) the isolation of few, if any, integrated transgenics from a single bombardment. The use of antibiotic resistance genes and their corresponding antibiotics for selection of transgenic *C. elegans* in biolistic bombardments might in principle address many of these challenges.

Recently, the selection of transgenic worms created by microinjection was reported using antibiotic resistance markers and selection for transformation by growth on antibiotics. Two antibiotics and their corresponding resistance genes were tested: i) G418 was used in combination with a neomycin resistance gene, and ii) puromycin, which required the use of a detergent (Triton-X 100) to permeabilize the worms for the antibiotic to be effective, was used in combination with a puromycin resistance gene [5,6].

The use of these antibiotic selection approaches for isolating transgenics from biolistic bombardments was not reported; furthermore, G418 selection was reported to be unsuccessful for selecting transformants via bombardment [6].

We reported the successful selection of transgenics following biolistic bombardment using a different antibiotic, hygromycin B, and the hygromycin B phosphotranspherase gene – encoding a kinase that inactivates hygromycin by phosphorylation – to select transgenic *C. elegans*, as part of our ongoing efforts to develop and apply methods to expand the genetic code of cells and animals[7–9]. Subsequently, the extensive characterization of a method for biolistic bombardment and enrichment of transgenics using a two-antibiotics plus detergent (puromycin, G418 and Triton-X 100) and a fluorescent protein gene, was reported [10]. Following bombardment, the dual antibiotic plus detergent method requires the removal of adults- which are not sufficiently sensitive to the antibiotic- by repeated gravity sedimentation, resuspension of the L1s and growth for 4 days in liquid media containing 0.5 mg ml^{-1} puromycin and G418 and 0.1% Triton-X 100 to enrich for transgenics. The antibiotic and detergent treatment is not sufficient to directly select transgenics and a further step, in which fluorescent transformants are manually isolated from a background of wild type animals, is required [10].

Antibiotic based selection has the potential to improve traditional bombardment approaches. However in the published dual antibiotic plus fluorescence approach 25% of all the bombardments in *C. elegans* reported fail to yield *any* transmitting transgenics. Moreover, in approximately half of the cases examined no integrated strains were identified using this approach. In most cases one independent line is isolated from a bombardment, and the maximum number of independent lines isolated by this approach is unknown. Here we develop and characterize a simple, rapid and cost-effective transformation and selection protocol to generate transgenic *C. elegans* using biolistic bombardment and hygromycin B selection and demonstrate the advantages of this approach.

Materials and Methods

Worm Strains and Maintenance

For bombardment either wild type *C. elegans* (N2 Bristol strain) or strains carrying the *smg-2*(e2008) allele were used. Worms were grown under standard conditions on NGM agar plates (1×NGM) [11] seeded with *E.coli* OP50. Post-bombardment selection was performed on 3×NGM agar plates (based on the recipe for 1×NGM, but with 7.5 g/l peptone, 0.1 mM CaCl₂, 0.5 mM MgSO₄) seeded with *E. coli* HB101, which give thicker bacterial lawns. Synchronized cultures for bombardments were grown in liquid medium containing: S-Basal (50 mM KPO_4, pH 6.0, 100 mM NaCl, 5 mg/L cholesterol), 10 mM potassium citrate (pH 6.0), trace metals solution (50 µM disodium EDTA, 25 µM $FeSO_4 \times 7H_2O$, 10 µM $MnCl_2 \times 4H_2O$, 10 µM $ZnSO_4 \times 7H_2O$, 1 µM $CuSO_4 \times 5H_2O$), 3 mM $CaCl_2$, 3 mM $MgSO_4$, supplemented with Antibiotic-Antimycotic 100× (GIBCO Life Technologies). Worms in liquid culture were fed with concentrated *E. coli* HB101 pellets.

Microscopy

Worm populations on NGM plates were imaged using a Rolera Bolt camera (QImaging) mounted on a Leica M165FC fluorescent stereo microscope with 16.5:1 zoom optics. When acquiring fluorescence images, plates were cooled to 4°C before imaging to reduce the movement of the animals and allow for longer exposure times.

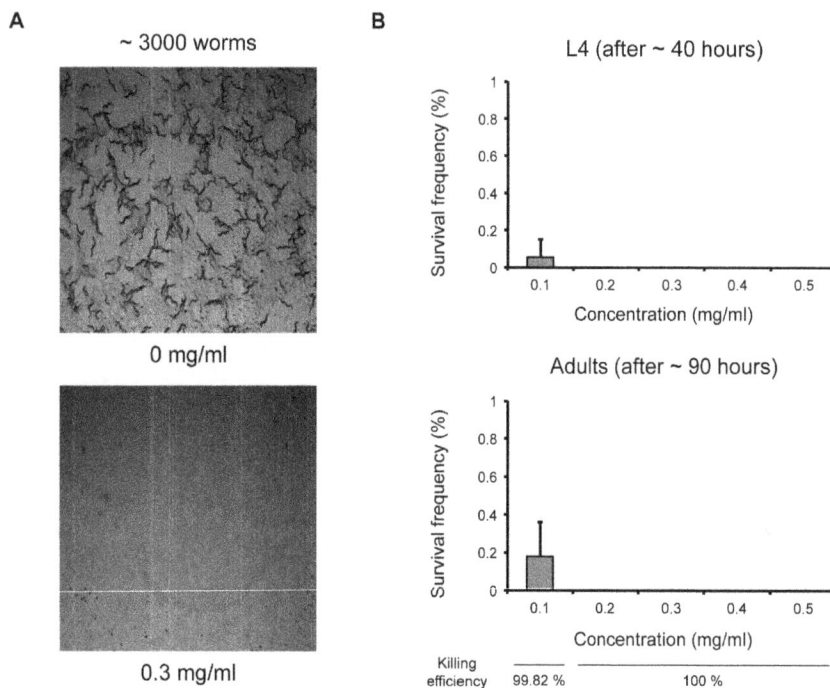

Figure 1. Survival of *C. elegans* in the presence of hygromycin B. a) 3000 synchronized L1 larvae were transferred to seeded 6 cm NGM plates and immediately treated with hygromycin B. Images of control and treated plates were acquired 40 hours later. b) Survival assay on hygromycin B dilution. 3000 synchronised L1 larvae were plated on seeded 6 cm plates and immediately treated with hygromycin B at the specified final concentrations. The plates were scored for L4 larvae after 40 hours and for adults after 90 hours. The experiment was performed in triplicate, error bars represent standard deviation. c) Selection of transgenic animals on hygromycin B. 10 synchronized L1 larvae carrying a hygromycin B resistance gene and expressing GFP/mCherry were mixed with the indicated number of wild type L1 larvae on a seeded 6 cm plate and immediately treated with hygromycin B (0.3 mg ml⁻¹). The number of adult transgenic (fluorescence-positive) and non-transgenic (fluorescence-negative) animals was scored after 90 hours. The experiment was performed with two independent hygromycin B resistant transgenic strains.

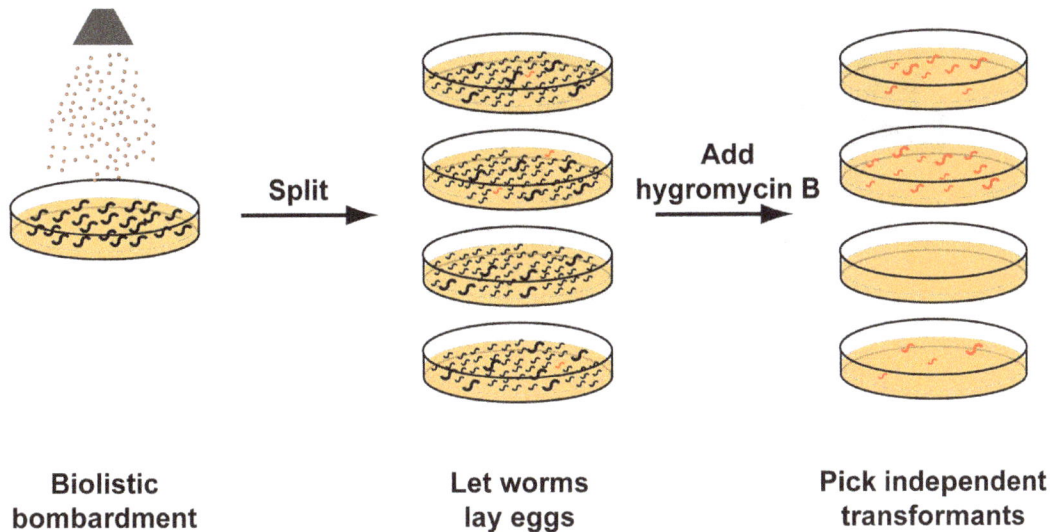

Figure 2. Hygromycin B selection of transgenic *C. elegans* created by biolistic bombardment. Gold beads coated with the DNA mixture of interest, carrying the hygromycin B resistance gene, are bombarded onto worms spread on an agar plate. Bombarded worms are split onto 10–20 selection plates and left to lay eggs for 2–3 days, at which point hygromycin B is added to the plates. After 4–7 days surviving transgenic worms can be picked.

Confocal images were acquired using a Zeiss 780 UV confocal inverted microscope equipped with a 488 nm laser to image GFP and a 561 nm laser to image mCherry. Animals were washed off plates, anaesthetized using 0.2 mM levamisole and mounted on 3% agar pads.

Survival Assays

Survival of N2 animals in the presence of various concentrations of hygromycin B was tested on NGM plates seeded with *E.coli* OP50. Worms were synchronized by bleaching and 300 (low density) or 3000 (high density) L1 larvae were added to seeded 6 cm plates and immediately treated with the indicated amounts of hygromycin B (diluted to 200 µl in M9 buffer). Plates were then incubated at 20°C and scored for L4 larvae after 40 hours and for adult worms after 60 and 90 hours. The experiment was performed in triplicate.

Survival of transgenic animals and selection efficiency was assayed by mixing 10 transgenic L1 larvae expressing both the hygromycin B resistance gene and GFP with 1000 or 10000 non-transgenic wild type L1s on a 6 cm seeded NGM plate. After transfer to the plate, hygromycin B was immediately added to a final concentration of 0.3 mg ml^{-1}. The number of surviving adults was scored after 90 hours of growth at 20°C.

Bombardment Protocol

Biolistic bombardment was performed using a He-1000 apparatus (Bio-Rad). The protocol was based on previously published protocols using rescue of the *unc-119* mutant phenotype as selection marker [3,4].

Bombardments were performed with synchronized populations of worms at the late L4/young adult stage using gold beads and rupture discs as indicated. The best efficiencies were achieved with 0.3–3 µm gold beads (microcarriers) (ChemPur) and 1100 psi rupture discs (Bio-Rad).

To synchronize worms, the animals were grown until the culture contained a large number of adults, at which point feeding was stopped to induce growth arrest of any larvae hatched subsequently. After 1–2 days L1 larvae were harvested through

removal of older animals by sedimentation and subsequent centrifugation of the supernatant for 4 min at 800 g and 4°C. The L1 larvae were used to start a fresh synchronized culture which was used for bombardment when the animals had reached the late L4/early adult stage.

For each bombardment 6 mg of gold beads were coated with 10 µg of DNA essentially as previously described [4]. Linearized DNA was not purified prior to coating, except when otherwise indicated. For co-bombardment of more than one vector, a minimum of 1 µg of resistance plasmid was used with 9 µg of co-bombarded plasmid(s).

60 mg of mixed beads (0.3–3 µm, ChemPur) were prepared for 10 bombardments (6 mg of beads/bombardment). The beads were first washed by vortexing for 30 min in 70% HPLC-grade ethanol (Sigma-Aldrich), pelleted briefly and then washed 3×with sterile H$_2$0 (for every wash, beads were vortexed for 1 min, soaked for 1 min, and spun down briefly). The beads were finally resuspended in 1 ml of 50% glycerol, and stored for a maximum of two weeks at 4°C if not immediately used.

For coating, 100 µl of beads in 50% glycerol (6 mg of beads) were transferred to DNA LoBind tubes (Eppendorf), centrifuged briefly, the supernatant removed and the beads resuspended in the DNA mixture, followed by the immediate addition of 20 µl fresh spermidine solution (100 mM in H$_2$0) and 50 µl CaCl$_2$ (2.5 M in H$_2$0), with brief vortexing after each addition. The samples were then either incubated on ice for about 30 minutes with occasional resuspension or vortexed gently at 4°C. The beads were then washed once with 300 µl 70% HPLC-grade ethanol, once with 1000 µl 100% HPLC-grade ethanol and then resuspended in 140 µl 100% HPLC-grade ethanol and kept on ice until bombardment. We found that it is possible to store the DNA coated beads in 100% ethanol at −20°C at least over night without reducing transformation efficiency. Biolistic macrocarriers (Bio-Rad) and rupture discs (Bio-Rad) were washed in isopropanol (Sigma-Aldrich) and left to dry. The macrocarriers were then placed into a hepta-adaptor and 20 µl of DNA coated beads in ethanol were spread onto each macrocarrier. The hepta-adaptor with macrocarriers was then placed into the vacuum chamber of

Brightfield	Psnb-1::GFP	Prps-0::mCherry	Merged

64.1.1

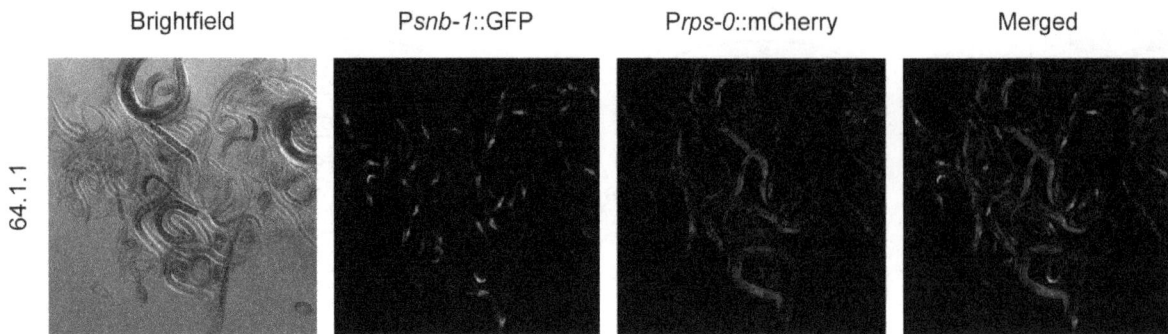

Figure 3. Hygromycin B independent transmission of integrated transgenes. Worms expressing the hygromycin B resistance gene and mCherry from the ubiquitous Prps-0 promoter and GFP from the pan-neuronal Psnb-1 promoter (Prps-0::HygB, Prps-0::mCherry, Psnb-1::GFP, bombardment 64, strain 64.1.1) were grown as described in Figure S3. Worms were washed off the plate and imaged using a confocal microscope as described in the methods section. The strain was grown and propagated in the absence of hygromycin B.

the He-1000 biolistic bombardment apparatus under vacuum until all ethanol had evaporated.

A Biolistic Hepta Stop Screen (Bio-Rad) and rupture disc were added and the adapter was mounted into the instrument. The plate with worms was placed at a distance of 6 cm from the stopping disk and the worms bombarded at 27.5 inches Hg vacuum.

Hygromycin B Selection

After bombardment, the animals were left to recover at either room temperature or 20°C for 30 min to 1 h. The worms were then either washed off the bombarded plate and transferred to 10–20 selection plates (3×NGM, seeded with *E. coli* HB101) or alternatively cut into chunks and the chunks then placed onto the surface of the selection plates. The animals were left to lay eggs for 2–3 days, at which point the plates were generally full of very young F1 larvae. Hygromycin B was then added onto the surface of the plates.

Selections were performed by adding hygromycin B (>85% pure, Invivogen) onto the surface of the plates to a final concentration of 0.3 mg ml^{-1} hygromycin B, without the addition of any permeabilizing agents. The final hygromycin B concentration was calculated taking into account the total volume of the agar in the plate. To help spread the hygromycin solution evenly and to reduce post-bombardment contamination, we diluted the hygromycin B with several other antibacterial and antifungal solutions. The standard 'selection mixture' added to a single 9 cm plate was: 100 µl hygromycin B (>85% pure hygromycin B, 100 mg/ml, InvivoGen), 600 µl Antibiotic-Antimycotic 100× (GIBCO Life Technologies), 80 µl Amphotericin B 250 µg/ml (GIBCO Life Techologies), 20 µl 1000×Kanamycin solution (final concentration 50 µg/ml, Kanamycindisulfate, Merck). Selection mixture was added also to a control plate of non-transformed worms. In cases where the food on the selection plates was depleted at the time of hygromycin B addition, we added 100–200 µl of the concentrated bacteria used for growing liquid cultures 1–2 days after adding hygromycin B. Plates were then screened for transformants after a minimum of 4–7 days, at which time lines with high transmission rates were beginning to form populations. Plates that appeared to contain no transgenic worms were rescreened after a further 1–2 weeks.

We note that dauer larvae generally appeared to be resistant to hygromycin B. They were, however, unable to exit the dauer stage in the presence of antibiotic even when the population density was low and food was abundant. The presence of dauer larvae did therefore not negatively influence selection efficiency.

Plasmid Construction

We created a 3-way Gateway pEXPR vector (SG120) carrying the hygromycin B resistance gene hygromycin B phosphotranspherase (*HygR*), which inactivates hygromycin B by phosphorylation, to be used for co-bombardment with any other vector of interest when creating transgenic strains [12]. For ubiquitous and strong expression of the gene, a 2.8 kb fragment upstream of *rps-0* was used [13].

New Improved *HygR* Plasmids

In order to reduce the number of co-bombarded vectors, we created a pDEST Gateway vector where the Gateway cloning cassette is free for cloning purposes and the *HygR* resistance cassette was introduced into the pDEST vector backbone. Also, two multiple cloning sites, MCS1 (*KasI/NarI/SfoI_SpeI*) and MCS2 (*KpnI_NheI_AvrII_AscI*), were introduced upstream and downstream of the *HygR* cassette, to facilitate vector linearization (Figure S2 in File S1, Table S3 in File S1). The resistance cassette (Prps-0::HygR::unc54) was assembled by overlap extension (OE) PCR using primers WP21, WP64, WP65, WP68, WP23 and WP39. All PCR reactions were carried out using Phusion Polymerase (Finnzymes). The PCR product was then recombined into the pDONR P4-P1R Gateway vector (IR84). The resistance cassette was amplified from IR84 using primers WP191 and WP195, and the pDEST R4-R1 plasmid backbone was amplified using primers WP189 and WP190. Both PCR products were digested with *SpeI* and *AscI* and ligated to create the new pDEST variant carrying the resistance cassette in the backbone (IR95).

To reduce the size of our constructs, we tested a much shorter version of Prps-0 than the one previously reported [7,13] (0.8 kb instead of 2.8 kb) and confirmed that it is fully functional *in vivo*. To increase expression levels of the resistance gene, we codon optimized the gene and introduced artificial introns [14] (resistance cassette assembled by PCR using primers WP62, WP64, WP66, WP69, WP23 and WP48, recombined into pDONR 221 to give IR87; resistance cassette amplified from IR87 using WP193 and WP195 to create a pDEST IR98). We furthermore created versions of the resistance vector co-expressing a fluorescent marker, GFP or mCherry, from the same promoter as the resistance gene. The resistance –fluorescence operon was assembled by overlap extension (OE) PCR using primers WP43 (WP62), WP64, WP65 (WP66), WP127 (WP76), WP77, WP80

A

B

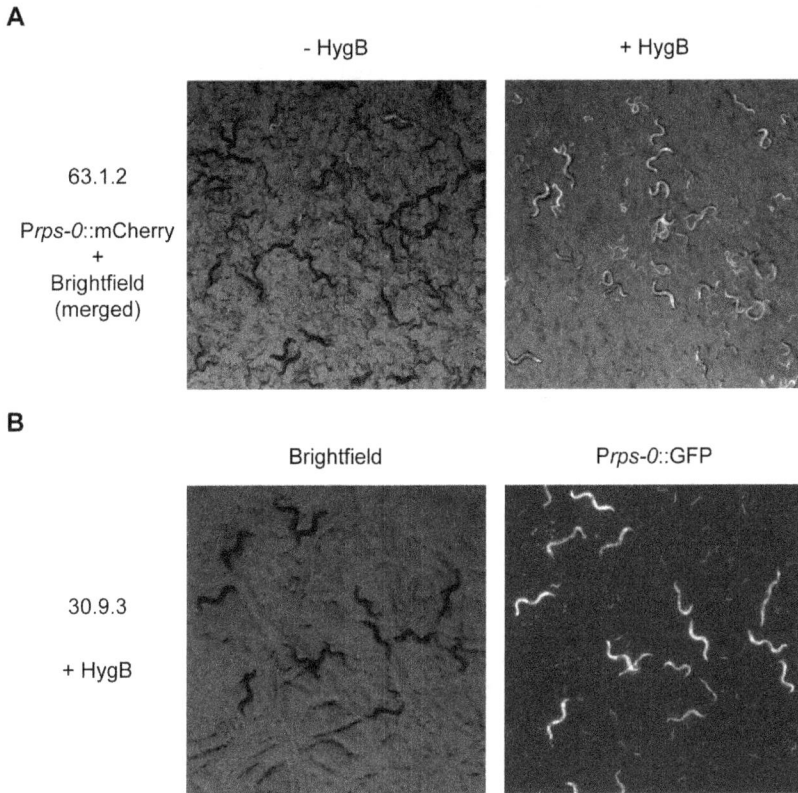

Figure 4. Hygromycin B allows long term maintenance of non-integrated transgenes. a) 3 transgenic worms from a non-integrated line transformed with the hygromycin B resistance and GFP (P*rps-0::HygB::unc-54*, P*rps-0*::GFP::*unc-54*, bombardment 63, strain 63.1.2) were grown on a seeded plate for 3 generations in the absence and presence of hygromycin B. Worms were imaged as described in the methods section. Exposure time: 700 ms. b) Worms with a non-integrated transgenic array carrying the hygromycin B resistance gene and GFP (P*rps-0::HygB::unc-54*, P*rps-0*::GFP::*unc-54*, bombardment 30, strain 30.9.3) were propagated for >30 generations in the presence of hygromycin B. Worms were imaged as described in the methods section. Exposure time: 700 ms.

(WP82), WP44 (WP78), WP84 (WP85), WP23 and WP48. The PCR product was then recombined into the pDONR 221 Gateway vector (IR83, IR88, IR89, IR90). The operon cassette was amplified from these vectors using primers WP191 (WP193) and WP195, digested with *SpeI* and *AscI* and ligated with the pDEST digested backbone (described above) to create the new pDEST resistance vectors carrying a resistance – fluorescence operon (IR99– IR105) (Figure S2 in File S1, Table S3 in File S1).

Results

We first investigated the efficiency of hygromycin B for killing wild-type *C. elegans*. We demonstrated that addition of low concentrations of hygromycin B to worms on the agar is sufficient to kill all L1 wild-type worms. For high worm densities (3000 worms per 6 cm plate) 0.2 mg ml^{-1} hygromycin B is sufficient to kill all non-transgenic worms (Figure 1, Figure S1 in File S1, Table S1 in File S1), at lower densities (300 worms per 6 cm plate) 0.1 mg ml^{-1} hygromycin B is sufficient to kill all non-transgenic worms (Figure S1 in File S1, Table S1 in File S1, Supplementary Note in File S1).

To demonstrate that the hygromycin B phosphotranspherase gene can rescue survival on hygromycin B we created several constructs (Figure S2 in File S1, Table S2 in File S1, Table S3 in File S1) for strong and ubiquitous expression of hygromycin B phosphotransferase from the P*rps-0* promoter in *C. elegans*. Final constructs used a *C. elegans* optimized *HygR* gene and a minimal P*rps-0* promoter to minimize vector size. We have created several Gateway compatible vectors carrying the *HygR* in the backbone, both with and without co-expressed fluorescent markers to quickly and easily generate constructs bearing genes of interest for bombardment. The vectors contain additional unique restriction sites, which should ensure that at least one site will be compatible with any given insert and thus facilitate linearization (Figure S2 in File S1, Table S2 in File S1, Table S3 in File S1). We demonstrated that transgenic worms bearing *HygR* are resistant to hygromycin B up to at least 1 mg ml^{-1} (data not shown) [7]. Moreover, we demonstrated in a model selection that when a small number of transgenic worms were placed in a 10^3 fold excess of wild type worms and exposed to hygromycin B the wild type worms were all killed off while the transgenic worms all survived and developed (Figure 1c).

In our optimized bombardment protocol (Figure 2) we grew a population of well-synchronized late L4s/young adults and used about 40 000–90 000 (0.5–1.5 ml settled) worms per bombardment. We found that using 0.3–3 µm gold beads was more cost effective than using 1 µm beads and did not impinge on efficiency (Table S4 in File S1, Table S5 in File S1). Furthermore, switching to 1100 psi rupture disks from the more commonly used 1350 psi disks increased the number of recovered transgenics (Table S4 in File S1, Table S5 in File S1). We demonstrated that DNA linearization prior to bombardment increased the number of transgenics obtained (Table S6 in File S1) [1].

Table 1. Efficiency of biolistic *C. elegans* transgenesis using hygromycin B selection.

Rupture disks	DNA linearized	Bombard-ments	Bombard-ments yielding a transgenic strain (%)	Bombard-ments yielding >=10 independent transgenic strains (%)	independent strains per bombard-ment (average)	Bombard-ments screened for integrants	Bombard-ments yielding an integrated transgenic strain (%)	Bombard-ments yielding >=2 independent integrated transgenic strains (%)	independent integrated strains per bombard-ment (average)
1350	no	24	23 (96%)	5 (21%)	5.75	0	n.d.	n.d.	n.d.
1100	no	11	11 (100%)	4 (36%)	8.09	0	n.d.	n.d.	n.d.
1100	yes	45	45 (100%)	38 (84%)	11.58	22	18 (82%)	15 (68%)	2.4

n.d. – not determined.

After bombardment, we split the worms onto 10 to 20 plates to allow the subsequent isolation of independent transgenic strains arising from each plate. We let the animals lay eggs for 2–3 days before adding hygromycin B directly onto the surface of the plates. The worms surviving after 4 to 7 days were collected. All worms isolated from this single step selection on hygromycin B were transgenic. A plate of non-transformed worms was routinely included in our selection, and was used to confirm that the selection conditions killed all non-transformed worms.

With the optimized approach, based on a large dataset, transgenic strains were obtained from every bombardment (100%) (Table 1). Ten or more independent transmitting transgenic strains were obtained from 84% of bombardments, with an average of 11.6 independent transgenic strains per bombardment (Table 1). The number of independent transformants identified was in many cases limited simply by the number of plates we chose to split the bombardment across and additional experiments demonstrated that the creation of up to 28 independent lines from a single optimized bombardment was possible (Table S6 in File S1). We examined twenty-two bombardments for genomic integration. We found that the large majority of the bombardments (82%) led to an integrated transgenic strain, with an average of 2.4 integrated independent strains per bombardment (Table 1, Figure 3, Figure S3 in File S1, Table S4 in File S1). Since it is likely that some integrated strains generated by biolistic bombardment will carry multiple copies of the transgene, they are likely to give higher expression levels than strains constructed using recently introduced methods for targeted single copy integration [15,16], which may in some cases be desirable. When working with non-integrated strains, which may be advantageous when high transgene expression levels are desired, we found that we could easily grow and maintain large transgenic cultures at high densities both on agar plates (Figure 4) and in liquid culture (data not shown) in the presence of hygromycin B. When co-bombarding multiple plasmids, we found co-expressing strains in every bombardment. For each bombardment the majority of strains show co-expression. However, we also observed strains that did not express one or more of the bombarded constructs (data not shown).

Discussion

We report the direct isolation of transgenic *C. elegans* via biolistic bombardment using hygromycin B, an antibiotic that quantitatively kills worms at low concentration. Unlike previous reports which use combinations of antibiotics and detergent to permeabilize the worms to the antibiotic our approach uses only a single antibiotic and does not require the use of detergents to permeabilize the worms or a subsequent fluorescent reporter screen to isolate transgenic lines.

In contrast to previous methods that use extensive liquid handling, centrifugations and separation steps the method we report uses a single agar based selection, allowing the direct and straightforward isolation of independent lines in a single step. The reduced hands-on time makes the method scalable, and we routinely perform bombardments and selections on ten or more constructs in parallel. Hygromycin B is approximately 15 times less expensive than puromycin, making the use of multiple selection plates, for the straightforward isolation of independent lines, economically feasible for many laboratories.

Crucially, the method provides substantial improvements in the success of bombardments for isolating transmitting strains, the isolation of multiple independent strains, and the isolation of integrated strains. While previous methods give transmitting strains in up to 80% of transformations, the method we developed yielded transmitting strains in 100% of bombardments. Moreover, our method allows us to routinely isolate multiple independent transmitting strains, indeed 84% of bombardments yielded greater than ten independent transmitting strains with bombardments yielding up to 28 independent transmitting strains. While previous methods yield an integrated strain from up to 70% of bombardments (Ben Lehner, personal communication) we find integrated strains in 82% of bombardments, and we find greater than two independent integrated strains in 68% of bombardments. Using our approach a single bombardment suffices to obtain both an integrated transgenic strain and a variety of extrachromosomal transgenic strains with a range of transgene copy number and transgene expression levels that can be stably maintained.

While we have focused on characterizing hygromycin B based selections for bombardment based transgenesis, the selection will be useful for marking the introduction or deletion of DNA into *C. elegans* by other methods. Indeed, the hygromycin B cassette we have reported here can also be used to select targeted knock-ins generated using a CRISPR-Cas9 system in *C. elegans* [17].

Supporting Information

File S1 Contains: Figure S1. Survival of *C. elegans* on hygromycin B. Synchronized L1 larvae were added to seeded 6 cm NGM plates and immediately treated with the indicated hygromycin B concentrations. Plates were imaged 40 hours after treatment. a) At 300 worms per plate no treated worms reached the L4 stage even at the lowest hygromycin B concentration. b) At 3000 worms per plate some animals reached the L4 stage at the lowest antibiotic concentration, whereas at higher concentrations no animals reached the L4 stage. **Figure S2.** Gateway-based vectors constructed for the selection of hygromycin B resistant transgenics. a) Basic hygromycin B resistance vectors. b) New optimized hygromycin B resistance vectors. *unc-54* 3'UTR –3' untranscribed downstream region of *unc-54*. *Amp* – ampicillin resistance gene. *ori* – origin of replication. attR4 & attR3– Gateway recombination sites. *ccdB* – toxicity gene. CAT - chloramphenicol acetyltransferase gene. MCS1 - *KasI/NarI/SfoI*, *SpeI* unique restriction sites. MCS2– *KpnI*, *NheI*, *AvrII*, *AscI* unique restriction sites. P*rps-0*_short – shorter upstream region (0.8 kb) of *rps-0*. HygR CeOPT – *C. elegans* optimized (codon optimization, introns) hygromycin B phosphotransferase gene. *gpd-2/gpd-3* outron – CEOPX036 operon intergenic region (outron) between *gpd-2* and *gpd-3* genes. GFP – green fluorescent protein gene (optimized for *C. elegans*). mCherry – mCherry fluorescent protein

gene (optimized for *C. elegans*). **Figure S3.** Hygromycin B independent transmission of integrated transgenes. Worms were judged to contain integrated transgenes using previously reported criteria: transmission frequency of 100%, no mosaicism in any worm [10]. An individual animal from an integrated transgenic strain was transferred to a single plate and a population grown until food was exhausted, at which time progeny were twice more transferred to fresh plates by chunking. The hygromycin B resistance gene and GFP were expressed under the control of the ubiquitous P*rps-0* promoter (P*rps-0*::HygR::*gpd-2/gpd-3*::GFP::*unc-54*, bombardment 61, strains 61.4.3, 61.10.2; P*rps-0*_short::HygR, P*rps-0*::GFP::*unc-54*, bombardment 65, strain 65.2.1). Worms were imaged as described in the methods section. Exposure time: 700 ms. All of the strains were grown and propagated in the absence of hygromycin B. **Table S1.** Survival of *C. elegans* on hygromycin B. 300 (low density) or 3000 (high density) synchronized L1 larvae were plated on 6 cm plates and immediately treated with hygromycin B. The plates were then scored for the number of of L4 larvae (after ~40 hours) and adult animals (after ~60 and ~90 hours). The experiment was repeated in triplicate. **Table S2.** Primers used for vector construction. **Table S3.** List of plasmids generated. **Table S4.** Detailed summary of all bombardments performed. **Table S5.** Optimization of bombardment conditions. The upper table shows the effect of rupture discs on transformation efficiency (rupture discs define the force at which gold beads are shot onto the worms). The lower table shows the effect of bead size in combination with different types of rupture discs. **Table S6.** Influence of DNA linearization on bombardment efficiency. To test the importance of DNA linearization on transformation efficiency, a single DNA mixture was prepared and then further processed in 3 different ways (non-linearized, linearized but not purified before coating, linearized and purified before coating). Purification was performed using a PCR purification kit (Qiagen). **Supplementary Note.**

Acknowledgments

We would like to thank Mario de Bono and his research group for discussions and technical help.

Author Contributions

Conceived and designed the experiments: IR SG JWC. Performed the experiments: IR SG. Analyzed the data: IR SG JWC. Wrote the paper: IR SG JWC.

References

1. Mello CC, Kramer JM, Stinchcomb D, Ambros V (1991) Efficient gene transfer in *C. elegans*: extrachromosomal maintenance and integration of transforming sequences. EMBO J 10: 3959–3970.
2. Mello C, Fire A (1995) DNA transformation. Methods Cell Biol 48: 451–482.
3. Praitis V, Casey E, Collar D, Austin J (2001) Creation of low-copy integrated transgenic lines in *Caenorhabditis elegans*. Genetics 157: 1217–1226.
4. Berezikov E, Bargmann CI, Plasterk RH (2004) Homologous gene targeting in *Caenorhabditis elegans* by biolistic transformation. Nucleic Acids Res 32: e40.
5. Semple JI, Garcia-Verdugo R, Lehner B (2010) Rapid selection of transgenic *C. elegans* using antibiotic resistance. Nat Methods 7: 725–727.
6. Giordano-Santini R, Milstein S, Svrzikapa N, Tu D, Johnsen R, et al. (2010) An antibiotic selection marker for nematode transgenesis. Nat Methods 7: 721–723.
7. Greiss S, Chin JW (2011) Expanding the genetic code of an animal. J Am Chem Soc 133: 14196–14199.
8. Chin JW (2012) Molecular biology. Reprogramming the genetic code. Science 336: 428–429.
9. Davis L, Chin JW (2012) Designer proteins: applications of genetic code expansion in cell biology. Nat Rev Mol Cell Biol 13: 168–182.
10. Semple JI, Biondini L, Lehner B (2012) Generating transgenic nematodes by bombardment and antibiotic selection. Nat Methods 9: 118–119.
11. Brenner S (1974) The genetics of *Caenorhabditis elegans*. Genetics 77: 71–94.
12. Gritz L, Davies J (1983) Plasmid-encoded hygromycin B resistance: the sequence of hygromycin B phosphotransferase gene and its expression in *Escherichia coli* and *Saccharomyces cerevisiae*. Gene 25: 179–188.
13. Hunt-Newbury R, Viveiros R, Johnsen R, Mah A, Anastas D, et al. (2007) High-throughput in vivo analysis of gene expression in *Caenorhabditis elegans*. PLoS Biol 5: e237.
14. Redemann S, Schloissnig S, Ernst S, Pozniakowsky A, Ayloo S, et al. (2011) Codon adaptation-based control of protein expression in *C. elegans*. Nat Methods 8: 250–252.
15. Robert V, Bessereau JL (2007) Targeted engineering of the *Caenorhabditis elegans* genome following Mos1-triggered chromosomal breaks. EMBO J 26: 170–183.
16. Frokjaer-Jensen C, Davis MW, Hopkins CE, Newman BJ, Thummel JM, et al. (2008) Single-copy insertion of transgenes in *Caenorhabditis elegans*. Nat Genet 40: 1375–1383.
17. Chen C, Fenk LA, de Bono M (2013) Efficient genome editing in *C. elegans* by CRISPR-targeted homologous recombination. Nucleic Acids Res. in press.

7

Genetic Background Affects Human Glial Fibrillary Acidic Protein Promoter Activity

Xianshu Bai, Aiman S. Saab, Wenhui Huang, Isolde K. Hoberg, Frank Kirchhoff, Anja Scheller*

Department of Molecular Physiology, University of Saarland, Homburg, Germany

Abstract

The human glial fibrillary acidic protein (hGFAP) promoter has been used to generate numerous transgenic mouse lines, which has facilitated the analysis of astrocyte function in health and disease. Here, we evaluated the expression levels of various hGFAP transgenes at different ages in the two most commonly used inbred mouse strains, FVB/N (FVB) and C57BL/6N (B6N). In general, transgenic mice maintained on the B6N background displayed weaker transgene expression compared with transgenic FVB mice. Higher level of transgene expression in B6N mice could be regained by crossbreeding to FVB wild type mice. However, the endogenous murine GFAP expression was equivalent in both strains. In addition, we found that endogenous GFAP expression was increased in transgenic mice in comparison to wild type mice. The activities of the hGFAP transgenes were not age-dependently regulated. Our data highlight the importance of proper expression analysis when non-homologous recombination transgenesis is used.

Editor: Roy A. Quinlan, University of Durham, United Kingdom

Funding: This work is supported by DFG SPP 1172, DFG SFB 894, European Commission FP7-202167 NeuroGLIA and FP7-People ITN-237956 EdU-Glia (http://www.eduglia.eu/).The funders had no role in study design, data collection and analysis, decision to publish, or preparation of the manuscript.

Competing Interests: The authors have declared that no competing interests exist.

* E-mail: anja.scheller@uks.eu

Introduction

Glial fibrillary acidic protein (GFAP) is the major intermediate filament protein in astrocytes, the main glia population of the brain, and has become the bona fide marker of astrocytes [1–3]. GFAP expression starts already during embryonic development in radial glia [3–5] and is highly sensitive to any kind of pathology such as acute brain injury (stroke, trauma), chronic neurodegeneration (Alzheimer's and Parkinson's disease) and aging [6–11].

In the past two decades, a 2.2 kb fragment 5′ upstream of the open reading frame of the human GFAP gene (hGFAP promoter) [12,13] has been frequently used to drive transgenic expression of several proteins (e.g. LacZ, GFP or Cre) selectively in astrocytes [14–16]. To study physiological properties of astrocytes, we used this promoter for transgenic expression of fluorescent proteins (FPs) and the tamoxifen-inducible Cre DNA recombinase CreERT2 (CT2, a fusion protein of the Cre DNA recombinase and the ligand-binding domain of the estrogen receptor) [17–20]. Transgenic mice were generated by injection of linearized vector DNA [21] into oocytes of the most commonly used inbred mouse strains, FVB/N (FVB) and C57BL/6N (B6N). FVB mice (white fur) with large litters and high reproductive capacity are widely used for gene transfer experiments owing to their large and prominent pronuclei [22]. B6N mice (black fur) represent the preferred mouse strain for behavioral experiments despite developing spontaneous auditory degeneration in young adulthood [23–25].

To generate a homogenous genetic background suitable for a wide range of behavioral experiments, we crossbred transgenic FVB mice (expressing ECFP, EGFP or CT2 under the control of hGFAP promoter) to B6N mice. Unexpectedly, we found that the

transgenic protein expression was strongly influenced by the genetic background of the mouse strain. Since hGFAP transgenic mice are widely used within the scientific community, we performed a quantitative comparison of transgene expression in both inbred strains.

Materials and Methods

Ethics Statement

This study was carried out at the University of Saarland in strict accordance with the recommendations to European and German guidelines for the welfare of experimental animals. Animal experiments were approved by the Saarland state's "Landesamt für Gesundheit und Verbraucherschutz" in Saarbrücken/Germany (animal license number: 72/2010).

Animals

FVB/NRj (FVB) and C57Bl/6NRj (B6N) wild type mice were used as wild type (WT) controls (purchased from Janvier, France). Transgenic mice TgN(hGFAP-ECFP)$_{GCFD}$ = hGFAP-ECFP$_{GCFD}$, TgN(hGFAP-EGFP)$_{GFEA/GFEC}$ = hGFAP-EGFP$_{GFEA/GFEC}$ and TgN(hGFAP-CreERT2)$_{GCTF}$ = hGFAP-CT2$_{GCTF}$ were originally generated by non-homologous recombination in the FVB background [18–20], expressing FPs and CT2 in astrocytes (Fig. 1A). After at least 12 generations of crossbreeding to B6N mice, they were considered as being of B6N background. TgN(hGFAP-AmCyan)$_{GCYM}$ = hGFAP-AmCyan$_{GCYM}$ mice were originally generated by injection of DNA in B6N oocytes [18]. We crossbred B6N(hGFAP-ECFP)$_{GCFD}$ and B6N(hGFAP-AmCyan)$_{GCYM}$ to FVB once to get B6NxFVB1 and twice to get B6NxFVB2 using transgenic males and wild type females (Fig. 1A). The four-letter indices represent distinct founder lines. For visualization of recombined cells in hGFAP-CT2$_{GCTF}$, mice were bred to TgH(Rosa26-CAG-loxP-stop-loxP-tdTomato) (R26tdTom) reporter mice (Jaxlab: B6; 129S6-Gt (ROSA) 26Sortm14 (CAG-tdTomato) Hze/J [26], in which CAG represents the ubiquitously active cytomegalovirus enhancer fused to the chicken beta-actin promoter. Litters of FVB (hGFAP-CT2)$_{GCTF}$ mice are of mixed background when crossed to R26tdTom reporter mice which are in a mixed C57BL/6J C57BL/6N background.

Real Time-PCR and Western Blot Analysis

Levels of messenger RNA (mRNA) and genomic DNA were detected by reverse transcriptase PCR (RT-PCR), levels of proteins were detected by sodium dodecyl sulfate polyacrylamide gel electrophoresis and subsequent Western blot analysis as described previously [19,27]. The cerebellum was homogenized (Precellys homogenizer, peqlab, Erlangen, Germany) and divided for RNA extraction (1/6) with RNeasy mini kit (QIAGEN, Hilden, The Netherlands) and for protein analysis (5/6). Primer sequences for RT-PCR were as follows (in 5′ to 3′ direction): GFAP-forward, TGG AGG AGG AGA TCC AGT TC; GFAP-reverse, AGC TGC TCC CGG AGT TCT; ExFP (EGFP and ECFP)-forward, GAA GCG CGA TCA CAT GGT; ExFP-reverse, CCA TGC CGA GAG TGA TCC; AmCyan-forward, GAG AAC CTT CAC CTA CGA GGA C; AmCyan-reverse, TCG AAG CAG TTG CCC TTC; Cre-forward, CCT GGA AAA TGC TTC TGT CCG; Cre-reverse, CAG GGT GTT ATA AGC AAT CCC; β-actin-forward, GGG TCA GAA GGA CTC CTA TG; β -actin-reverse, GGT CTC AAA CAT GAT CTG GG.

After gel separation, proteins were transferred to nitrocellulose membrane and probed with polyclonal rabbit anti-GFAP (1:1000, Dako cytomation, Glostrup, Denmark), polyclonal rabbit anti-GFP (1:1000, abcam, Cambridge, England), polyclonal rabbit anti-human estrogen receptor α (1:200, Santa Cruz, Santa Cruz, USA) or monoclonal mouse anti-α-Tubulin (1:10000, Sigma, St. Louis, USA) antibodies.

Analysis of Transgenic Copy Number

Transgene copy number was determined by quantitative RT-PCR as described previously with slight modifications [28]. Briefly, plasmids of hGFAP-ECFP, hGFAP-EGFP, hGFAP-CT2 and hGFAP-AmCyan were used to establish a copy number standard curve. Genomic DNA was extracted from respective mouse tails with the Spin Tissue Mini kit (Stratec Molecular, Berlin, Germany). We selected heterozygous and homozygous NG2-EYFP [29] and NG2-CreERT2 (provided by Wenhui Huang, unpublished) knock-in mice as copy number controls for ECFP/EGFP and CT2, respectively. For the hGFAP-AmCyan transgene, we extracted genomic DNA from primary astrocytes [30]. The primers for ExFPs, CT2 and AmCyan were the same as the one used for cDNA PCR.

Tamoxifen Treatment

To induce DNA recombination in hGFAP-CT2$_{GCTF}$ × R26tdTom reporter mice, tamoxifen (10 mg/ml corn oil, Sigma, St. Louis, USA) was intraperitoneally injected into seven-week-old mice for three consecutive days (100 mg/kg body weight). Ten days after the first injection, mice were perfused and analyzed.

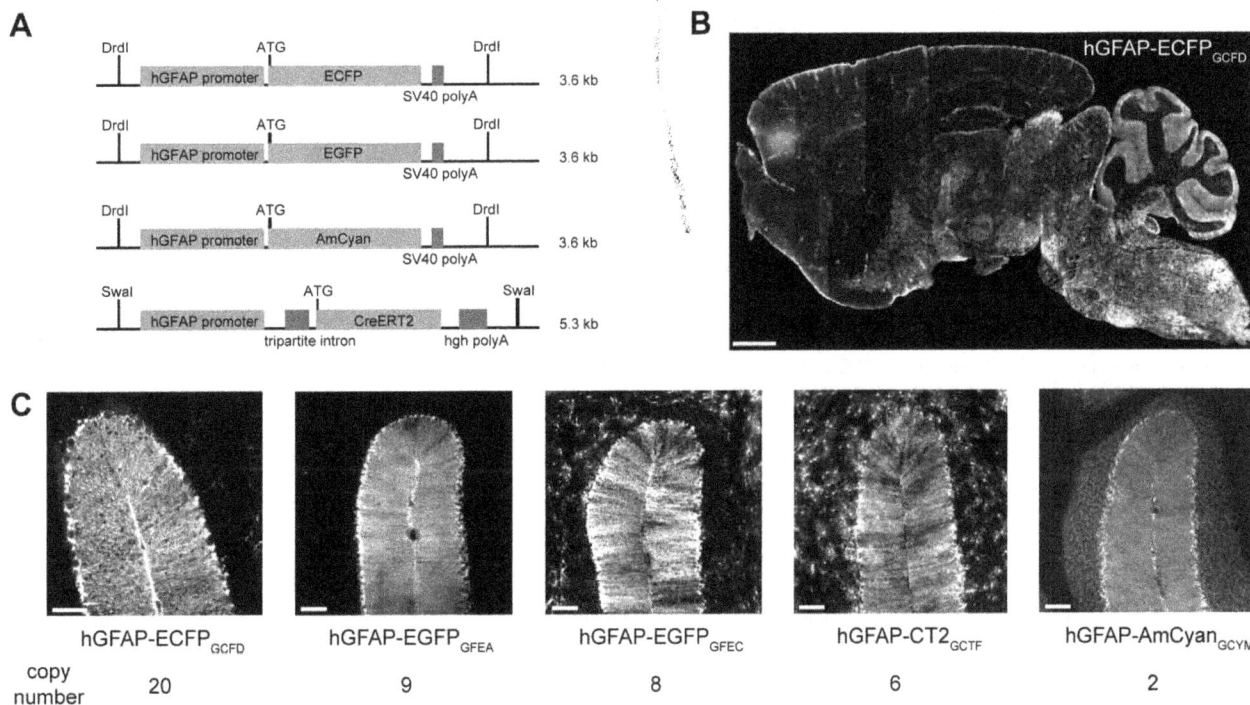

Figure 1. hGFAP promoter controlled transgene expression in five different mouse lines. (A) Transgenic constructs used for oocyte injection. (B) Widespread expression of ECFP in FVB(hGFAP-ECFP)$_{GCFD}$ mice with high levels in the cerebellum. Scale bar indicates 1 mm. (C) Abundant fluorescent signals from Bergmann glia of FVB(hGFAP-ECFP)$_{GCFD}$, FVB(hGFAP-EGFP)$_{GFEA/C}$, B6N(hGFAP-AmCyan)$_{GCYM}$ and FVB(hGFAP-CT2$_{GCFT}$ × R26tdTom). Transgene copy numbers are indicated below the respective mouse lines. Scale bars indicate 100 µm.

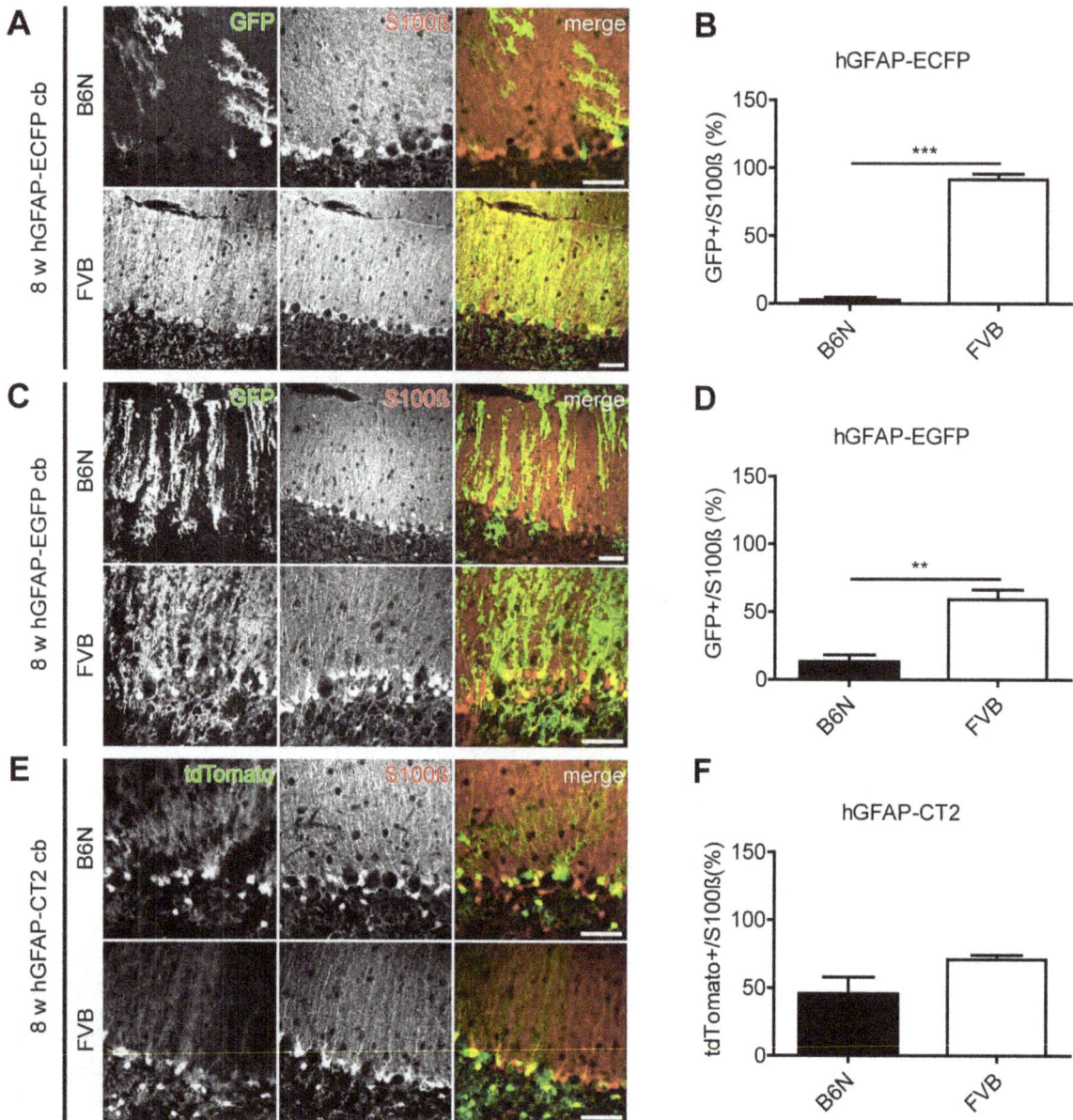

Figure 2. Immunohistochemical analysis of reporter protein expression in different transgenic mouse lines showed lower expression in the B6N background when compared to FVB. Cerebellar vibratome slices (cb) of 8-week-old mice were immunolabeled with anti-GFP (A and C) and anti-S100β antibodies (A, C and E), endogenous fluorescence of tdTomato in E. Upper panels depict transgene expression in B6N, lower panels in FVB. The S100β staining indicates all Bergmann glia. Results of comparative analysis in B6N and FVB mice are presented as percentage of transgene expressing Bergmann glia (S100β positive cells) (B, D and F). ***: $p<0.001$, **: $p<0.01$. Scale bars indicate 50 μm.

Immunohistochemical Analysis of Transgenic Protein Expression

After perfusion and post-fixation with 4% formaldehyde in 0.1 M phosphate buffer (pH 7.4), free-floating vibratome brain slices were generated, blocked and permeabilized as described previously [31]. Slices were incubated with polyclonal goat anti-GFP (for ECFP and EGFP, 1:1000, Rockland, Gilbertsville, USA) and/or polyclonal rabbit anti-S100β (1:500, abcam, Chambridge, England) followed by incubation with Alexa488-conjugated anti-goat IgG/Alexa555-conjugated anti-rabbit IgG (1:2000, Invitrogen, Grand Island NY, USA).

Statistical Analysis

Three animals of every experimental age group and every strain were studied in three independent experiments. In RT-PCR experiments, cerebella of pups (one week old, 1 w) and adult mice (eight weeks old, 8 w) were investigated. We compared always mice of the same gender in both backgrounds, mostly males.

Statistical differences were analyzed using the two-tailed t-test for two-grouped data and one-way Anova for three-grouped data. Data are shown as mean+SEM.

Results

Description of hGFAP Transgenic Mouse Lines

Transgenic mice used in this study were generated by non-homologous recombination with different transgene copy numbers (TCN). Their detailed expression patterns have already been described previously [17–20]. The transgenic mouse lines used for comparison of transgene activity are categorized in three groups (Fig. 1A): (1) hGFAP-ECFP$_{GCFD}$ (TCN = 20) and hGFAP-EGFP$_{GFEA/GFEC}$ (TCN = 9 and 8, respectively) are based on the same vector with a SV40 polyA site and injected into FVB oocytes. (2) hGFAP-AmCyan$_{GCYM}$ (TCN = 2) is based on the same vector, but injected into B6N oocytes. (3) The vector to produce hGFAP-CT2$_{GCTF}$ (TCN = 6) contained additional regulatory elements (a generic intron in front of the ATG start codon and the polyA site of the human growth hormone instead of the SV40 polyA [32]. Vector DNA was injected into FVB oocytes. In groups 1 and 2, the expression of the FPs was directly controlled by the hGFAP promoter, while the hGFAP-CT2 mouse line required crossbreeding to a Cre-reporter line. For that purpose we used the R26tdtom mouse line, in which the final expression level in astrocytes was controlled by a ubiquitously active promoter (CAG) [26].

All transgenic mice were fertile and could be crossed to homozygosity without overt pathological phenotype. Genetically modified mice, generated by non-homologous recombination, are known for line-dependent transgene expression patterns [33,34]. In the CNS of our mouse lines we also observed a region dependent pattern of transgene expression. Only 10 to 30% of cortical astrocytes expressed ECFP, while 60 to 90% of all Bergmann glia in the cerebellum and astrocytes in the brainstem expressed ECFP in the hGFAP-ECFP$_{GFCD}$ mouse line (Fig. 1B). However, within the progeny of a given line, in the same inbred strain, the expression pattern did not change.

To evaluate the impact of genetic background on transgene expression, we focused on Bergmann glia (Fig. 1C) in the following lines: hGFAP-ECFP$_{GCFD}$, hGFAP-EGFP$_{GFEC}$, hGFAP-CT2$_{GCTF}$ and hGFAP-AmCyan$_{GCYM}$. The highly organized distribution of Bergmann glia facilitated the quantitative analysis by cell-counting.

We analyzed the sensitivity of the hGFAP promoter (gfa2) [12,13] in FVB and B6N strains by comparing protein and mRNA levels.

Transgenic Mice in B6N Backgrounds Displayed Diminished Transgene Expression

The extent of FP expression (ECFP and EGFP) in FVB and B6N mice was evaluated by cell counting after immunohisto-chemistry (Fig. 2) and Western blot analysis (Fig. 3).

To compensate for differences in physical fluorescence properties we enhanced the signal by using anti-GFP antibodies for both hGFAP-ECFP$_{GCFD}$ and hGFAP-EGFP$_{GFEC}$ mouse lines in FVB and B6N backgrounds. Immunohistochemistry data revealed that nearly all Bergmann glial cells expressed ECFP (91.5±4.0%) (Fig. 2A and B) in FVB(hGFAP-ECFP)$_{GCFD}$ mice, however, ECFP was hardly detectable in B6N(hGFAP-ECFP)$_{GCFD}$ mice (2.8±1.8%, Fig. 2A and B). A reduction in transgene expression was also observed in another transgenic mouse line: in FVB(hGFAP-EGFP)$_{GFEC}$, in which 59.2±7.1% of Bergmann glia were EGFP-positive, while markedly less EGFP expressing Bergmann glia (13.2±4.9%) were observed in B6N(hGFAP-EGFP)$_{GFEC}$ mice (Fig. 2C and D). We further analyzed a third transgenic mouse line, hGFAP-CT2$_{GCTF}$, in which the inducible Cre DNA recombinase CT2 was expressed under the control of the same hGFAP promoter. To activate CT2, we injected

tamoxifen to hGFAP-CT2 × R26tdTom female mice for three consecutive days to induce recombination and subsequent expression of the red fluorescent reporter protein tdTomato in astrocytes. Ten days after the first tamoxifen injection, reporter expression was not significantly different in FVB mice (71.1±3.2%) compared to B6N mice (45.8±12.0%) (Fig. 2E and F).

We confirmed the higher expression of FPs and CT2 in cerebella of FVB mice by Western blot analysis. Young FVB mice showed significantly higher expression of transgenes than B6N mice in all three examined lines (left panel in Fig. 3A) (ratios: FVB$_{ECFP}$/B6N$_{ECFP}$ = 37.3; FVB$_{EGFP}$/B6N$_{EGFP}$ = 8; FVB$_{CT2}$/B6N$_{CT2}$ = 2.81). No or only a weak protein signal could be detected in B6N(hGFAP-ECFP)$_{GCFD}$ and B6N(hGFAP-EGFP)$_{GFEC}$ adult mice. In contrast, in B6N(hGFAP-CT2)$_{GCTF}$ adult mice CT2 expression was clearly detectable (Fig. 3A) but still significantly lower compared to FVB mice.

We then investigated whether the endogenous GFAP expression varies in WT and transgenic mouse lines. The expression of the hGFAP transgenes appeared to be independently regulated from the endogenous mouse GFAP gene (Fig. 3B). In WT as well as in hGFAP-ECFP$_{GCFD}$ mice the level of endogenous GFAP was not different between B6N and FVB mice of the same age. However, we found that WT mice expressed less GFAP than transgenic mice in all the transgenic mouse lines that we studied in this work (Fig. 3C).

Cell counting after immunohistochemical labeling as well as Western blot analysis revealed higher levels of FP expression in FVB compared to B6N mice. CT2 protein expression was significantly higher in FVB cerebellar homogenates; however, tdTomato reporter expression analysis was similar in transgenic FVB and B6N mice.

Lower Transgenic FP and CT2 mRNA Levels in B6N than in FVB Mice

To investigate whether the different expression levels of transgenic proteins were caused by transcriptional/posttranslational regulation or protein degradation, we studied mRNA levels in WT and transgenic FVB and B6N mice. Both B6N and FVB WT mice showed equal levels of endogenous GFAP mRNA at the same age, consistent with the protein data (Fig. 3B). However, endogenous GFAP mRNA levels dropped significantly from young (1 w) to adult (8 w) WT mice (B6N vs. FVB: 1.0±0.06 vs. 1.0±0.04 at 1 w; 0.64±0.05 vs. 0.60±0.03 at 8 w) (Fig. 4A). After comparing the endogenous mouse GFAP mRNA levels in WT mice, we quantified mRNA levels of FPs and CT2 controlled by transgenic hGFAP promoters in both inbred strains. In line with the protein data, almost no ECFP mRNA was detectable in B6N(hGFAP-ECFP)$_{GCFD}$ mice at any age (Fig. 4B). Also only low levels of EGFP mRNAs were detected in B6N(hGFAP-EGFP)$_{GFEC}$ young mice, however, those increased in the adult (Fig. 4C), again consistent with our Western blot data (Fig. 3A). CT2 mRNA levels were significantly lower in B6N compared to FVB at both ages (Fig. 4D), also consistent with the protein data from the Western blots (Fig. 3A).

In young hGFAP-ECFP$_{GCFD}$ mice (with the highest difference in protein levels), ECFP mRNA levels were about 3500 times higher in FVB than in B6N (Fig. 4B), however, in hGFAP-CT2$_{GCTF}$ mice of the same age, the CT2 mRNA levels were almost comparable (FVB/B6N = 1.25, Fig. 4D). Furthermore, ECFP was down-regulated with age in FVB(hGFAP-ECFP) (37% remaining, Fig. 4B), while the other two lines displayed either similar (FVB(hGFAP-EGFP$_{GFEC}$), Fig. 4C) or increased

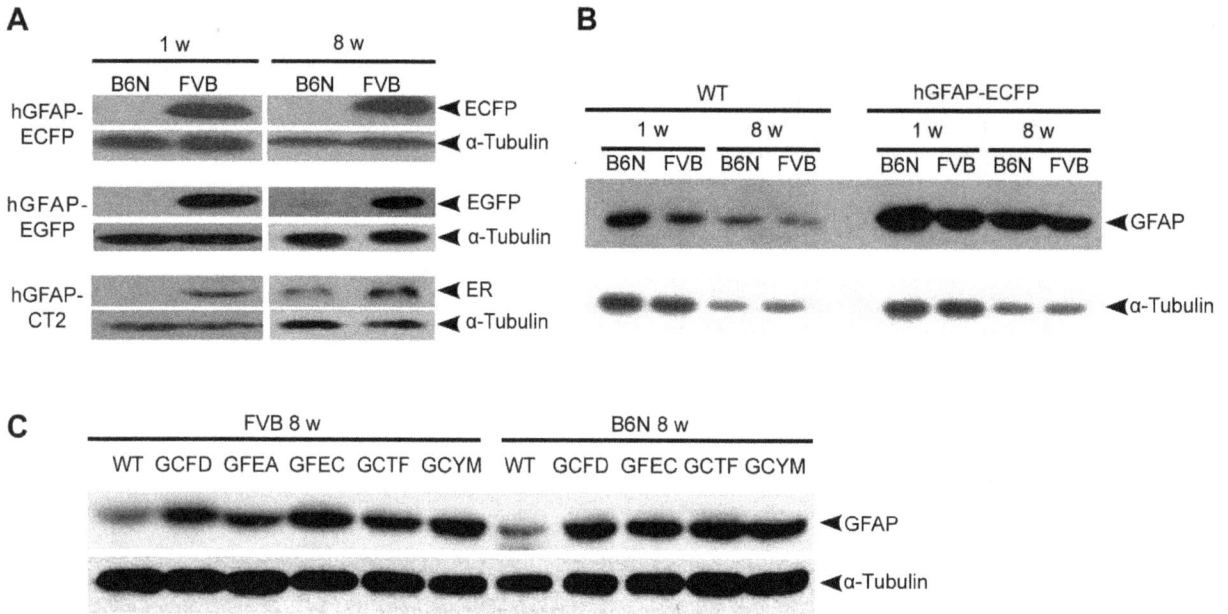

Figure 3. Comparative Western blot analysis of endogenous GFAP and transgenic proteins in B6N and FVB mice. Cerebellar homogenates of transgenic and wild type mice (1 w and 8 w) were probed with anti-GFP (to detect ECFP or EGFP), anti-human estrogen receptor α (ER α, recognizing CT2), and anti-GFAP and anti-α-tubulin antibodies. (A) Western blot analysis of transgene expression. (B) Western blot analysis of endogenous GFAP expression in WT and transgenic mice (hGFAP-ECFP)$_{GCFD}$. (C) Western blot analysis of endogenous GFAP expression in WT and five transgenic mouse lines (hGFAP-ECFP$_{GCFD}$; hGFAP-EGFP$_{GFEA}$; hGFAP-EGFP$_{GFEC}$; hGFAP-CT2$_{GCTF}$; hGFAP-AmCyan$_{GCYM}$) in both FVB and B6N background.

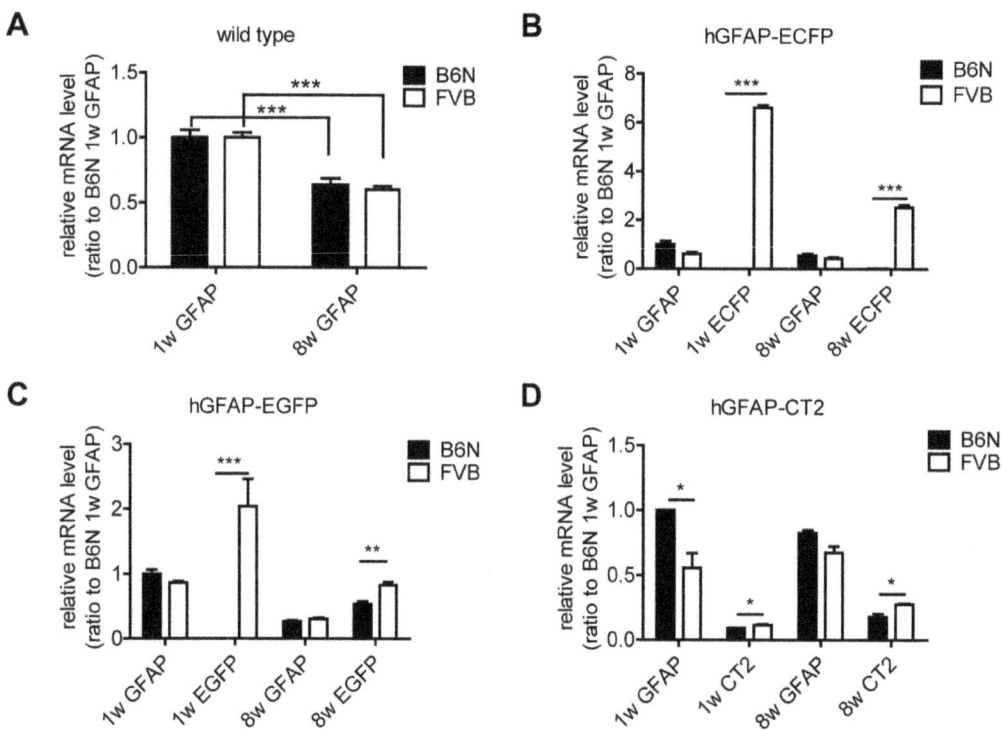

Figure 4. Quantitative RT-PCR analysis of transgene and endogenous GFAP mRNA levels in FVB and B6N mice. (A) Cerebellar GFAP mRNA levels in wild type B6N and FVB mice (1 w and 8 w). (B-D) Transgenic mRNA levels compared to endogenous GFAP mRNA levels in the cerebellum of B6N and FVB mice (1 w and 8 w). (B) hGFAP-ECFP$_{GCFD}$. (C) hGFAP-EGFP$_{GFEC}$. (D) hGFAP-CT2$_{GCTF}$. Relative expression is normalized to GFAP mRNA level in 1 w B6N mice. *: p<0.05, **: p<0.01, ***: p<0.001. Data are obtained from three independent experiments with samples from three mice (n = 3) in every experiment.

(B6N(hGFAP-EGFP$_{GFEC}$ and hGFAP-CT2) and FVB(hGFAP-CT2), Fig. 4C and D) levels of mRNA in adult mice.

Additionally, one crossbreeding of the second EGFP-expressing transgenic line, FVB(hGFAP-EGFP)$_{GFEA}$ (Fig. 1C), to B6N WT mice significantly reduced EGFP mRNA level in FVBxB6N1 littermates compared with FVB mice (FVB:FVBxB6N1 ratio = 6.75), while endogenous GFAP mRNA levels were equal as shown for the other transgenic lines before (data not shown).

Taken together, quantitative RT-PCR results confirmed that FVB mice showed higher hGFAP promoter activity as confirmed by higher transgene mRNA level. However, the endogenous GFAP mRNA levels were equal among all the WT and transgenic mouse lines except in hGFAP-CT2 line.

We also noted that the activity of the hGFAP promoter is highly variable. For instance, in young FVB(hGFAP-ECFP)$_{GCFD}$ mice the ECFP mRNA levels were about 11 times higher than the endogenous GFAP mRNA levels, while in young FVB(hGFAP-CT2)$_{GCTF}$ transgenic mice CT2 mRNA levels were decreased to 18.5% (Fig. 4B–D).

Crossbreeding of B6N Mice to FVB Increases Transgene Expression

So far we could demonstrate that crossbreeding FVB transgenic mice to B6N results in a severe down-regulation of transgene expression. Therefore, we wanted to know whether the reverse experiment, backcrossing of transgenic B6N mice to FVB, could enhance low transgene expression (Fig. 5). For this purpose, we selected the hGFAP-ECFP$_{GCFD}$ mouse line, because it exhibited the strongest difference in transgene expression between B6N and FVB (Fig. 2A, B and 5A). Strikingly, one single backcrossing (B6NxFVB1) already significantly reactivated ECFP mRNA 22-fold (B6N vs. B6NxFVB1: 0.03±0.01 vs. 0.65±0.16) and protein expression 13-fold (B6N vs. B6NxFVB1: 2.8±1.8% vs. 36.2±5.5%) in B6NxFVB1 mice compared to B6N (Fig. 5). These results again indicate that FVB mice have higher hGFAP promoter activity than B6N. This observation provides additional proof that the genetic background has a clear impact on hGFAP promoter activity.

We further tested the effect of backcrossing to FVB in an additional mouse line. But this time we chose a mouse line, i.e. hGFAP-AmCyan$_{GCYM}$, which was originally generated by injection of the transgenic vector into B6N oocytes and maintained in a B6N background. After crossing for only one generation to FVB (Fig. 6A) we realized that AmCyan expression in the cerebellum, e.g. the number of reporter-positive Bergmann glia, was not significantly enhanced (B6N vs. B6NxFVB1: 56.6±4.5 vs. 64.0±0.5%). The mRNA levels of AmCyan were comparable as well (B6N vs. B6NxFVB1: 0.98±0.09 vs.0.96±0.07). An additional backcrossing (B6NxFVB2) showed a significantly higher level of AmCyan mRNA when compared to B6N and B6NxFVB1 (B6NxFVB2: 1.5±0.08) (Fig. 6C).

However, when analyzing the cortex, a brain region where AmCyan expression levels were initially very low (16.2±6.4%), we could detect a more than two-fold increase of AmCyan-expressing astrocytes already in B6NxFVB1 (36.7±3.4%) when compared to B6N mice (Fig. 6B).

These results suggest that a low hGFAP promoter activity in B6N mice can be increased by crossbreeding to the FVB background.

Discussion

In the current study, we investigated in different transgenic mouse lines the hGFAP promoter-controlled expression of FPs or CT2. We found that the activity of this promoter was strongly dependent on the chosen inbred strain, FVB or B6N.

(1) Transgene Activity in Inbred Strains

FVB mice expressed higher FP and CT2 levels than B6N, especially the hGFAP-ECFP$_{GCFD}$ and hGFAP-EGFP$_{GFEC}$ lines (Fig. 2, 3 and 4). In addition, a single backcrossing of B6N to FVB rescued silenced FP expression (Fig. 5) or increased the existing expression (Fig. 6). All our data demonstrate a stronger activity of the hGFAP promoter in FVB than in B6N mice.

Three different mechanisms have been reported to regulate the transcription of the *Gfap* gene: DNA methylation; histone methylation and acetylation; as well as spatial positioning.

DNA Methylation Influences Transcription Factor Binding

Epigenetic studies showed that in early stages of embryonic development, the methylation of the GFAP promoter at CpG islands represses transcription by preventing the binding of STAT3 (signal transducer and activator of transcription 3) in a complex with Smad1/4 (signal transducer and transcriptional modulator) and p300 at the corresponding promoter element [35–37]. During late embryogenesis, enhanced demethylation of the GFAP promoter and subsequently increased GFAP expression are characteristic properties of astroglial differentiation [36]. Similarly, in human malignant gliomas, the GFAP expression is controlled by methylation. Here, however, an enhanced methylation of the promoter causes a silencing of the *Gfap* gene [38]. Similarly, for the imprinted transgene RSVIgmyc higher levels of methylation were found in C57BL/6J than in FVB [39], indicating higher methylation activities in B6N. Since the *Gfap* gene transcription occurs monoallelically in the cerebral cortex [40], different methylation conditions could be a very potent mechanism to cause the observed different transgene expression levels in B6N and FVB.

Histone Methylation and Acetylation Affect Chromatin Structure

Similar to DNA methylation, histone methylation represents another mechanism of transcriptional silencing or activation. Growth factors (basic fibroblast growth factor 2) positively affect the binding of the STAT/CBP complex with the GFAP promoter by inducing H3K4 (lysine 4 at histone 3) methylation and suppression of H3K9 (lysine 9 at histone 3) methylation around the STAT3-binding site, leading to an increased GFAP expression in developing astrocytes [37]. Histone acetylation can be positively related to transcriptional activity as well [41]. At the GFAP promoter, binding of STAT3 to the CBP/p300 complex activates the intrinsic histone acetyltransferase of the coactivators CBP and p300 and subsequent relaxing of the chromatin structure, resulting in enhanced transcription [42]. Unfortunately, the GFAP promoter difference in histone acetylation/methylation between inbred strains has not yet been investigated.

Spatial Positioning as a Mean to Regulate Transcription

The spatial positioning of gene loci within the nucleus has been discussed as a mechanism of transcriptional regulation. In cultured astrocytes the active *Gfap* alleles appear preferentially positioned towards the center of the nucleus while inactive alleles are more frequently found at the periphery as it could be shown by fluorescence *in situ* hybridization [40,41]. Transcription-preferring localization within the nuclear architecture appears as an effective mean to regulate gene expression, and it is tempting to speculate

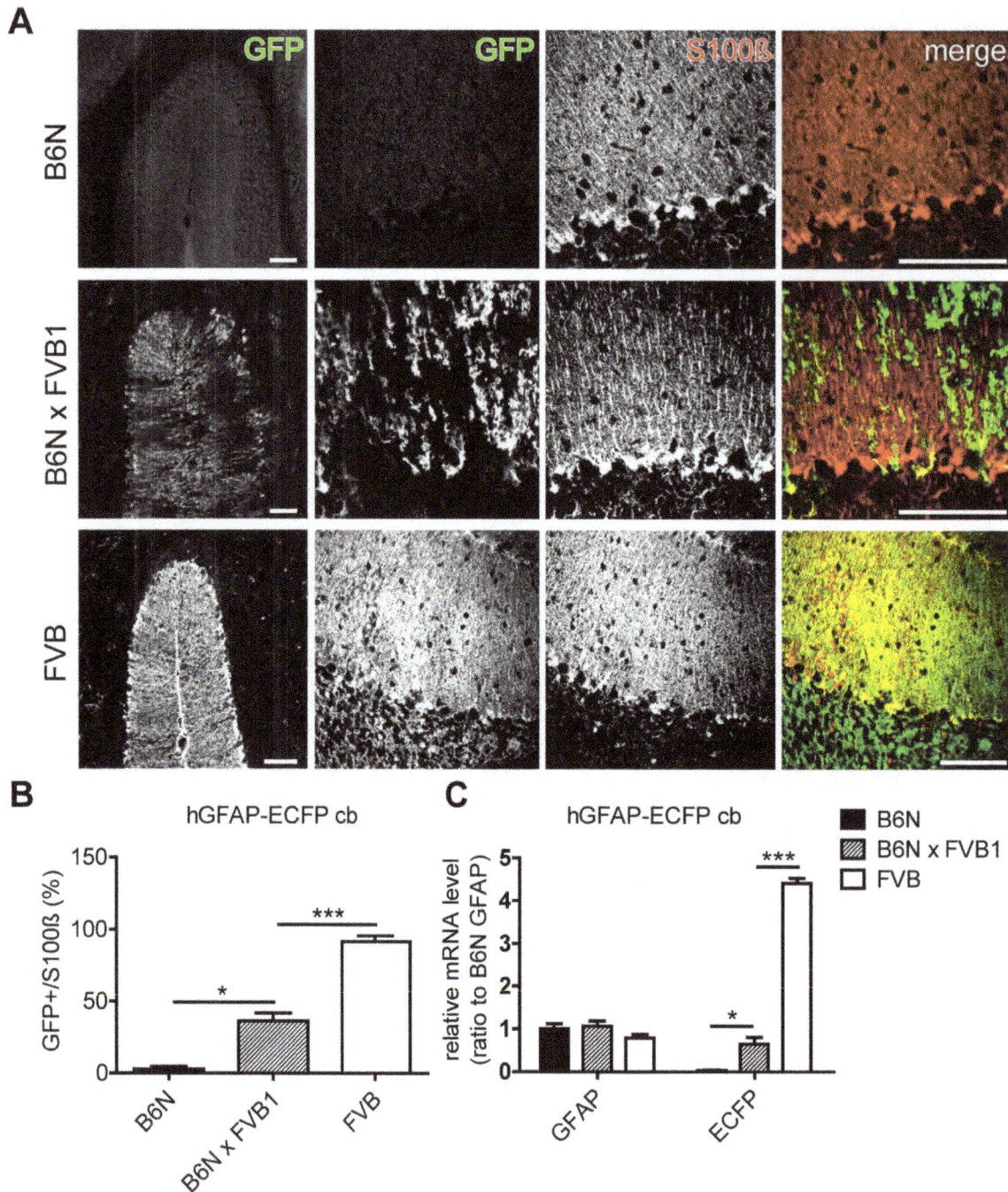

Figure 5. Backcrossing of B6N (hGFAP-ECFP)GCFD mice to FVB for a single generation re-activated transgenic ECFP expression. (A and B) Cerebellar slices of 8-week-old mice were immunostained with anti-GFP and anti-S100β antibodies and analyzed. Only single ECFP expressing Bergmann glia (S100β positive cells) were detected in B6N(hGFAP-ECFP)$_{GCFD}$ mice (A, upper panel), while ~91.5% of Bergmann glia were ECFP positive in FVB(hGFAP-ECFP)$_{GCFD}$ mice (A, lower panel). Backcrossing of B6N(hGFAP-ECFP)$_{GCFD}$ for one generation with FVB WT mouse led to increased ECFP expression in B6NxFVB1 littermates (A, middle panel). (C) GFAP and ECFP mRNA levels in B6N, FVB and B6NxFVB1 mice (8 w). Relative expression is normalized to GFAP mRNA level in B6N mice. *: p<0.05 and ***: p<0.001. Data are obtained from three independent experiments with samples from three mice (n=3) in every experiment. Scale bars indicate 100 μm.

that such chromatin remodeling mechanisms are subject to the genetic background of inbred strains.

(2) Differences in Transgenic Constructs

The variable composition of the transgenic plasmids used to generate the analyzed mouse lines, seems to affect the transgenic

expression pattern. For hGFAP-ECFP$_{GCFD}$, hGFAP-EGFP$_{GFEA/GFEC}$ and hGFAP-AmCyan$_{GCYM}$ lines, the simplest cloning strategy has been used: A fragment of the hGFAP promoter (gfa2) [12], a Kozak sequence (TCG CCA CCA TG, [43]) followed by the open reading frame (ORF) of the transgenic protein and termination by the SV40 polyadenylation (polyA)

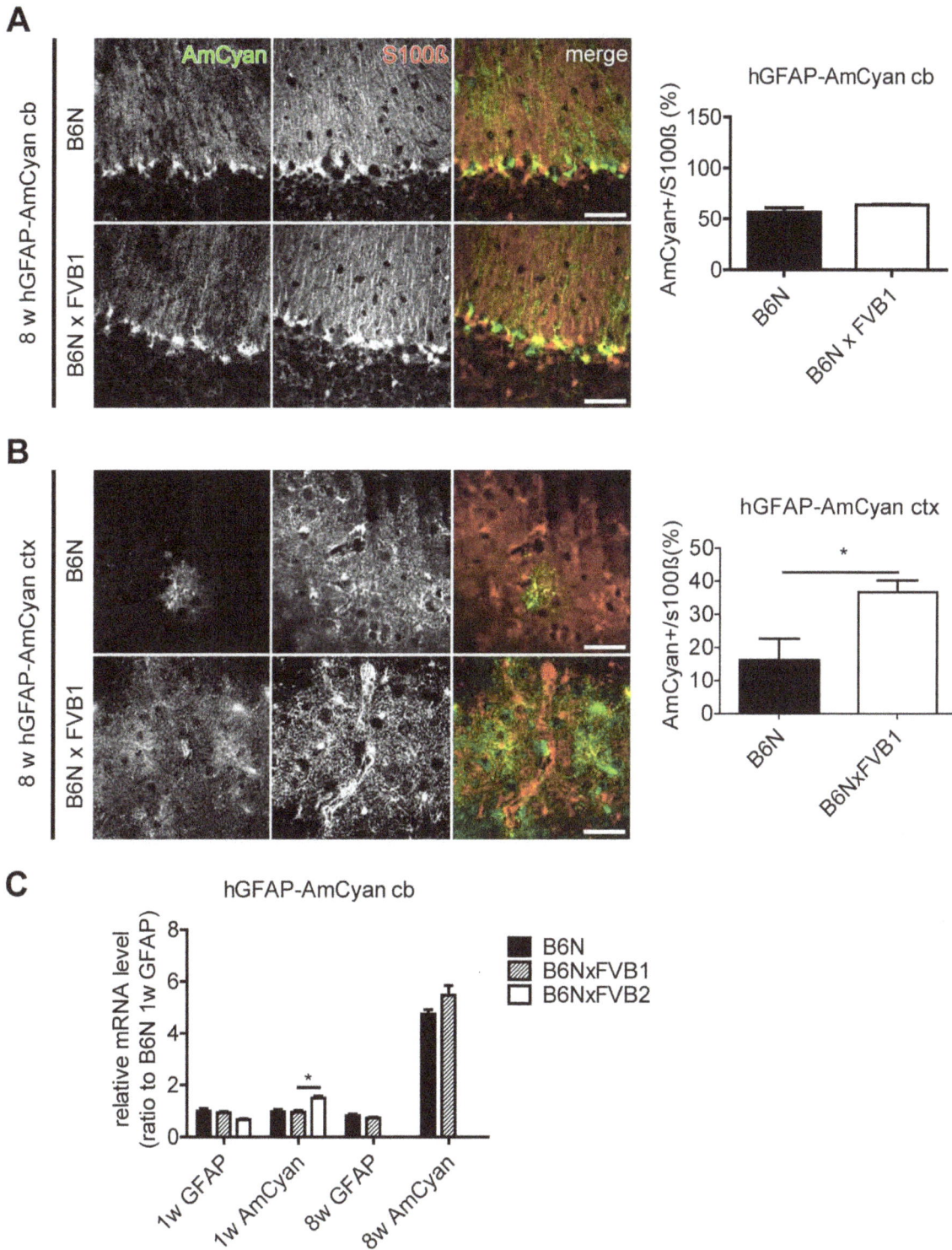

Figure 6. Backcrossing of B6N(hGFAP-AmCyan)GCYM mice to FVB increased transgenic AmCyan expression. Cerebellar (A) and cortical (B) brain slices of 8-week-old mice were immunostained with anti-S100β antibodies. Backcrossing of B6N(hGFAP-AmCyan)$_{GCYM}$ (upper panels in A and B) for one generation to FVB (lower panels in A and B) did not significantly enhance AmCyan expression in cerebellum, but caused higher levels in the cortex of B6NxFVB1 littermates when compared to B6N. Results of comparative analysis (right panels in A and B) in B6N and B6NxFVB1 mice are provided as percentage of transgene expressing Bergmann glia (A) and cortical astrocytes (B) (S100β positive cells). (C) GFAP and AmCyan mRNA levels in 1-week-old (B6N, B6NxFVB1 and B6NxFVB2) and 8-week-old (B6N and B6NxFVB1) mouse cerebellum. A second backcrossing resulted in enhanced transgene levels. Relative expression is normalized to GFAP mRNA level in 1-week-old B6N mice. *: p<0.05. Data are obtained from three independent experiments with samples from three mice (n = 3) in every experiment. Scale bars indicate 50 µm.

sequence [44,45] (Fig. 1C). For the generation of hGFAP-CT2 transgenic mice, the construct was modified by insertion of a generic intron to stabilize primary transcripts [46]. In addition, the viral polyA sequence was exchanged with an eukaryotic polyA sequence (hgh polyA, human growth hormone, [32]). For transgenesis several hundred linearized DNA molecules were injected into a single fertilized oocyte, which usually integrate as concatemers into the genome [47].

The protein and mRNA data suggest that CT2 mice are less affected by the change of inbred strains compared with FP-transgenic mice (hGFAP-ECFP$_{GCFD}$ and hGFAP-EGFP$_{GFEC}$). The levels of expressed CT2 protein and mRNA were still lower in B6N than FVB, but the overall difference was strikingly lower than in the lines with FP expression (Fig. 2, 3 and 4). Since splicing is known as an mRNA stabilizing mechanism [47–49], we assume that the additional splicing induced by the generic intron in the CT2 construct reduces the transcriptional variability between the inbred strains as prominently observed with the FP constructs. In addition, the recombination frequency (the functional readout of the CT2 enzyme activity) was not affected by the genetic background. Besides the improved stability of the mRNA this might also be due to the low number of enzyme molecules that are required for recombination of loxP sites and resulting in reporter protein expression (tdTomato) in FVB and B6N after CT2 induction (Fig. 2E and F).

(3) Transgene Copy Number

Previous reports have shown that transgene copy number (TCN) affects the level of transgene expression in the mammalian system due to the concatemeric integration [50]. While lower copy numbers lead to higher transgene expression, high copy numbers have the opposite effect. Here, we observed that mouse lines with higher TCN (hGFAP-ECFP$_{GCFD}$ = 20 copies) showed an overall high transgene expression (FVB: more than 90% of Bergmann glia) compared to mouse lines with smaller TCN (hGFAP-EGFP$_{GFEC}$ = 8 copies) (FVB: about 60% of Bergmann glia). Compared to background changes the mouse lines with higher TCN have shown higher sensitivities to inbred strain changes while mouse lines with smaller TCN were less sensitive (Fig. 3A). For hGFAP-AmCyan$_{GCYM}$ (2 copies) this could additionally explain why we could not detect a significant difference in Bergmann glial AmCyan expression after a single backcross to FVB (B6N vs. B6NxFVB1: both around 60%, Fig. 6). In contrast, crossing of B6N(hGFAP-ECFP)$_{GCFD}$ mice to FVB led to a remarkable 13-fold increase.

However, also the design of the construct can reduce the impact of TCN on transgenic protein expression or transgenic mRNA levels. This could be shown by the CT2 construct, where the differences between B6N and FVB were remarkably smaller than in the FP lines (Fig. 3 and 4) while the copy numbers were comparable (hGFAP-EGFP$_{GFEC}$ = 8 and hGFAP-CT2$_{GCTF}$ = 6).

(4) Endogenous GFAP

In all analyzed transgenic mouse lines the Western Blot analysis of cerebellar homogenates indicated an upregulation of endogenous GFAP protein (Fig. 3C) compared with both WT strains.

However, further analysis by qPCR revealed no difference in endogenous GFAP promoter activity: the mRNA levels did not change between the background strains. Also the number of GFAP positive cells was comparable in B6N and FVB (data not shown). Previous studies using hGFAP transgenes did not report an upregulation of the endogenous GFAP level [4,5,17–19,51–57]. We assume that the increase in GFAP protein might be harder to detect when using immunofluorescence detection techniques that are most frequently exerted. Although the increase in GFAP could be an early indicator of a slight pathology, we did not observe behavioral abnormalities in our mouse lines [17–19].

(5) Developmental Regulation of GFAP

GFAP mRNA expression is developmentally regulated. Endogenous mRNA levels peak at the first and second postnatal week and decrease into adulthood [58,59], an observation we could confirm in WT mice (Fig. 4A). However, transgenic mRNAs were differently regulated, with decreases of FP mRNAs in FVB(hGFAP-ECFP)$_{GCFD}$ and FVB(hGFAP-EGFP)$_{GFEC}$ in line with the endogenous GFAP mRNA, while mRNA levels of the FP in B6N(hGFAP-EGFP)$_{GFEC}$ and of CT2 in both backgrounds of hGFAP-CT2$_{GCTF}$ increased with age (Fig. 4D and 6C), thereby indicating the presence of different regulatory mechanisms.

Conclusion

The random transgene insertion site underlies local influences of cis-acting regulatory elements that thereby affect the strength of transgenic expression and the high variability of expression patterns among individual founders [33,34,47,60,61]. Additionally, the copy number of transgene insertion could influence the stability of transgenic expression [50]. Here, we show that also changing the inbred strain strongly modulates the activity of the human GFAP promoter. FVB mice showed always higher transgenic activity than B6N mice at the same age. By extended crossing into the FVB or B6N background and vice versa the level of transgene expression could be reversibly (in the time span of generations) modulated.

Although all our mouse lines showed weaker expression in B6N, it is hard to extrapolate whether this occurs in all hGFAP mouse lines. Since this promoter is frequently used to study astrocyte function, we recommend a careful control of the genetic background.

Acknowledgments

The authors are grateful to Frank Rhode for excellent technical assistance, to Daniel Rhode for animal husbandry and to Hongkui Zeng (Allen Institute for Brain Science, Seattle, Washington, USA) for providing R26tdTom reporter mice. We thank Dr. Yvonne Schwarz (Department of Molecular Neurophysiology, University of Saarland, Homburg, Germany) for providing primary astrocyte cultures.

Author Contributions

Conceived and designed the experiments: A. Scheller. Performed the experiments: XB. Analyzed the data: XB. Contributed reagents/materials/analysis tools: A. Saab WH IKH. Wrote the paper: XB A. Scheller FK.

References

1. Bignami A, Eng LF, Dahl D, Uyeda CT (1972) Localization of the glial fibrillary acidic protein in astrocytes by immunofluorescence. Brain Res 43: 429–435.
2. Eng LF, Vanderhaeghen JJ, Bignami A, Gerstl B (1971) An acidic protein isolated from fibrous astrocytes. Brain Res 28: 351–354.
3. Middeldorp J, Hol EM (2011) GFAP in health and disease. Prog Neurobiol 93: 421–443.
4. Malatesta P, Hack MA, Hartfuss E, Kettenmann H, Klinkert W, et al. (2003) Neuronal or glial progeny: regional differences in radial glia fate. Neuron 37: 751–764.
5. Casper KB, McCarthy KD (2006) GFAP-positive progenitor cells produce neurons and oligodendrocytes throughout the CNS. Mol Cell Neurosci 31: 676–684.

6. Goss JR, Finch CE, Morgan DG (1991) Age-related changes in glial fibrillary acidic protein mRNA in the mouse brain. Neurobiol Aging 12: 165–170.
7. Yoshida T, Goldsmith SK, Morgan TE, Stone DJ, Finch CE (1996) Transcription supports age-related increases of GFAP gene expression in the male rat brain. Neurosci Lett 215: 107–110.
8. Eng LF, Ghirnikar RS (1994) GFAP and astrogliosis. Brain Pathol 4: 229–237.
9. Ransom B, Behar T, Nedergaard M (2003) New roles for astrocytes (stars at last). Trends Neurosci 26: 520–522.
10. Pekny M, Nilsson M (2005) Astrocyte activation and reactive gliosis. Glia 50: 427–434.
11. Parpura V, Heneka MT, Montana V, Oliet SH, Schousboe A, et al. (2012) Glial cells in (patho)physiology. J Neurochem 121: 4–27.
12. Besnard F, Brenner M, Nakatani Y, Chao R, Purohit HJ, et al. (1991) Multiple interacting sites regulate astrocyte-specific transcription of the human gene for glial fibrillary acidic protein. J Biol Chem 266: 18877–18883.
13. Masood K, Besnard F, Su Y, Brenner M (1993) Analysis of a segment of the human glial fibrillary acidic protein gene that directs astrocyte-specific transcription. J Neurochem 61: 160–166.
14. Brenner M, Kisseberth WC, Su Y, Besnard F, Messing A (1994) GFAP promoter directs astrocyte-specific expression in transgenic mice. J Neurosci 14: 1030–1037.
15. Zhuo L, Sun B, Zhang CL, Fine A, Chiu SY, et al. (1997) Live astrocytes visualized by green fluorescent protein in transgenic mice. Dev Biol 187: 36–42.
16. Zhuo L, Theis M, Alvarez-Maya I, Brenner M, Willecke K, et al. (2001) hGFAP-cre transgenic mice for manipulation of glial and neuronal function in vivo. Genesis 31: 85–94.
17. Nolte C, Matyash M, Pivneva T, Schipke CG, Ohlemeyer C, et al. (2001) GFAP promoter-controlled EGFP-expressing transgenic mice: a tool to visualize astrocytes and astrogliosis in living brain tissue. Glia 33: 72–86.
18. Hirrlinger PG, Scheller A, Braun C, Quintela-Schneider M, Fuss B, et al. (2005) Expression of reef coral fluorescent proteins in the central nervous system of transgenic mice. Mol Cell Neurosci 30: 291–303.
19. Hirrlinger PG, Scheller A, Braun C, Hirrlinger J, Kirchhoff F (2006) Temporal control of gene recombination in astrocytes by transgenic expression of the tamoxifen-inducible DNA recombinase variant CreERT2. Glia 54: 11–20.
20. Lalo U, Pankratov Y, Kirchhoff F, North RA, Verkhratsky A (2006) NMDA receptors mediate neuron-to-glia signaling in mouse cortical astrocytes. J Neurosci 26: 2673–2683.
21. Gordon JW, Scangos GA, Plotkin DJ, Barbosa JA, Ruddle FH (1980) Genetic transformation of mouse embryos by microinjection of purified DNA. Proc Natl Acad Sci U S A 77: 7380–7384.
22. Taketo M, Schroeder AC, Mobraaten LE, Gunning KB, Hanten G, et al. (1991) FVB/N: an inbred mouse strain preferable for transgenic analyses. Proc Natl Acad Sci U S A 88: 2065–2069.
23. Li HS, Borg E (1991) Age-related loss of auditory sensitivity in two mouse genotypes. Acta Otolaryngol 111: 827–834.
24. Li HS (1992) Influence of genotype and age on acute acoustic trauma and recovery in CBA/Ca and C57BL/6J mice. Acta Otolaryngol 112: 956–967.
25. Willott JF, Aitkin LM, McFadden SL (1993) Plasticity of auditory cortex associated with sensorineural hearing loss in adult C57BL/6J mice. J Comp Neurol 329: 402–411.
26. Madisen L, Zwingman TA, Sunkin SM, Oh SW, Zariwala HA, et al. (2010) A robust and high-throughput Cre reporting and characterization system for the whole mouse brain. Nat Neurosci 13: 133–140.
27. Saab AS, Neumeyer A, Jahn HM, Cupido A, Šimek AA, et al. (2012) Bergmann glial AMPA receptors are required for fine motor coordination. Science 337: 749–753.
28. Shepherd CT, Moran Lauter AN, Scott MP (2009) Determination of transgene copy number by real-time quantitative PCR. Methods Mol Biol 526: 129–134.
29. Karram K, Goebbels S, Schwab M, Jennissen K, Seifert G, et al. (2008) NG2-expressing cells in the nervous system revealed by the NG2-EYFP-knockin mouse. Genesis 46: 743–757.
30. Kim HJ, Magrané J (2011) Isolation and culture of neurons and astrocytes from the mouse brain cortex. Methods Mol Biol 793: 63–75.
31. Hirrlinger J, Scheller A, Hirrlinger PG, Kellert B, Tang W, et al. (2009) Split-cre complementation indicates coincident activity of different genes in vivo. PLoS One 4: e4286.
32. Seeburg PH (1982) The human growth hormone gene family: nucleotide sequences show recent divergence and predict a new polypeptide hormone. DNA 1: 239–249.
33. Feng G, Mellor RH, Bernstein M, Keller-Peck C, Nguyen QT, et al. (2000) Imaging neuronal subsets in transgenic mice expressing multiple spectral variants of GFP. Neuron 28: 41–51.
34. Heim N, Garaschuk O, Friedrich MW, Mank M, Milos RI, et al. (2007) Improved calcium imaging in transgenic mice expressing a troponin C-based biosensor. Nat Methods 4: 127–129.
35. Fukuda S, Taga T (2005) Cell fate determination regulated by a transcriptional signal network in the developing mouse brain. Anat Sci Int 80: 12–18.
36. Takizawa T, Nakashima K, Namihira M, Ochiai W, Uemura A, et al. (2001) DNA methylation is a critical cell-intrinsic determinant of astrocyte differentiation in the fetal brain. Dev Cell 1: 749–758.
37. Namihira M, Kohyama J, Abematsu M, Nakashima K (2008) Epigenetic mechanisms regulating fate specification of neural stem cells. Philos Trans R Soc Lond B Biol Sci 363: 2099–2109.
38. Restrepo A, Smith CA, Agnihotri S, Shekarforoush M, Kongkham PN, et al. (2011) Epigenetic regulation of glial fibrillary acidic protein by DNA methylation in human malignant gliomas. Neuro Oncol 13: 42–50.
39. Weichman K, Chaillet JR (1997) Phenotypic variation in a genetically identical population of mice. Mol Cell Biol 17: 5269–5274.
40. Takizawa T, Gudla PR, Guo L, Lockett S, Misteli T (2008) Allele-specific nuclear positioning of the monoallelically expressed astrocyte marker GFAP. Genes Dev 22: 489–498.
41. Takizawa T, Meshorer E (2008) Chromatin and nuclear architecture in the nervous system. Trends Neurosci 31: 343–352.
42. Cheng PY, Lin YP, Chen YL, Lee YC, Tai CC, et al. (2011) Interplay between SIN3A and STAT3 mediates chromatin conformational changes and GFAP expression during cellular differentiation. PLoS One 6: e22018.
43. Kozak M (1987) An analysis of 5'-noncoding sequences from 699 vertebrate messenger RNAs. Nucleic Acids Res 15: 8125–8148.
44. Proudfoot N, O'Sullivan J (2002) Polyadenylation: a tail of two complexes. Curr Biol 12: R855–857.
45. Proudfoot NJ, Brownlee GG (1976) 3' non-coding region sequences in eukaryotic messenger RNA. Nature 263: 211–214.
46. Choi T, Huang M, Gorman C, Jaenisch R (1991) A generic intron increases gene expression in transgenic mice. Mol Cell Biol 11: 3070–3074.
47. Gama Sosa MA, De Gasperi R, Elder GA (2010) Animal transgenesis: an overview. Brain Struct Funct 214: 91–109.
48. Buchman AR, Berg P (1988) Comparison of intron-dependent and intron-independent gene expression. Mol Cell Biol 8: 4395–4405.
49. Huang MT, Gorman CM (1990) Intervening sequences increase efficiency of RNA 3' processing and accumulation of cytoplasmic RNA. Nucleic Acids Res 18: 937–947.
50. Garrick D, Fiering S, Martin DI, Whitelaw E (1998) Repeat-induced gene silencing in mammals. Nat Genet 18: 56–59.
51. Lee Y, Messing A, Su M, Brenner M (2008) GFAP promoter elements required for region-specific and astrocyte-specific expression. Glia 56: 481–493.
52. Lee Y, Su M, Messing A, Brenner M (2006) Astrocyte heterogeneity revealed by expression of a GFAP-LacZ transgene. Glia 53: 677–687.
53. de Leeuw B, Su M, ter Horst M, Iwata S, Rodijk M, et al. (2006) Increased glia-specific transgene expression with glial fibrillary acidic protein promoters containing multiple enhancer elements. J Neurosci Res 83: 744–753.
54. Messing A, Brenner M (2003) GFAP: functional implications gleaned from studies of genetically engineered mice. Glia 43: 87–90.
55. Brenner M, Messing A (1996) GFAP Transgenic Mice. Methods 10: 351–364.
56. Ganat YM, Silbereis J, Cave C, Ngu H, Anderson GM, et al. (2006) Early postnatal astroglial cells produce multilineage precursors and neural stem cells in vivo. J Neurosci 26: 8609–8621.
57. Su M, Hu H, Lee Y, d'Azzo A, Messing A, et al. (2004) Expression specificity of GFAP transgenes. Neurochem Res 29: 2075–2093.
58. Tardy M, Fages C, Riol H, LePrince G, Rataboul P, et al. (1989) Developmental expression of the glial fibrillary acidic protein mRNA in the central nervous system and in cultured astrocytes. J Neurochem 52: 162–167.
59. Lewis SA, Cowan NJ (1985) Temporal expression of mouse glial fibrillary acidic protein mRNA studied by a rapid in situ hybridization procedure. J Neurochem 45: 913–919.
60. Elder GA, Friedrich VL, Liang Z, Li X, Lazzarini RA (1994) Enhancer trapping by a human mid-sized neurofilament transgene reveals unexpected patterns of neuronal enhancer activity. Brain Res Mol Brain Res 26: 177–188.
61. Dobie K, Mehtali M, McClenaghan M, Lathe R (1997) Variegated gene expression in mice. Trends Genet 13: 127–130.

8

Gut Microbiota Contributes to the Growth of Fast-Growing Transgenic Common Carp (*Cyprinus carpio* L.)

Xuemei Li[1,4,9], Qingyun Yan[1,2,9], Shouqi Xie[2], Wei Hu[2], Yuhe Yu[1]*, Zihua Hu[3]*

1 Key Laboratory of Aquatic Biodiversity and Conservation of Chinese Academy of Sciences, Institute of Hydrobiology, Chinese Academy of Sciences, Wuhan, China, **2** State Key Laboratory of Freshwater Ecology and Biotechnology, Institute of Hydrobiology, Chinese Academy of Sciences, Wuhan, China, **3** Center for Computational Research, New York State Center of Excellence in Bioinformatics and Life Sciences, Department of Ophthalmology, Department of Biostatistics, Department of Medicine, State University of New York at Buffalo, Buffalo, New York, United States of America, **4** Key Laboratory of Freshwater Biodiversity Conservation, Ministry of Agriculture of China, Yangtze River Fisheries Research Institute, Chinese Academy of Fishery Sciences, Wuhan, China

Abstract

Gut microbiota has shown tight and coordinated connection with various functions of its host such as metabolism, immunity, energy utilization, and health maintenance. To gain insight into whether gut microbes affect the metabolism of fish, we employed fast-growing transgenic common carp (*Cyprinus carpio* L.) to study the connections between its large body feature and gut microbes. Metagenome-based fingerprinting and high-throughput sequencing on bacterial 16S rRNA genes indicated that fish gut was dominated by Proteobacteria, Fusobacteria, Bacteroidetes and Firmicutes, which displayed significant differences between transgenic fish and wild-type controls. Analyses to study the association of gut microbes with the fish metabolism discovered three major phyla having significant relationships with the host metabolic factors. Biochemical and histological analyses indicated transgenic fish had increased carbohydrate but decreased lipid metabolisms. Additionally, transgenic fish has a significantly lower Bacteroidetes:Firmicutes ratio than that of wild-type controls, which is similar to mammals between obese and lean individuals. These findings suggest that gut microbiotas are associated with the growth of fast growing transgenic fish, and the relative abundance of Firmicutes over Bacteroidetes could be one of the factors contributing to its fast growth. Since the large body size of transgenic fish displays a proportional body growth, which is unlike obesity in human, the results together with the findings from others also suggest that the link between obesity and gut microbiota is likely more complex than a simple Bacteroidetes:Firmicutes ratio change.

Editor: Kostas Bourtzis, International Atomic Energy Agency, Austria

Funding: This work was supported by the Major State Basic Research Development Program of China (2009CB118705), the National Natural Science Foundation of China (31172084, 31071896), the State Key Laboratory of Freshwater Ecology and Biotechnology (2012FB03), the Knowledge Innovation Program of the Chinese Academy of Sciences (Y15E04), and the Youth Innovation Promotion Association, CAS (Y22Z07). The funders had no role in study design, data collection and analysis, decision to publish, or preparation of the manuscript.

Competing Interests: The authors have declared that no competing interests exist.

* E-mail: yhyu@ihb.ac.cn (YY); zihuahu@ccr.buffalo.edu (ZH)

⍑ These authors contributed equally to this work.

Introduction

Microbes, which colonize animal gut, function collectively as an extra 'organ' for the host. Their community structure is shaped by the combining effects of host genotype, lifestyle, living environments, and selective pressures from gut habitats [1–4]. The genomes of these microbes (microbiome) exceed the size of the host nuclear genome by a few orders of magnitude, contributing to a broad range of functions which have not evolved wholly on the host [5–7]. The genomes of the microbes, unlike its host genome, can dynamically change the configuration of their components to fulfill the needs of the community as a whole and of the host. A growing number of studies have shown gut microbiota has a tight and coordinated connection with host metabolism, energy utilization and storage, immunity and nutritional status, and health maintenance [8–11].

Obesity, which results from the accumulation of excess adipose tissue, presents a good example for illustrating the potential interactions between the mammalian host and its dynamic symbionts. Also, the dynamics of microbial genomic and metabolic diversity are key factors maintaining host's health [8]. The causes driving obesity appear to be complex. A consensus hypothesis is a heterogeneous group of conditions with multiple causes, including behavioral and environmental factors such as a sedentary lifestyle and excessive consumption of energy-dense foods [12]. It has recently been proposed that gut microbiota, as an environmental factor, may shape the host immune network and metabolic activity which in turn alters energy metabolism accompanying the obese state. The potential mechanisms underlying this relationship include increased nutrient absorption from the diet, prolonged intestinal transit time, altered bile acid entero-hepatic cycle, increased cellular uptake of circulating triglycerides, and altered tissue composition of biologically active polyunsaturated fatty acid [13].

Although there have been about 30 bacterial phyla described to date, the development of obesity has most often been associated with significant changes to Bacteroidetes and Firmicutes levels. Obese mice resulting from a high-fat/high-sugar western diet, as compared with mice receiving a low-fat/high-polysaccharide diet, display enrichment in Firmicutes at the expense of Bacteroidetes [14]. Similar to animal models, obese people have a relatively higher proportion of Firmicutes, when compared to lean people. Surgically- or diet-induced weight loss can reduce the proportion of Firmicutes [15–18]. However, this finding is inconsistent from other studies. Duncan et al. have showed proportions of Bacteroidetes and Firmicutes among fecal bacteria have no association in human obesity [19]. In another study, overweight and obese subjects have a ratio of Bacteroidetes to Firmicutes in favor of Bacteroidetes [20]. Recently, Jumpertz et al. investigated dynamic changes of gut microbiota by applying pyrosequencing to examine bacterial 16S rRNA genes and reported no phylum level difference between fecal microbiota of obese and lean subjects [21]. Therefore, the link between obesity and the microbiota is likely more sophisticated than the simple phylum-level Bacteroidetes:-Firmicutes ratio change.

In this report we conducted a series studies to reveal the relationships between gut microbes and the fast-growing feature of transgenic common carp, which was modified with an 'all-fish' growth hormone gene [22,23]. Although the transgenic fish has a large body size stimulated by the recombinant grass carp (*Ctenopharyngodon idellus*) growth hormore gene (*gcGH*) when compared to wild-type controls, unlike obesity in human it displays an isometric body growth. In fact, significantly lower levels of growth hormone receptor (*GHR*) mRNA have been found in adipose tissues of obese human subjects, as compared with the lean human counterparts [24]. Furthermore, mice with growth hormone receptor deficiency (*GHR−/−*) have a greater percent fat mass but with no significant differences in absolute fat mass through the life, and animals with lean mass show an opposite trend [25]. It has been found that the organization of fish intestine is similar to that of mammals, and more importantly many homologous genes which are regulated by gut microbes in mammals show similar expression responses in fish [26]. Therefore the fast-growing transgenic fish provides a good model not only to study the impact of gut microbial communities on the growth of fish but also to investigate if the increase of Firmicutes at the expense of Bacteroidetes is unique to obesity.

By using metagenome-based methods, we found significant differences of gut microbiota composition between transgenic fish and wild-type controls during a two-year field study, while both displayed high degree of similarities within each group. The results were further confirmed by high-throughput sequencing on bacterial 16S rRNA genes. We further extended our study to reveal the association of gut microbes with fish metabolism and discovered three major phyla (Proteobacteria, Bacteroidetes, and Firmicutes) had significant relationships with the host metabolic factors. Furthermore, transgenic fish had increased carbohydrate but decreased lipid metabolisms, which were evidenced by both biochemical and histological analyses. Additionally, we observed that the fast-growing transgenic fish had a significantly lower Bacteroidetes:Firmicutes ratio than that of wild-type controls. The results indicate that the Bacteroidetes:Firmicutes ratio change is not unique to obesity. The results also suggest that the relative abundance of Firmicutes over Bacteroidetes could be one of the factors contributing to the fast growth of transgenic fish, although Bacteroidetes and Firmicutes account for only a small proportion of its gut microbiota.

Materials and Methods

Animals and Ethics Statement

Individuals of transgenic fish (*Cyprinus carpio* L.) at different developmental stages (from larvae to adults) were sampled, and counterparts from wild-type fish were used as controls. All experiments involving animals were performed under protocols approved by the Institutional Animal Care and Use Committee of Institute of Hydrobiology, Chinese Academy of Sciences (Approval ID: keshuizhuan 08529).

Experimental Design and Fish Husbandry

Transgenic fish and wild-type controls were reared with the same commercial feed in ponds at Guanqiao Experimental Station. Individuals sampled from larval stage to adult animals (from April 2009 to March 2011) were used to study the structure and dynamics of gut microbiota. To study nutrient metabolism related to the gut microbiota, four different diets (Table S1) were given to both transgenic and wild-type fish (during the stages of 2-month and 5-month) under laboratory condition. Briefly, the transgenic fish and wild-type controls were transferred to the laboratory at the stage of 2-month, and then acclimated to laboratory condition with a practical diet twice a day (09:00 AM and 16:00 PM) for the first 3 weeks and an equal mixture of the four experimental diets (Table S1) for the 4th week. At the beginning of the laboratory growth experiment, acclimated transgenic fish were weighed after one day of food deprivation and then randomly distributed into 12 tanks (20 individuals for each tank, totally about 60 g). Three tanks were randomly assigned as replicates for each dietary treatment. After a growth trial for six weeks, fish in each tank were also weighed after one day of food deprivation and then were randomly selected for analysis as described in the following sections.

Intestine Sampling Procedures and Bacterial dna Preparation

For larval stages (3- and 6-day post-incubation), intestines were removed aseptically under a dissecting microscope, and three replicated samples for both transgenic fish and wild-type controls were used for investigating the diversity and dynamics of gut microbiota. For each of the late stages five (for individuals at the stages of 2–5 month) or three individuals (at the stages of 8–23 month) from both transgenic fish and wild-type controls were randomly selected and subject to the following procedures. The intestine was first carefully removed under sterile environments. Whole intestinal tract (for individuals at 2–5 month stages) or part of foregut, mid gut, and hind gut (for individuals at of 8–23 month stages) was then collected for subsequent DNA extraction. For fish reared in the laboratory condition three individuals from each diet treatment (one individual from each of triplicate tanks) were randomly selected and whole intestine of each individual were collected as described above.

DNA preparation was performed by incubating intestinal homogenates in 1 ml lysis solution (30 mM EDTA, 10 mMTris-HCl, 0.5% sodium dodecyl sulfate (SDS), 0.1 mg proteinase K, 0.05 mg RNase A) at 55°C bath for 10 h, followed by standard phenol/chloroform extraction and precipitating with cold ethanol as previously described [27].

PCR-DGGE and Sequencing

To amplify bacterial 16S ribosomal RNA gene, PCR reactions (25 µL) were prepared, each containing approximately 1 ng/µl DNA templates, 1×buffer (without $MgCl_2$), 2 mM $MgCl_2$, 0.06 unit/µl*Taq* DNA polymerase, 80 µM of deoxynucleotide triphosphate, and 0.25 µM of each universal bacterial target primer 357F-GC and 518R [28] (Table S2). Touchdown PCRs were performed on a S1000^{TM} thermal cycler (Bio-Rad) with the following conditions: 5 min at 94°C, followed by 10 cycles of 30 sec at 94°C, 30 sec at 66–57°C, and 60 sec at 72°C. This procedure was followed by 20 cycles of 30 sec at 94°C, 30 sec at

56°C and 60 sec at 72°C with a post-amplification extension of 10 min at 72°C. All PCR products were confirmed by agarose gel electrophoresis.

Approximately equal amounts of PCR products were separated by denaturing gradient gel electrophoresis (DGGE) using 9.0% polyacrylamide gel with a 45–70% denaturing gradient. Electrophoresis was performed at 60°C with 100 V for 12 h according to the method described previously [29]. Gels were then stained in 1×TAE buffer containing 1×SYBR Gold (Molecular Probes) for 30 min, followed by photographing with a Gel Doc^{TM} XR imaging system (Bio-Rad). DGGE band types were originally assigned and matched using the Quantity One® software (Bio-Rad, version 4.6.9), and the banding patterns were then manually checked.

Dominant bacterial operation taxonomic units (OTUs) were recovered from DGGE profiles by excising the bands with relatively high density and re-amplified using the same primer pairs without GC-clamp (357F and 518R). The resulting products were visualized using 1.8% agarose gels. Target 16S rRNA gene fragments were excised and purified using agarose gel DNA extraction kit (Axygen), cloned into pMD18-T vector (TaKaRa), and then transformed *Escherichia coli* (DH-5a) with a plasmid. Two positive clones for each OTU were sequenced.

All partial 16S rRNA gene sequences were compared with those in the public Ribosomal Database Project II [30] to ascertain their closest relatives. Neighbor-joining phylogenetic trees were calculated using ClustalX in combination with MEGA (4.0) package [31]. Bootstrap (1000) was performed to evaluate the phylogenetic tree.

Quantifying Firmicutes and Bacteroidetes by Q-PCR

Real-time quantitative PCR (Q-PCR) was used to quantify the relative abundance of gut Firmicutes and Bacteroidetes [32] using standards constructed with known amounts of plasmid DNA. In brief, PCR products of 16S rRNA genes were gel-purified, cloned into pMD18-T vector, and then transformed into *Escherichia coli* cells. After confirming by sequencing, plasmid DNA containing cloned 16S rRNA gene was extracted. The resulting DNA concentrations were determined by spectrophotometry with serial dilutions. Standard curves were then established using diluted plasmid DNA in Q-PCR. The abundance of Firmicutes, Bacteroidetes, and total bacteria in each intestinal sample was evaluated.

The Q-PCR was performed on an ABI 7500 FAST system (Applied biosystems). Each PCR (25 µL) contains 1×SYBR Green qPCR master mix (Shanghai Ruian), 0.2 µM of each primer (Table S2), and 2 µL DNA templates. PCR cycling included an initial denaturation for 2 min at 95°C, followed by 40 cycles of 94°C for 10 sec, 60°C for 40 sec. Fluorescence readings were taken at each extension step, and a final melting analysis was performed to check nonspecific product formation. Three replicates were analyzed for each sample.

Bacterial 16S rRNA Gene Pyrosequencing

The V1-V3 regions, which have more related variations for 16S rRNA gene than shorter sequences or the full-length sequence [33], were amplified using the bacterial primers 27F and 534R (Table S2) with Pyrobest^{TM} DNA polymerase (Takara). The sample-unique 10-base bar-code was add to each primer for sorting of PCR amplicons into different samples, and the underlined text indicates universal bacterial primers. PCR products were purified with the QIAquick Gel Extraction Kit (Qiagen), after quantifying by the Qubit^{TM} Quantitation Platform (Invitrogen), 200 ng product from each sample was pooled for

pyrosequencing by a 454 GS FLX Titanium system (454 Life Sciences/Roche Applied Science) according to the manufacturer's instructions. After pyrosequencing all reads were scored for quality filtering and the sequences that passed quality control were used to pick operational taxonomic units (OTUs). The representative sequence of each OTU was used for taxonomy assignment and generating phylogenetic tree. Alpha- and beta-diversity were also calculated for comparing bacterial communities, clustering and PCA were also performed to visually depict the differences between samples. All these analyses were performed according to the procedures described elsewhere [34]. The pyrosequencing dataset was deposited into European Nucleotide Archive under the accession number ERP002333.

Statistical Analyses

Statistical analyses were performed with the software SPSS, R package, and XLSTAT. The pyrosequencing results were analyzed using the pipeline of QIIME [35] and the Fast UniFrac online toolkit (http://bmf2.colorado.edu/fastunifrac/) [36]. A binary matrix from DGGE band matching data was used to calculate Sørensen similarities for unweighted pair-group method with arithmetic average (UPGMA) clustering. The band patterns were also analyzed using the Raup and Crick probability-based index of similarity (S_{RC}), which provides a measurement of statistically significant similarity and dissimilarity at the 95% confidence level [37]. The similarity index is the probability that the randomized similarity would be less than or equal to the observed similarity, and S_{RC} values above 0.95 or below 0.05 signifies the similarity or differences [38]. The S_{RC} was calculated using the PAST program. In addition, canonical ordination of redundancy analysis (RDA) was performed with Canoco for Windows 4.5 to screen microbial phyla that could significantly predict metabolic characters and to explore the potential relationships between intestinal microbes and host's metabolism. One way ANOVA and two-tailed Student's t-test were performed to assess the differences between transgenic fish and wild-type controls. For the time-series and multiple diet treatments the statistical significance between transgenic and wild-type fish was evaluated using one-side Wilcoxon signed-rank test and one-side paired t-test.

Results

Gut Microbiota Composition Differs between Transgenic Fish (Cyprinus carpio L.) And Wild-type Controls

A two-year field study was performed to explore the similarities and differences in gut microbiota composition between transgenic fish (represented by T in Figures and Tables) and wild-type controls (represented by C in Figures and Tables). Individual fish reared in ponds with the same commercial feed from larval stage to adult animals were used for the comparison of gut microbiota composition. Sørensen similarity based on DGGE patterns of 16S rRNA genes (V3 region) indicated that gut microbiotas in transgenic fish were different from those in wild-type controls as shown in Figure 1, where UPGMA clustering classifies samples of transgenic fish and wild-type controls into two distinct groups in all developmental stages from the field study (only the 17-month developmental stage showed some exception). Analyses were also performed to compare the mean value of Sørensen index either between transgenic fish and wild-type controls (between-group) or within each group of transgenic fish and wild-type controls (within-group). While both within-groups displayed high degree of similarities (0.81±0.07 and 0.74±0.12), the lowest similarities (0.43±0.30) was observed in the between-groups (Figure 2a).

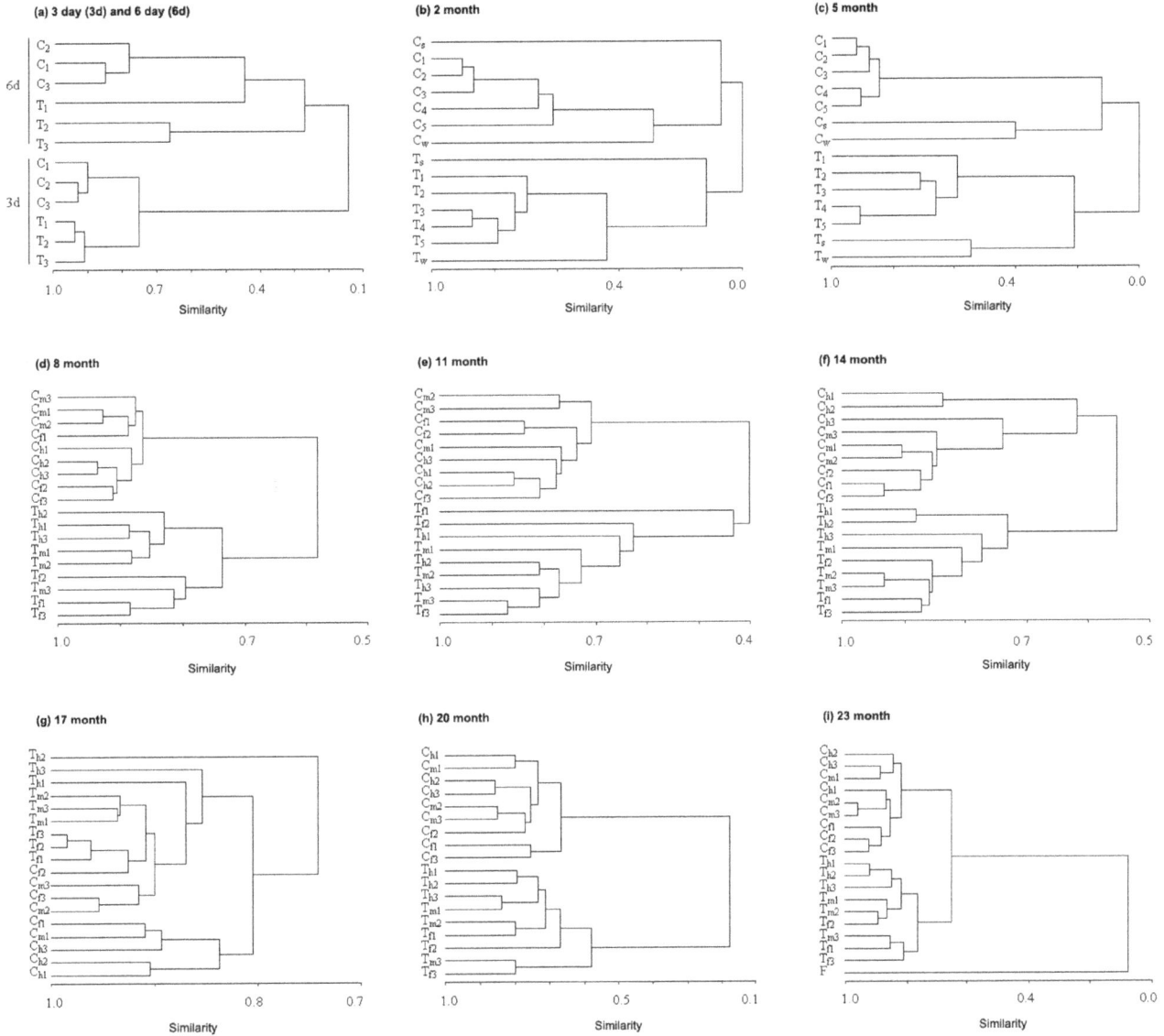

Figure 1. UPGMA clustering over Sørensen similarity of gut microbiota composition between transgenic fish and wild-type controls across different developmental stages. The similarity matrix was calculated using the binary data, and clustering was performed using the unweighted pair-group method with arithmetic average (UPGMA). T_i and C_i indicate the ith replication of transgenic fish and wild-type control, respectively; T_{fi}, T_{mi}, and T_{hi} represent the ith foregut, midgut, and hindgut samples from transgenic fish, respectively, and C_{fi}, C_{mi}, and C_{hi} from the corresponding part of controls; T_s and T_w represent the sediment and water samples collected from the pond where transgenic fish were reared, and C_s and C_w represent controls; F indicates food sample.

Statistical analyses using one-side Wilcoxon signed-rank test revealed significant differences ($p<0.001$) regarding Sørensen index between the within-groups and the between-group comparisons.

The degree of similarity was further assessed using Raup and Crick similarity index (S_{RC}) [37,38]. This probability-based similarity index tells whether the samples are significantly similar ($S_{RC} \geq 0.95$), significantly dissimilar ($S_{RC} \leq 0.05$), or have no significant difference ($0.05 < S_{RC} < 0.95$). Similar to the above findings S_{RC} was ≥ 0.95 for most within-group comparisons except the 17-month developmental stage for transgenic fish and the 14-month as well as 17-month stages for wild-type controls (Figure 2b), suggesting these within-group similarities were mainly driven by deterministic considerations. On the other hand, S_{RC} displayed low similarity between transgenic fish and wild-type controls with

a mean value of 0.42 ± 0.34. Significantly low S_{RC} (<0.05) was observed for the 2-, 5-, and 20-month developmental stages. Further statistical analyses using one-side Wilcoxon singed-rank test revealed significant differences ($p<0.01$) between the within-groups and the between-group. The S_{RC} values between environmental samples (sediment and water samples from the ponds) and corresponding gut samples ranged from 0.05 and 0.95 with a mean value of 0.42 ± 0.30. These findings suggest that gut microbiota composition is not significantly similar to the environments.

To confirm the findings from the DNA fingerprinting approaches and to further study gut microbe phylotypes, 454-pyrosequencing was applied to analyze 16S rRNA genes (V1-V3 regions). Gut samples collected from fish raised under laboratory condition with four different diets (in Figures: C represents control

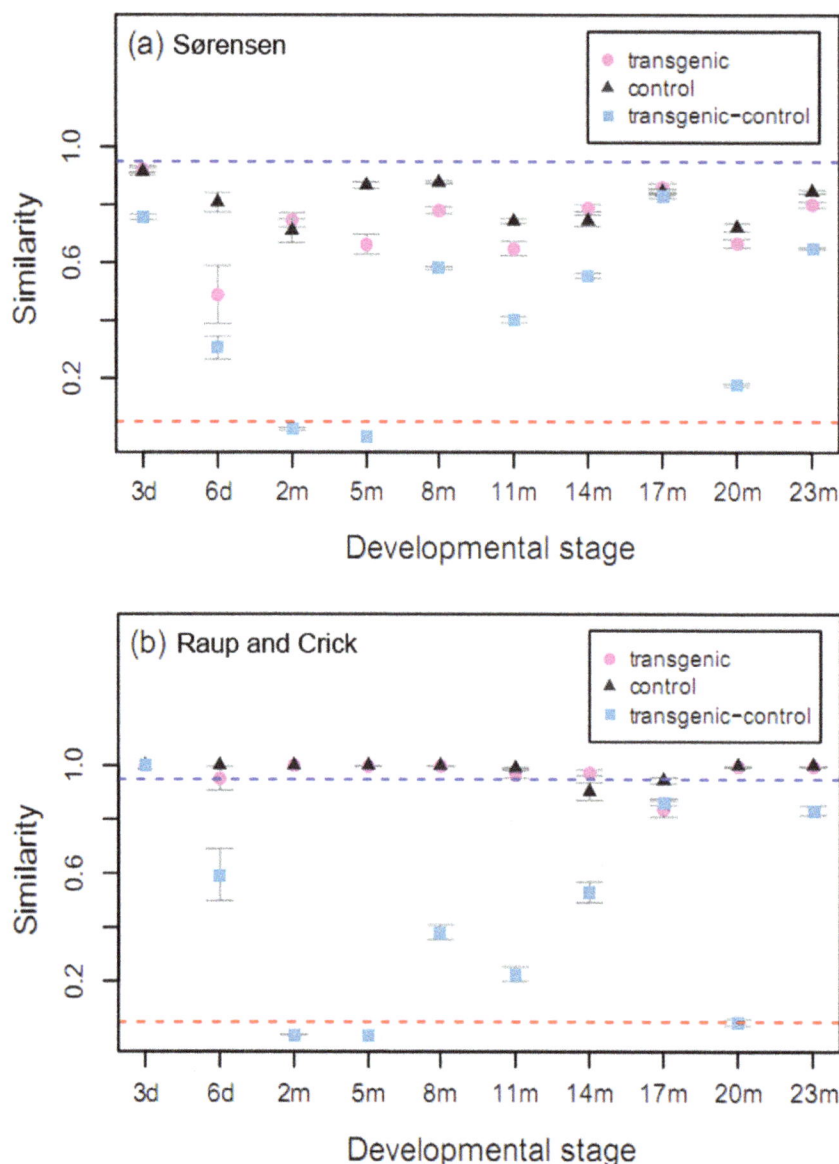

Figure 2. Comparison of similarities and differences for gut microbiota composition between transgenic fish and wild-type controls across different developmental stages. (a) Comparison of the average Sørensen index obtained from DGGE patterns of 16S rRNA genes between transgenic fish and wild-type controls or within each group of transgenic fish and wild-type controls. (b) Comparison of Raup and Crick similarity index (S_{RC}) obtained from DGGE patterns of 16S rRNA genes between transgenic fish and wild-type controls or within each group of transgenic fish and wild-type controls. Dashed lines indicate significant cutoff for difference (low line) and similarity (upper line). Error bars represent the standard error of the mean.

diet, *HP* represents high protein diet, *HC* represents high carbohydrate diet, and *HL* represents high lipid diet, Table S1) were used for this analysis. A total of 621,110 valid bacterial 16S rRNA gene reads were obtained. OTUs at 97% homology cutoff indicated the gut microbiota was dominated by Proteobacteria (59%–87%), Fusobacteria (6%–19%), Bacteroidetes (5%–16%), and Firmicutes (1%–3%). Principal component analysis (PCA plot with UniFrac scaled axis) indicated transgenic and control samples in general showed relatively higher similarities within each group than those of between groups (Figure 3). The difference of gut microbiota composition was also observed from the average OTU counts of the dominating bacterial phyla (Figure S1). The number of members unique to transgenic fish was between 18% (Proteobacteria) and 46% (Actinobacteria).

Alpha-diversity analysis of OTUs from different diet treatments revealed no significant differences ($p>0.05$) between wild-type controls and transgenic fish regarding the ACE, Chao1, and Shannon diversity, although wild-type controls had comparatively larger values than samples of transgenic fish except the high protein diet group (Table 1).

Transgenic Fish has a Low Bacteroidetes:Firmicutes Ratio

Previous studies indicated that obesity in humans and animals might be associated with decreased gut Bacteroidetes:Firmicutes ratio [14–18,39,40]. This characteristic has also been explored here to see whether it is unique to obesity.

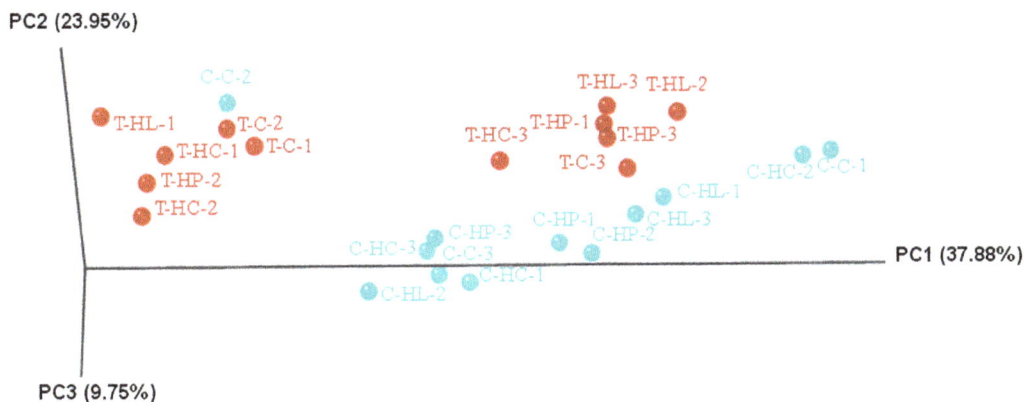

Figure 3. Principal Component Analysis (PCA plot with UniFrac scaled axis) of individual samples with different diet treatments. For each sample code, the first letter *T* represents transgenic fish, and *C* wild-type controls; the middle letter(s) indicates the diet treatments (*C*: control diet, *HP*: high protein diet, *HC*: high carbohydrate diet, *HL*: high lipid diet), and the last numbers represent replicate samples.

Q-PCR quantification of 16S rRNA genes copies for these two bacterial phyla indicated that at the initial developmental stages (3- and 6-day) both transgenic fish (Figure 4a) and wild-type controls (Figure 4b) had very high Bacteroidetes:Firmicutes ratio. The relative Bacteroidetes abundance in transgenic fish decreased dramatically to less than 2% at the 2-month and the 5-month developmental stages. The trend of lower Bacteroidetes:Firmicutes ratio was retained through the rest of the late developmental stages, and the largest Bacteroidetes abundance only accounted for 33% (at the 14-month developmental stage). By contrast, wild-type fish displayed a relatively high Bacteroidetes:Firmicutes ratio. This was especially true at the developmental stages of 2-month, 5-month, and 20-month, for which the relative Bacteroidetes abundance was more than 71%, 46%, and 94%, respectively. One-side paired t-test indicated that the proportion of Bacteroidetes in wild-type controls was significantly larger than those in transgenic fish ($p = 0.04$). For fish raised in laboratory condition the relative abundance of Bacteroidetes and Firmicutes was detected by high-throughput sequencing based on 16S rRNA genes (V1-V3 regions). In agreement with the findings from fish raised in natural conditions, transgenic fish had a much lower Bacteroidetes:Firmicutes ratio than that of wild-type controls (data not shown).

The results indicate that the Bacteroidetes:Firmicutes ratio change is not unique to obesity. Although Bacteroidetes and Firmicutes account for only a small proportion of gut microbiota in fish, the finding from this study suggests that the relative abundance of Firmicutes over Bacteroidetes could be one of the factors contributing to the fast growth of transgenic common carp. The result together with the findings from others 19–21] also

suggest that the link between obesity and gut microbiota is likely more complex than a simple change of Bacteroidetes:Firmicutes ratio.

Proteobacteria, Bacteroidetes, and Firmicutes Display Significant Relationship with Metabolism of the Investigated Common Carp

Canonical ordination of redundancy analysis (RDA) was employed to explore the potential relationships between gut microbiota and host's metabolism. In this analysis, the host metabolic factors resulting from biochemical analysis were used as response variables and gut microbial groups, which were measured by high-throughput sequencing of 16S rRNA genes, as explanatory variables. As shown in Figure 5, three major phyla (Proteobacteria, Bacteroidetes, and Firmicutes) display significant relationships (Monte Carlo test $p<0.05$) with the host metabolic factors, and 96.7% of the response-explanatory variable relation can be significantly explained by the first two axes ($p<0.05$) (Table 2). Moreover, the Bacteroidetes and Firmicutes display a close correlation in predicting the host's metabolism, which is best evidenced by a small angle between these two variables. On the other hand, Proteobacteria is not correlated with Bacteroidetes and Firmicutes in explaining the host's metabolism, as large angles exist between Firmicutes and Proteobacteria as well as between Bacteroidetes and Proteobacteria.

Table 1. Alpha-diversity of gut microbiota calculated according to the composition and relative abundance of OTUs with 97%-identity.

	Control diet treatment		High protein diet treatment		High carbohydrate diet treatment		High lipid diet treatment	
	T	**C**	**T**	**C**	**T**	**C**	**T**	**C**
Chao1	1522.31±236.06	2074.92±604.71	3355.92±790.68	2182.72±498.22	2116.61±44.06	2455.50±349.88	2814.63±443.53	2924.76±656.02
ACE	1597.77±244.93	2205.18±599.29	3408.50±800.13	2155.29±424.72	2108.88±53.83	2568.71±349.81	2888.23±481.03	3026.54±649.95
Shannon	5.83±0.32	6.41±0.69	6.26±0.68	5.82±0.44	5.59±0.45	6.52±0.67	5.67±0.88	6.18±0.31

(a) transgenic

(b) control

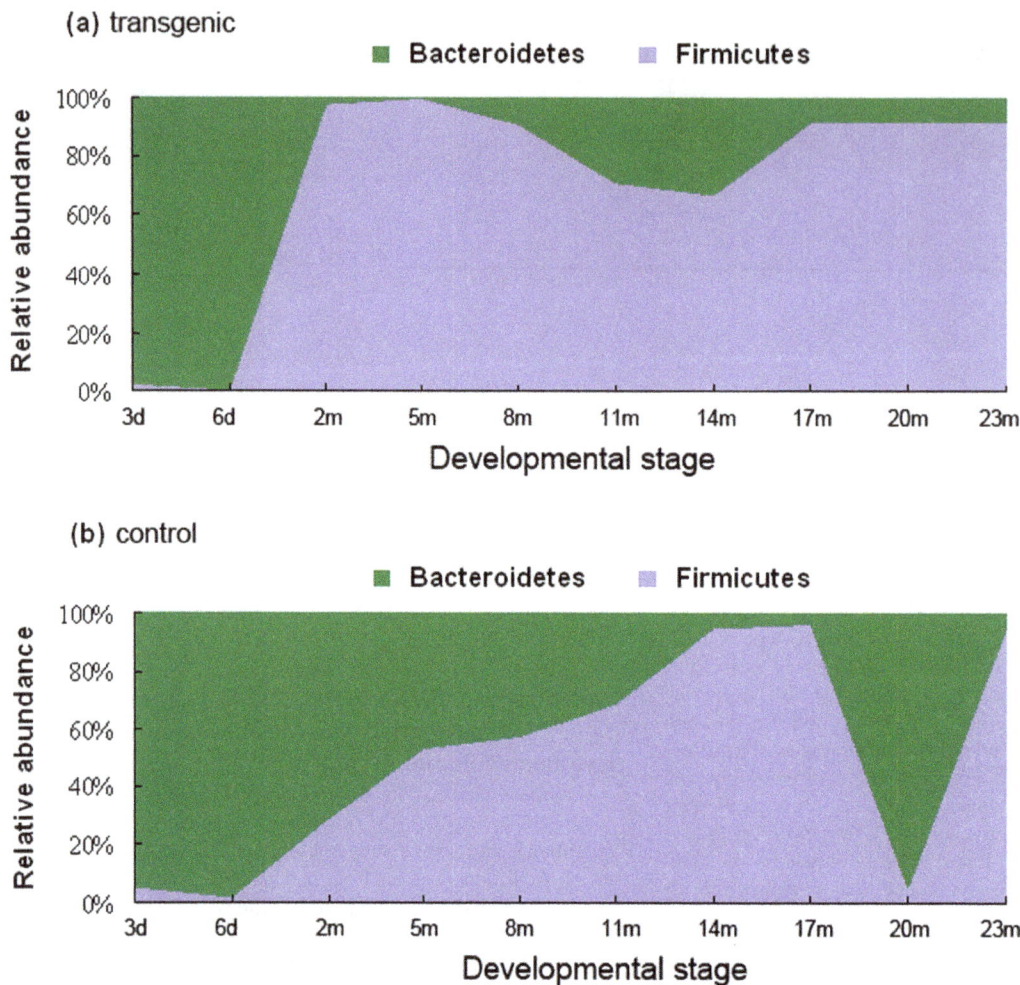

Figure 4. Comparison of relative Bacteroidetes and Firmicutes abundance at different developmental stages. Real-time quantitative PCR (Q-PCR) was used to quantify the abundance of gut Firmicutes and Bacteroidetes based on the 16S rRNA genes (V3 region). (a) Relative abundance of Firmicutes and Bacteroidetes in transgenic fish. (b) Relative abundance of Firmicutes and Bacteroidetes in wild-type controls.

Transgenic Fish Displays Increased Carbohydrate but Decreased Lipid Metabolisms

Physiological and biochemical analyses were performed to study the metabolic differences between transgenic fish and wild-type controls. The results (Table 3) indicated that transgenic fish had significantly lower concentration of glucose (GLU) than that of wild-type controls across all diet treatments and that high carbohydrate diet significantly increased the amount of gut amylase in transgenic fish, suggesting an increased carbohydrate digestion. On the contrary, the lipid-related metabolic parameters such as serum alanine aminotransferase (ALT), aspartate aminotransferase (AST), and triglyceride (TG), which reflect a certain degree of liver dysfunction and represent lipid deposits in the liver, were generally higher in transgenic fish than those of wild-type controls ($p<0.05$), indicating lipid metabolism in transgenic fish was partially disturbed. However, no significant difference was observed for protein metabolism between transgenic fish and wild-type controls. This was evidenced by similar levels of trypsin across 4 types of diets and of all serum indices (except GLU). Supporting these findings, histological analyses of liver tissues revealed that samples from transgenic fish had enlarged cells with increased

amount of lipid droplets, while liver tissues from wild-type fish harbored more glycogen deposits (Figure 6).

Discussion

Gut microbiota is a complex community of microorganisms that colonizes the gastrointestinal tract of animals. A growing number of studies have shown gut microbiotas, which are involved in energy harvest and storage as well as in a variety of metabolic functions such as fermenting and absorbing undigested carbohydrates [41], are especially important for host's metabolism [8–11]. Changes in gut microbiota composition are associated with many diseases such as celiac disease [42], austim [43], and obesity [15,17]. The latter is most likely associated with the changes of two microbiota divisions Bacteroidetes and Firmicutes.

The present study on the difference of gut microbiota composition between wild-type and fast-growing transgenic fish allows us to: i) investigate the associations between gut microbes and the metabolism of fish and ii) address the question whether the change of microbiota Bacteroidetes over Firmicutes is unique to obesity in human. The comparison of gut microbiota between transgenic and wild-type fish provides an analogy but different comparison to obese and lean people in terms of body weight and

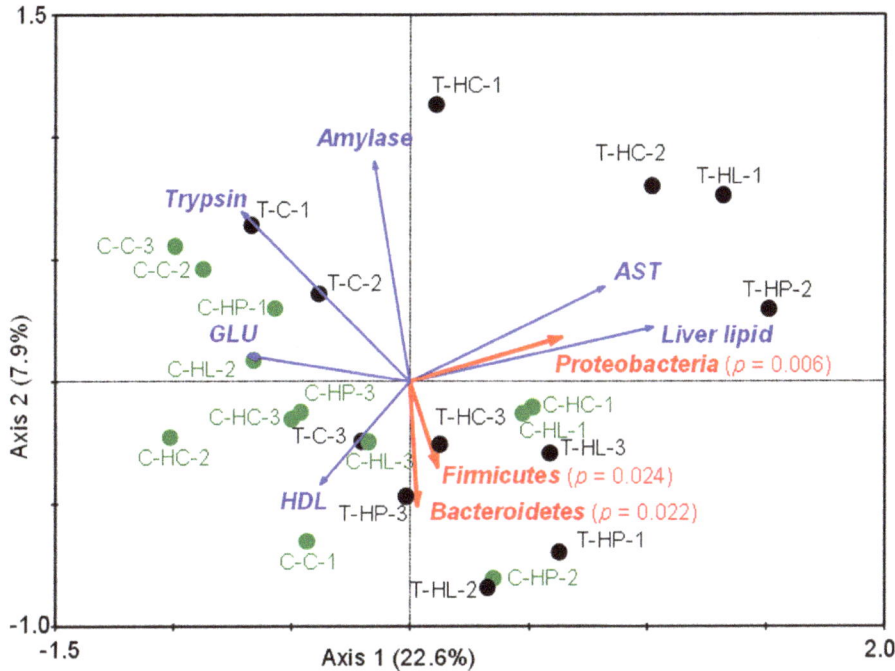

Figure 5. Triplot of the redundancy analysis (RDA) showing significant relationship between metabolic-related factors (response variables) and microbial groups (explanatory variables). First and second ordination axes were plotted, representing 22.6% and 7.9% of the variability in the data set, respectively. P-values obtained by Monte Carlo test were reported. For each sample code, the first letter *T* represents transgenic fish, and *C* wild-type controls; the middle letter(s) indicates the diet treatments (*C*: control diet, *HP*: high protein diet, *HC*: high carbohydrate diet, *HL*: high lipid diet), and the last numbers represent replicated samples. GLU represents glucose, AST aspartate aminotransferase, and HDL high-density lipoprotein.

size. Unlike obesity in humans the large body size of transgenic fish displays a proportional body growth, which has been demonstrated by previous reports [22,23] and was also confirmed in this study (Figure S2). Different factors might contribute to the fast-growing nature of transgenic fish such as expression of the integrated growth hormone gene [44], increased food intake [45], improved feed efficiency [46], and altered fish behaviors [47,48].

A core microbiota comprised of Proteobacteria, Fusobacteria, Bacteroidetes, and Firmicutes was identified herein. Proteobacteria accounted for the largest proportion of gut microbiota, which is in agreement with the results from other fish [49,50]. We, however, observed significant differences of gut microbiota

Table 2. Summary statistics of redundancy analysis (RDA) showing the relationships between gut microbiota and host metabolism by canonical axes with associated *p* values from Monte Carlo test.

Axes	1	2	3	4
Eigenvalues	0.226	0.079	0.010	0.279
Response-explanatory variable correlation	0.700	0.558	0.337	0.000
Cumulative percentage variance				
of response data	22.6	30.5	31.5	59.4
of response-explanatory variables relation	71.8	96.7	100.0	0.0
Significance test of the first canonical axis (p value)	0.012			
Significance test of all canonical axes (p value)	0.006			

composition between transgenic fish and wild-type controls. It is also very important to note that the largest differences between transgenic fish and wild-type controls are during the early developmental stages (during 2-month and 5-month), for which both Sørensen and S_{RC} indices display significant differences. The results were further confirmed by high-throughput sequencing on bacterial 16S rRNA genes. These findings suggest that gut microbiota composition is most likely filtered to be different between transgenic fish and controls at particular developmental stages.

One purpose of this study is to explore the association between gut microbes and fish metabolism, in turn, their growth. Transgenic fish have a significantly faster growth rate, when compared to wild-type controls. This is true for fish reared in both nature (ponds) and laboratory conditions (tanks). This fast-growing characteristic is correlated with gross energy intake and growth energy, both of which are significantly more than those from wild-type fish (Figure S2). This leads to an interesting question as to whether gross energy intake for fast-growing transgenic fish is linked to gut microbiotas. We found 3 out of the 4 core microbiotas were associated with host's metabolism. This has been illustrated by canonical ordination of redundancy analysis, from which three microbiotas Proteobacteria, Bacteroidetes, and Firmicutes present significant relationships (Monte Carlo test $p<0.05$) with host metabolic characteristics. We also noticed that Bacteroidetes displayed a close correlation with Firmicutes in predicting host metabolism. Therefore, other than the important role played by the dominating phylum Proteobacteria, both Bacteroidetes and Firmicutes could have some impact on fish metabolism. Further analyses are needed to reveal whether the

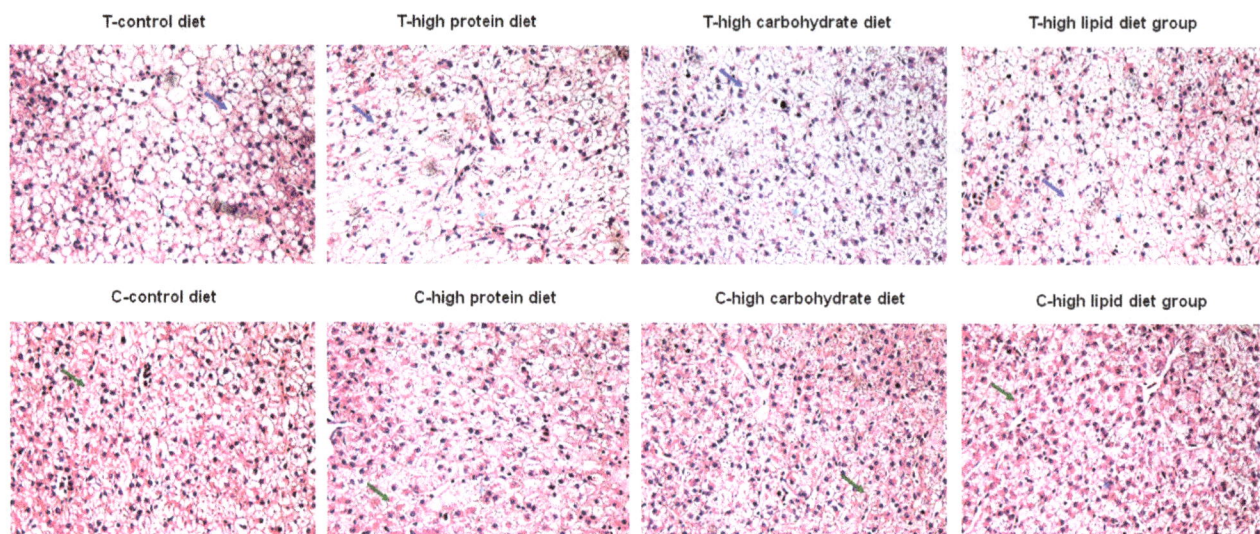

Figure 6. Histological analyses of liver tissues. Liver tissues from 4 different diet treatments were formaldehyde-fixed, followed by staining with Hematoxylin and eosin. The results (75×) from transgenic fish are shown in the upper panels, and wild-type controls with corresponding diet treatments in the lower panels. Arrows in upper panels indicate lipid droplet (blue) and those in lower panels (green) show glycogen deposits.

large body size of transgenic fish is the consequence of the gut microbiota change or the cause for the change of gut microbiota.

These findings not only indicate microbiotas play very important roles in host metabolism but also suggest a close connection between gross energy intake and fast-growing feature

of transgenic fish. The large gross energy intake in transgenic fish might come from increased carbohydrate metabolisms, which was demonstrated by both biochemical and histological analyses. While transgenic fish had significantly lower concentrations of glucose than those of wild-type fish, high carbohydrate diet

Table 3. Comparison of metabolic differences between transgenic and wild-type fish.

	Control diet	High protein	High carbohydrate	High lipid
Whole-fish lipid (wet weight)	0.04 (T>C)	0.02 (T>C)	NS	0.01 (T>C)
Whole-fish protein (wet weight)	NS	NS	0.01 (T<C)	NS
Whole-fish ash (wet weight)	NS	NS	NS	NS
Whole-fish water	0.02 (T<C)	0.03 (T<C)	NS	0.02 (T<C)
Whole-fish energy	NS	NS	NS	NS
Muscle lipid (wet weight)	NS	0.04 (T<C)	NS	NS
Muscle protein (wet weight)	0.00 (T<C)	NS	NS	NS
Muscle ash (wet weight)	NS	0.01 (T<C)	NS	NS
Muscle water	0.01 (T>C)	0.02 (T>C)	0.04 (T>C)	0.02 (T>C)
Muscle energy	NS	NS	NS	NS
Liver lipid (wet weight)	0.04 (T>C)	0.01 (T>C)	NS	0.01 (T>C)
Amylase	NS	NS	0.04 (T>C)	NS
Trypsin	NS	NS	NS	NS
Serum glucose	0.02 (T<C)	0.01 (T<C)	0.02 (T<C)	0.02 (T<C)
Serum cholesterol	NS	NS	NS	NS
Serum triglyceride	NS	NS	0.01 (T>C)	NS
Serum alanine aminotransferase	0.00 (T>C)	NS	0.00 (T>C)	0.00 (T>C)
Serum aspartate aminotransferase	0.03 (T>C)	NS	0.00 (T>C)	0.04 (T>C)
Serum high-density lipoprotein	0.01 (T<C)	NS	0.04 (T<C)	0.03 (T<C)
Serum low-density lipoprotein	0.04 (T<C)	NS	0.00 (T<C)	NS

The numbers represent the significant p values, and the subsequent parenthesis show in which directions the differences are. NS indicates not significant. The letter T in parenthesis represents transgenic fish, and C wild-type control.

significantly increased the amount of gut amylase. Histological analysis of liver sections also revealed liver tissues from wild-type fish harbored more glycogen deposits. These results indicated transgenic fish could take the advantage of carbohydrate diets for the fast growth. This may be partially attributed from the high efficiency of polysaccharide fermentation by Firmicutes [15,51,52], as more Firmicutes were detected in the gut of transgenic individuals.

Another important question is whether the increase of Firmicutes at the expense of Bacteroidetes is unique to mammalian obesity. Although previous studies have shown that the increased proportion of Firmicutes have a direct connection with the development of obesity [14–18], other studies have differing conclusions [19–21]. The present study therefore could shed light on these contradicted findings, as the gut microbiota between transgenic fish and wild-type control provide an analogy but different comparison to obese and lean mammalian individual in terms of body weight and size. Similar to the results from mammals, both RTQ-PCR quantification of 16S rDNA (V3 region) copies and high-throughput sequencing on 16S rDNA (V1-V3 regions) reveal lower Bacteroidetes:Firmicutes ratio in transgenic fish. These results demonstrate that the change of microbiota Bacteroidetes over Firmicutes is also true to the fast-growing transgenic fish. Since transgenic fish has a proportional body growth which is unlike obese individual in mammals, the finding therefore suggests that the link between obesity and microbiota is likely more complex than the simple Bacteroidetes:Firmicutes ratio change [17].

Further evidence comes from similar mechanism for excessive energy harvest in both transgenic fish and obese mammals, for which gut microbiota affects body weight by increasing energy harvest from dietary fibers. Metagenomic and biochemical analyses have revealed that mouse gut microbiota is enriched with bacterial genes capable of fermenting dietary fibers [52]. The notion of changing energy harvest by gut microbiota has also been explored in human by Jumpertz et al. [21], who tested whether microbiota in lean and obese individuals were correlated with the efficiency of dietary energy harvest. They found that the changes of gut microbiota were directly correlated with stool energy loss in lean individuals and that a 20% increase in Firmicutes and a corresponding decrease in Bacteroidetes were associated with an increased energy harvest. Therefore, excessive calories from fiber by microbiota metabolism could be one of the important factors contributing to obese state, which is in good agreement with gut microbes affecting the metabolism of vertebrate fish from this study. These results indicate that the relative abundance of Firmicutes over Bacteroidetes could be one of the factors contributing to the fast growth of transgenic common carp, even though they account for only a small proportion of the total gut microbiota.

References

1. Ley RE, Hamady M, Lozupone C, Turnbaugh PJ, Ramey RR, et al. (2008) Evolution of mammals and their gut microbes. Science 320: 1647–1651.
2. Li X, Yu Y, Feng W, Yan Q, Gong Y (2012) Host species as a strong determinant of the intestinal microbiota of fish larvae. J Microbiol 50: 29–37.
3. Rawls JF, Mahowald MA, Ley RE, Gordon JI (2006) Reciprocal gut microbiota transplants from zebrafish and mice to germ free recipients reveal host habitat selection. Cell 127: 423–433.
4. Yan Q, van der Gast CJ, Yu Y (2012) Bacterial community assembly and turnover within the intestines of developing zebrafish. PLoS ONE 7: e30603.
5. Bäckhed F, Manchester JK, Semenkovich CF, Gordon JI (2007) Mechanisms underlying the resistance to diet-induced obesity in germ-free mice. Proc Natl Acad Sci U S A 104: 979–984.
6. O'Hara AM, Shanahan F (2006) The gut flora as a forgotten organ. EMBO Rep 7: 688–693.

Supporting Information

Figure S1　Venn diagrams displaying similarity and difference for all phyla and 4 major bacterial phyla between transgenic fish and wild-type controls. The number of shared members is listed in the middle, the number of members unique to transgenic fish is shown on the left, and that unique to wild-type controls is indicated on the right. T represents transgenic fish and C wild-type controls.

Figure S2　Comparison of body weight and energy intake between the fast-growing transgenic fish (T) and wild-type controls (C). (a) Body weight comparison between transgenic fish and wild-type controls reared in ponds. (b) Body weight comparison between transgenic fish and wild-type controls raised in laboratory tanks. (c) Comparison of condition factor ($100 \times$ weight/length^3) between transgenic fish and wild-type controls reared in ponds. (d) Comparison of condition factor between transgenic fish and wild-type controls raised in laboratory tanks. (e) Gross energy intake comparison between transgenic fish and wild-type controls raised in laboratory tanks. (f) Growth energy comparison between transgenic fish and wild-type controls raised in laboratory tanks. I_E and G_E represent gross energy intake and growth energy, respectively. W_i stands for the initial value of weight and W_f the final weight. Asterisks indicate significant differences for the comparisons obtained from two-tailed Student's t-test (* stands for $p<0.05$ and ** $p<0.005$).

**Table S1　**Formulation and chemical composition of the experimental diet.

**Table S2　**Oligonucleotide sequences of all PCR primers used in the study.

Acknowledgments

We thank Drs. Christopher J. van der Gast (NERC Centre for Ecology and Hydrology) and Gail M. Seigel (State University of New York at Buffalo) for critical reading and comments to the manuscript. We also thank Xiaoming Zhu, Yunxia Yang and Junyan Jin (Institute of Hydrobiology, Chinese Academy of Sciences) for their kind help in the laboratory fish husbandry. Thanks also to Jiajia Ni, Lili Dai, Shu Wu, Jinjin Li, Hongjuan Hu and Chun Wang for their help in the experiment.

Author Contributions

Conceived and designed the experiments: YY QY XL. Performed the experiments: XL QY. Analyzed the data: QY ZH XL. Contributed reagents/materials/analysis tools: QY XL SX WH. Wrote the paper: QY ZH.

7. Qin J, Li R, Raes J, Arumugam M, Burgdorf KS, et al. (2010) A human gut microbial gene catalogue established by metagenomic sequencing. Nature 464: 59–65.
8. Bäckhed F (2011) Programming of host metabolism by the gut microbiota. Ann Nutr Metab 58: 44–52.
9. Nicholson JK, Holmes E, Wilson ID (2005) Gut microorganisms, mammalian metabolism and personalized health care. Nat Rev Microbiol 3: 431–438.
10. Stevens CE, Hume ID (1998) Contributions of microbes in vertebrate gastrointestinal tract to production and conservation of nutrients. Physiol Rev 78: 393–427.
11. Velagapudi VR, Hezaveh R, Reigstad CS, Gopalacharyulu P, Yetukuri L, et al. (2010) The gut microbiota modulates host energy and lipid metabolism in mice. J Lipid Res 51: 1101–1112.
12. Friedman JM (2004) Modern science versus the stigma of obesity. Nat Med 10: 563–569.

13. Musso G, Gambino R, Cassader M (2011) Interactions between gut microbiota and host metabolism predisposing to obesity and diabetes. Annu Rev Med 62: 361–380.

14. Turnbaugh PJ, Bäckhed F, Fulton L, Gordon JI (2008) Diet-induced obesity is linked to marked but reversible alterations in the mouse distal gut microbiome. Cell Host Microbe 3: 213–223.

15. Ley RE, Turnbaugh PJ, Klein S, Gordon JI (2006) Microbial ecology: human gut microbes associated with obesity. Nature 444: 1022–1023.

16. Santacruz A, Marcos A, Wärnberg J, Marti A, Martin-Matillas M, et al. (2009) Interplay between weight loss and gut microbiota composition in overweight adolescents. Obesity 17: 1906–1915.

17. Turnbaugh PJ, Hamady M, Yatsunenko T, Cantarel BL, Duncan A, et al. (2009) A core gut microbiome in obese and lean twins. Nature 457: 480–484.

18. Zhang H, DiBaise JK, Zuccolo A, Kudrna D, Braidotti M, et al. (2009) Human gut microbiota in obesity and after gastric bypass. Proc Natl Acad Sci U S A 106: 2365–2370.

19. Duncan SH, Lobley GE, Holtrop G, Ince J, Johnstone AM, et al. (2008) Human colonic microbiota associated with diet, obesity and weight loss. Int J Obes 32: 1720–1724.

20. Schwiertz A, Taras D, Schafer K, Beijer S, Bos NA, et al. (2010) Microbiota and SCFA in lean and overweight healthy subjects. Obesity 18: 190–195.

21. Jumpertz R, Le DS, Turnbaugh PJ, Trinidad C, Bogardus C, et al. (2011) Energy-balance studies reveal associations between gut microbes, caloric load, and nutrient absorption in humans. Am J Clin Nutr 94: 58–65.

22. Zhu Z (1992) Growth hormone gene and the transgenic fish. In: You CB, Chen ZL, editors. Agricultural Biotechnology. Beijing: China Science and Technology Press. 106–116.

23. Zhu Z (1992) Generation of fast growing transgenic fish: Methods and mechanisms. In: Hew CL, Fletcher GL, editors. Transgenic Fish. Singapore: World Scientific Publishing. 92–119.

24. Erman A, Veilleux A, Tchernof A, Goodyer CG (2011) Human growth hormone receptor (GHR) in obesity: I. GHR mRNA expression in omental and subcutaneous adipose tissues of obese women. Int J Obes 35: 1511–1519.

25. Berryman DE, List EO, Palmer AJ, Chung MY, Wright-Piekarski J, et al. (2010) Two-year body composition analyses of long-lived GHR null mice. J Gerontol A Biol Sci Med Sci 65: 31–40.

26. Rawls JF, Samuel BS, Gordon JI (2004) Gnotobiotic zebrafish reveal evolutionarily conserved responses to the gut microbiota. Proc Natl Acad Sci U S A 101: 4596–4601.

27. Li XM, Yu YH, Xie SQ, Yan QY, Chen YH (2011) Effect of chitosan on intestinal bacteria of allogynogenetic crucian carp, Carassius auratus gibelio, as depicted by polymerase chain reaction-denaturing gradient gel electrophoresis. J World Aquacult Soc 42: 539–548.

28. Muyzer G, de Waal EC, Uitterlinden AG (1993) Profiling of complex microbial populations by denaturing gradient gel electrophoresis analysis of polymerase chain reaction-amplified genes coding for 16S rRNA. Appl Environ Microbiol 59: 695–700.

29. Yu Y, Yan Q, Feng W (2008) Spatiotemporal heterogeneity of plankton communities in Lake Donghu, China, as revealed by PCR-denaturing gradient gel electrophoresis and its relation to biotic and abiotic factors. FEMS Microbiol Ecol 63: 328–337.

30. Cole JR, Chai B, Farris RJ, Wang Q, Kulam SA, et al. (2005) The Ribosomal Database Project (RDP-II): sequences and tools for high-throughput rRNA analysis. Nucleic Acids Res 33: D294–296.

31. Tamura K, Dudley J, Nei M, Kumar S (2007) MEGA4: Molecular Evolutionary Genetics Analysis (MEGA) software version 4.0. Mol Biol Evol 24: 1596–1599.

32. Guo X, Xia X, Tang R, Zhou J, Zhao H, et al. (2008) Development of a real-time PCR method for Firmicutes and Bacteroidetes in faeces and its application to quantify intestinal population of obese and lean pigs. Lett Appl Microbiol 47: 367–373.

33. Schloss PD (2010) The effects of alignment quality, distance calculation method, sequence filtering, and region on the analysis of 16S rRNA gene-based studies. PLoS Comput Biol 6: e1000844.

34. Caporaso JG, Lauber CL, Walters WA, Berg-Lyons D, Lozupone CA, et al. (2011) Global patterns of 16S rRNA diversity at a depth of millions of sequences per sample. Proc Natl Acad Sci U S A 108: 4516–4522.

35. Caporaso JG, Kuczynski J, Stombaugh J, Bittinger K, Bushman FD, et al. (2010) QIIME allows analysis of high-throughput community sequencing data. Nat Methods 7: 335–336.

36. Hamady M, Lozupone C, Knight R (2010) Fast UniFrac: facilitating high-throughput phylogenetic analyses of microbial communities including analysis of pyrosequencing and PhyloChip data. ISME J 4: 17–27.

37. Raup DM, Crick RE (1979) Measurement of faunal similarity in paleontology. Journal Paleontol 53: 1213–1227.

38. Rowan AK, Snape JR, Fearnside D, Barer MR, Curtis TP, et al. (2003) Composition and diversity of ammonia-oxidising bacterial communities in wastewater treatment reactors of different design treating identical wastewater. FEMS Microbiol Ecol 43: 195–206.

39. Costello EK, Gordon JI, Secor SM, Knight R (2010) Postprandial remodeling of the gut microbiota in Burmese pythons. ISME J 4: 1375–1385.

40. Semova I, Carten JD, Stombaugh J, Mackey LC, Knight R, et al. (2012) Microbiota regulate intestinal absorption and metabolism of fatty acids in the zebrafish. Cell Host & Microbe 12: 277–288.

41. Gill SR, Pop M, Deboy RT, Eckburg PB, Turnbaugh PJ, et al. (2006) Metagenomic analysis of the human distal gut microbiome. Science 312: 1355–1359.

42. Elinav E, Strowig T, Kau AL, Henao-Mejia J, Thaiss CA, et al. (2011) NLRP6 inflammasome regulates colonic microbial ecology and risk for colitis. Cell 145: 745–757.

43. Robinson CJ, Bohannan BJ, Young VB (2010) From structure to function: the ecology of host-associated microbial communities. Microbiol Mol Biol Rev 74: 453–476.

44. Wang YP, Hu W, Wu G, Sun YH, Chen SP, et al. (2001) Genetic analysis of 'all-fish' growth hormone gene transferred carp (Cyprinus carpio L.) and its F1 generation. Chin Sci Bull 46: 1175–1179.

45. Fu C, Li D, Hu W, Wang Y, Zhu Z (2007) Growth and energy budget of F2 'all-fish' growth hormone gene transgenic common carp. J Fish Biol 70: 347–361.

46. Guan B, Hu W, Zhang TL, Wang YP, Zhu ZY (2008) Metabolism traits of 'all-fish' growth hormone transgenic common carp. Aquaculture 284: 217–223.

47. Duan M, Zhang TL, Hu W, Li ZJ, Sundström LF, et al. (2011) Behavioral alterations in GH transgenic common carp may explain enhanced competitive feeding ability. Aquaculture 317: 175–181.

48. Li DL, Fu CZ, Hu W, Zhong S, Wang YP, et al. (2007) Rapid growth cost in "all-fish" growth hormone gene trainsgenic carp: reduced critical swimming speed. Chin Sci Bull 52: 1501–1506.

49. Roeselers G, Mittge EK, Stephens WZ, Parichy DM, Cavanaugh CM, et al. (2011) Evidence for a core gut microbiota in the zebrafish. ISME J 5: 1595–1608.

50. Wu S, Wang G, Angert ER, Wang W, Li W, et al. (2012) Composition, diversity, and origin of the bacterial community in grass carp intestine. PLoS ONE 7: e30440.

51. Bäckhed F, Ding H, Wang T, Hooper LV, Koh GY, et al. (2004) The gut microbiota as an environmental factor that regulates fat storage. Proc Natl Acad Sci U S A 101: 15718–15723.

52. Turnbaugh PJ, Ley RE, Mahowald MA, Magrini V, Mardis ER, et al. (2006) An obesity-associated gut microbiome with increased capacity for energy harvest. Nature 444: 1027–1031.

9

Detection of a Soluble Form of CD109 in Serum of CD109 Transgenic and Tumor Xenografted Mice

Hiroki Sakakura[1,2], **Yoshiki Murakumo**[1,3], **Shinji Mii**[1], **Sumitaka Hagiwara**[2], **Takuya Kato**[1], **Masato Asai**[1], **Akiyoshi Hoshino**[1], **Noriyuki Yamamoto**[2], **Sayaka Sobue**[4], **Masatoshi Ichihara**[4], **Minoru Ueda**[2], **Masahide Takahashi**[1]*

1 Department of Pathology, Nagoya University Graduate School of Medicine, Nagoya, Aichi, Japan, 2 Department of Oral and Maxillofacial Surgery, Nagoya University Graduate School of Medicine, Nagoya, Aichi, Japan, 3 Department of Pathology, Kitasato University School of Medicine, Sagamihara, Kanagawa, Japan, 4 Department of Biomedical Sciences, College of Life and Health Sciences, Chubu University, Kasugai, Aichi, Japan

Abstract

CD109, a glycosylphosphatidylinositol-anchored glycoprotein, is expressed at high levels in some human tumors including squamous cell carcinomas. As CD109 is reportedly cleaved by furin and its soluble form is secreted into culture medium *in vitro*, we hypothesized that CD109 could serve as a tumor marker *in vivo*. In this study, we investigated CD109 as a novel serum tumor marker using transgenic mice that overexpress mouse CD109 (mCD109-TG mice) and tumor xenografted mice inoculated with human CD109 (hCD109)-overexpressing HEK293 cells. In sera and urine of mCD109-TG mice, mCD109 was detected using western blotting. In xenografted mice, hCD109 secreted from inoculated tumors was detected in sera, using western blotting and CD109 ELISA. Concentrations of tumor-secreted CD109 increased proportionally as tumors enlarged. Concentrations of secreted CD109 decreased notably by 17 h after tumor resection, and became undetectable 48 h after resection. The half-life of tumor-secreted CD109 was about 5.86 ± 0.17 h. These results indicate that CD109 is present in serum as a soluble form, and suggest its potential as a novel tumor marker in patients with cancers that express CD109.

Editor: Pierre Busson, Institute of Cancerology Gustave Roussy, France

Funding: This work was supported by Grants-in-Aid for Global Center of Excellence (GCOE) research, Scientific Research (A) commissioned by the Ministry of Education, Culture, Sports, Science and Technology (MEXT) of Japan (to MT), by Scientific Research (C) commissioned by MEXT of Japan (to YM). The funders had no role in study design, data collection and analysis, decision to publish, or preparation of the manuscript.

Competing Interests: The authors have declared that no competing interests exist.

* E-mail: mtakaha@med.nagoya-u.ac.jp

Introduction

CD109, a glycosylphosphatidylinositol (GPI)-anchored cell-surface glycoprotein, is a member of the α2-macroglobulin/C3,C4,C5 family [1–4]. CD109 was identified as a cell-surface antigen expressed in KG1a acute myeloid leukemia cells, fetal and adult CD34$^+$ bone marrow mononuclear cells, activated platelets, activated T lymphoblasts, leukemic megakaryoblasts, mesenchymal stem cell subsets and endothelial cells [5–7]. We previously reported that, whereas CD109 is expressed in only limited cell types in normal human and mouse tissues, including myoepithelial cells of the breast, salivary, lacrimal, and bronchial secretary glands; basal cells of the prostate and bronchial epithelia; and basal to suprabasal layers of epidermis [8–13], its expression is frequently detected in several tumor tissues, including squamous cell carcinomas (SCCs) of the oral cavity, esophagus, lung and uterus, basal-like breast carcinoma, malignant melanoma of the skin, and urothelial carcinoma of the bladder, using immunohistochemical studies with anti-CD109 antibody [8–12], [14–16]. High expression of CD109 is also frequently detected in premalignant squamous epithelial lesions, and was associated with differentiation of SCCs in the oral cavity [14]. Reportedly, CD109 is a component of the TGF-β1 receptor system, and negatively regulates TGF-β1 signaling [17]. CD109 has been shown to be cleaved by furin into two forms, a 180-kDa soluble form and a 25-kDa membrane-attached form; the 180-kDa soluble form is secreted from the cell surface into culture medium *in vitro*. Processing CD109 into 180-kDa and 25-kDa proteins is necessary in regulating TGF-β1 signaling [18]. Moreover, release of CD109 from cell surfaces, or the addition of recombinant CD109, downregulates TGF-β1 signaling and TGF-β1 receptor expression in human keratinocytes [19]. Proteomics study indicates that CD109 is released from some tumor cell lines and related to TGF-β signaling *in vitro* [20]. Thus, if a soluble form of CD109 secreted from tumors is detectable in body fluid, it could be a novel marker for malignant and premalignant lesions.

In this study, we used transgenic mice that express exogenous CD109 and xenografted mice inoculated with HEK293 cells that overexpress CD109, and found a soluble form of serum CD109 that increases proportionally with the volume of xenografted tumors. Our findings suggest that CD109 is a potential tumor marker.

Materials and Methods

Ethics Statement

All animal protocols were approved by the Animal Care and Use Committee of Nagoya University Graduate School of Medicine (Approval ID number: 25004).

Antibodies

Anti-CD109-C-9 mouse monoclonal antibody (mAb), which detects 180-kDa N-terminal fragment of human and mouse CD109, was purchased from Santa Cruz Biotechnology (Santa Cruz, CA, USA; Fig. 1A). Anti-CD109-11H3 mAb, which detects 25-kDa C-terminal fragment of human CD109 was kindly provided by Immuno-Biological Laboratories Co., Ltd. (IBL, Gunma, Japan; Fig. 1A). Anti-CD109-6G1 mAb and anti-CD109-8H1 mAb, which detect 180-kDa N-terminal fragment of human CD109, but not mouse CD109, and were used for CD109 Enzyme-Linked Immunosorbent Assay (ELISA) (kindly provided by IBL). Anti-FLAG M2 mAb, anti-FLAG rabbit polyclonal antibody (pAb) and anti-β-actin mAb were purchased from Sigma (St Louis, MO, USA).

Cells and cell culture conditions

HEK293 (derived from human embryonic kidney) cells were maintained in DMEM supplemented with 8% FBS at 37°C in 5% CO_2 condition.

Vector construction and generation of stable transfectants

FLAG-tagged human *CD109* (*FLAG-hCD109*) cDNA was cloned into pcDNA3.1(+) and used in transfection experiments as described previously [10]. Figure 1A shows a schematic illustration of FLAG-hCD109. The stable transfectants were obtained after Geneticin (Invitrogen, Carlsbad, Ca, USA) selection (HEK293-FLAG-hCD109 and HEK293-VC).

Animals

All mice were housed in a specific pathogen-free facility. Their cages contained hardwood chip bedding at 25°C on a 12-h light/dark cycle.

Generation of mouse *CD109* transgenic (mCD109-TG) mice

The pCAGGS vector was kindly provided by Dr. J. Miyazaki, Osaka University Graduate School of Medicine [21]. Mouse *CD109* (*mCD109*) cDNA that lack the native leader sequence was inserted into pSRαCHFX vector to generate *FLAG*- and *His*-tagged *mCD109* cDNA with sequences encoding CD8α signal peptide [8], [22]. This epitope-tagged *mCD109* was amplified by PCR with primers (forward : 5'-TTTTGGCAAAGAATTCTG-CAGGCCACCATGGCCTT-3' and reverse : 5'-CCTGAG-GAGTGAATTCTCAATGTTGCACAAAGTAC-3') and inserted between the two *Eco*RI sites of pCAGGS possessing the *CAG* promoter using In-Fusion HD Cloning Kit (Clontech, Palo Alto, CA, USA) (Fig. 1B). The construct was digested with *Sal*I and *Hind*III, purified and microinjected into fertilized eggs of (C57BL/6×DBA/2) F1 (BDF1) mice in Research Laboratory for Molecular Genetics of Yamagata University School of Medicine. We established three lines of mCD109-TG mice (TG-203, 206 and 213). The incorporation of the transgene was screened by PCR using primers (forward: 5'-GGCCGGATTACAAGGACGAT-3' and reverse : 5'-GAGTGTGAGCACCCGAAACTT -3'). The transgenic mice were back-crossed for more than seven generations into the C57BL/6 strain and used for all experiments. Their nontransgenic littermates (WT-mice) were used as controls.

Establishment of HEK293 xenografted mice

Female BALB/c-*nu/nu* mice were obtained from Charles River Japan (Kanagawa, Japan). Eight to 10 week-old-mice were used for HEK293 xenografted mice experiments. HEK293-FLAG-hCD109 and HEK293-VC cells were both adjusted to a concentration of 1.0×10^7 cells suspended in 100 μl serum-free DMEM. The cell suspensions with 80 μl matrigel (Becton Dickinson, Bedford, MA, USA) were then injected subcutaneously into right flanks of BALB/c-*nu/nu* mice (HEK293-FLAG-hCD109, N = 7; -VC, N = 7).

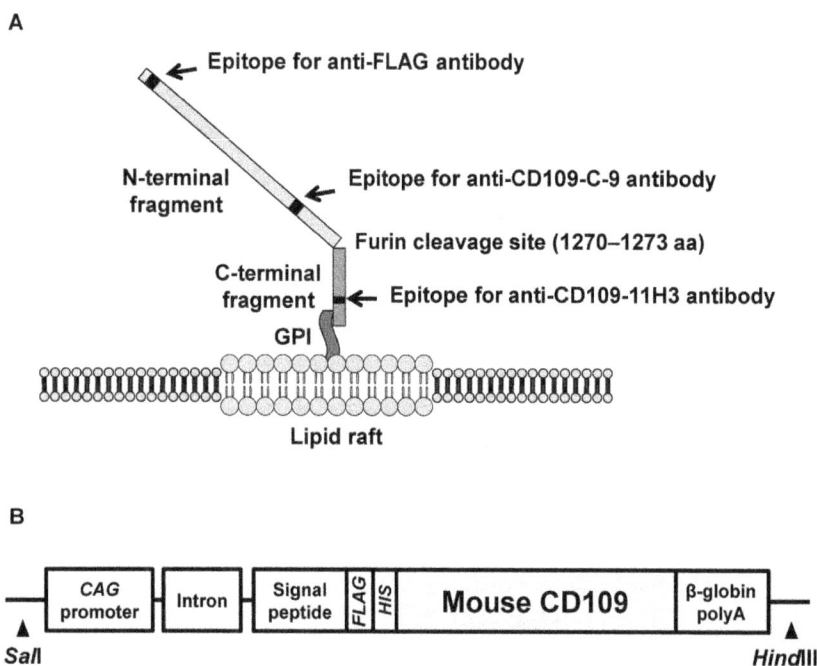

A

B

Figure 1. Structure of CD109 cell-surface glycoprotein. A, Schematic illustration of FLAG-tagged human CD109 (FLAG-hCD109) structure on cytoplasmic membrane. Anti-CD109-C-9 mAb, anti-FLAG mAb and anti-FLAG pAb can detect 180-kDa N-terminal fragments; anti-CD109-11H3 mAb can detect 25-kDa C-terminal fragments. B, Schematic illustration of construction of *FLAG*-tagged mouse *CD109* (*FLAG-mCD109*) transgene.

Tumor development was followed in individual mice every 7 days by sequential caliper measurements of length (L) and width (W). Tumor volume was calculated as volume $= L \times W^2 \times \pi/6$. Developed tumors were resected 42 days after xenografts. Using general anesthesia (sevoflurane) and local anesthesia (0.2% lidocaine), tumor tissue was excised with skin; wounds were then sutured. Resected tissues were cut into 5-mm^3 specimens and quickly frozen for protein extraction or fixed in 10% neutral-buffered formalin for histological analysis.

Preparation of serum samples

Blood from mice was collected from tail veins or retro-orbital sinuses under general anesthesia. Blood collection was performed every 7 days before tumor resection and 17, 48, 72 and 168 h after tumor resection in xenografted mice. Sera was separated by centrifugation ($2000 \times$ g for 15 min) at 4°C and stored at −80°C until analysis.

Immunoprecipitation of CD109 in the sera of xenografted mice

Sera of xenografted mice were diluted at 1:50 in TBS buffer (50 mM Tris-HCl, 150 mM NaCl, pH 7.6). Samples were incubated for 4 h at 4°C with 20 μl Anti-FLAG M2 Affinity Gel beads (Sigma). Beads were washed 3 times with TBS buffer, bead-bound immune complexes were resuspended in 50 μl of $2 \times$ SDS sample buffer (62.5 mM Tris-HCl, pH 6.8, 2% SDS, 25% glycerol, 20 μg/ml bromophenol blue) containing 2% β-mercaptoethanol, and boiled at 100°C for 2 min. After removing beads by centrifugation, samples were subjected to western blotting.

Western blotting

Frozen tissues were homogenized by TissueRuptor (QIAGEN, Hilden, Germany) in $5 \times$ SDS sample buffer (175 mM Tris-HCl, pH 6.8, 5% SDS, 25% glycerol, 50 μg/ml bromophenol blue) and sonicated until no longer viscous. After measuring protein concentration using the DC protein Assay Kit (Bio-Rad Laboratories, Hercules, CA, USA), lysates were boiled at 100°C for 2 min in the presence of 2% β-mercaptoethanol. Serum samples of mCD109-TG and xenografted mice were diluted at 1:10 with PBS; equal volumes of $2 \times$ SDS sample buffer were then added to the diluted samples. They were boiled at 100°C for 2 min in the presence of 2% β-mercaptoethanol and subjected to western blotting. Urine samples were collected from mCD109-TG mice and desalted by Bio-Spin 6 column (Bio-Rad Laboratories) following manufacturer's instructions. Five×SDS sample buffer was added to the desalted samples, which were then boiled at 100°C for 2 min in the presence of 2% β-mercaptoethanol. Samples containing equal amounts of protein were applied to SDS-PAGE and transferred to polyvinylidene fluoride membranes (Millipore Corporation, Bedford, MA, USA). Membranes were blocked with Blocking One (Nacalai Tesque, Kyoto, Japan) for 1 h at RT, and then incubated with primary antibodies at 4°C overnight. After washing three times with TBST buffer (20 mM Tris-HCl, pH 7.6, 137 mM NaCl, 0.1% Tween 20), membranes were incubated with secondary antibodies conjugated to horseradish peroxidase (HRP) (Dako, Kyoto, Japan) for 1 h at RT. After washing membranes three times with TBST, antigen-antibody reaction was visualized using the ECL Detection Kit (GE Healthcare, Buckinghamshire, UK).

Immunohistochemistry

Tumor tissues of xenografted mice were resected as described above. The tissues were fixed in 10% neutral-buffered formalin, dehydrated and embedded in paraffin. Sections (4-μm thick) were prepared for hematoxylin and eosin (HE) staining and immuno-histochemistry. For immunohistochemistry, slides were deparaffinized in xylene and rehydrated in a graded ethanol series. For antigen retrieval, they were immersed into Target Retrieval Solution, pH 9.0 (Dako) and heated for 30 min at 98°C in a water bath. Non-specific binding was blocked with 10% normal goat serum for 20 min at RT. Sections were incubated with primary antibodies for 90 min at RT. Endogenous peroxidase was inhibited with 3% hydrogen peroxide in PBS for 15 min. Slides were incubated with secondary antibody conjugated to HRP-labeled polymer (EnVision+; Dako) for 25 min at RT. Reaction products were visualized with diaminobenzidine (Dako). Nuclei were counterstained with hematoxylin.

Cell proliferation assay

Equal numbers of cells were plated in 96-well plates (2×10^3 cells per well) with 100 μl of DMEM with 4% FBS. Cell proliferation assay was conducted 24 h after seeding using WST-1 Reagent (Roche Diagnostics, Mannheim, Germany) according to manufacturer's protocol. Absorbance was measured at 440 nm every 24 h using a PowerScan4 microplate reader (DS Pharma Medical Co. Ltd., Osaka, Japan).

CD109 ELISA

Serum levels of CD109 from xenografted mice were assessed using the CD109 ELISA kit (kindly provided from IBL) according to manufacturer's instructions. The ELISA kit detects human CD109 but not mouse CD109. All serum samples were diluted at 1:50 in dilution buffer; 100 μl of the individual samples and each standard were added to anti-human-CD109-6G1 mAb-coated testing plate wells and incubated at 4°C overnight. After incubation, samples were aspirated and washed 4 times with washing buffer. Then 100 μl anti-CD109-8H1 mAb conjugated to HRP-labeled solution was added and incubated for 1 h at 4°C. Wells were then washed 5 times with washing buffer, and 100 μl tetramethyl benzidine solution was added to each well and incubated for 30 min at room temperature. Finally, 100 μl stop solution was added to each well. After the reaction, absorbance of each sample was measured at 450 nm/570 nm using a Power-Scan4 microplate reader (DS Pharma Medical Co.). Concentration of CD109 was calculated on the basis of a standard curve.

Calculation of half-life of CD109 in sera of HEK293-FLAG-hCD109 xenografted mice

Half-life ($T_{1/2}$) of CD109 in sera of HEK293-FLAG-hCD109 xenografted mice was calculated using the following equations:

$$C_1 = C_B \exp(-k \times t)$$

$$T_{1/2} = \ln(2)/k$$

in which C_1: concentration of CD109 at 17 h after tumor resection; C_B: concentration of CD109 before tumor resection; k: rate constant of elimination; and t: time after tumor resection (17 h in this case) [23].

Statistical analysis

Data are presented as mean ± SE. Statistical significance was determined with Tukey–Kramer's HSD test of one-factor factorial ANOVA using KaleidaGraph 4.0 for Windows (Synergy Software, Reading, PA, USA). $P < 0.05$ was considered significant.

Results

CD109 is present as a soluble molecule *in vivo*

After establishment of mCD109-TG mice, we assessed their tissue distribution of mCD109 expression. Whole lysates prepared from various organs of TG and WT siblings of mCD109-TG (TG-206) mice were analyzed for endogenous and exogenous mCD109 expression by western blotting with anti-FLAG pAb and anti-CD109-C-9 mAb. In lysates from a WT mouse, two bands were detected of ~160 kDa and ~190 kDa from testis and skin, using anti-CD109-C-9 mAb but not anti-FLAG pAb (Fig. 2A). The 160-kDa and 190-kDa bands are thought to represent N-terminal fragments of CD109 cleaved by furin and full-length CD109

before cleavage, respectively. Two bands with similar molecular masses were detected in lysates of heart, lung, esophagus, stomach, colon, spleen, pancreas, bladder, testis, ovary, uterus and skin of TG mice using both anti-CD109-C-9 mAb and anti-FLAG pAb, indicating expression of exogenous FLAG-mCD109 in mCD109-TG mice. (Fig. 2A).

We then analyzed the presence of soluble CD109 in sera and urine of TG mice, using western blotting. In serum samples from TG mice, both anti-FLAG pAb and anti-CD109-C-9 mAb recognized a major band of ~160 kDa (Fig. 2B). However, expression of the same band was very low in samples from WT mice, as detected by anti-CD109-C-9 mAb. In TG mice urine samples, 160- and 190-kDa bands were detected with both anti-

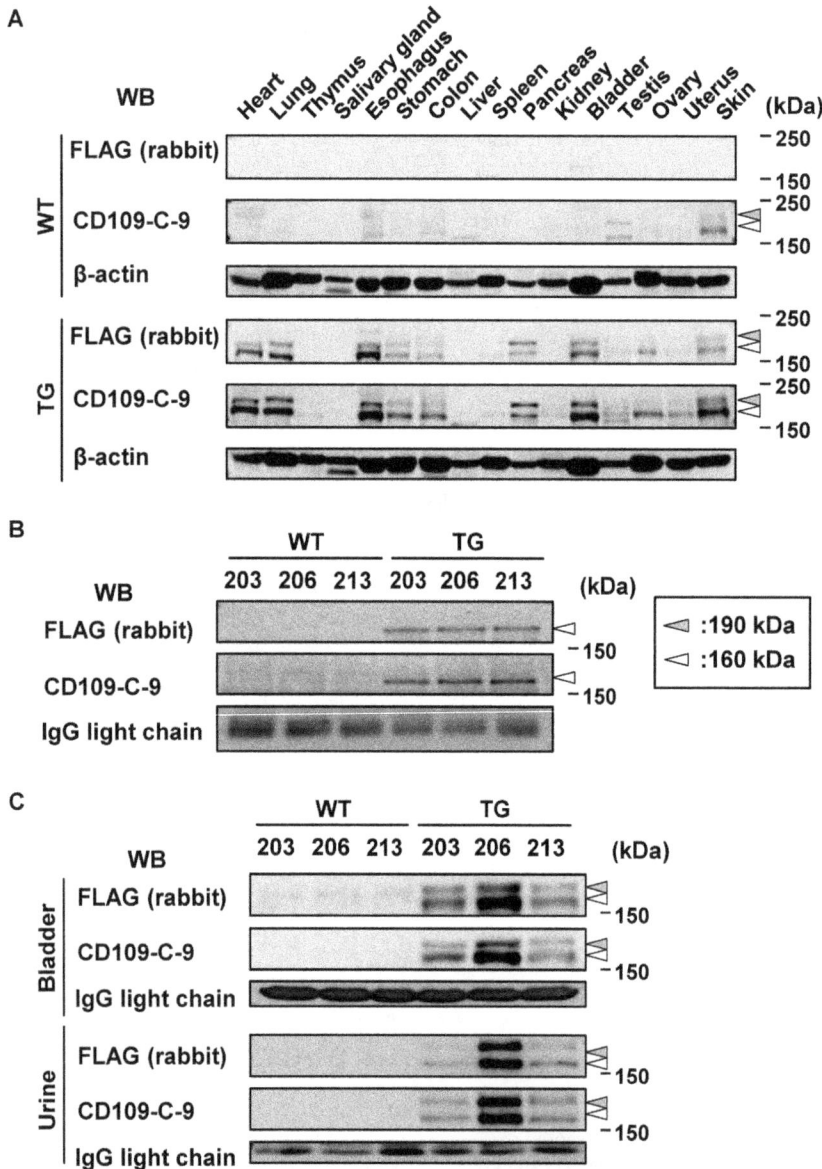

Figure 2. Detection of soluble CD109 in serum and urine of mCD109-TG mice. A, Expression of CD109 in various tissues of mCD109-TG (TG-206) mice. Blots of anti-β-actin antibody are shown as internal controls. White and gray arrowheads: 160- and 190-kDa bands, respectively. B, Detection of soluble CD109 in serum of mCD109-TG mice. White arrowheads:160-kDa bands. C, Expression of CD109 in bladder, and soluble CD109 in urine, of mCD109-TG mice. White and gray arrowheads:160- and 190-kDa bands, respectively. Sera, bladder tissues and urine of WT and TG siblings of 3 lines of mCD109-TG mice (TG-203, 206, and 213) were analyzed using western blots with the indicated antibodies; blots of anti-β-actin antibody and IgG light chain were used as internal controls.

FLAG pAb and anti-CD109-C-9 mAb, and in bladder lysate, but these bands could not be detected in samples from WT mice (Fig. 2C). Both 160-kDa and 190-kDa CD109 appear to be secreted from bladder tissue. These findings indicated that CD109 is a soluble molecule in serum and urine.

Characterization of HEK293 stable transfectants

Before starting the xenografted mouse experiments, we characterized HEK293-FLAG-hCD109 and -VC cell lines, which were used for the xenografted mouse experiments. Expression of hCD109 in cell lysates and amount of secreted hCD109 in the culture media of these cells was assessed using western blotting with anti-FLAG, -CD109-C-9, and -CD109-11H3 antibodies. Two bands of ~180 kDa and ~190 kDa were detected in lysate of HEK293-FLAG-hCD109 cells using anti-FLAG and -CD109-C-9 mAbs. As previously reported, the 180-kDa band represents N-terminal fragments of CD109 cleaved by furin (Fig. 1A), and the 190-kDa band represents an immature glycosylated form of full-length CD109 [18]. The difference in molecular mass between mouse (160 kDa) and human (180 kDa) CD109 may be due to differences of glycosylation. A single 180-kDa band, which represents N-terminal fragment after the cleavage, was detected in culture medium using the same antibodies (Fig. 3A). A small amount of 180-kDa protein was also detected in cell lysate of HEK293-VC cells using anti-CD109-C-9 mAb, which was thought to be the endogenous CD109 protein. Membrane-attached 25-kDa CD109 C-terminal fragment (Fig. 1A) was detected in lysates of HEK293-FLAG-hCD109 and -VC cells using anti-CD109-11H3 mAb; interestingly, a 25-kDa CD109 fragment was also faintly detected in culture medium of HEK293-FLAG-hCD109, suggesting that small amounts of C-terminal 25-kDa CD109 fragment was released from cell surfaces to culture medium. We also evaluated cell proliferation of HEK293 transfectants (Fig. 3B). No significant difference in cell proliferation was observed between HEK293-FLAG-hCD109 and -VC cell lines.

Establishment of HEK293-FLAG-hCD109 xenografted mice

To verify whether CD109 secreted from tumors is detectably present in body fluids, we established xenografted mice using HEK293-FLAG-hCD109 and -VC cells. The schedule of the xenografted mouse experiment is schematically illustrated in Figure 4A. Xenografted cells injected in BALB/c nude mice were allowed to grow for 42 days; tumors derived from the xenografted cells were resected. The developed tumors were round, movable and sharply marginated. No significant difference was found between tumors from HEK293-FLAG-hCD109 and -VC cells in gross pathological findings or growth rates (Fig. 4B,C). No apparent adhesion or invasion to adjacent tissues was observed in any tumors derived from either cell line. It is unclear why there was no difference in tumor growth between HEK293-FLAG-hCD109 and -VC xenografts, although CD109 is reportedly a negative regulator of TGF-β signaling.

The resected tumor tissue was evaluated by H&E staining and immunohistochemistry. The HEK293-FLAG-hCD109 tumors showed strong expression of exogenous FLAG-hCD109 when immunostained with anti-FLAG pAb, anti-CD109 C-9 and -CD109-11H3 mAbs. However, no histological differences between HEK293-FLAG-hCD109 and -VC tumors were observed in the H&E-stained sections (Fig. 4D). Cell lysates of tumor tissues were analyzed by western blotting with anti-FLAG pAb, anti-CD109-C-9 and -CD109-11H3 mAb, which confirmed high CD109 expression in HEK293-FLAG-hCD109 tumor tissues compared with that in HEK293-VC tumor tissues (Fig. 4E).

Tumor-secreted CD109 is detectable in serum and associated with tumor volume in xenografted mice

Next, we evaluated whether FLAG-hCD109 protein released from developed tumors is detectable in serum of xenografted mice. Serum samples from xenografted mice were subjected to western blotting with anti-FLAG pAb and -CD109-C-9 mAb. Tumor-secreted FLAG-hCD109 protein in sera of xenografted mice was detected by western blotting, as was FLAG-mCD109 protein in sera of mCD109-TG mice (Fig. 5A), whereas FLAG-hCD109 could not be detected in the urine (data not shown). Tumor volume and serum CD109 concentrations were analyzed over time. Serum CD109 concentration assessed by western blotting with anti-CD109-C-9 mAb increased after Day 28, whereas assessment by western blotting combined with immunoprecipitation using Anti-FLAG M2 Affinity Gel showed increased FLAG-hCD109 after Day 14 (Fig. 5B). CD109 ELISA, which is specific for human CD109, used for quantitative estimation of serum FLAG-hCD109, detected logarithmic increase of hCD109 concentration after Day 14, when the xenografted tumors were also logarithmically growing (Figs. 4C, 5C). These results indicate that tumor-secreted CD109 in serum proportionally increases with tumor volume (Fig. 5D).

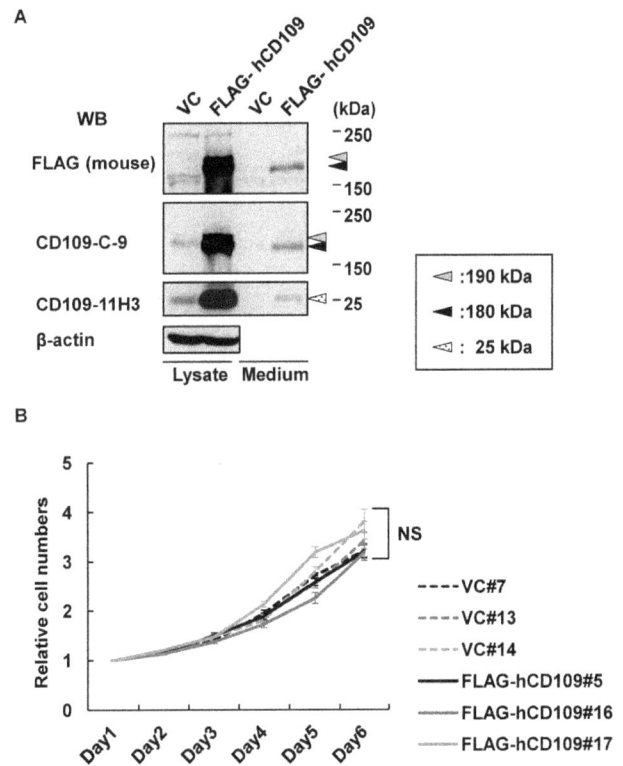

Figure 3. Characterization of HEK293-FLAG-hCD109 and -VC cell lines. A, Expression of CD109 in total cell lysates and detection of soluble CD109 in culture media of HEK293-FLAG-hCD109 and -VC cell line. Total cell lysates and culture media were subjected to western blotting using the indicated antibodies; blots of anti-β-actin antibody shown as internal controls. B, Cell proliferation analysis of HEK293-FLAG-hCD109 and -VC cell lines. Absorbance values at day 1 are defined as 1.0. NS: not significant, using the Tukey–Kramer's HSD test.

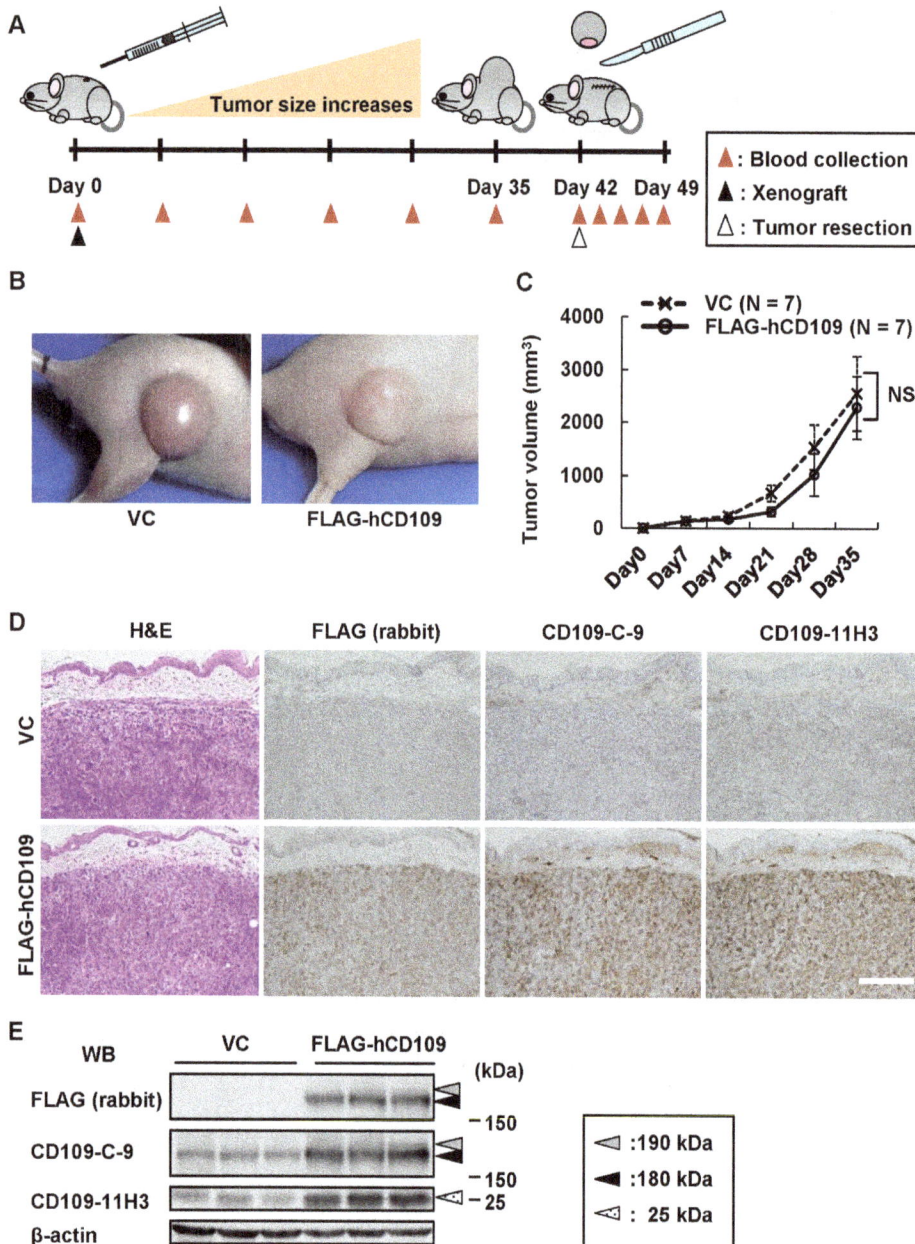

Figure 4. Characterization of tumors developed in xenografted mice. A, Schematic illustration of the schedule of xenografted mice experiments using HEK293-FLAG-hCD109 and -VC cell lines. Blood was collected every 7 days after the xenograft until tumor resection, and 17, 48, 72 and 168 h after tumor resection. B, Gross appearance of xenografted tumors in mice. C, Growth curves of xenografted tumors. Seven xenografted tumors of HEK293-FLAG-hCD109 and seven xenografted tumors of HEK293-VC were analyzed for tumor volume, as indicated in Materials and Methods. NS: not significant, using Tukey–Kramer's HSD test. D, Histopathological appearance of xenografted tumors. Sections of xenografted tumors resected at Day 42 were subjected to H&E staining, and immunohistochemical staining with the indicated antibodies. Scale bar: 200 μm. E, Expression of CD109 in xenografted tumors. Total cell lysates from three of each xenografted tumor groups of HEK293-FLAG-hCD109 and -VC cell lines were analyzed for CD109 expression by western blotting. Dotted-white, black and gray arrowheads: 25-, 180- and 190-kDa bands, respectively. Blots of anti-β-actin antibody shown as internal control.

Tumor-secreted CD109 in the serum rapidly decreases after tumor resection

Next, we evaluated whether tumor-secreted CD109 in the serum of HEK293-FLAG-hCD109 xenografted mice decreases after tumor resection. Forty-two days after the xenograft, tumor resections were performed as described in Materials and Methods, and serum samples were analyzed for CD109 by western blotting and ELISA. Serum CD109 concentration notably decreased 17 h

after resection compared with that before resection. CD109 was undetectable 48, 72 and 168 h after the operation (Fig. 6A,B). Average concentrations of CD109 before the operation and 17 h afterwards were 521.9 ± 81.4 ng/ml and 70.1 ± 11.5 ng/ml, respectively. Assuming that CD109 has a monophasic elimination pattern, the half-life of tumor-secreted CD109 calculated from these results was 5.86 ± 0.17 h, which indicates that tumor-secreted

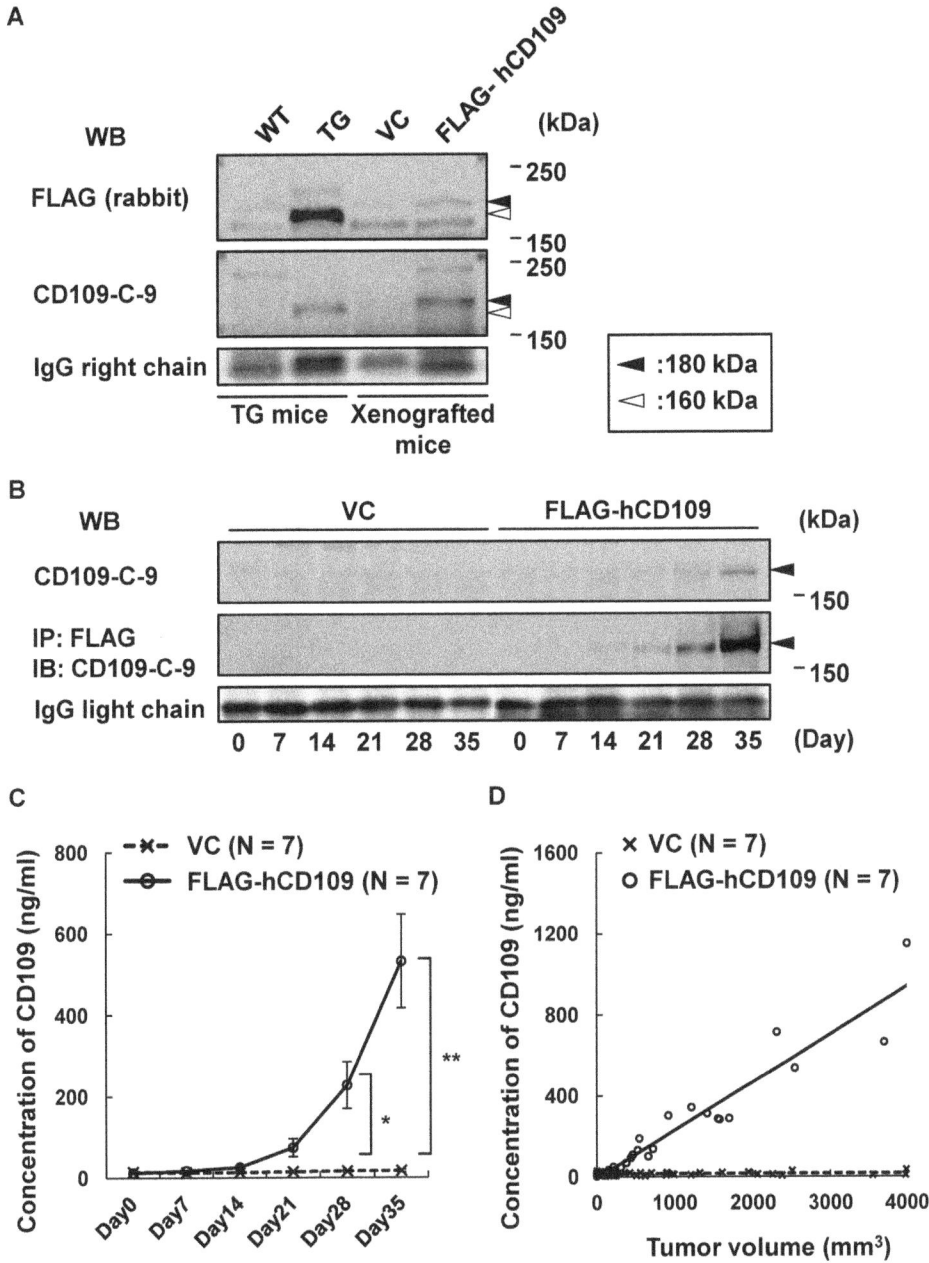

Figure 5. Proportional increase of soluble CD109 in serum with tumor volume in HEK293-FLAG-hCD109 xenografted mice. A, Detection of soluble CD109 in serum of HEK293-FLAG-hCD109 xenografted mice. Sera of mCD109-TG (TG-206) mice and xenografted mice were analyzed for CD109 by western blotting with the indicated antibodies. Molecular weight of mCD109: ~160 kDa (white arrowheads); that of FLAG-hCD109: ~180 kDa (black arrowheads); western blot for IgG light chain is indicated as an internal control. B, Analysis over time for soluble CD109 in serum of xenografted mice. Serum samples were analyzed for CD109 by western blotting with anti-CD109-C-9 mAb (upper panel). They were also analyzed for tumor-secreted FLAG-hCD109 by immunoprecipitation with Anti-FLAG M2 Affinity Gel, followed by western blotting with anti-CD109-C-9 mAb (middle panel); western blot for IgG light chain is indicated as an internal control. C, CD109 concentration in sera of xenografted mice (HEK293-FLAG-hCD109, N = 7; -VC, N = 7). Tumor-secreted FLAG-hCD109 in sera was assessed by CD109 ELISA, which recognize human CD109, but not mouse CD109. *$P<0.01$, **$P<0.0001$, using Tukey–Kramer's HSD test. D, Relationship between concentration of tumor-secreted FLAG-hCD109 in sera with tumor volumes in xenografted mice. Results in C and Fig. 4C were combined and graphically summarized.

CD109 is rapidly washed out from the serum after tumor resection.

Discussion

In the present study, we investigated the availability of CD109 as a tumor marker, using mCD109-TG mice and HEK293

xenografted mice. CD109 is highly expressed in some tumors, especially in SCCs of lung, esophagus, uterus and oral cavity, whereas it is expressed by very limited cells in normal tissues such as myoepithelial cells of the mammary, salivary and lacrimal glands, and basal cells of the prostate and bronchial epithelia. Reportedly, CD109 is cleaved and released from cell surfaces to

A

WB   (kDa)

CD109-C-9

IP: FLAG
IB: CD109-C-9   -150

IgG light chain   ◄ :180 kDa

BO 17 48 72 168 (h)

B

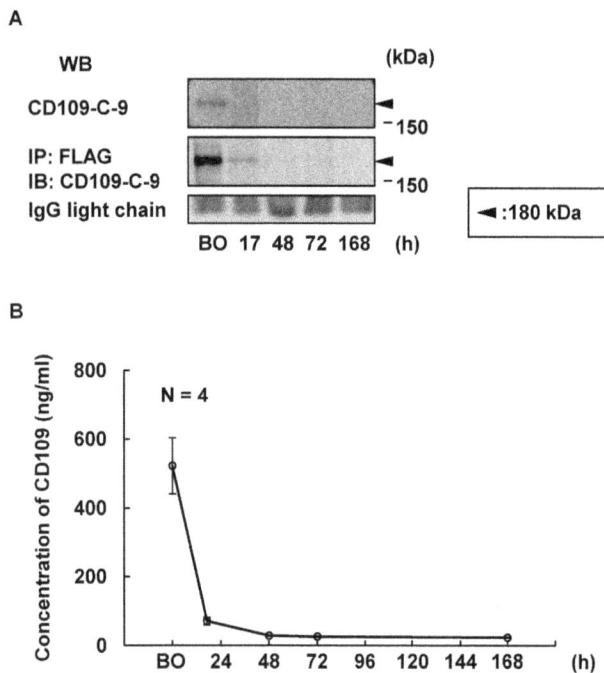

Figure 6. Rapid decrease of tumor-secreted FLAG-hCD109 in serum of HEK293-FLAG-hCD109 xenografted mice after tumor resection. A, Analysis over time of tumor-secreted FLAG-hCD109 in sera of HEK293-FLAG-hCD109 xenografted mice after tumor resection. Serum samples were analyzed for CD109 by western blotting with anti-CD109-C-9 mAb (upper panel). They were also analyzed for tumor-secreted FLAG-hCD109 by immunoprecipitation with Anti-FLAG M2 Affinity Gel, followed by western blotting with anti-CD109-C-9 mAb (middle panel). Black arrowheads: 180-kDa bands; western blot of IgG light chain is indicated as an internal control. B, Quantitative assessment of concentration of serum tumor-secreted FLAG-hCD109 in HEK293-FLAG-hCD109 xenografted mice after tumor resection using CD109 ELISA (N = 4). BO: Before operation.

culture media in *in vitro* studies [18–20]. These findings suggest the potential for CD109 in cancer management, as a novel tumor marker for SCCs or other CD109-expressing tumors. Thus, we first investigated whether the soluble form of CD109 is present in the body fluids using transgenic mice that overexpress mCD109. Our results indicate that the soluble form of CD109 is present in serum and urine of mCD109-TG mice. As CD109 was overexpressed in heart and lung of mCD109-TG mice, CD109 might be secreted by the cardiovascular system in mCD109-TG mice. Exogenous CD109 was also highly expressed in bladders of mCD109-TG mice, suggesting that soluble CD109 could be secreted from bladder epithelia. On the other hand, CD109 was not detected in urine of HEK293-FLAG-hCD109 xenografted mice (data not shown), which indicates that serum CD109 is not efficiently excreted in the urine.

In the xenografted mouse experiments, we investigated the association between tumor volume and concentration of tumor-secreted CD109 in the serum. Tumor-secreted CD109 was detected in sera of HEK293-FLAG-hCD109 xenografted mice; quantitative assessment by CD109 ELISA showed a logarithmic increase in parallel to that of xenografted tumor volume. These results suggested that serum CD109 concentration reflects amounts of CD109

secreted from tumor cells, especially from tumor tissues. CD109 is expressed in both malignant and premalignant lesions in the oral squamous epithelia. Therefore, CD109 is a new candidate among serum tumor markers for premalignant and malignant lesions. In addition, serum CD109 immediately decreased after tumor resections in this study. Seventeen hours after the operation, the concentration of CD109 was reduced to one-seventh to one-eighth, indicating that half-life of tumor-secreted CD109 is about 5.86 ± 0.17 h. The half-life of α2-macroglobulin, a structural family protein of CD109, is reported to be several hours, which is similar to that of CD109 [24]. Compared with other tumor markers, the half-life of CD109 is longer than those of SCC antigen and CYFRA (2.2 h and 1.5 h, respectively), and is shorter than those of CEA and CA19-9 (1.5 days and 12 h, respectively) [23]. Therefore, CD109 may be a good marker for monitoring tumor progression and response to surgical treatment, which proportionally increases with tumor volume and rapidly decreases after tumor resection. On the other hand, CD109 was detected in urine of mCD109-TG mice, which exhibit high levels of exogenous FLAG-mCD109 in the bladder. CD109 did not appear to be excreted from serum to urine in the xenografted mouse experiment, suggesting that urine CD109 was secreted from bladder tissues. We previously reported that CD109 is expressed in urothelial carcinomas of the bladder in an immunohistochemical study [16]. Thus, we propose that CD109 could become a urine-based marker for urothelial carcinoma of the urinary tract.

The CD109 ELISA used in this study detects only human CD109, not mouse CD109, and clearly detected an exponential increase in tumor-produced human CD109 concentration. However, endogenous human CD109 could be present in sera of human subjects, which may confound quantitative assessments of tumor-produced CD109 in human serum, as previous publications have reported CD109 expression in activated T-cells, activated platelets, and endothelial cells; this implies that serum CD109 is secreted from hemocytes (including T-cells), platelets or endothelial cells [1–7]. Investigations of the utility of secretory proteins such as sIL-2, sVEGFR, sEGFR, and many other proteins as tumor markers have been widely reported [25–27]. However, these proteins can be difficult to use clinically because they appear in low levels in normal individuals [28]. Thus, detailed investigation of CD109 concentration in serum of normal individuals and cancer patients is required, and specific tools for detecting tumor-derived CD109 by eliminating normal tissue-derived CD109 would be quite useful for application of CD109 as a tumor marker.

Taken together, our results indicate that CD109 is a potential tumor marker. Further investigations using clinical samples from patients with carcinomas such as squamous cell carcinomas are needed to verify the usefulness of CD109 in cancer management.

Acknowledgments

We thank Mr. K. Imaizumi, Mr. K. Uchiyama, Mrs. K. Ushida, and Mrs. A. Itoh for their technical assistance.

Author Contributions

Conceived and designed the experiments: HS YM. Performed the experiments: HS YM SM SH. Analyzed the data: HS AH NY. Contributed reagents/materials/analysis tools: HS YM SM TK MA. Wrote the paper: HS YM SM MI MT. Responsible for the project design: YM MU MT. Generated transgenic mice: SS MI.

References

1. Sutherland DR, Yeo E, Ryan A, Mills GB, Bailey D, et al. (1991) Identification of a cell-surface antigen associated with activated T lymphoblasts and activated platelets. Blood 77: 84–93.

2. Haregewoin A, Solomon K, Hom RC, Soman G, Bergelson JM, et al. (1994) Cellular expression of a GPI-linked T cell activation protein. Cell Immunol 156: 357–370.

3. Smith JW, Hayward CP, Horsewood P, Warkentin TE, Denomme GA, et al. (1995) Characterization and localization of the Gov^{a/b} alloantigens to the glycosylphosphatidylinositol-anchored protein CD109 on human platelets. Blood 86: 2807–2814.

4. Lin M, Sutherland DR, Horsfall W, Totty N, Yeo E, et al. (2002) Cell surface antigen CD109 is a novel member of the α_2 macroglobulin/C3, C4, C5 family of thioester-containing proteins. Blood 99: 1683–1691. doi: 10.1182/blood.V99.5.1683.

5. Kelton JG, Smith JW, Horsewood P, Humbert JR, Hayward CPM, et al. (1990) Gov^{a/b} alloantigen system on human platelets. Blood 75: 2172–2176.

6. Murray LJ, Bruno E, Uchida N, Hoffman R, Nayar R, et al. (1999) CD109 is expressed on a subpopulation of CD34+ cells enriched in hematopoietic stem and progenitor cells. Exp Hematol 27: 1282–1294. doi: 10.1016/S0301-472X(99)00071-5.

7. Giesert C, Marxer A, Sutherland DR, Schuh AC, Kanz L, et al. (2003) Antibody W7C5 defines a CD109 epitope expressed on CD34+ and CD34− hematopoietic and mesenchymal stem cell subsets. Ann NY Acad Sci 996: 227–230. doi: 10.1111/j.1749-6632.2003.tb03250.x.

8. Hashimoto M, Ichihara M, Watanabe T, Kawai K, Koshikawa K, et al. (2004) Expression of CD109 in human cancer. Oncogene 23: 3716–3720. doi:10.1038/sj.onc.1207418.

9. Zhang JM, Hashimoto M, Kawai K, Murakumo Y, Sato T, et al. (2005) CD109 expression in squamous cell carcinoma of the uterine cervix. Pathol Int 55: 165–169. doi: 10.1111/j.1440-1827.2005.01807.x.

10. Sato T, Murakumo Y, Hagiwara S, Jijiwa M, Suzuki C, et al. (2007) High-level expression of CD109 is frequently detected in lung squamous cell carcinomas. Pathol Int 57: 719–724. doi: 10.1111/j.1440-1827.2007.02168.x.

11. Hasegawa M, Hagiwara S, Sato T, Jijiwa M, Murakumo Y, et al. (2007) CD109, a new marker for myoepithelial cells of mammary, salivary, and lacrimal glands and prostate basal cells. Pathol Int 57: 245–250. doi: 10.1111/j.1440-1827.2007.02097.x.

12. Hasegawa M, Moritani S, Murakumo Y, Sato T, Hagiwara S, et al. (2008) CD109 expression in basal-like breast carcinoma. Pathol Int 58: 288–294. doi: 10.1111/j.1440-1827.2008.02225.x.

13. Mii S, Murakumo Y, Asai N, Jijiwa M, Hagiwara S, et al. (2012) Epidermal hyperplasia and appendage abnormalities in mice lacking CD109. Am J Pathol 181: 1180–1189. doi: 10.1016/j.ajpath.2012.06.021.

14. Hagiwara S, Murakumo Y, Sato T, Shigetomi T, Mitsudo K, et al. (2008) Up-regulation of CD109 expression is associated with carcinogenesis of the squamous epithelium of the oral cavity. Cancer Sci 99: 1916–1923. doi: 10.1111/j.1349-7006.2008.00949.x.

15. Ohshima Y, Yajima I, Kumasaka MY, Yanagishita T, Watanabe D, et al. (2010) CD109 expression levels in malignant melanoma. J Dermatol Sci 57: 140–142. doi: 10.1016/j.jdermsci.2009.11.004.

16. Hagikura M, Murakumo Y, Hasegawa M, Jijiwa M, Hagiwara S, et al. (2010) Correlation of pathological grade and tumor stage of urothelial carcinomas with CD109 expression. Pathol Int 60: 735–743. doi: 10.1111/j.1440-1827.2010.02592.x.

17. Finnson KW, Tam BY, Liu K, Marcoux A, Lepage P, et al. (2006) Identification of CD109 as part of the TGF-β receptor system in human keratinocytes. FASEB J 20: 1525–1527. doi: 10.1096/fj.05-5229fje.

18. Hagiwara S, Murakumo Y, Mii S, Shigenomi T, Yamamoto N, et al. (2010) Processing of CD109 by furin and its role in the regulation of TGF-β signaling. Oncogene 29: 2181–2191. doi:10.1038/onc.2009.506.

19. Litvinov IV, Bizet AA, Binamer Y, Jones DA, Sasseville D, et al. (2011) CD109 release from the cell surface in human keratinocytes regulates TGF-β receptor expression, TGF-β signalling and STAT3 activation: relevance to psoriasis. Exp Dermatol 20: 627–632. doi: 10.1111/j.1600-0625.2011.01288.x.

20. Caccia D, Zanetti Domingues L, Miccichè F, De Bortoli M, Carniti C, et al. (2011) Secretome compartment is a valuable source of biomarkers for cancer-relevant pathways. J Proteome Res 9: 4196–4207. doi: 10.1021/pr200344n.

21. Niwa H, Yamamura K, Miyazaki J (1991) Efficient selection for high-expression transfectants with a novel eukaryotic vector. Gene 108: 193–199.

22. Ichihara M, Hara T, Kim H, Murate T, Miyajima A (1997) Oncostatin M and leukemia inhibitory factor do not use the same functional receptor in mice. Blood 90: 165–173.

23. Yoshimasu T, Maebeya S, Suzuma T, Bessho T, Tanino H, et al. (1999) Disappearance curves for tumor markers after resection of intrathoracic malignancies. Int J Biol Markers 14: 99–105.

24. Imber MJ, Pizzo SV (1981) Clearance and binding of two electrophoretic "fast" forms of human α_2-macroglobulin. J Biol Chem 256: 8134–8139.

25. Janik JE, Morris JC, Pittaluga S, McDonald K, Raffeld M, et al. (2004) Elevated serum soluble interleukin-2 receptor levels in patients with anaplastic large cell lymphoma. Blood 104: 3355–3357.

26. Ebos JM, Lee CR, Bogdanovic E, Alami J, Van Slyke P, et al. (2008) Vascular endothelial growth factor-mediated decrease in plasma soluble vascular endothelial growth factor receptor-2 levels as a surrogate biomarker for tumor growth. Cancer Res 68: 521–529. doi: 10.1158/0008-5472.CAN-07-3217.

27. Kim SC, Park HM, Lee SN, Han WS (2004) Expression of epidermal growth factor receptor in cervical tissue and serum in patients with cervical neoplasia. J Low Genit Tract Dis 8: 292–297.

28. Vaidyanathan K, Vasudevan DM (2012) Organ specific tumor markers: what's new? Ind J Clin Biochem 27: 110–20.

10

Development of Transgenic Minipigs with Expression of Antimorphic Human Cryptochrome 1

Huan Liu[1,3,9], **Yong Li**[1,3,9], **Qiang Wei**[1,3,9], **Chunxin Liu**[1,3], **Lars Bolund**[1,4], **Gábor Vajta**[1,5], **Hongwei Dou**[2,3], **Wenxian Yang**[2,3], **Ying Xu**[2,3], **Jing Luan**[2,3], **Jun Wang**[1,6,7], **Huanming Yang**[1], **Nicklas Heine Staunstrup**[2,4]*, **Yutao Du**[1,2,3]*

1 BGI-Shenzhen, Shenzhen, Guangdong, China, 2 BGI Ark Biotechnology, BGI-Shenzhen, Shenzhen, Guangdong, China, 3 ShenZhen Engineering Laboratory for Genomics-Assisted Animal Breeding, BGI-Shenzhen, Shenzhen, Guangdong, China, 4 Department of Biomedicine, University of Aarhus, Aarhus C, Denmark, 5 Central Queensland University, Rockhampton, Queensland, Australia, 6 Department of Biology, University of Copenhagen, Copenhagen, Denmark, 7 King Abdulaziz University, Jeddah, Saudi Arabia

Abstract

Minipigs have become important biomedical models for human ailments due to similarities in organ anatomy, physiology, and circadian rhythms relative to humans. The homeostasis of circadian rhythms in both central and peripheral tissues is pivotal for numerous biological processes. Hence, biological rhythm disorders may contribute to the onset of cancers and metabolic disorders including obesity and type II diabetes, amongst others. A tight regulation of circadian clock effectors ensures a rhythmic expression profile of output genes which, depending on cell type, constitute about 3–20% of the transcribed mammalian genome. Central to this system is the negative regulator protein Cryptochrome 1 (CRY1) of which the dysfunction or absence has been linked to the pathogenesis of rhythm disorders. In this study, we generated transgenic Bama-minipigs featuring expression of the Cys414-Ala antimorphic human Cryptochrome 1 mutant (hCRY1^{AP}). Using transgenic donor fibroblasts as nuclear donors, the method of handmade cloning (HMC) was used to produce reconstructed embryos, subsequently transferred to surrogate sows. A total of 23 viable piglets were delivered. All were transgenic and seemingly healthy. However, two pigs with high transgene expression succumbed during the first two months. Molecular analyzes in epidermal fibroblasts demonstrated disturbances to the expression profile of core circadian clock genes and elevated expression of the proinflammatory cytokines IL-6 and TNF-α, known to be risk factors in cancer and metabolic disorders.

Editor: Nicholas S Foulkes, Karlsruhe Institute of Technology, Germany

Funding: This work was supported by research grants from ShenZhen Engineering Laboratory for Genomics-Assisted Animal Breeding in Shenzhen, China and the National Basic Research Program of China (973 Program 2011CB944201). The funders had no role in study design, data collection and analysis, decision to publish, or preparation of the manuscript.

Competing Interests: The authors have declared that no competing interests exist.

* E-mail: duyt@genomics.cn (YD); nhs@hum-gen.au.dk (NHS)

❾ These authors contributed equally to this work.

Introduction

Upholding an entrained biorhythm is pivotal for the timing of metabolic processes (reviewed in [1,2]). Hence, a highly developed hierarchical system of circadian clocks beginning with a master clock residing in the suprachiasmatic nucleus (SCN) of the anterior hypothalamus coordinates rhythmic cell behavior in all peripheral metabolic tissues [3,4]. Cryptochrome (CRYs) proteins belong to a class of light-sensitive flavoproteins that play a core role in the molecular pathways underlying circadian rhythms in mammalian cells (reviewed in [1]). The heterodimeric transcription factor composed of the circadian locomotor output cycles kaput (CLOCK) and the brain and muscle aryl hydrocarbon receptor nuclear translator (ARNT)-like protein 1 (BMAL1) controls the expression of E-box containing genes including those coding for CRY and periodic circadian (PER) proteins. In turn, CRY/PER heterodimers inhibit CLOCK/BMAL1 transactivation activity through a negative transcriptional/translational feed-back loop (TTFL). Timely post-translational modifications of CRY and

PER, such as phosphorylation by casein kinase I, target them for proteosomal degradation thereby releasing inhibition of CLOCK/BMAL1 in the auto-regulatory cycle [5,6].

Regulation of several physiological systems including the endocrine, rest-activity cycle and metabolic system, is heavily intertwined with the circadian clock. In fact, 3–20% of all peripherally active genes are expressed with a periodicity of 24-hours and many of these are involved in metabolic processes [1,7,8,9]. Naturally, recent evidence suggests that disturbance of circadian rhythms increases the risk of metabolic disorders such as obesity and type II diabetes [10,11]. Moreover, mice with the sole modification of overexpressing an autologous but mutated CRY1 share key characteristics with type II diabetes mellitus patients [12].

Minipigs have become important biomedical models for human ailments due to the well described anatomical and physiological resemblance to humans. Hence, minipigs as an animal model for at least forty human ailments including cardiovascular and metabolic disorders have been described [13,14,15,16].

Creation of transgenic animals by somatic cell nuclear transfer (SCNT) has been described for a variety of animals including sheep [17], mouse [18,19], cow [20,21] and pig [22,23,24]. In brief, the nuclei of genetically manipulated somatic donor cells are introduced into enucleated oocytes. Subsequently, the reconstructed embryos are implanted in the uterus of recipient sows. Recently a simplified procedure has been devised, termed handmade cloning (HMC) [25,26], which has been applied in a number of cases including the generation of genetically engineered pigs [27]. Hence, minipigs expressing a mutated APP gene associated with early onset of Alzheimer's disease [28] have been produced by this method. More recently minipigs ectopically expressing human integrins as a model for skin inflammation [29], minipigs expressing a human PCSK9 gain-of-function mutant, as a model for early onset atherosclerosis [30], or minipigs expressing the nematode fat-1 gene for improved nutritional value of pigs [31] have also been produced.

Here we describe HMC-based generation of transgenic Bama-minipigs harboring an antimorphic hCRY1 mutation (Cys414-Ala), which previously has been shown to entail disturbance of the circadian rhythm but also phenotypes reminiscent of type II diabetes, such as hyperglycemia and polydipsia [32]. Twenty-three alive·and healthy piglets were delivered and all animals presented peripheral exogenous Cry1 expression. Moreover, altered expression patterns of circadian clock and proinflammatory genes were evident in cultured fibroblasts, suggesting a disturbance of peripheral oscillators.

Materials and Methods

Ethics statement

Animal experiment procedures were approved by the life ethics and biological safety review committee of BGI-Research.

Plasmids and hCry1^{AP} vector construction

The CMV promoter in pcDNA3.1 (Invitrogen, Paisley, UK) was released by BglII/NheI and exchanged with the CMV early enhancer and chicken beta-actin (CAG) promoter PCR amplified from pEGFP-N1 (Clontech, CA, USA). The new construct was denoted pCAG-Neo. A multiple cloning site was, subsequently, inserted in EcoRI and NotI relaxed pCAG-Neo. The first intron of the chicken beta-actin gene was inserted into NheI/EcoRI. Human Cry1 (hCry1) cDNA was purchased from FulenGen (ID:M0114, FulenGen, Guangzhou, GD, China). The Cys414-Ala mutation was introduced by site-directed mutagenesis (Quik-Change, Stratagene, Santa Clara, CA, USA) employing 5′-CAGTTTTTTTCACTGCTATGCCCCTGTTGGTTTTGGTAGG-3′ and 5′-CCTACCAAAACCAACAGGGGCATAGCAGTGAAAAAACTG-3′ as forward and reverse PCR primers, respectively. The PCR product was digested with DpnI and inserted into the pMD18-T plasmid (TaKaRa, Shiga, Japan) and subsequently transformed into competent DH5α cells. The purified plasmid was digested with AgeI/NotI to release the hCry1^{AP} CDS fragment, which was then inserted into EcoRI/NotI digested pCAG-intron-Neo producing the final construct designated pCAG-intron-hCry1^{AP}.SV40-Neo.

Generation of transgenic donor cells for HMC

Porcine fetal fibroblasts (PFF) derived from a 32 d old Bama (BM) pig fetus were cultured in DMEM (Gibco, Invitrogen, Paisley, UK) supplemented with 15% FBS (HyClone, Logan, UT, USA), 1% L-glutamine, and 1% NEAA and maintained in 5% CO_2 atmosphere at 37°C. Approximately 1.5×10^6 BM-PFF cells were transfected, using an Amaxa Nucleofector kit (Lonza,

Verviers, Belgium) according to manufacturer's directions and seeded into 6-well cell plates (JET Biofil, Guangzhou, China). One day after transfection, cells were trypsinized and reseeded into six 10 cm culture dishes (Becton Dickinson, Lincoln Park, NJ, USA) in complete DMEM medium containing 500 µg/mL G418 (Invitrogen). After 10 days of drug selection, G418-resistant colonies were picked and expanded.

Handmade cloning and embryo transfer

The specific HMC procedure used for pig cloning has previously been described [25,31]. Briefly, cumulus–oocyte complexes (COCs) collected from porcine ovaries were washed and incubated in 4-well plates for in vitro maturation (IVM) at 38.5°C in 5% CO_2 humidified atmosphere. After 41–42 h cumulus cells were separated from the matured oocytes by hyaluronidase treatment. Following partial digestion of the zonae pellucidae, nuclei were removed by bisection. Each cytoplast was transferred into a fusion chamber containing fusion medium and fused to a single transgenic fibroblast with a single 100 V direct current (DC) impulse of 2.0 kV/cm for 9 µs. After one hour, each cytoplast-fibroblast pair was fused with another cytoplast in activation medium using a DC impulse of 0.8 kV/cm for 80 µs. Hereafter, the fused cells were incubated for 4–6 h in porcine zygote medium 3 (PZM-3) supplemented with 5 µg/ml cytochalasin B and 10 µg/ml cyclohexinmide at 38.5°C, 5%CO_2, 5% O_2, and 90% N_2 at maximum humidity. Reconstructed embryos developed into transgenic blastocysts during an additional 6 days of culture in PZM-3 medium and were eventually transferred surgically into the uterine horns of recipient sows [33]. Pregnancies were diagnosed by ultrasonography on day 28 post-transfer and monitored every 2 weeks afterwards [25]. Twenty-three live, transgenic minipigs were eventually obtained.

Identification of hCry1 in the transgenic animals

Genomic DNA was phenol-chloroform extracted from transfected primary fetal fibroblasts or tail clips from the 23 transgenic minipigs and a non-transgenic control (#321-1^{wt}) at seven days of age. Transgene identification was carried out in PCR reactions of 20 ng genomic DNA, 20pmol of each primer (cACTB intron forward 5′-TTCATACCTCTTATCTTCCTCCCA-3′ and CRY1 reverse 5′-CTTCCACTGCTGCTACAACCTG-3′) and 0.5 unit of rTaq polymerase (Takara). PCR conditions were as follows; 5 min at 95°C; 35 cycles of 30 sec. at 95°C, 30 sec. at 60°C, 40 sec. at 72°C; and final extension at 72°C for 5 min. Finally, 4 µL PCR product was electrophoresed on a 1.0% agarose gel.

For relative measurements of hCry1^{AP} mRNA levels total RNA was Trizol extracted from transfected primary fetal fibroblasts as well as from tail clips of the 23 transgenic minipigs and the non-transgenic control (#321-1^{wt}). Subsequently, 1 µg RNA was in one step DNase treated and reverse transcribed using the RevertAid First Strand cDNA Synthesis Kit (Takara) according to manufacturer's instructions. From the total cDNA, 1 µL was used for quantification using an ABI 7500 Real-Time PCR machine (ABI, CA, USA), Power SYBR Green (Takara) and hCry1 specific primers (**Table 1**). Conditions were as follows: 30 sec. at 95°C; 35 cycles of 5 sec. at 95°C, 34 sec at 60°C, and 40 sec. at 72°C. GAPDH was used as internal control (**Table 1**).

Transcription analysis of cytokine and circadian clock genes

Epidermal fibroblasts from three transgenic pigs (#175-3^{AP}, #175-5^{AP}, and #377-3^{AP}) and a non-transgenic pig (#321-1^{wt})

Table 1. Quantitative RT-PCR primer-sequences.

Gene name	Forward primer	Reverse primer
pPer2	ACACCCAGAAGGAGGAGCAGAGC	CGAGGCTTGACCCGTTTGGACTT
pBaml1	TTTGTCGTAGGATGTGACCGAGGGA	CGCCGTGCTCCAGAACATAATCG
pCry1	CTTCTTGCGTCAGTGCCATCTAA	ATGATGCTCTGCGTGTCCTCTTC
hCry1	TTCATACCTCTTATCTTCCTCCCA	CTTCCACTGCTGCTACAACCTG
pTNF-α	CCACGCTCTTCTGCCTACTG	GAGGTACAGCCCATCTGTCG
pIL-6	AGGGAAATGTCGAGGCTGTG	CTCAGGCTGAACTGCAGGAA
pClock	GAACAATAGACCCAAAGGAACCA	CCCAGAACTTCAAATGGCAAATA

were expanded from ear-derived skin biopsies. The established cell cultures were seeded in 12-well plates and maintained until 100% confluency at which the cells were synchronized with 50% FBS in DMEM for 2 hours and successively recovered in serum-free medium for 6 hours. At the given time points the cells were harvested and total RNA Trizol extracted. Quantitative RT-PCR for the gene-transcripts of porcine Per2, Cry1, Clock, Bmal1, Interleukin-6 (IL-6), Tumor-necrosis-factor alpha (TNF-α), and the transgenic human Cry1 (**Table 1**) was performed as described above.

Body temperature measurements

Three transgenic (#208-2^{AP}, #377-1^{AP}, #377-6^{AP}) and three non-transgenic (#321-1^{wt}, #321-2^{wt}, #321-3^{wt}) minipigs were housed individually with 11:13 h light-dark conditions. They were fed three times a day (8:00 am, 2:00 pm, and 6:00pm) with a standard swine feed and had unlimited access to water. Body temperature was measured using an infrared thermometer (DT-8806H, CEM, Hong Kong) every second hour through 98 hours.

Statistical analysis

P-values were calculated by a two-tailed Student's t-test or a one-sided ANOVA where appropriate to test the null hypothesis of no difference between the compared groups. The assumption of equal variances was tested by F-test; if significantly different Welch's correction was applied. The assumption of equal SD was tested by Bartlett-test; if significantly different the non-parametric Dunn's Multiple Comparisons Test was used. Assumption of normality was tested using the Kolmogorov and Smirnov method. In all statistical analyses, p-values <0.05 were considered significant.

Results

Establishment of transgenic donor cells for HMC

With the objective to selectively express an antimorphic human CRY1 (hCRY1) mutant in cultured cells, we firstly constructed a vector comprising two modules; (i) a CAG-intron-hCry1^{wt} expression cassette composed of the CAG promoter, the first intron of chicken beta-actin (cACTB) gene and the wt hCRY1 gene, and (ii) a downstream SV40-Neo selection cassette composed of the SV40 promoter and the neomycin (Neo) resistance gene. Hereafter, the hCry1 Cys414-Ala antimorphic mutation (hCry1^{AP}) was introduced by site-directed mutagenesis generating the final vector pCAG-intron-hCry1^{AP}.SV40-Neo (**Figure 1A**). Cys414 is part of a well conserved CP dipeptide motif across CRY-proteins and phyla, indicating its importance for CRY1 protein function-ality. Based on the solved crystal-structure of murine CRY2, the

CP motif appears situated within a co-factor binding pocket regulating CRY1 stability through the interaction with the ubiquitin ligase complex SCF^{FBXL3}, flavin adenine dinucleotide (FAD) and PERs [34]. Although both the catalytic function with PER proteins and repression of the CLOCK/BMAL1 complex appears unaffected by the Cys414-Ala point mutation [12,32], circadian rhythms and cellular function (in e.g. pancreatic beta-cells) are disturbed. Thus, suggesting that Cry1^{AP} affects kinetic and dynamic properties or influences non-canonical circadian pathways.

In order to generate hCRY1^{AP} expressing donor cells, we stably transfected Bama porcine fetal fibroblasts (BM-PFFs) with pCAG-intron-hCry1^{AP}.SV40-Neo and applied a selection pressure with G-418 for 10 days. Subsequently, 25 individual G-418 resistant colonies were isolated and expanded. Genomic DNA extracted from the 25 clones was subjected to PCR analysis with Cry1-Neo specific primers revealing 12 positive clones (data not shown). Subsequently, RT-qPCR on total mRNA extracted from the 12 positive clones specific for hCry1 transcripts demonstrated that hCry1 was expressed in all clones at varying intensities (**Figure 1B**). Thus, relative to hCry1^{AP} clone # 35, the three clones designated # 6, # 12 and # 15 exhibited the highest expression levels, with hCry1 mRNA levels 17, 10 and 8 fold above that of clone # 35, respectively. These high-expressing clones were chosen for HMC. Importantly, the hCry1 transcript was undetectable in un-transfected BM-PFFs.

Generation of transgenic cloned pigs by HMC

The selected hCry1^{AP} clones (# 6, # 12 and # 15) were individually applied as donor cells for HMC. The blastocyst rate was 45.5%, and a total of 691 reconstituted embryos were surgically transferred to the uteri of six naturally cycling recipient sows, with 90–135 reconstituted embryos transferred per recipient (**Table 2**). Of the six recipient sows, five became pregnant from which four went to term delivering a total of 32 piglets of which six were stillborn and two were mummified. Of the living 24 piglets one presented with malformations and was euthanized. Further two succumbed within the first months. The surviving 21 piglets remain healthy and gain weight according to the growth curve of the herd (**Figure 2A**).

Identification of hCry1^{AP} expression in the twenty-three transgenic minipigs

All cloned minipigs were evaluated for transgene integration and expression. Hence, transgenic status was demonstrated in all 23 cloned animals by PCR amplification of hCry1 on tail clip extracted genomic DNA obtained from the cloned minipigs at the age of one week. A PCR fragment with the expected size of 746 bp

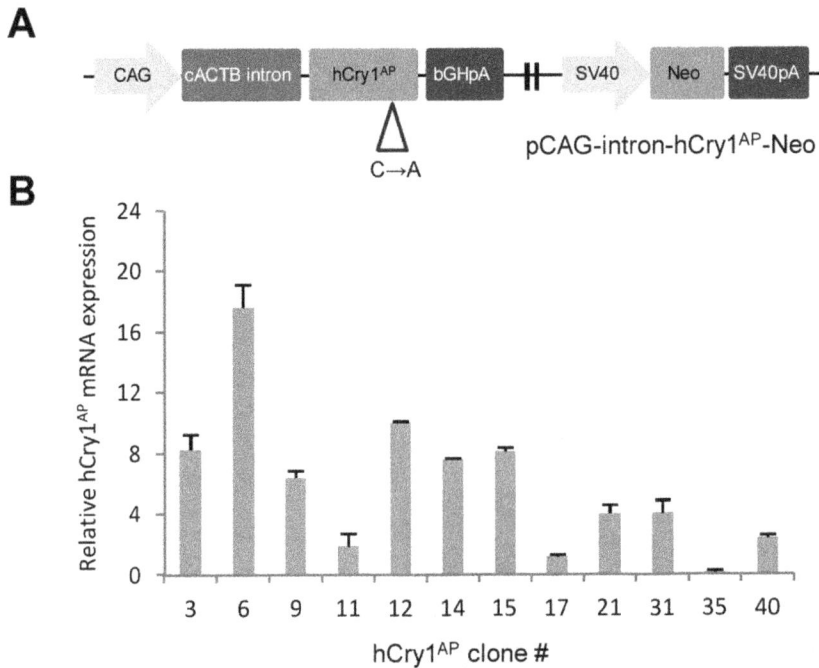

Figure 1. Functional analysis of the antimorphic human Cry1 expression vector. (A) Graphic illustration of the pCAG-intron-hCry1^{AP}.SV40-Neo expression vector. Expression of the antimorphic human Cry1 (hCry1^{AP}) is driven by the CMV enhancer chicken beta-actin (CAG) chimeric promoter and terminated with the bovine growth hormone polyadenylation site (bGHpA). A selection cassette consisting of a Simian virus 40 (SV40) promoter, a Neomycin (Neo) gene and SV40 polyadenylation signal is placed downstream. **(B)** Quantitative RT-PCR analysis of hCry1^{AP} expression in transgenic Bama-minipigs fetal fibroblasts (BM-PFFs). Reverse transcribed total RNA was used for amplification with hCry1-specific exon-exon primers and normalized to endogenous GAPDH. The depicted expression levels are relative to cell clone # 17. The experiment is performed in triplicate and data are presented as mean values ± standard deviation.

appeared in the lanes for each of the cloned minipigs but not for an age-matched non-transgenic (NT) control (**Figure 2B, upper**). Assessment of transgene expression was achieved by conducting RT-PCR on total RNA extracted from the tail clips. Again an easily detectable band corresponding to hCry1 cDNA was observed for all transgenic minipigs but not for the NT control (**Figure 2B, lower**).

Analysis of the expression level of Cry1 showed that the relative expression level in the transgenic minipigs varied up to 7-fold compared to that in the lowest expressing transgenic animal (#351-4^{AP}) (**Figure 2C**). A general higher expression level was detected in animals originating from donor cell # 6 (sow #377 and #351) suggesting a correlation between the expressions observed in the donor cells and in the transgenic minipigs. Interestingly, the expression appears to be more heterogeneous in piglets arising from donor cell # 6 compared to the expression pattern among piglets originating from donor cell # 12 (sow #175 and #208). This suggests that non-inherited positional effects work on the transgene - especially in animals originating from donor cell # 6. Pig #377-1^{AP} and pig #208-2^{AP} were the animals displaying the highest relative hCry1^{AP} expression. Notably, they perished at the age of about 35 days, hinting a possible concentration dependent toxicity.

Altered oscillation of circadian gene expression in the transgenic minipigs

Clock-gene expression levels inherently oscillate in well defined phases with time in intact peripheral tissue. In cell culture this feature is lost but can, however, be reinitiated and synchronized through serum-shock. Hence, in *ex vivo* fibroblasts this leads to a

cell-autonomous oscillation in clock gene expression for at least 48 hours [35,36]. Furthermore, functional CRY1 is required for a sustained circadian rhythm in dissociated murine fibroblasts [37]. In order to evaluate if the introduced point mutation in hCry1 would exert disruption to the cell-autonomous oscillation, we transfected primary porcine fibroblasts with a construct expressing the functional hCRY1 protein (pCAG-intron-hCry1.SV40-Neo) or the antimorphic variant (pCAG-intron-hCry1^{AP}.SV40-Neo). Following antibiotic selection and RT-qPCR analysis, clones expressing the functional (clone # 3) or antimorphic hCry1 (clone # 3) at comparable intensity and displaying similar growth rates were selected and further expanded (**Figure 3A**). At confluency the cells were serum shocked for 2 h after which they were returned to serum-free medium for an additional 6 h. Subsequently, cells were harvested for RT-qPCR analysis every 4 hours through 48 hours.

The mRNA level of pPer2 in the hCry1 transfected cells appears to have a first zenith at Zeitgeber time (ZT) 16 h and a second at ZT 36 h (**Figure 3B**). On the other hand, the zeniths of pCry1 mRNA expression are slightly offset occurring at ZT 20 h and 44 h (**Figure 3C**). This slight offset but synchronous oscillation of Per2 and Cry1 has previously been documented in SCN as well as peripheral tissue [38]. Interestingly, the rhythmic pattern is only partially maintained in the hCry1^{AP} transfected cells and, moreover, displays oscillations of greater magnitude, suggesting that the regulation of pPer2 and pCry1 is unbalanced in the presence of the antimorphic hCRY1^{AP}.

With the intent to examine circadian gene expression patterns in NT as compared to transgenic minipigs we cultured epidermal fibroblasts from transgenic minipigs (#175-3^{AP}, #175-5^{AP} and

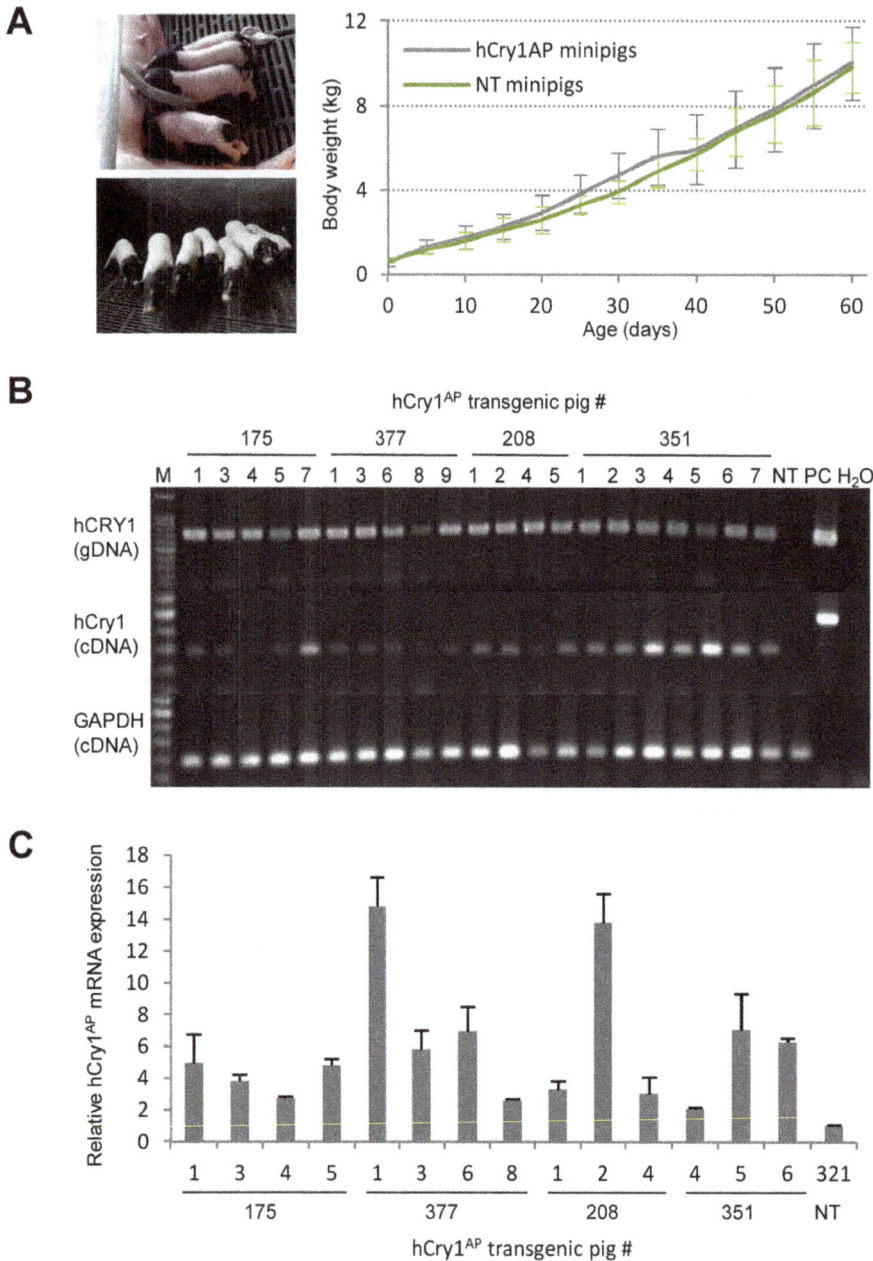

Figure 2. Demonstration of transgenesis in cloned minipigs produced by HMC. (**A**) Pictures of cloned Bama-minipigs at the age of five and 60 days. Curve indicates mean weight increase over the first 60 days of the 21 transgenic (hCry1^{AP}) and five non-transgenic (NT) animals (#321-1^{wt}, #321-2^{wt}, #321-3^{wt}, #321-5^{wt}, and #321-6^{wt}). (**B**) PCR and RT-PCR analysis on gDNA and total RNA, respectively, isolated from tail clips of the 23 cloned minipigs born from three recipient sows (#175, #377 and #208) as well as from one non-transgenic (NT) control. The PCR analysis employing hCry1 specific primers revealed genomic integration of the transgenic cassette in all the cloned animals (upper panel). RT-PCR analysis using exon-exon primers for hCry1 and porcine GAPDH showed robust expression of hCry1^{AP} in the transgenic animals with no detectable band in the lane corresponding to the NT control (lower two panels); PC, plasmid control; M, 100 bp marker. (**C**) Quantitative RT-PCR performed on cDNA from 14 of the 23 transgenic animals and a NT control (#321-1^{wt}). Total mRNA extracted from tail clips was reverse transcribed and used for quantification of hCry1 normalized to endogenous ACTB. The expression values are relative to the NT control. The experiment is performed in triplicate and data are presented as mean values ± standard deviation.

#377-3^{AP}) and from a NT control (#321-1^{wt}). At confluency the cells were serum shocked for 2 h and then allowed to recover in serum-free medium for another 6 h. Afterwards the cells were harvested for RT-qPCR analysis every 4 hours through 52 hours. Firstly, we validated hCry1^{AP} mRNA expression in all the three cell populations. After an initial increase in expression level in pig #175-5^{AP}, all mRNA levels (relative to pig #175-3^{AP}) remained

within a stable range with only minor yet rhythmic fluctuations. Importantly, no call above background was observed in the control (**Figure 4A**). Secondly, we examined porcine Cry1, Per2, Clock and Bmal1 mRNA levels in the cultured cells following synchronization. Given the nature of the feedback regulatory effect of CRY1/PER2 on CLOCK/BMAL1 the corresponding two sets of genes are expected to have peak expression levels occurring in

Table 2. Summary of cloning efficiencies obtained with hCRY1^{AP}-transgenic fibroblasts.

Donor sow ID	Blastocysts transferred	Donor cell	Delivered piglets	Alive	Presenting malformations	Stillborn	Mummified
S1-377	135	hCRY1^{dn}-6	9	5		2	2
S2-351	111	hCRY1^{dn}-6	10	8		2	
S3-175	135	hCRY1^{dn}-12	7	6		1	
S4-208	134	hCRY1^{dn}-12	6	4	1	1	
S5-584	93	hCRY1^{dn}-15	Not pregnant				
S6-180	83 (17PA)	hCRY1^{dn}-15	Abortion				
Total	691		32	23	1	6	2

antiphase to each other under normal entrainment. In fibroblasts from the NT control pig #321-1^{wt}, pPer2 and pCry1 transcript levels displayed around 24 h oscillations with a first zenith around ZT 15 h for pPer2 and ZT 17 h for pCry1 (**Figure S1**). Notably, the oscillations of Per2 and Cry1 appeared slightly displaced and with a circadian rhythm similar to the one observed in the transfected porcine fibroblasts. Conversely pBaml1 transcript levels displayed a first nadir around ZT 12 and 36, thus, being inversely correlated to pPer2 and, to a lesser extent, pCry1 expression. On the other hand, pClock mRNA levels presented a much weaker circadian pattern although a weak zenith is detectable a ZT 20 and 44. However, this seems to be consistent with previous findings in murine peripheral tissues [39]. In the transgenic animals the expression curves of hCry1, pPer2, pCry1, pClock, and pBmal1 displayed key differences and similarities compared to the NT control (**Figure 4 A–E**). Oscillation in the mRNA levels of hCry1 follows that of pCry1, indicating that hCry1 is susceptible to the same post-transcriptional regulation otherwise governing the levels of pCry1. Supporting this notion, is the observation that the 3′ untranscribed region (UTR) of mouse Cry1 mRNA contains a destabilizing element to which heterogeneous nuclear ribonucleoprotein D (hnRNP D) binds facilitating mRNA degradation [40]. As in the NT control, expression levels of pPer2, pClock and pBaml1 in the transgenic animals follow a rhythmic pattern. However, in the latter the amplitude is increased and the oscillation period decreased. Hence, the ZT for the first zenith is comparable between the control and the transgenic animals. However, cells from the latter seem to complete two cycles within 24 h compared to a single cycle in the NT control.

Group means and analysis of variance between group means at each time point is summarized in **Table S1 and S2**.

Looking specifically at the acrophases of pCry1 and pPer2 within the first 24 hours, there is a high concordance between the cells derived from the transgenic and NT animals (**Figure S2**). Hence for pCry1 the acrophase is at 18.9 h ($\pm$0.6 h) in the transgenic cells and at 18.4 h in the NT control cells. Similar, for pPer2 the acrophase occurred at 14.1 h ($\pm$0.2 h) in the transgenic cells and at 13.6 h in the NT control cells. This rhythmicity is maintained in the NT cells producing pCry1 and pPer2 acrophases on day 2 at 16.4 h and 13.6 h, respectively. Interestingly, however, data from the transgenic cells only fitted the applied model if subdivided into 12 h time frames. Giving acrophases at 7 h ($\pm$0.2 h) and 19.7 h ($\pm$1.2 h) for pCry1 and 5.8 h ($\pm$0.2 h) and 16.4 h ($\pm$1.4 h) for pPer2. Taken together, the redundancy in the TTFLs seemingly ensures rhythmicity in the transgenic fibroblasts but with an altered circadian period (τ) and amplitude. A slight intra-individual variation in τ seems evident

but falls in line with previous reports on circadian rhythms in human fibroblasts [41].

Increased activation of proinflammatory markers in the hCry1^{AP} transgenic minipigs

Next we wanted to determine whether the altered circadian rhythms and/or the expression of the antimorphic hCry1 in the transgenic animals would lead to activation of proinflammatory cytokines. The level of IL-6 and TNF-α mRNA was quantified in cDNA originating from serum shocked hCry1^{AP} transgenic fibroblasts (derived from pig #377-1^{AP}, #208-4^{AP}, and #175-5^{AP}) and NT fibroblasts (derived from pig #321-1^{wt}). An initial analysis revealed a sharp and linear increase in the mRNA level of IL-6 in the transgenic animals reaching a 7–10 fold increment at ZT 14 compared to ZT 0. Notably, this observation was absent in the NT control (**Figure 5A**). To confirm these findings we repeated the experiment with epidermal fibroblasts derived from pig #175-5^{AP} and a NT control (#321-1^{wt}) with measurements every fourth hour through 28 hours (**Figure 5B**). The more detailed time course confirmed the initial steep increase in IL-6 mRNA levels until ZT 12 after which the level dropped off reaching a nadir at ZT 20. Comparing the expression levels of IL-6 at nadir and zenith, minipig #175-5^{AP} presents a level about 3-fold higher relative to the NT control. Notably, however, the IL-6 mRNA level in both the transgenic and NT animal revolted with a τ of approximately 12 hours. In a parallel experiment assessing TNF-α mRNA levels in the same pool of cells derived from pig #175-5^{AP} and the NT control showed significant increase in TNF-α expression in the transgenic minipig, that is, reaching levels 3–9 fold higher relative to the levels observed in the control (**Figure 5C**). This lends further support to the notion of proinflammatory induction in the transgenic cells. Interestingly, a complete circadian cycle was completed in 16 hours in cells from the NT control animal whereas in only took 8 hours for a complete cycle in the cells derived from the transgenic animal.

Circadian oscillation of body temperature

In mammals, body temperature (T_b) shows circadian oscillation. Therefore, recording T_b fluctuation is considered an effective way of determining circadian rhythm perpetuations [42]. Body temperature was recorded every second hour through 4 days in five transgenic minipigs (#175-5^{AP}, #208-4^{AP}, #351-3^{AP}, #351-6^{AP}, and #377-2^{AP}) and three NT minipigs (#321-1^{wt}, #321-3^{wt}, and #C-28^{wt}) (**Figure 6A**). Animals in both groups displayed normal T_b curves with a steady increase during daylight and a steady decrease during nighttime. Three of the five hCry1^{AP} and two of the three NT pigs showed rhythmic T_b oscillations with τ

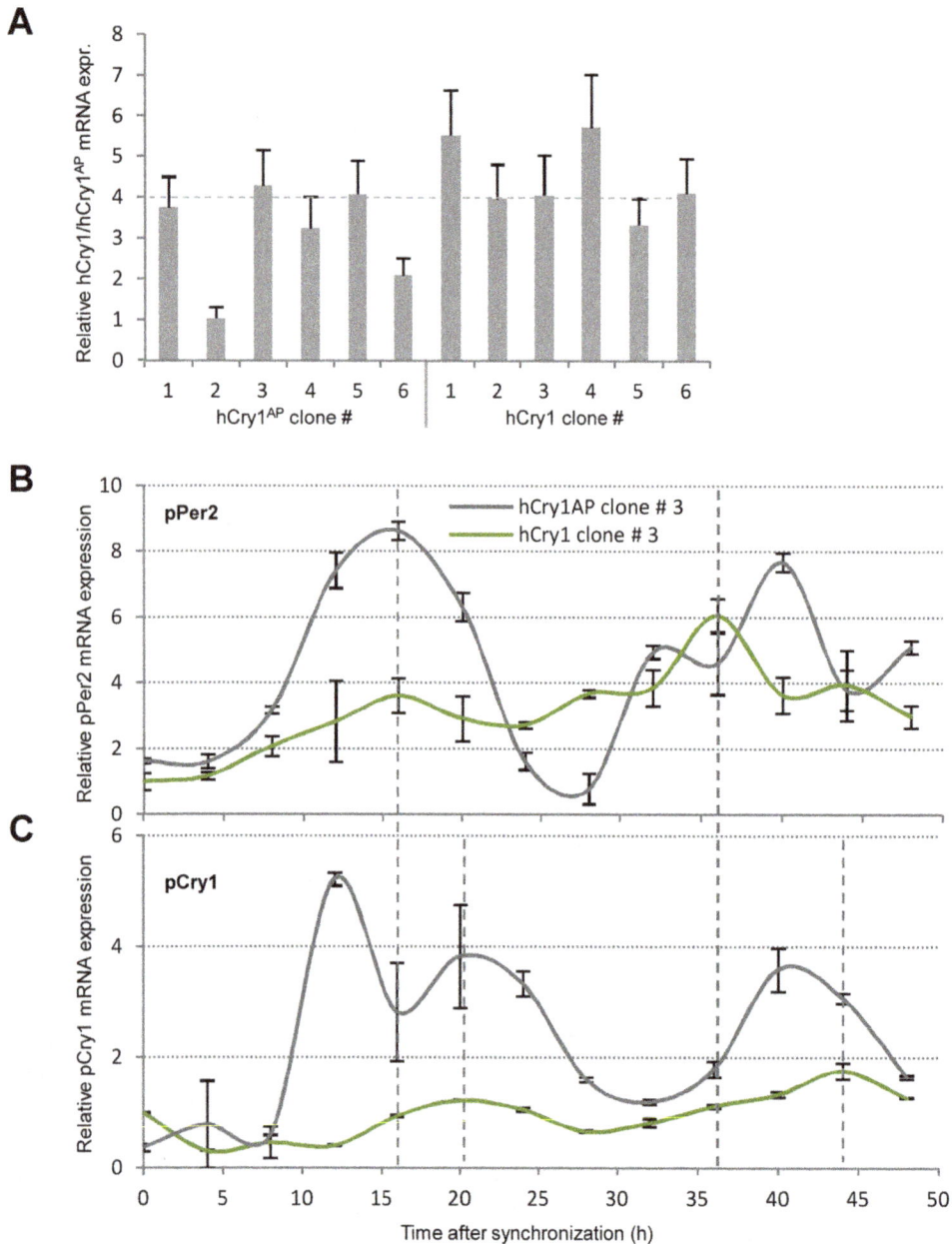

Figure 3. Expression of the antimorphic but not the wt hCry1 instigate altered expression profiles of pPer2 and pCry1. Primary porcine fibroblasts were stably transfected with pCAG-intron-hCry1.SV40-Neo or pCAG-intron-hCry1[AP].SV40-Neo and relative expression were assessed by RT-qPCR with exon-exon spanning hCry1 primers. (**B–C**) Serum-shocked cells were harvested every fourth hour through 48 hours. Total RNA was extracted and used for RT-qPCR with exon-exon spanning primers targeting pPer2 or pCry1. GAPDH normalized data relative to hCry1 clone # 3 is depicted as a function of time. The experiment is performed in triplicate and data are presented as mean values with smoothened curves. Grey dashed lines indicate the first and second zenith of pPer2 and pCry1 mRNA expression in hCry1 containing cells.

around 23 hours as calculated by the chi-square periodogram (**Table S3**). Among the remaining three animals, especially the transgenic minipig #208-4[AP] exhibited an abnormal T_b curve (**Figure 6B**). In summary, these results may imply that expression of the antimorphic mutant hCry1[AP] potentially leads to disturbance of the circadian rhythm of T_b in the transgenic minipigs.

Discussion

Minipigs have become an attractive mammalian model for research in human ailments including metabolic, cardio-vascular

and degenerative disorders. Small body sizes as well as physiological and anatomical resemblance to humans has raised the interest in minipigs and so basic as well as clinical studies have already been conducted. Furthermore, the complete sequencing of the pig genome has further spawned a general interest in genetically modified porcine disease-models [43]. In this report we describe the generation of Bama-minipigs holding a genomically integrated transcription cassette from which a heterologous and antimorphic CRY1 is expressed. The transgene is under the control of a CAG promoter which has shown almost ubiquitous activity in all tissues including the brain and skin [44,45]. Expression of the transgene

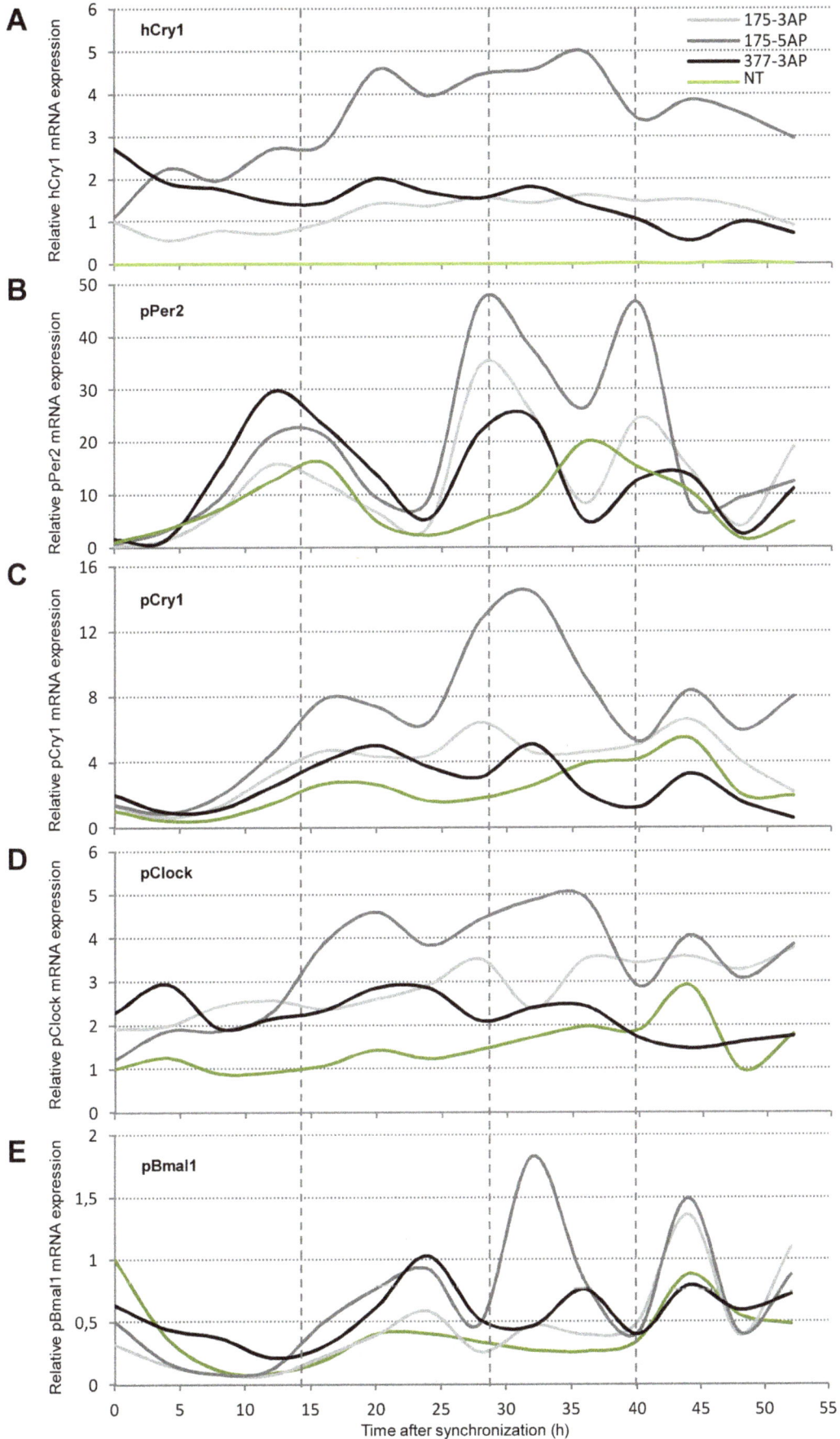

Figure 4. Altered oscillation patterns of key clock transcripts in fibroblasts from transgenic minipigs. Fibroblasts were expanded from skin-biopsies obtained from three transgenic minipigs (#175-3[AP], #175-5[AP], and #377-3[AP]) and one NT control (#321-1[wt]). The cells were serum-shocked after which total RNA was extracted every fourth hour through 52 hours. Quantitative RT-PCR was performed with exon-exon spanning primers targeting (**A**) hCry1 (**B**) pPer2 (**C**) pCry1 (**D**) pClock and (**E**) pBmal1 normalized to endogenous ACTB. Expression is relative to minipig #175-3[AP] in (**A**) and to the NT control in (**B–E**) and is depicted as a function of time. The experiment is performed in triplicate and data are presented as mean values with smoothened curves. Grey dashed lines indicate the first, second and third zenith of pPer2 mRNA expression in cells from minipig #175-5[AP].

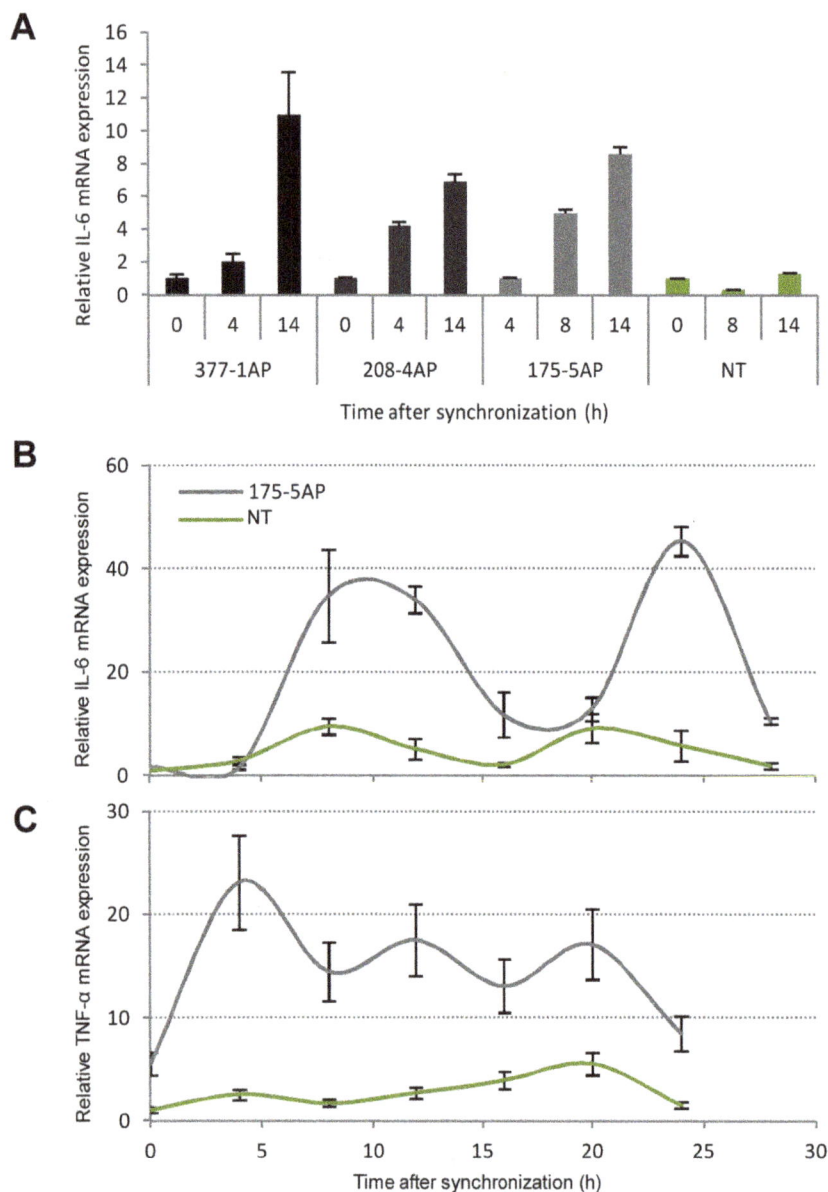

Figure 5. Induction of the proinflammatory cytokines IL-6 and TNF-α in fibroblast from the hCry1[AP] transgenic animals. (**A**) Fibroblasts derived from skin-biopsies obtained from three transgenic minipigs (#377-1[AP], #175-5[AP], and #208-4[AP]) and one NT control (#321-1[wt]) were serum-shocked and allowed to recover. Subsequently, total RNA was extracted at three time points within 14 hours post-recovery. Quantitative RT-PCR was performed with exon-exon spanning primers targeting porcine IL-6 and normalized to endogenous ACTB. Values are relative to the first measurement time point. (**B**) Relative IL-6 RT-qPCR on total RNA extracted from serum-shocked fibroblasts originating from transgenic minipig #175-5[AP] and NT control pig #321-1[wt] as above. Cells were harvested every fourth hour through 24–28 hours. Values are relative to the first measurement time point (ZT 0). (**C**) Relative TNF-α mRNA expression in fibroblasts from transgenic minipig #175-5[AP] and NT control pig #321-1[wt] as above. All experiments are performed in triplicate and data are presented as mean values ± standard deviation.

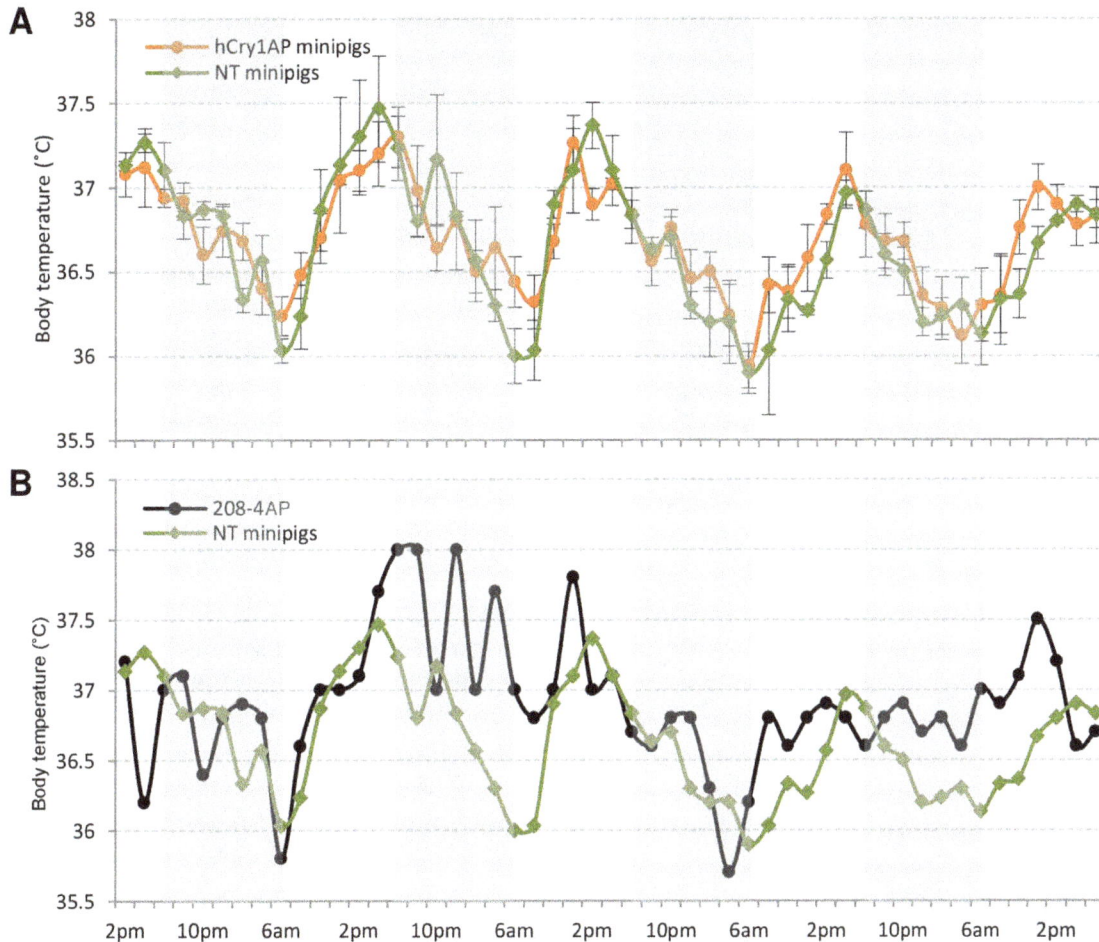

Figure 6. Discrete circadian oscillation of body temperature in hCry1^{AP} transgenic compared to NT control minipigs. (A) Body temperature was measured using an infrared thermometer in five transgenic minipigs (#175-5^{AP}, #377-2^{AP}, #351-3^{AP}, #351-6^{AP}, and #208-4^{AP}) and three NT control animals (#321-1^{wt}, #321-2^{wt}, and #321-3^{wt}) every second hour over a time course of 4 days. Mean body temperature is plotted as a function of time ± standard deviation. **(B)** Body temperature fluctuations of minipig #208-4^{AP} compared to the mean from the three NT control minipigs (#321-1^{wt}, #321-2^{wt}, and #321-3^{wt}) obtained as described above.

in the 23 live-born, seemingly healthy, cloned animals was verified by PCR and RT-qPCR on DNA and RNA, respectively, extracted from skin biopsies. Curiously, despite originating from only two different donor-cell populations the expression pattern in the transgenic minipigs appeared rather heterogeneous. As the cycle of the external cues (light and feed) are highly controlled it is unlikely that the circadian clocks among the pigs are out of synchronization. However, stochastic events, genomic rearrangements or epigenetic changes during embryogenesis could explain the variation in expression.

Next we wanted to assess to what extent the introduced antimorphic hCry1^{AP} would be expressed in cultured fibroblasts from the pigs and whether this would alter the autonomous peripheral oscillations of other key clock components. It is well known that several clock effectors including Per2 and Cry1 contain 3'UTRs prone for mRNA degradation [40,46]. Interestingly, the mRNA level of hCry1^{AP} itself oscillated with a periodicity comparable to pCry1, suggesting that the transgenic Cry1 is subject to the same post-transcriptional regulation as the endogenous Cry1. Furthermore, there appears to be consistency between expression levels of hCry1^{AP} and the clock genes, hence,

minipig #175-5^{AP} presented the highest relative expression in all assays.

Whereas peripheral oscillations of core clock gene transcripts demonstrated an approximately 24 hours periodicity in epidermal fibroblasts from the NT control pig, the same gene transcripts oscillated with a frequency of about 12 hours in cells from the three transgenic animals tested (#175-3^{AP}, #175-5^{AP}, and #377-3^{AP}). Importantly, previous reports demonstrated coherence between peripheral oscillations *in vitro* and *in vivo*. Furthermore, deviations in the circadian rhythm of clock gene expression were found to be retained *in vitro* [47,48]. These results, thus, appear to be in line with the abnormal circadian behavior observed in mouse Cry1^{AP} transgenic mice [32]. This indicates that the regulatory function of CRY1 cannot be rescued by CRY2 or a second regulatory loop, involving the orphan nuclear receptors Rev-Erbα and RORα.

Oscillations in the mRNA levels of the clock driving genes pClock and pBaml1 are quantitatively small compared to those of the regulatory arm, pCry1 and pPer2. Importantly, however, there is an anti-phase correlation between the former and pPer2, suggesting that regulation of the cell-autonomous circadian clock is maintained but with a τ of only 12 hours. Notably, oscillation of pPer2 is both very consistent and robust across all transgenic fibroblast cultures. This supports not only the notion of pPer2 being the primary intrinsic Zeitgeber of rhythmicity in peripheral tissue [49,50] but also substantiating the τ alternating effect of hCRY1^{AP}. Furthermore, the expression profiles of Cry1 and pPer2 are synchronized but slightly staggered. This lag in time has previously been described in both the SCN and peripheral tissue [39,51], supporting the concept of PER2 induced stabilization of CRY1 by prevention of ubiquitylation and proteosomal degradation [34,52]. Together with the likewise adopted model of CRY1 induced stabilization of PER2 through similar mechanisms [52,53], we propose a working model in which the hCRY1^{AP} protein enjoys increased stability leading to augmented pPER2 stability which downstream leads to increased stability of endogenous CRY1. Further supporting this view is the marked difference in expression amplitude between hCry1 and hCry1^{AP} transfected porcine fibroblasts.

Sleep-loss studies have identified physiological connection between the circadian clock and the immune system [54]. As a chronic state of metainflammation is known to be a substantial risk factor in metabolic disorders and cancer [55,56,57] and because the absence of CRY in knockout mice has been shown to entail increased expression levels of proinflammatory cytokines including TNF-α and IL-6 [58], we sought to determine if the antimorphic variant potentially could lead to a higher steady-state of inflammation in the transgenic animals. A considerable difference was apparent with average levels of IL-6, 4-fold ($p = 0.047$), and TNF-α, 5-fold ($p = 0.0025$), higher in fibroblasts from minipig #175-5^{AP} relative to the control. Notably, the oscillation of IL-6 in pig #175-5^{AP} displayed a periodicity of about 12 hours which is coherent with the expression profile of hCry1^{AP}. However, IL-6 in the NT control followed the same pattern although pCry1 had a τ of about 24 hours.

CRY proteins have shown to bind adenylyl cyclase (ADCY) leading to a decrease in cyclic AMP production thereby preventing the activation of NF-κB and downstream expression of IL-6 - with the opposite effect taken place in the absence of CRY protein. Our data suggests that hCRY1^{AP} competes with endogenous CRY proteins for binding to ADCY but that it lacks repressor activity.

Body temperature (T_b) control is a very important projection emanating from the SCN as it plays a role in the resetting of SCN itself but also as an auxiliary Zeitgeber of peripheral oscillators [51,59]. Core T_b displays a clear circadian rhythm in mammals with fluctuations of 1–4°C [60]. We therefore wanted to assess whether the presence of hCRY1^{AP} protein would interfere with T_b control in the transgenic animals as previously observed in Cry1$^{-/-}$/Cry2$^{-/-}$ double knockout mice [61]. The average τ and amplitude of the T_b rhythm was unchanged ($\tau \approx 23$ hours, $p \leq 0.005$) in the transgenic pigs compared to the controls under normal light-dark and feeding conditions. However, there were a few outliers and especially pig #208-4^{AP} presented an arrhythmic T_b oscillation, suggesting, that the central circadian clock, in at least this animal, is disturbed even under normal entrainment. It has previously been implied that CRY1 is dispensable for persistent SCN rhythmicity [37]. However, it is conceivable that the antimorphic hCRY1 mutant interferes with the circadian clock in a non-canonical fashion.

In summary we have generated 23 Bama-minipigs transgenic for a tissue unspecific antimorphic human Cry1. We have shown stable expression of the transgene in the skin of all cloned minipigs and demonstrated that this gives rise to altered circadian patterns of key peripheral oscillators as well as amplified induction of proinflammatory cytokines. Further analyses are currently undertaken and we anticipate that such porcine models will contribute to a better understanding of the link between the pathogenesis of circadian and metabolic disorders.

Supporting Information

Figure S1 Circadian rhythmicity in expression patterns of key circadian regulatory genes. Fibroblasts expanded from skin-biopsies obtained from a non-transgenic minipig (#321-1) were serum-shocked after which total RNA was extracted every fourth hour through 52 hours. Quantitative RT-PCR was performed with exon-exon spanning primers targeting hCRY1, pPer2, pCry1, pClock, and pBmal1 normalized to endogenous ACTB. Expression is relative to the first measurement time point (ZT 0) and depicted as a function of time. The experiment is performed in triplicate and data are presented as means. Grey dashed lines indicate a complete 24 h oscillation (ZT 15 and 36 h, respectively).

Figure S2 Comparison of acrophases of pCry1 and pPer2 expression in synchronized transgenic and non-transgenic epidermal fibroblasts. The mRNA expression values of pPer2 and pCry1 depicted in figure 4 B–C were subdivided into windows of 12 or 24 hours. The acrophase was calculated using the free software program Acro (www.periodogram.org). The acrophase in ZT as well as the 95% confidence interval (CI) and a measure of goodness-of-fit is shown.

Table S1 Mean values and standard deviation (in brackets) of the relative mRNA expression levels shown in figure 4.

Table S2 One-way analysis of variance (ANOVA) comparing means of the relative mRNA expression level observed in the non-transgenic animal to the levels observed in the three transgenic animals as shown in figure 4. # Dunn's Multiple Comparisons Test (non-parametric analysis as SD's are not identical); ns not significant; * $p < 0.05$; ** $p < 0.01$; *** $p < 0.001$.

Table S3 Chi-square periodogram data output (www.periodogram.org) using 51 body temperature entries per animal over a four day period.

Acknowledgments

We are grateful to all participants of the transgenic group at BGI and the cloning group at BGI Ark Biotechnology (BAB).

Author Contributions

Conceived and designed the experiments: HL YL LB YD NHS. Performed the experiments: HL QW CL. Analyzed the data: HL NHS. Contributed reagents/materials/analysis tools: JW YD HY. Wrote the paper: HL NHS. Cloned the animals: HD WY YX JL. Helped finalize the manuscript: LB GV.

References

1. Green CB, Takahashi JS, Bass J (2008) The meter of metabolism. Cell 134: 728–742.

2. Maury E, Ramsey KM, Bass J (2010) Circadian rhythms and metabolic syndrome: from experimental genetics to human disease. Circ Res 106: 447–462.

3. Yamazaki S, Numano R, Abe M, Hida A, Takahashi R, et al. (2000) Resetting central and peripheral circadian oscillators in transgenic rats. Science 288: 682–685.

4. Yoo SH, Yamazaki S, Lowrey PL, Shimomura K, Ko CH, et al. (2004) PERIOD2::LUCIFERASE real-time reporting of circadian dynamics reveals persistent circadian oscillations in mouse peripheral tissues. Proc Natl Acad Sci U S A 101: 5339–5346.

5. Lee C, Etchegaray JP, Cagampang FR, Loudon AS, Reppert SM (2001) Posttranslational mechanisms regulate the mammalian circadian clock. Cell 107: 855–867.

6. Eide EJ, Vielhaber EL, Hinz WA, Virshup DM (2002) The circadian regulatory proteins BMAL1 and cryptochromes are substrates of casein kinase Iepsilon. J Biol Chem 277: 17248–17254.

7. McCarthy JJ, Andrews JL, McDearmon EL, Campbell KS, Barber BK, et al. (2007) Identification of the circadian transcriptome in adult mouse skeletal muscle. Physiol Genomics 31: 86–95.

8. Reddy AB, Karp NA, Maywood ES, Sage EA, Deery M, et al. (2006) Circadian orchestration of the hepatic proteome. Curr Biol 16: 1107–1115.

9. Storch KF, Lipan O, Leykin I, Viswanathan N, Davis FC, et al. (2002) Extensive and divergent circadian gene expression in liver and heart. Nature 417: 78–83.

10. Eckel-Mahan KL, Patel VR, Mohney RP, Vignola KS, Baldi P, et al. (2012) Coordination of the transcriptome and metabolome by the circadian clock. Proc Natl Acad Sci U S A 109: 5541–5546.

11. Bouatia-Naji N, Bonnefond A, Cavalcanti-Proenca C, Sparso T, Holmkvist J, et al. (2009) A variant near MTNR1B is associated with increased fasting plasma glucose levels and type 2 diabetes risk. Nat Genet 41: 89–94.

12. Okano S, Hayasaka K, Igarashi M, Iwai H, Togashi Y, et al. (2010) Non-obese early onset diabetes mellitus in mutant cryptochrome 1 transgenic mice. Eur J Clin Invest 40: 1011–1017.

13. Xi S, Yin W, Wang Z, Kusunoki M, Lian X, et al. (2004) A minipig model of high-fat/high-sucrose diet-induced diabetes and atherosclerosis. Int J Exp Pathol 85: 223–231.

14. Oxenhandler RW, Adelstein EH, Haigh JP, Hook RR, Jr., Clark WH, Jr. (1979) Malignant melanoma in the Sinclair miniature swine: an autopsy study of 60 cases. Am J Pathol 96: 707–720.

15. Forster R, Ancian P, Fredholm M, Simianer H, Whitelaw B (2010) The minipig as a platform for new technologies in toxicology. J Pharmacol Toxicol Methods 62: 227–235.

16. Spurlock ME, Gabler NK (2008) The development of porcine models of obesity and the metabolic syndrome. J Nutr 138: 397–402.

17. Wilmut I, Schnieke AE, McWhir J, Kind AJ, Campbell KH (1997) Viable offspring derived from fetal and adult mammalian cells. Nature 385: 810–813.

18. Wakayama T, Shinkai Y, Tamashiro KL, Niida H, Blanchard DC, et al. (2000) Cloning of mice to six generations. Nature 407: 318–319.

19. Wakayama T, Yanagimachi R (1999) Cloning of male mice from adult tail-tip cells. Nat Genet 22: 127–128.

20. Cibelli JB, Stice SL, Golueke PJ, Kane JJ, Jerry J, et al. (1998) Cloned transgenic calves produced from nonquiescent fetal fibroblasts. Science 280: 1256–1258.

21. Kato Y, Tani T, Sotomaru Y, Kurokawa K, Kato J, et al. (1998) Eight calves cloned from somatic cells of a single adult. Science 282: 2095–2098.

22. Betthauser J, Forsberg E, Augenstein M, Childs L, Eilertsen K, et al. (2000) Production of cloned pigs from in vitro systems. Nat Biotechnol 18: 1055–1059.

23. Onishi A, Iwamoto M, Akita T, Mikawa S, Takeda K, et al. (2000) Pig cloning by microinjection of fetal fibroblast nuclei. Science 289: 1188–1190.

24. Polejaeva IA, Chen SH, Vaught TD, Page RL, Mullins J, et al. (2000) Cloned pigs produced by nuclear transfer from adult somatic cells. Nature 407: 86–90.

25. Du Y, Kragh PM, Zhang Y, Li J, Schmidt M, et al. (2007) Piglets born from handmade cloning, an innovative cloning method without micromanipulation. Theriogenology 68: 1104–1110.

26. Kragh PM, Vajta G, Corydon TJ, Purup S, Bolund L, et al. (2004) Production of transgenic porcine blastocysts by hand-made cloning. Reprod Fertil Dev 16: 315–318.

27. Schmidt M, Kragh PM, Li J, Du Y, Lin L, et al. (2010) Pregnancies and piglets from large white sow recipients after two transfer methods of cloned and transgenic embryos of different pig breeds. Theriogenology 74: 1233–1240.

28. Kragh PM, Nielsen AL, Li J, Du Y, Lin L, et al. (2009) Hemizygous minipigs produced by random gene insertion and handmade cloning express the Alzheimer's disease-causing dominant mutation APPsw. Transgenic Res 18: 545–558.

29. Staunstrup NH, Madsen J, Primo MN, Li J, Liu Y, et al. (2012) Development of transgenic cloned pig models of skin inflammation by DNA transposon-directed ectopic expression of human beta1 and alpha2 integrin. PLoS One 7: e36658.

30. Al-Mashhadi RH, Sorensen CB, Kragh PM, Christoffersen C, Mortensen MB, et al. (2013) Familial hypercholesterolemia and atherosclerosis in cloned minipigs created by DNA transposition of a human PCSK9 gain-of-function mutant. Sci Transl Med 5: 166ra161.

31. Zhang P, Zhang Y, Dou H, Yin J, Chen Y, et al. (2012) Handmade cloned transgenic piglets expressing the nematode fat-1 gene. Cell Reprogram 14: 258–266.

32. Okano S, Akashi M, Hayasaka K, Nakajima O (2009) Unusual circadian locomotor activity and pathophysiology in mutant CRY1 transgenic mice. Neurosci Lett 451: 246–251.

33. Yoshioka K, Suzuki C, Tanaka A, Anas IM, Iwamura S (2002) Birth of piglets derived from porcine zygotes cultured in a chemically defined medium. Biol Reprod 66: 112–119.

34. Xing W, Busino L, Hinds TR, Marionni ST, Saifee NH, et al. (2013) SCF(FBXL3) ubiquitin ligase targets cryptochromes at their cofactor pocket. Nature 496: 64–68.

35. Nagoshi E, Saini C, Bauer C, Laroche T, Naef F, et al. (2004) Circadian gene expression in individual fibroblasts: cell-autonomous and self-sustained oscillators pass time to daughter cells. Cell 119: 693–705.

36. Welsh DK, Yoo SH, Liu AC, Takahashi JS, Kay SA (2004) Bioluminescence imaging of individual fibroblasts reveals persistent, independently phased circadian rhythms of clock gene expression. Curr Biol 14: 2289–2295.

37. Liu AC, Welsh DK, Ko CH, Tran HG, Zhang EE, et al. (2007) Intercellular coupling confers robustness against mutations in the SCN circadian clock network. Cell 129: 605–616.

38. Glossop NR, Hardin PE (2002) Central and peripheral circadian oscillator mechanisms in flies and mammals. J Cell Sci 115: 3369–3377.

39. Zvonic S, Ptitsyn AA, Conrad SA, Scott LK, Floyd ZE, et al. (2006) Characterization of peripheral circadian clocks in adipose tissues. Diabetes 55: 962–970.

40. Woo KC, Ha DC, Lee KH, Kim DY, Kim TD, et al. (2010) Circadian amplitude of cryptochrome 1 is modulated by mRNA stability regulation via cytoplasmic hnRNP D oscillation. Mol Cell Biol 30: 197–205.

41. Brown SA, Fleury-Olela F, Nagoshi E, Hauser C, Juge C, et al. (2005) The period length of fibroblast circadian gene expression varies widely among human individuals. PLoS Biol 3: e338.

42. Sei H, Oishi K, Morita Y, Ishida N (2001) Mouse model for morningness/eveningness. Neuroreport 12: 1461–1464.

43. Groenen MA, Archibald AL, Uenishi H, Tuggle CK, Takeuchi Y, et al. (2012) Analyses of pig genomes provide insight into porcine demography and evolution. Nature 491: 393–398.

44. Schonig K, Weber T, Frommig A, Wendler L, Pesold B, et al. (2012) Conditional gene expression systems in the transgenic rat brain. BMC Biol 10: 77.

45. Vandermeulen G, Richiardi H, Escriou V, Ni J, Fournier P, et al. (2009) Skin-specific promoters for genetic immunisation by DNA electroporation. Vaccine 27: 4272–4277.

46. Woo KC, Kim TD, Lee KH, Kim DY, Kim W, et al. (2009) Mouse period 2 mRNA circadian oscillation is modulated by PTB-mediated rhythmic mRNA degradation. Nucleic Acids Res 37: 26–37.

47. Hasan S, Santhi N, Lazar AS, Slak A, Lo J, et al. (2012) Assessment of circadian rhythms in humans: comparison of real-time fibroblast reporter imaging with plasma melatonin. FASEB J 26: 2414–2423.

48. Pagani L, Semenova EA, Moriggi E, Revell VL, Hack LM, et al. (2010) The physiological period length of the human circadian clock in vivo is directly proportional to period in human fibroblasts. PLoS One 5: e13376.

49. Kornmann B, Schaad O, Bujard H, Takahashi JS, Schibler U (2007) System-driven and oscillator-dependent circadian transcription in mice with a conditionally active liver clock. PLoS Biol 5: e34.

50. Fan Y, Hida A, Anderson DA, Izumo M, Johnson CH (2007) Cycling of CRYPTOCHROME proteins is not necessary for circadian-clock function in mammalian fibroblasts. Curr Biol 17: 1091–1100.

51. Brown SA, Zumbrunn G, Fleury-Olela F, Preitner N, Schibler U (2002) Rhythms of mammalian body temperature can sustain peripheral circadian clocks. Curr Biol 12: 1574–1583.

52. Yagita K, Tamanini F, Yasuda M, Hoeijmakers JH, van der Horst GT, et al. (2002) Nucleocytoplasmic shuttling and mCRY-dependent inhibition of ubiquitylation of the mPER2 clock protein. EMBO J 21: 1301–1314.

53. Langmesser S, Tallone T, Bordon A, Rusconi S, Albrecht U (2008) Interaction of circadian clock proteins PER2 and CRY with BMAL1 and CLOCK. BMC Mol Biol 9: 41.

54. Bollinger T, Bollinger A, Oster H, Solbach W (2010) Sleep, immunity, and circadian clocks: a mechanistic model. Gerontology 56: 574–580.

55. Ben-Neriah Y, Karin M (2011) Inflammation meets cancer, with NF-kappaB as the matchmaker. Nat Immunol 12: 715–723.

56. Gregor MF, Hotamisligil GS (2011) Inflammatory mechanisms in obesity. Annu Rev Immunol 29: 415–445.

57. Sun B, Karin M (2012) Obesity, inflammation, and liver cancer. J Hepatol 56: 704–713.

58. Narasimamurthy R, Hatori M, Nayak SK, Liu F, Panda S, et al. (2012) Circadian clock protein cryptochrome regulates the expression of proinflammatory cytokines. Proc Natl Acad Sci U S A 109: 12662–12667.

59. Buhr ED, Yoo SH, Takahashi JS (2010) Temperature as a universal resetting cue for mammalian circadian oscillators. Science 330: 379–385.

60. Refinetti R, Menaker M (1992) The circadian rhythm of body temperature. Physiol Behav 51: 613–637.

61. Nagashima K, Matsue K, Konishi M, Iidaka C, Miyazaki K, et al. (2005) The involvement of Cry1 and Cry2 genes in the regulation of the circadian body temperature rhythm in mice. Am J Physiol Regul Integr Comp Physiol 288: R329–335.

11

A Critical Regulatory Role for Macrophage Migration Inhibitory Factor in Hyperoxia-Induced Injury in the Developing Murine Lung

Huanxing Sun[1], Rayman Choo-Wing[1], Angara Sureshbabu[1], Juan Fan[2], Lin Leng[2], Shuang Yu[2], Dianhua Jiang[3], Paul Noble[3], Robert J. Homer[4], Richard Bucala[2], Vineet Bhandari[1]*

1 Department of Pediatrics, Yale University, New Haven, Connecticut, United States of America, 2 Department of Medicine, Yale University, New Haven, Connecticut, United States of America, 3 Department of Medicine, Duke University School of Medicine, Durham, North Carolina, United States of America, 4 Department of Pathology, Yale University, New Haven, Connecticut, United States of America

Abstract

Background: The role and mechanism of action of MIF in hyperoxia-induced acute lung injury (HALI) in the newborn lung are not known. We hypothesized that MIF is a critical regulatory molecule in HALI in the developing lung.

Methodology: We studied newborn wild type (WT), MIF knockout (MIFKO), and MIF lung transgenic (MIFTG) mice in room air and hyperoxia exposure for 7 postnatal (PN) days. Lung morphometry was performed and mRNA and protein expression of vascular mediators were analyzed.

Results: MIF mRNA and protein expression were significantly increased in WT lungs at PN7 of hyperoxia exposure. The pattern of expression of Angiopoietin 2 protein (in MIFKO>WT>MIFTG) was similar to the mortality pattern (MIFKO>WT>MIFTG) in hyperoxia at PN7. In room air, MIFKO and MIFTG had modest but significant increases in chord length, compared to WT. This was associated with decreased expression of Angiopoietin 1 and Tie 2 proteins in the MIFKO and MIFTG, as compared to the WT control lungs in room air. However, on hyperoxia exposure, while the chord length was increased from their respective room air controls, there were no differences between the 3 genotypes.

Conclusion: These data point to the potential roles of Angiopoietins 1, 2 and their receptor Tie2 in the MIF-regulated response in room air and upon hyperoxia exposure in the neonatal lung.

Editor: Peter Chen, University of Washington, United States of America

Funding: Supported in part by grants 0755843T (VB) from the American Heart Association, ATS-07-005 (VB) from the American Thoracic Society; HL-74195 (VB), HL-085103 (VB) from the National Institutes of Health (NIH) National Heart, Lung, and Blood Institute (NHLBI), and AI042310 (RB) from the National Institute of Allergy and Infectious Diseases (NIAID) of the NIH. The funders had no role in study design, data collection and analysis, decision to publish, or preparation of the manuscript.

Competing Interests: The authors have declared that no competing interests exist.

* E-mail: vineet.bhandari@yale.edu

Introduction

Hyperoxia-induced lung injury is characterized by an influx of inflammatory cells, increased pulmonary permeability, and endothelial and epithelial cell injury/death [1,2]. A variety of molecular mediators, including cytokines have been implicated as having a critical role in these processes [1,2]. There are significant differences in the response to HALI in the developing and mature lungs, and it is important to identify and understand such unique responses in order to assess potential therapeutic approaches to HALI in such circumstances [1,3]. We have recently reported that loss of macrophage migration inhibitory factor (MIF) in the mouse leads to impaired lung maturation, akin to a respiratory distress syndrome (RDS) in premature newborns, and increased mortality [4]. Neonates with RDS are frequently exposed to high concentrations of oxygen for a prolonged duration as part of their therapeutic regimen. Surprisingly, we found no information

about MIF expression on exposure to hyperoxia, or the role, if any, in HALI in the developing lung.

Hence, we hypothesized that MIF is a critical regulatory molecule in HALI in the newborn lung. To address this hypothesis, we studied newborn (NB) mice genetically deficient in MIF (MIF null mutant or knock out: MIFKO). We generated a novel lung-targeted MIF overexpressing TG (MIFTG) mouse and examined the phenotype. These genetically defined mouse strains were further subjected to a HALI model, with sustained exposure to hyperoxia for 7 postnatal (PN) days. Given previous reports by us [4] and others [5–13] about VEGF being involved in MIF signaling and the potential role of VEGF in alveolar simplification in the developing lung, we also evaluated NB MIFKO mice crossbred with the vascular endothelial factor (VEGF) overexpressing TG mice.

Our goal was to study alterations in pulmonary phenotype and the expression of vascular mediators in the varied mouse models. Specifically, we evaluated lung morphometry and the expression of vascular endothelial growth factor (VEGF), its receptors (R1–3), angiopoietin (Ang), and its receptor (Tie2) in the lung.

Materials and Methods

Animals

MIFKO mice were backcrossed for this study into the C57BL/6 genetic background (generation N10) [4]. A transgenic mouse overexpressing MIF in Type II lung alveolar cells, which produce MIF under basal conditions [14], was created by a previously established methodology [15]. Briefly, a 0.4 Kb DNA fragment containing the complete mouse MIF cDNA [16] was inserted into the Xbal/Spel site of an expression plasmid under the control of the rat CC10, which is a Type II epithelial-cell and conducting airway-cell specific promoter in the developing lung [17]. Plasmid transfection into Bruce4 ES cells permitted immediate establishment of founder mice in the pure C57BL/6 background. Immunohistochemistry analysis at 6 weeks confirmed MIF overexpression in alveolar epithelium and quantitative, real-time PCR analysis of MIF mRNA in lung tissue and MIF protein levels in bronchoalveolar lavage fluid revealed a 2.4 and 4.3 fold increase in MIF mRNA transcripts and MIF protein, respectively (**figure S1**).

Lung-targeted and regulatable VEGF TG mice overexpressing $VEGF_{165}$, a kind gift from Jack Elias, MD, were generated and characterized as previously described [18–20], and were cross-bred with MIF KO mice. The VEGF TG mice express $VEGF_{165}$ in the lung with maternal exposure to doxycycline (dox) in the drinking water, leading to transmammary activation in the TG (+) pups [20]. Maternal exposure to dox was performed from PN1 to PN7.

All animal work was approved by the Institutional Animal Care and Use Committee at the Yale University School of Medicine.

Oxygen exposure

For the NB animals, exposure to hyperoxia (along with their mothers) was performed by placing mice in cages in an airtight Plexiglass chamber ($55 \times 40 \times 50$ cm), as described previously [20,21]. For the NB survival experiments, exposure to 100% oxygen was initiated on PN1 and continued until PN7. Two lactating dams were used and alternated in hyperoxia and room air (RA) every 24 h. The litter size was limited to 10–12 pups per dam to control for the effects of litter size on nutrition and growth.

Throughout the experiment, mice were given free access to food and water, and oxygen levels were continually monitored. The inside of the chamber was maintained at atmospheric pressure and mice were exposed to a 12 hr light-dark cycle.

Analysis of mRNA

RNA was isolated from frozen lungs using TRIzol Reagent (Invitrogen Corporation, Carlsbad, CA) and treated by DNase. RNA samples were then purified by RNeasy kit (Qiagen Sciences, Maryland, USA) according to the manufacturer's instructions. RNA samples were subjected to real time PCR and semi-quantitative RT-PCR. The primers used for real time PCR:

MIF-F: 5′-CCA GAA CCG CAA CTA CAG TAA -3′
MIF-R: 5′ -CCG GTG GAT AAA CAC AGA AC -3′
The primers used for semi-quantitative RT-PCR are as follows:
β-Actin-F, 5′-GTGGGCCGCTCTAGGCACCA -3′
β-Actin-R, 5′-TGGCCTTAGGGTTCAGGGGG -3′
VEGF-A-F, 5′- GACCCTGGCTTTACTGCTGTA -3′
VEGF-A-R, 5′- GTGAGGTTTGATCCGCATGAT -3′
VEGF-C-F, 5′- AACGTGTCCAAGAAATCAGCC -3′
VEGF-C-R, 5′- AGTCCTCTCCCGCAGTAATCC -3′

VEGF-R1-F, 5- CACCACAATCACTCCAAAGAAA -3′
VEGF-R1-R, 5- CACCAATGTGCTAACCGTCTTA -3′
VEGF-R2-F, 5- ATTGTAAACCGGGATGTGAAAC -3′
VEGF-R2-R, 5- TACTTCACAGGGATTCGGACTT -3′
VEGF-R3-F, 5- GCTGTTGGTTGGAGAGAAGC -3′
VEGF-R3-R, 5- TGCTGGAGAGTTCTGTGTGG -3′
Ang1-F, 5′- AGGCTTGGTTTCTCGTCAGA -3′
Ang1-R, 5′- TCTGCACAGTCTCGAAATGG -3′
Ang2-F, 5′- GAACCAGACAGCAGCACAAA -3′
Ang2-R, 5′- AGTTGGGGAAGGTCAGTGTG -3′
Tie2-F, 5′- GGACAGTGCTCCAACCAAAT -3′
Tie2-R, 5′- TTGGCAGGAGACTGAGACCT -3′
mRNA band densities were measured by densitometry using NIH image J and expressed in Arbitrary Densitometric Units (ADU), as previously described [21].

Histology

Lung tissues obtained from the NB mice from the RA experiments at PN7 were subjected to a standard protocol for lung inflation (25 cm) and fixed overnight in 10% buffered formalin [22]. After washing in fresh PBS, fixed tissues were dehydrated, cleared, and embedded in paraffin by routine methods. Sections (5 μm) were collected on Superfrost Plus positively charged microscope slides (Fisher Scientific Co., Houston, Texas, USA), deparaffinized, and stained with hematoxylin & eosin, as described previously [20].

Lung Morphometry

Alveolar size was estimated from the mean chord length of the airspace, as described previously [20]. Chord length increases with alveolar enlargement. Alveolar septal wall thickness was estimated using Image J software, adapting the method described previously for bone trabecular thickness, for lung [23].

Bronchoalveolar lavage (BAL) fluid total and differential cell counts and IL-6 levels

BAL fluid was obtained and total and differential cell counts enumerated, as previously reported [24,25]. IL-6 levels were measured using ELISA (R&D Systems, Minneapolis, MN), as per manufacturer's instructions.

Quantitative measurements of Pulmonary artery hypertension (PAH)-induced right ventricular hypertrophy (RVH)

We assessed quantitative measurements of PAH-induced RVH by RV/left ventricle (LV) and RV/(LV + interventricular septum or IVS) ratios, using the methodology as described previously [26].

Western Blotting

We detected MIF, Ang1, Ang2, Tie2, VEGF-A, and VEGFR1 protein with β-actin as control from lung lysates using Western analysis, as described previously [20,25]. Anti-MIF specific antibody was purchased from Abcam (Cambridge, MA). Ang1 and Ang2 antibodies were purchased from Millipore (Billerica, MA). VEGF-A and β-actin antibodies were purchased from Santa Cruz Biotechnology, Santa Cruz, CA. VEGFR1 and Tie2 antibodies were obtained from Cell Signaling, Danvers, MA.

Statistical Analyses

Values are expressed as mean ± SEM. Groups were compared with the Student's two-tailed unpaired t test or the logrank test (for the survival analysis), using GraphPad Prism 3.0 (GraphPad

Software, Inc., San Diego, CA), as appropriate. A $P<0.05$ was considered statistically significant.

Results

Effect of hyperoxia on MIF expression in NB WT, MIFKO and MIFTG mice lungs

Using the hyperoxia-exposed NB WT mouse model, we noted significantly increased expression of lung MIF mRNA (**figure 1A**). MIF protein expression data from NB WT, MIFKO and MIFTG mice lungs at PN7 in RA and after exposure to 100% O_2 at PN7 has been shown in **figure 1B**.

Effect of hyperoxia on survival in NB WT, MIFKO and MIFTG mice

There was no statistically significant difference in survival in RA (up to 7 days) of MIFKO mice (MIFKO; n = 24; survival = 82.6%) or lung-targeted, over-expressing MIFTG mice (MIFTG; n = 12; survival = 91.7%) when compared to WT mice (n = 26; survival = 92.3%). However, when we evaluated the survival of MIFKO

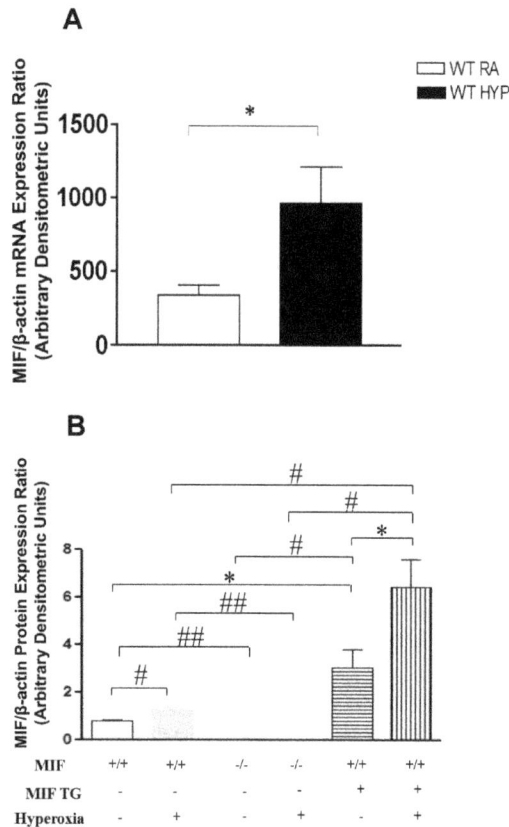

Figure 2. Effect of hyperoxia on survival in NB MIF KO and MIF TG mice at PN7. NB MIF KO, MIF TG and WT control mice were exposed to 100% O_2 and survival was assessed. The noted values represent the number of animals in each group. NB: newborn; WT: wild type; MIF KO: macrophage migration inhibitory factor knock out; MIF TG: macrophage migration inhibitory factor (over-expressing) transgenic; HYP: hyperoxia. $\#P \leq 0.01$.

mice and MIFTG mice after hyperoxia for 7 days, there was a significant decrease in survival of the MIFKO mice when compared to WT controls at PN7 (p≤0.01) (**figure 2**). Notably, the MIFTG mice had significantly increased survival compared to the MIFKO mice at PN7 (p≤0.01) (**figure 2**). Similarly, there was a trend towards increased survival of MIFTG, compared to WT mice (p = 0.08) (**figure 2**).

Effect of hyperoxia on lung architecture in NB MIFKO and MIFTG mice

Surprisingly, both the MIFKO and MIFTG mouse lungs exhibited simplified alveoli with a reduction in secondary crests when compared to the WT control mice in RA (**figure 3A**). After hyperoxia exposure from PN1-7, there was a significant alveolar simplification in NB WT, MIFKO and MIFTG mice when compared to their respective RA controls (**figure 3A**) However, the alveolar simplification appeared to be of a similar degree in the NB WT, MIFKO and MIFTG mice lungs exposed to hyperoxia till PN7(**figure 3A**). These results were confirmed by lung morphometry on chord length measurements (**figure 3B**). Septal thickness was higher in NB WT mice exposed to hyperoxia, compared to RA (**figure 3C**). NB MIFKO mice lungs had the highest septal thickness, which decreased on hyperoxia exposure, but with values similar to that of NB WT in hyperoxia (**figure 3C**). NB MIFTG mice lungs in RA or hyperoxia-exposed also had similar values to that of NB WT in hyperoxia (**figure 3C**).

Taken together, these data suggest that either a lack or an excess of MIF in the developing lung in RA results in significant alterations in pulmonary architecture.

Effect of hyperoxia on BAL total cell counts, total protein and IL-6 levels in NB MIFKO and MIFTG mice

To gain insight into potential mechanisms underlying alveolar simplification and the increased survival of the NB MIFTG mice on hyperoxia exposure, we evaluated cell counts in BAL fluid obtained from MIFKO, MIFTG, and WT mice subjected to RA or hyperoxia. In RA, MIFKO animals had the lowest cell numbers among the experimental groups (**figure 3D**). As anticipated, exposure to hyperoxia resulted in a significant increase in the total BAL cell count at PN7 in WT animals (**figure 3D**). A similar pattern was noted in the lungs of MIFKO and MIFTG mice that survived exposure to hyperoxia (**figure 3D**). On exposure to hyperoxia, WT animals had the most robust increase in BAL total

Figure 1. Effect of hyperoxia on MIF expression in NB WT, MIF KO and MIF TG mice lungs at PN7. (A) Quantitative real-time PCR of MIF mRNA expression in lungs of NB mice exposed to hyperoxia (as described in "Methods" section) at PN7. Control lungs were exposed to room air for 7 days. The figure is representative of n = 3–4 mice per group. **(B)** MIF protein expression data from WT, NB MIF KO and MIF TG mice lungs at PN7 in RA and exposed to 100% O2 at PN7. The figure is representative of n = 4 mice per group. NB: newborn; WT: wild type; MIF KO: macrophage migration inhibitory factor knock out; MIF TG: macrophage migration inhibitory factor (over-expressing) transgenic; β: beta; HYP: hyperoxia; PN: post natal; RA: room air. $*P \leq 0.05$; $\#P \leq 0.01$; $\#\#P < 0.0001$.

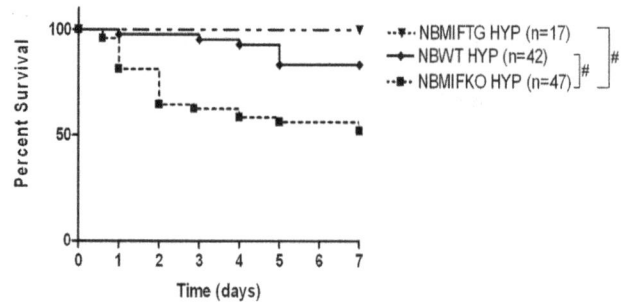

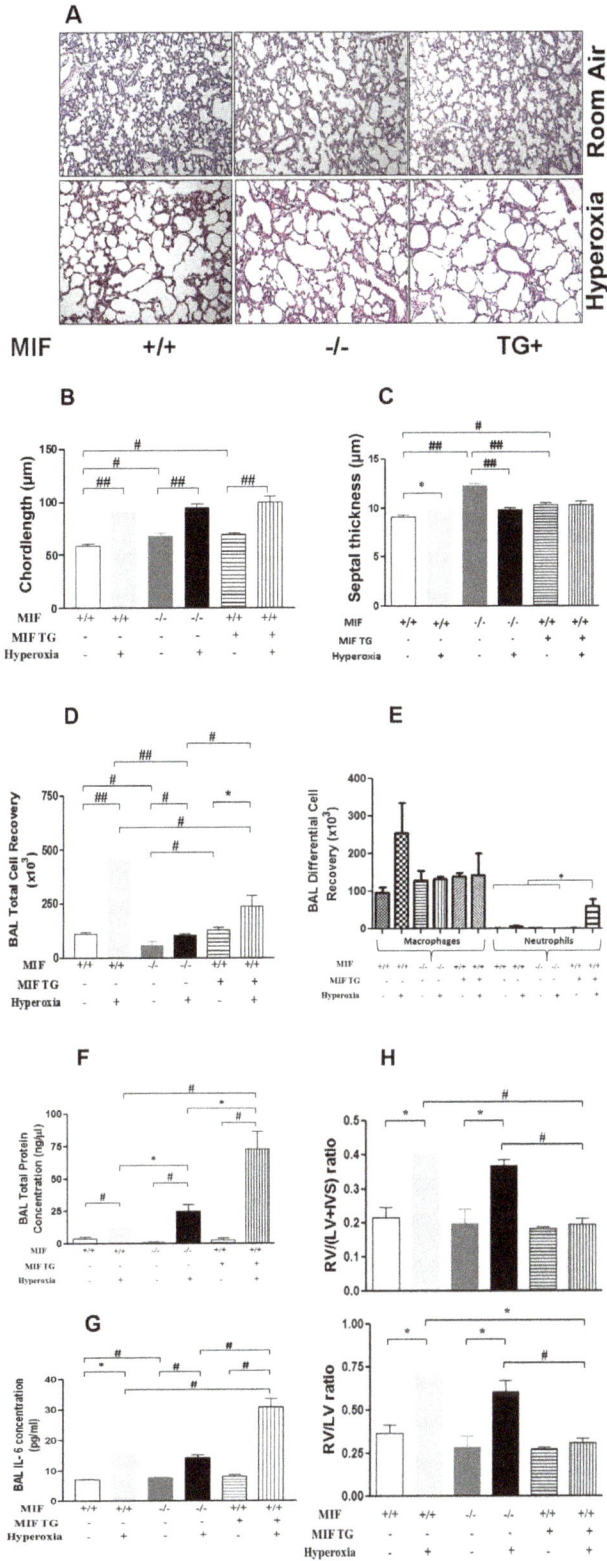

thickness, confirmed features noted on lung histology (3B and 3C). Each bar represents the mean ± SEM of a minimum of 5 animals. BAL total and differential cell recovery of NB MIF KO, MIF TG and WT control mice exposed to RA or survived 100% O_2 at PN7 (3D and 3E). Each bar represents the mean ± SEM of a minimum of 12 animals. BAL total protein levels of NB MIF KO, MIF TG and WT control mice exposed to RA or survived 100% O_2 at PN7 (3F). Each bar represents the mean ± SEM of a minimum of 4 animals. BAL IL-6 levels of NB MIF KO, MIF TG and WT control mice exposed to RA or survived 100% O_2 at PN7 (3G). Quantitative measurements of RV hypertrophy (3H).[§]Each bar represents the mean ± SEM of a minimum of 3 animals. MIF +/+: wild type; MIF −/−: macrophage migration inhibitory factor knock out; MIF TG +: macrophage migration inhibitory factor (over-expressing) transgenic; BAL: bronchoalveolar lavage; IL-6: interleukin-6; RV: right ventricle; LV: left ventricle; IVS: interventricular septum. [§]Note: MIF KO assessments were done at PN6, due to their increased mortality in hyperoxia at PN7. ##$P<0.0001$, #$P≤0.01$, *$P<0.05$.

Figure 3. Effect of hyperoxia on lung architecture, BAL total and differential cell counts, BAL total protein concentrations and IL-6 levels in NB MIF KO and MIF TG mice at PN7. Representative photomicrographs of lung histology (H&E stain) of NB MIF KO, MIF TG and WT control mice exposed to RA or survived 100% O_2 at PN7 (3A). The figures are illustrative of a minimum of 8 animals in each group. Alveolar size, as measured by chord length and septal

cell counts, and it was significantly higher than the cell counts observed in the MIFTG, which, in turn, was higher than in MIFKO mice (**figure 3D**). In the differential cell counts, the majority of the cells in RA or hyperoxia were macrophages. Interestingly, there was a marked increase in neutrophils in the MIFTG mice in hyperoxia (**figure 3E**).

BAL total protein concentrations were significantly increased in the hyperoxia-exposed animals (**figure 3F**), compared to their respective RA controls. The BAL total protein was significantly higher in the MIFKO vs. the WT mice exposed to hyperoxia. However, the highest increase of BAL total protein concentration was noted in the MIFTG mice.

BAL IL-6 levels were significantly increased in the hyperoxia-exposed animals (**figure 3G**). The highest increase of BAL IL-6 levels was noted in the MIFTG mice, with similar concentrations noted in the WT and KO mice exposed to hyperoxia.

Measurements of PAH-induced RVH revealed that these values were significantly increased in the WT and MIFKO mice exposed to hyperoxia, but significantly decreased in the MIFTG mice exposed to hyperoxia (**figure 3H**).

This suggests that enhanced inflammation, as assessed by BAL fluid cell counts, total protein concentration and IL-6 levels, could account for the differences in lung architecture after hyperoxia exposure when compared to RA. Interestingly, the MIFTG mice exposed to hyperoxia appeared to be protected against PAH-induced RVH. However, an explanation for the alveolar simplification noted in RA in the MIFKO and MIFTG mice, and the increased survival of the NB MIFTG mice on hyperoxia exposure was still required.

Effect of hyperoxia on mRNA and protein expression of vascular growth factors and their receptors in MIFKO and MIFTG mice

Since alveolar simplification is associated with impaired vasculogenesis [27], we next studied the mRNA expression of vascular endothelial growth factors (VEGF) -A and –C, and their receptors -R1, -R2 and -R3. In addition, we measured Angiopoietin (Ang)-1 and 2, and their common receptor, Tie2 (**figures 4A**–**B**). As shown in **figure 4A**, there was a significant reduction in mRNA expression of VEGF-A in the MIF KO mice lungs in RA and after hyperoxia when compared to the WT mice in RA at PN7. No differences were observed in the expression levels of VEGF-C and VEGF-R2 mRNA (data not shown). Interestingly, MIFTG mice exposed to hyperoxia had consistently decreased expression of VEGF -R1 and -R3 mRNA, compared to WT or MIFTG mice in RA, as well as WT or MIFKO in

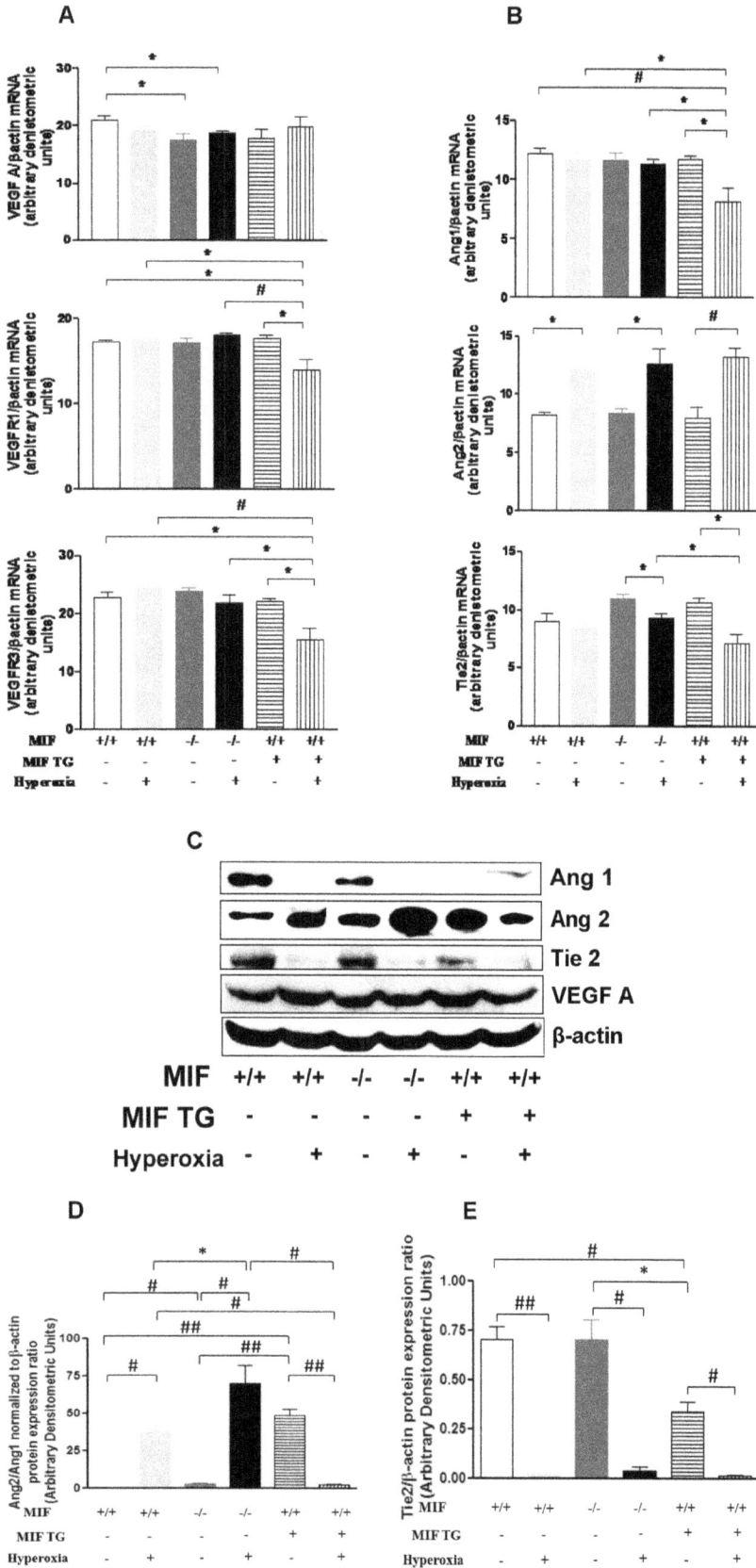

Figure 4. Effect of hyperoxia on RNA expression of vascular mediators and their receptors in NB MIF KO and MIF TG mice at PN7.
The ratios of mRNA of VEGF -A, and the receptors -R1 and -R3 (4A) as well as Ang1 and Ang2 and their receptor Tie2 (4B) with β-actin were quantified by densitometry. Ang1, Ang2, Tie2, and VEGF-A proteins, with β-actin as controls, were detected by western blotting (4C). The densitometric values

of Ang2/Ang1 ratio and Tie2 protein expression are noted in 4D and 4E, respectively. The noted values represent assessments in a minimum of 4 animals in each group. MIF +/+: wild type; MIF −/−: macrophage migration inhibitory factor knock out; MIF TG +: macrophage migration inhibitory factor (over-expressing) transgenic; VEGF: vascular endothelial growth factor; R: receptor; Ang: angiopoietin; Tie2: Tyrosine kinase with Ig and EGF homology domain 2 (receptor for Ang1 and 2); β-actin: beta actin. ##$P<0.0001$, #$P\leq0.01$, *$P<0.05$.

hyperoxia (**figure 4A**). A similar pattern of decreased expression of Ang1 was also observed in MIF TG mice lungs exposed to hyperoxia when compared to WT or MIFKO mice (**figure 4B**). However, the expression of Ang2 mRNA was significantly increased in the hyperoxia-exposed mouse lungs when compared to their respective RA controls (**figure 4B**). Finally, Tie2 mRNA expression was decreased in the MIFKO and MIFTG mouse lungs after hyperoxia when compared to their respective RA controls (**figure 4B**) and MIFTG mouse lungs showed a significant decrease in the expression of Tie2 after hyperoxia when compared to hyperoxia-exposed MIFKO mouse lungs (**figure 4B**).

We sought to confirm the modest differences in mRNA at the protein level (**figure 4C**). VEGF-A protein expression appeared to be similar in the WT, MIFKO, and MIFTG mice. Ang1 and Tie2 expression were lower in the MIFKO and MIFTG lungs, compared to WT. Ang1 and Tie2 protein expression decreased further in all the mouse strains after hyperoxia exposure. Although Ang2 protein expression decreased in the MIFTG group after hyperoxia exposure, it was increased in the WT and MIFKO groups, which contrasted with the results obtained by RNA expression. This pattern may reflect either impaired release of the protein or its uptake and retention by hyperoxia-exposed cells in WT and MIFKO groups. The protein expression has been quantified in figures **4D and 4E**.

In summary, the lack of MIF in the murine developing lung appears not to influence VEGF-A protein expression while a lack or excess of MIF is associated with a significant reduction in Ang1 and Tie2 proteins. Exposure to hyperoxia led to a further decrease in Ang1 protein, and an increase in Ang2 protein, except in the case of excess MIF expressed by MIFTG lungs, where a decrease in Ang2 protein was noted (**4C and 4D**).

Effect of lung-targeted VEGF overexpression in MIF KO mice lungs

Our data did not show a difference in VEGF-A protein expression in the WT and MIFKO mice in RA (**figure 4C**). However, earlier reports have suggested VEGF involvement in MIF signaling [7,9,12,28] and alveolar simplification in the developing lung [4]. To test if enhanced VEGF signaling downstream of MIF could impact on lung architecture in the NB lung, we studied an externally-regulatable, lung-targeted VEGF$_{165}$ TG mouse cross-bred with the MIFKO mouse in RA. In our earlier study of VEGFTG mice in RA, we had noted thinning of the mesenchyme and larger alveoli that, in conjunction with increased expression and production of surfactant, suggested an enhancement of lung maturation [20]. Given the impaired lung maturation of the MIF KO mice [4], we wondered if provision of "exogenous" VEGF could enhance lung maturation in such a scenario. An examination of VEGF TG × MIF KO mice in RA revealed no significant recovery of the pulmonary phenotype and no difference in the BAL total cell counts (**figures 5A–C**) suggesting that VEGF and MIF function through non-overlapping pathways of lung development. This result was confirmatory of our data noting no effect of VEGF-A protein expression in the MIFKO, compared to WT (**figure 4C**).

Discussion

We have reported earlier that loss of MIF led to lung immaturity, a condition akin to RDS in premature NB [4]. Recently, in a preterm lamb model, it was suggested that MIF might be an important maturational and protective factor for NB lungs [29]. Since RDS in NB infants is almost always managed with high concentrations of supplemental oxygen, the present studies were undertaken to better define MIF's role in the developing lung in RA and in response to hyperoxia. We employed both genetic loss- and gain-of-function strategies with MIFKO and lung MIFTG mice, respectively.

Sustained and prolonged exposure to hyperoxia to the WT NB mouse lung in the saccular (PN1-4) and alveolar (>PN5) stages of development led to a significantly increased expression of MIF – both at the mRNA and protein level. This could be a damaging or a protective response. To test this, we exposed NB MIFKO, and a novel lung-targeted overexpressing MIFTG mouse, along with WT controls to hyperoxia. In accord with loss of MIF being harmful, there was significantly increased mortality in the MIFKO mice, compared to WT (and MIFTG) mice. Importantly, MIFTG mice had the best survival of all 3 groups. This would suggest that enhanced expression of MIF in the WT mouse on exposure to hyperoxia was a protective response. It is interesting to speculate that given the fact that NB mice are inherently resistant to HALI, vis-à-vis their adults counterparts (of the same strain) [1–3], MIF could be responsible, in part, for this enhanced survival response to HALI.

Next, we sought to establish the reason for the increased survival of the MIFTG mice in hyperoxia. The BAL cell count, though lower in the MIFTG vs. WT, was significantly higher than that of the MIFKO mice. Hence, while we cannot rule out a different inflammatory response in the lung parenchyma, we believe the degree of inflammation (as assessed by BAL cell counts) did not sufficiently explain the protective effects in hyperoxia. If the BAL total protein concentrations is taken as a surrogate marker for alveolar-capillary integrity [30,31], the increased amounts in the MIFKO mice vs. WT in hyperoxia would suggest that there may be more vascular leak/pulmonary edema in the MIFKO mice; thus, providing 1 possible explanation for their increased mortality. However, the highest BAL total protein concentration was noted in in the MIFTG mice lungs in hyperoxia, which had the best survival among the 3 groups. This would suggest that despite significant vascular leak/pulmonary edema, increased release/secretion of MIF itself and/or some other factor/protein was possibly contributing to the protective response in hyperoxia [32,33].

Hence, we also measured BAL IL-6 levels, and to our surprise, the highest concentrations were noted in the MIFTG mice exposed to hyperoxia. The marked increase in neutrophils in the MIFTG mice in hyperoxia appears to correlate with the increased IL-6 levels in the same mice. While traditionally, IL-6 is considered "pro-inflammatory", high levels of IL-6 in a lung-targeted mouse model of HALI was protective [21]. However, the same IL-6 TG mouse model in the NB period was exquisitely sensitive to HALI, and had significantly increased mortality [21]. It is possible that there is a threshold effect of IL-6 levels in the developing lungs exposed to hyperoxia in terms of it being

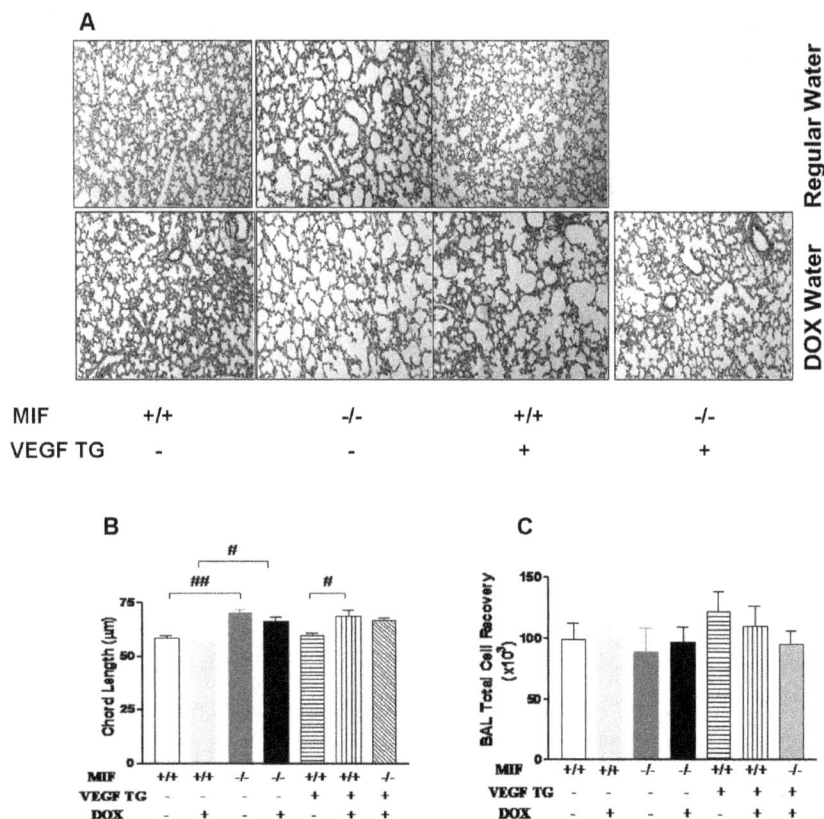

Figure 5. Effect of lung-targeted VEGF overexpression in NB MIF KO mouse lungs. Representative photomicrographs of lung histology (H&E stain) of NB MIF KO, MIF KO VEGF TG and WT litter-mate controls, on regular or dox (to activate VEGF165 in the VEGF TG mice) water mice in RA at PN7 (**5A**). The figures are illustrative of a minimum of 4 animals in each group. Alveolar size, as measured by chord length, confirmed features noted on lung histology (**5B**). Each bar represents the mean ± SEM of a minimum of three animals. BAL total cell count of NB MIF KO, MIF KO VEGF TG and WT litter-mate controls, on regular or dox (to activate VEGF165 in the VEGF TG mice) water mice in RA at PN7 (**5C**). Each bar represents the mean ± SEM of a minimum of three animals. VEGF: vascular endothelial growth factor; MIF +/+: wild type; MIF −/−: macrophage migration inhibitory factor knock out; VEGF TG+: vascular endothelial growth factor (over-expressing) transgenic; dox: doxycycline; BAL: bronchoalveolar lavage. ##$P<0.0001$, #$P≤0.01$.

damaging or protective, but additional research would be needed to sort this out.

The MIFKO mice appeared to have a similar degree of increased PAH-induced RVH as the WT mice exposed to hyperoxia; the latter response has also been noted in neonatal rats [26]. However, the MIFTG mice exposed to hyperoxia appeared to be protected against PAH-induced RVH. These data are supportive of their increased survival on hyperoxia exposure. Further research will be required to understand the mechanism of this differential response.

Since angiogenic agents have been implicated in the beneficial response to HALI in developing lung models [1,20], we also evaluated multiple factors and their receptors in our mouse models of HALI. It was striking to note that the expression of Ang2 (MIFKO>WT>MIFTG) paralleled the mortality pattern (MIF-KO>WT>MIFTG). While both Ang1 and 2 are known to act via their receptor Tie2, they have opposite effects. For example, in the context of HALI, Ang2 increases vascular permeability, while Ang1 tends to stabilize the blood vessels and decreases vascular permeability. Our previously published data [25,34] suggest that higher levels of Ang2 in response to hyperoxia exposure in the lung are detrimental in the context of HALI. Such effects in HALI would be further compounded by decreased Ang1. In other words, an increase in Ang2, with a concomitant decrease in Ang1 would

enhance HALI. In the WT mice lungs exposed to hyperoxia, Ang2 is increased, while Ang1 is decreased. In the MIFKO mice lungs exposed to hyperoxia, Ang2 is markedly increased with not much change in Ang1. The net effect of the Ang2/Ang1 expression ratio is increased Ang2, which we speculate could be contributing to enhanced HALI and increased mortality in the MIFKO mice. Since increased Ang2 has been shown in mouse models and human subjects (including premature NB with RDS) to be associated with increased mortality and/or adverse pulmonary outcomes [25,34,35], we speculate that a decreased concentration of Ang2 could be playing a role in the protective response of the MIFTG mice in hyperoxia.

On assessing the phenotype of the MIFKO and MIFTG mice lungs in RA, we noted a modest, but significant, increase in chord length and septal thickness, compared to WT RA controls. We observed a significant reduction in the protein expression of Ang1 and Tie2, compared to WT-RA controls. This result potentially suggests decreased Ang1 signaling, and given the importance of the Ang1-Tie2 axis in vascular integrity [36], and interaction in vascular and alveolar development [37,38], this pathway may be responsible for the alveolar simplification phenotype in the MIFKO and MIFTG mouse lungs in RA.

MIFKO and lung MIFTG mice in RA at PN7 showed altered alveolar architecture that did not worsen further upon hyperoxia exposure (based on chord length). This reiterates two points: first, an optimal amount of MIF is required for normal lung development, and second, the severity of impaired alveolarization with lack or excess of MIF is equivalent to that caused by hyperoxia in the WT developing lung, with no additive effect of hyperoxia in the MIFKO and MIFTG mouse lung phenotype. Hyperoxia exposure in these circumstances did not exacerbate the histological changes noted in RA in the MIFKO or MIFTG mouse lungs, suggesting that there is a limiting or "plateau" effect of MIF expression–either absent or excessive, on lung development. We suggest that the need for "just the right amount" of MIF is functionally critical, and this (the "Goldilocks effect") has been recognized by other investigators in relation to lung [39] and inner ear [40] development.

Taken together, these data suggest that restoring MIF activity within the lung to "just the right amount" is beneficial to the development of normal alveolar architecture.

VEGF-A has been noted to have a potential beneficial role in lung maturation in murine models [41], and we previously reported specifically on the role of $VEGF_{165}$ in lung maturation using the lung-targeted, externally-regulatable TG approach [20] that was utilized in the present study. However, when we cross-bred the $VEGF_{165}$ TG mouse onto a MIF KO background in RA, we did not find any recovery of the MIFKO pulmonary phenotype or impact on total BAL cell counts. This result was supported by our data of lack of effect on VEGF-A protein expression in the MIFKO vs. WT mice in RA. An alternative explanation could be that the increase in chord length noted in the VEGFTG mice at PN7 may have "negated" any potential recovery in lung architecture in the VEGFTG x MIFKO mouse lungs.

Another important aspect to highlight is the developmental regulation of the responses in HALI. For example, the decreased expression of Tie2 mRNA and protein, on exposure to hyperoxia, contrasts with our earlier data from adult mice [25] and could reflect the longer duration of oxygen exposure and/or developmental regulation in the NB lungs. The significant differences, and sometimes completely opposite responses in the NB vs. the adult to HALI have been noted by us [1,3,20,21], and others [42,43]. It is critical that both the developmental stage of the lung and the

severity/duration of hyperoxia be taken into account when comparing studies [44]. Since the functioning of these molecules are context-dependent, studies need to be pursued in developmentally-appropriate pulmonary models.

In summary, we noted that lack of MIF is harmful, while an excess of MIF in the lung is somewhat protective in the developing lung, in terms of survival on hyperoxia exposure. However, a lack or excess of MIF in the developing lung both lead to an alveolar simplification pulmonary phenotype in RA, which was not further worsened by hyperoxia exposure in the survivors. These effects were associated with alterations in the protein expression of the Ang1, Ang2 and Tie2 angiogenic factors. We speculate that the Ang1-Ang2-Tie2 axis signaling pathway mediates the pulmonary effects of MIF in the developing lung.

We suggest that restoring or enhancing MIF activity in the developing lung, without reaching supra-physiological levels, has the potential to improve impaired alveolarization in the infants at risk for adverse outcome secondary to HALI. Additional research is needed to ascertain the ideal circumstances for augmenting MIF action in the lung, potentially via the systemic or intra-pulmonary application of small molecule MIF modulators [45] that would be protective of HALI in the developing lung.

Supporting Information

Figure S1 *Left*: Quantitative, real-time PCR analysis of MIF mRNA in lung tissue in 6 week-old mice from the MIF-TG2 lines compared to wild-type (WT) controls. Methods and primers from [46]. *Right*: MIF protein levels in bronchoalveolar lavage fluid from mice measured by specific ELISA n = 3 mice per group. #$P<0.01$.

Author Contributions

Conceived and designed the experiments: HS RB VB. Performed the experiments: HS RC-W AS JF LL SY DJ RH. Analyzed the data: HS PN RH RB VB. Contributed reagents/materials/analysis tools: JF LL SY DJ PN RH RB VB. Wrote the paper: HS RC-W AS JF LL SY DJ PN RH RB VB.

References

1. Bhandari V, Elias JA (2006) Cytokines in tolerance to hyperoxia-induced injury in the developing and adult lung. Free Radic Biol Med 41: 4–18.
2. Bhandari V (2008) Molecular mechanisms of hyperoxia-induced acute lung injury. Front Biosci 13: 6653–6661.
3. Bhandari V (2002) Developmental differences in the role of interleukins in hyperoxic lung injury in animal models. Front Biosci 7: d1624–1633.
4. Kevill KA, Bhandari V, Kettunen M, Leng L, Fan J, et al. (2008) A role for macrophage migration inhibitory factor in the neonatal respiratory distress syndrome. J Immunol 180: 601–608.
5. Bondza PK, Metz CN, Akoum A (2008) Postgestational effects of macrophage migration inhibitory factor on embryonic implantation in mice. Fertil Steril 90: 1433–1443.
6. Ren Y, Law S, Huang X, Lee PY, Bacher M, et al. (2005) Macrophage migration inhibitory factor stimulates angiogenic factor expression and correlates with differentiation and lymph node status in patients with esophageal squamous cell carcinoma. Ann Surg 242: 55–63.
7. Ren Y, Chan HM, Li Z, Lin C, Nicholls J, et al. (2004) Upregulation of macrophage migration inhibitory factor contributes to induced N-Myc expression by the activation of ERK signaling pathway and increased expression of interleukin-8 and VEGF in neuroblastoma. Oncogene 23: 4146–4154.
8. White ES, Flaherty KR, Carskadon S, Brant A, Iannettoni MD, et al. (2003) Macrophage migration inhibitory factor and CXC chemokine expression in non-small cell lung cancer: role in angiogenesis and prognosis. Clin Cancer Res 9: 853–860.
9. Veillat V, Carli C, Metz CN, Al-Abed Y, Naccache PH, et al. (2010) Macrophage migration inhibitory factor elicits an angiogenic phenotype in human ectopic endometrial cells and triggers the production of major angiogenic factors via CD44, CD74, and MAPK signaling pathways. J Clin Endocrinol Metab 95: E403–412.
10. Xu X, Wang B, Ye C, Yao C, Lin Y, et al. (2008) Overexpression of macrophage migration inhibitory factor induces angiogenesis in human breast cancer. Cancer Lett 261: 147–157.
11. Bondza PK, Metz CN, Akoum A (2008) Macrophage migration inhibitory factor up-regulates alpha(v)beta(3) integrin and vascular endothelial growth factor expression in endometrial adenocarcinoma cell line Ishikawa. J Reprod Immunol 77: 142–151.
12. Ren Y, Tsui HT, Poon RT, Ng IO, Li Z, et al. (2003) Macrophage migration inhibitory factor: roles in regulating tumor cell migration and expression of angiogenic factors in hepatocellular carcinoma. Int J Cancer 107: 22–29.
13. Kim HR, Park MK, Cho ML, Yoon CH, Lee SH, et al. (2007) Macrophage migration inhibitory factor upregulates angiogenic factors and correlates with clinical measures in rheumatoid arthritis. J Rheumatol 34: 927–936.
14. Donnelly SC, Haslett C, Reid PT, Grant IS, Wallace WA, et al. (1997) Regulatory role for macrophage migration inhibitory factor in acute respiratory distress syndrome. Nat Med 3: 320–323.
15. DiCosmo BF, Geba GP, Picarella D, Elias JA, Rankin JA, et al. (1994) Airway epithelial cell expression of interleukin-6 in transgenic mice. Uncoupling of airway inflammation and bronchial hyperreactivity. J Clin Invest 94: 2028–2035.
16. Mitchell R, Bacher M, Bernhagen J, Pushkarskaya T, Seldin MF, et al. (1995) Cloning and characterization of the gene for mouse macrophage migration inhibitory factor (MIF). J Immunol 154: 3863–3870.
17. Perl AK, Wert SE, Loudy DE, Shan Z, Blair PA, et al. (2005) Conditional recombination reveals distinct subsets of epithelial cells in trachea, bronchi, and alveoli. Am J Respir Cell Mol Biol 33: 455–462.
18. Lee CG, Link H, Baluk P, Homer RJ, Chapoval S, et al. (2004) Vascular endothelial growth factor (VEGF) induces remodeling and enhances TH2-mediated sensitization and inflammation in the lung. Nat Med 10: 1095–1103.

19. Bhandari V, Choo-Wing R, Chapoval SP, Lee CG, Tang C, et al. (2006) Essential role of nitric oxide in VEGF-induced, asthma-like angiogenic, inflammatory, mucus, and physiologic responses in the lung. Proc Natl Acad Sci U S A 103: 11021–11026.

20. Bhandari V, Choo-Wing R, Lee CG, Yusuf K, Nedrelow JH, et al. (2008) Developmental regulation of NO-mediated VEGF-induced effects in the lung. Am J Respir Cell Mol Biol 39: 420–430.

21. Choo-Wing R, Nedrelow JH, Homer RJ, Elias JA, Bhandari V (2007) Developmental differences in the responses of IL-6 and IL-13 transgenic mice exposed to hyperoxia. Am J Physiol Lung Cell Mol Physiol 293: L142–150.

22. Harijith A, Choo-Wing R, Cataltepe S, Yasumatsu R, Aghai ZH, et al. (2011) A role for matrix metalloproteinase 9 in IFNgamma-mediated injury in developing lungs: relevance to bronchopulmonary dysplasia. Am J Respir Cell Mol Biol 44: 621–630.

23. Doube M, Klosowski MM, Arganda-Carreras I, Cordelieres FP, Dougherty RP, et al. (2010) BoneJ: Free and extensible bone image analysis in ImageJ. Bone 47: 1076–1079.

24. Bhandari V, Choo-Wing R, Homer RJ, Elias JA (2007) Increased Hyperoxia-Induced Mortality and Acute Lung Injury in IL-13 Null Mice. J Immunol 178: 4993–5000.

25. Bhandari V, Choo-Wing R, Lee CG, Zhu Z, Nedrelow JH, et al. (2006) Hyperoxia causes angiopoietin 2-mediated acute lung injury and necrotic cell death. Nat Med 12: 1286–1293.

26. de Visser YP, Walther FJ, Laghmani el H, Steendijk P, Middeldorp M, et al. (2012) Phosphodiesterase 4 inhibition attenuates persistent heart and lung injury by neonatal hyperoxia in rats. Am J Physiol Lung Cell Mol Physiol 302: L56–67.

27. Thebaud B, Abman SH (2007) Bronchopulmonary dysplasia: where have all the vessels gone? Roles of angiogenic growth factors in chronic lung disease. Am J Respir Crit Care Med 175: 978–985.

28. Munaut C, Boniver J, Foidart JM, Deprez M (2002) Macrophage migration inhibitory factor (MIF) expression in human glioblastomas correlates with vascular endothelial growth factor (VEGF) expression. Neuropathol Appl Neurobiol 28: 452–460.

29. Dani C, Corsini I, Burchielli S, Cangiamila V, Romagnoli R, et al. (2011) Natural surfactant combined with beclomethasone decreases lung inflammation in the preterm lamb. Respiration 82: 369–376.

30. Tang F, Yue S, Luo Z, Feng D, Wang M, et al. (2005) Role of N-methyl-D-aspartate receptor in hyperoxia-induced lung injury. Pediatr Pulmonol 40: 437–444.

31. You K, Xu X, Fu J, Xu S, Yue X, et al. (2012) Hyperoxia disrupts pulmonary epithelial barrier in newborn rats via the deterioration of occludin and ZO-1. Respir Res 13: 36.

32. James ML, Ross AC, Bulger A, Philips JB 3rd, Ambalavanan N (2010) Vitamin A and retinoic acid act synergistically to increase lung retinyl esters during normoxia and reduce hyperoxic lung injury in newborn mice. Pediatr Res 67: 591–597.

33. White CW, Greene KE, Allen CB, Shannon JM (2001) Elevated expression of surfactant proteins in newborn rats during adaptation to hyperoxia. Am J Respir Cell Mol Biol 25: 51–59.

34. Bhandari V, Choo-Wing R, Harijith A, Sun H, Syed MA, et al. (2012) Increased hyperoxia-induced lung injury in nitric oxide synthase 2 null mice is mediated via angiopoietin 2. Am J Respir Cell Mol Biol 46: 668–676.

35. Aghai ZH, Faqiri S, Saslow JG, Nakhla T, Farhath S, et al. (2008) Angiopoietin 2 concentrations in infants developing bronchopulmonary dysplasia: attenuation by dexamethasone. J Perinatol 28: 149–155.

36. Saharinen P, Eklund L, Miettinen J, Wirkkala R, Anisimov A, et al. (2008) Angiopoietins assemble distinct Tie2 signalling complexes in endothelial cell-cell and cell-matrix contacts. Nat Cell Biol 10: 527–537.

37. Hato T, Kimura Y, Morisada T, Koh GY, Miyata K, et al. (2009) Angiopoietins contribute to lung development by regulating pulmonary vascular network formation. Biochem Biophys Res Commun 381: 218–223.

38. Thebaud B, Ladha F, Michelakis ED, Sawicka M, Thurston G, et al. (2005) Vascular endothelial growth factor gene therapy increases survival, promotes lung angiogenesis, and prevents alveolar damage in hyperoxia-induced lung injury: evidence that angiogenesis participates in alveolarization. Circulation 112: 2477–2486.

39. Cohen JC, Lundblad LK, Bates JH, Levitzky M, Larson JE (2004) The "Goldilocks effect" in cystic fibrosis: identification of a lung phenotype in the cftr knockout and heterozygous mouse. BMC Genet 5: 21.

40. Frenz DA, Liu W, Cvekl A, Xie Q, Wassef L, et al. Retinoid signaling in inner ear development: A "Goldilocks" phenomenon. Am J Med Genet A 152A: 2947–2961.

41. Compernolle V, Brusselmans K, Acker T, Hoet P, Tjwa M, et al. (2002) Loss of HIF-2alpha and inhibition of VEGF impair fetal lung maturation, whereas treatment with VEGF prevents fatal respiratory distress in premature mice. Nat Med 8: 702–710.

42. Yang G, Abate A, George AG, Weng YH, Dennery PA (2004) Maturational differences in lung NF-kappaB activation and their role in tolerance to hyperoxia. J Clin Invest 114: 669–678.

43. Yang G, Madan A, Dennery PA (2000) Maturational differences in hyperoxic AP-1 activation in rat lung. Am J Physiol Lung Cell Mol Physiol 278: L393–398.

44. Rogers LK, Tipple TE, Nelin LD, Welty SE (2009) Differential responses in the lungs of newborn mouse pups exposed to 85% or >95% oxygen. Pediatr Res 65: 33–38.

45. Jorgensen WL, Gandavadi S, Du X, Hare AA, Trofimov A, et al. (2010) Receptor agonists of macrophage migration inhibitory factor. Bioorg Med Chem Lett 20: 7033–7036.

12

Ubiquilin-1 Overexpression Increases the Lifespan and Delays Accumulation of Huntingtin Aggregates in the R6/2 Mouse Model of Huntington's Disease

Nathaniel Safren[1,2,3,9]**, Amina El Ayadi**[4,9]**, Lydia Chang**[2,3]**, Chantelle E. Terrillion**[1,5]**, Todd D. Gould**[1,3,5,6]**,
Darren F. Boehning[4]**, Mervyn J. Monteiro**[1,2,3]*

1 Neuroscience Graduate Program, School of Medicine, University of Maryland, Baltimore, Maryland, United States of America, **2** Center for Biomedical Engineering and Technology, School of Medicine, University of Maryland, Baltimore, Maryland, United States of America, **3** Department of Anatomy and Neurobiology, School of Medicine, University of Maryland, Baltimore, Maryland, United States of America, **4** Department of Neuroscience and Cell Biology, University of Texas Medical Branch, Galveston, Texas, United States of America, **5** Department of Psychiatry, School of Medicine University of Maryland, Baltimore, Maryland, United States of America, **6** Department of Pharmacology, School of Medicine University of Maryland, Baltimore, Maryland, United States of America

Abstract

Huntington's Disease (HD) is a neurodegenerative disorder that is caused by abnormal expansion of a polyglutamine tract in huntingtin (htt) protein. The expansion leads to increased htt aggregation and toxicity. Factors that aid in the clearance of mutant huntingtin proteins should relieve the toxicity. We previously demonstrated that overexpression of ubiqulin-1, which facilitates protein clearance through the proteasome and autophagy pathways, reduces huntingtin aggregates and toxicity in mammalian cell and invertebrate models of HD. Here we tested whether overexpression of ubiquilin-1 delays or prevents neurodegeneration in R6/2 mice, a well-established model of HD. We generated transgenic mice overexpressing human ubiquilin-1 driven by the neuron-specific Thy1.2 promoter. Immunoblotting and immunohistochemistry revealed robust and widespread overexpression of ubiquilin-1 in the brains of the transgenic mice. Similar analysis of R6/2 animals revealed that ubiquilin is localized in huntingtin aggregates and that ubiquilin levels decrease progressively to 30% during the end-stage of disease. We crossed our ubiquilin-1 transgenic line with R6/2 mice to assess whether restoration of ubiquilin levels would delay HD symptoms and pathology. In the double transgenic progeny, ubiquilin levels were fully restored, and this correlated with a 20% increase in lifespan and a reduction in htt inclusions in the hippocampus and cortex. Furthermore, immunoblots indicated that endoplasmic reticulum stress response that is elevated in the hippocampus of R6/2 animals was attenuated by ubiquilin-1 overexpression. However, ubiquilin-1 overexpression neither altered the load of htt aggregates in the striatum nor improved motor impairments in the mice.

Editor: David Blum, Inserm U837, France

Funding: The authors have no support or funding to report.

Competing Interests: The authors have declared that no competing interests exist.

* E-mail: monteiro@umaryland.edu

9 These authors contributed equally to this work.

Introduction

Huntington's disease (HD) is an autosomal dominant, progressive neurodegenerative disorder characterized by chorea, psychiatric disturbances, and cognitive impairment [1,2]. Despite intensive investigation there is still no treatment to delay or prevent HD.

HD is caused by an expansion of a CAG trinucleotide repeat in exon 1 of the huntingtin (htt) gene [3]. Unaffected individuals have between 14 and 34 CAG repeats in this region, while those afflicted with HD have over 35 repeats [4,5]. Upon translation, this expansion leads to an aberrantly long tract of polyglutamine (polyQ) residues, which is believed to cause htt protein to misfold and acquire toxic properties [6]. In fact, there appears to be a direct correlation between the length of the CAG expansion and htt protein misfolding/aggregation and toxicity [7,8]. The accumulation of misfolded huntingtin poses a challenge to cellular proteostasis networks, and impairments in the ubiquitin protea-some system [9–11], endoplasmic reticulum associated degradation (ERAD) [12–14], and autophagy [15,16–18] has been reported. Efforts to restore these systems are being explored as potential therapy for HD.

Ubiquilins are a conserved family of proteins found in all eukaryotes. Humans and mice each possess four ubiquilin genes, each of which encodes a protein of about 600 amino acids. Ubiquilin proteins function to facilitate protein disposal through the proteasome and lysosomal degradation pathways, the same systems that appear compromised in HD [19–24]. Indeed there is growing evidence that links ubiquilin proteins and their genes to a number of neurodegenerative diseases. Ubiquilin proteins are found in the neuropathologic lesions that characterize HD, amyotrophic lateral sclerosis (ALS), Parkinson's disease, and Alzheimer's disease [25–28]. Mutations in ubiquilin-2 are linked to an aggressive X-linked form of ALS with dementia [29]. Similarly, mutations in ubiquilin-1 and 4 genes were recently linked to Brown-Vialetto-Van Laere syndrome and ALS, respectively [30,31]. Overexpression of ubiquilin-1 suppresses polyglutamine toxicity in cell culture and *Caenorhabditis elegans* models of HD, leading to a decrease in htt inclusions and cell death [19]. By contrast, knockdown of ubiquilin expression increases cell death, increases htt aggregates, induces endoplasmic reticulum (ER) stress, impairs autophagosome formation and maturation, and reduces lifespan of flies and nematodes [19,22,32,33].

Here we report on the effects of increasing ubiquilin-1 expression on HD progression in the R6/2 mouse model of HD [34]. We demonstrate that ubiquilin levels decrease progressively and dramatically during late-stages of disease in R6/2 mice. We produced transgenic mice overexpressing human ubiquilin-1 in neurons, which we crossed with R6/2 mice in order to test the hypothesis that restoration of ubiquilin levels would be protective in HD as a potential therapy for HD. We found that ubiquilin-1 overexpression dramatically increased lifespan, delayed formation of htt inclusions and attenuated ER stress in the hippocampus, but it did not improve motor deficits.

Materials and Methods

Animal research

This study was carried out in strict accordance with the recommendations in the Guide for the Care and Use of Laboratory Animals of the National Institutes of Health. The protocol was approved by the IACUC committee of the University of Maryland Baltimore.

Generation of ubiquilin-1 transgenic mice

In order to generate ubiquilin-1 overexpressing transgenic mice an ~1.8 kb cDNA fragment encoding human ubiquilin-1 fused with FLAG-tag at its N-terminus was inserted into the Thy1.2 expression cassette. The transgenic construct was then linearized with EcoRI and PvuI and used for pronuclear injection into fertilized eggs of the hybrid strain B6C3F2 and inbred strain C57BL/6J. Founder mice were then identified using Southern blotting and polymerase chain reaction (PCR). Two transgenic founder mice, line 48 and 62 were found to carry different copy numbers of the injected transgene. Immunoblots indicated line 62 had higher levels of the ubiquilin-1 overexpression and were backcrossed in a C57BL/6J background for 7 generations.

Animal husbandry and crosses

Line 62 ubiquilin-1 males were then crossed with ovary-transplanted R6/2 females (strain 006494) carrying 120 ± 5 CAG repeats from Jackson Labs (Bar Harbor, Maine). We used ovary-transplanted females because R6/2 mice are poor breeders. Mice were weaned and tail snipped at postnatal day 21. Following genotyping, mice were regrouped into experimental cages at week 4 containing four mice each, with one of each genotype. The following number of animals for each genotype were used for behavior: WT = 4, UBQ1 = 12, R6/2 = 10, R6/2-UBQ1 = 7. For survival studies, fourteen R6/2 mice were used, of which three were euthanized at 112, 130, and 136 days, and seven R6/2-UBQ1 mice were used, of which three were euthanized at 141, 142, and 164 days as they reached endpoint criteria. The rest were found dead. Four WT and fourteen UBQ1 mice were used for these survival studies, all of which survived the entire duration of the experiment. Data of the behavioral and pathological changes are reported for female animals to avoid potential confounds that can occur when using males [35]. All animals were house in a pathogen free facility at the University of Maryland animal facility and given regular mouse chow and fresh water. Only one person besides the personnel involved in cage maintenance handled the animals for the entire study. The person was blinded to the genotypes. The animals were monitored daily at the beginning of the study and twice daily when animals were close to reaching endpoint criteria. A veterinarian also ensured the health of animals. The criterion for euthanasia was when mice were unable to initiate movement when placed on their side for 20 seconds [36]. No drugs were used for the entire study except for tail biopsies for genotyping purposes. For this procedure the animals were anesthetized by brief inhalation with isoflurane and a small segment of the tail was removed using a razor blade. For euthanasia, mice were placed in an enclosed chamber and exposed to CO_2 from a regulated tank to effect, followed by cervical dislocation. This same procedure was used for animals used for all biochemical and immunohistochemical analyses.

Genotyping and CAG Repeat Sequencing

Genomic DNA was extracted from tail tissue using the Puregene Core Kit A (Qiagen, Hilden, Germany). Transgenic progeny were identified by Southern blotting and PCR analysis. For Southern blotting, mouse genomic DNA was digested with EcoRV enzyme and hybridized with ^{32}P-labelled with an EcoRV-XhoI fragment, spanning human ubiquilin-1 and 3′ sequences in the Thy1.2 transgenic cassette. The ubiquilin-1 transgene was screened using primers containing sequences specific to the Thy1.2 expression cassette and the ubiquilin-1 cDNA. The transgene was amplified using the sense primer (5′-TCTGAGTGGCAAAG-GACCTTA-3′) and the antisense primer (5′-GCTCTAGAC-TAAGACAAAAGTTGTCGCTGCATCTGACT) at the following cycling conditions: 98°C for 2 min, then 30 cycles (composed of 95°C for 10 sec, 62°C for 15 sec, and 72°C for 90 sec), followed by 72°C for 8 min, and terminating with 4°C. Mice containing the polyglutamine expansion were screened as previously described [35]. The CAG repeat length of R6/2 carriers was determined by Laragen (Los Angeles, CA).

Body Weight and Survival

Body weight was measured on a weekly basis to the nearest 0.1 g beginning at 8 weeks of age. Survival was measured as reported previously [56].

Rotarod

Rotarod analysis was performed on a custom built rotarod, set to accelerate from 4 to 40 rpm over a 530-second period. The rod was fitted with a bicycle inner tube to increase traction [35]. Latency to fall was recorded on three trials, with each trial separated by a 30-minute rest period. The trial with the longest latency was scored. At 7 weeks of age mice trained on three consecutive days. Only data from the last day of training was used in statistical analyses. Performance was measured weekly from 7 to 12 weeks of age.

Grip Strength

Forelimb grip strength was measured weekly from 8 weeks to 13 weeks of age using a Ugo Basile (Varese, Italy) 47106 grip strength meter. Mice were held in front of a grasping bar, which they instinctually gripped onto. Mice were pulled by the tail until their grip strength was overcome, and they lost grip of the bar. Their peak pull force was recorded in each of five trials. The three trials with the highest recorded force were then averaged.

Open Field

At 10 and 12 weeks of age locomotor activity was assessed using the open field test. To best measure the activity of the mice, trials

A

| a | b | c | d |
| EM48 | Ubqln | EM48 + Ubqln | DAPI |

B

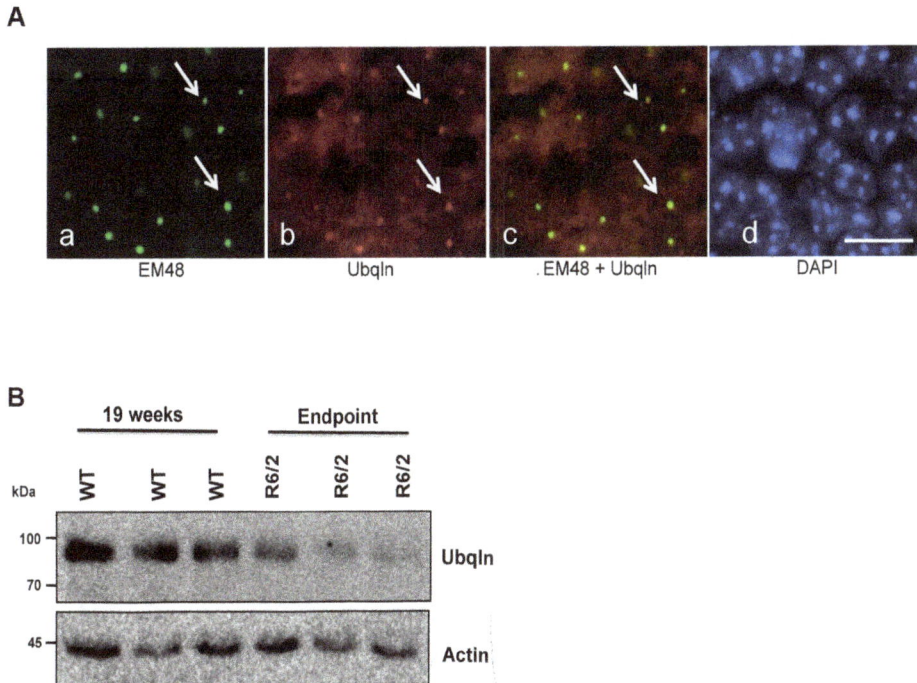

Figure 1. Ubiquilin is found in htt inclusions and its levels decrease in the R6/2 model of HD. (A) Confocal staining of a 15-week old end-stage R6/2 brain section of the hippocampus showing staining with anti-ubiquilin (red), anti-htt EM48 (green) and DAPI (blue). The merged image of the red and green fluorescence images shows ubiquilin colocalizes with htt inclusions (arrows). Bar = 15 µm. (B) Equal amounts of protein in whole brain homogenates from three end-stage R6/2 mice and three 19 week-old WT mice were immunoblotted for ubiquilin and for actin. Note the decline in ubiquilin levels in end-stage R6/2 animals.

were conducted one hour after the beginning of their dark cycle. Mice were placed into a 50×50 cm open field box arena, with one mouse from each genotype tested at the same time in different arenas. Total distance traveled was measured over the period of 15 minutes. Between trials, the floor of each chamber was washed with 70% ethanol followed by a ten-minute period where the smell of ethanol was allowed to dissipate. This was done to minimize any odors left by mice on previous trials that could potentially affect exploratory behavior.

Immunoblotting

Brains were dissected immediately following euthanization. One hemisphere was then homogenized in protein lysis buffer (PLB: 50 mM Tris pH 6.8, 150 mM NaCl, 20 mM EDTA, 1 mM EGTA, 0.5% SDS, 0.5% NP40, sarkosyl 0.5%, 1 mM prefabloc, 10 mM orthovanadate, 2.5 mM sodium fluoride) [37,38]. Total protein concentrations were then determined using the bicinch-oninic acid assay (Thermo Scientific, Rockford, IL). Brain homogenates and lysates were both stored in aliquots at −20°C.

Freshly thawed brain homogenates and cell lysates were each diluted to 15 µg/µL and loaded onto either 8.5 or 10% SDS PAGE gels, transferred to 0.45 µm PVDF membranes (Immobi-lon-P, Millipore, Billeria, MA) and probed with the following primary antibodies: mouse anti-FLAG (#F3165 Sigma Aldrich, St. Louis, MO), mouse anti-ubiquilin (clone 3D5E2, Invitrogen, Carlsbad, CA) which recognizes ubiquilin proteins, rat anti-BiP (sc-13539 Santa Cruz Biotechnology, Santa Cruz, CA), and goat anti-actin (Santa Cruz Biotechnology), rabbit anti-PDI (#2446 Cell Signaling, Danvers, MA), mouse anti-CHOP (#2895 Cell Signaling). Secondary antibodies used were horse radish peroxi-dase conjugated goat anti-mouse (#31430), goat anti-rabbit

(#31460), goat anti-rat (NA935V GE Healthcare) and rabbit anti-goat (#31492 Thermo Scientific).

Immunohistochemistry

Immediately following mouse euthanasia one hemisphere of the brain was removed and flash-frozen using dry ice. Hemispheres were stored at −80°C until they were retrieved for sectioning. 25 µm thick coronal sections were cut using a crysostat (Leica Biosystems, Buffalo Grove, IL). Immunohistochemistry was performed as previously described [39]. The following primary antibodies were used at the indicated concentrations: EM48 [40] (1:100; Millipore, Billerica, MA), mouse anti-ubiquilin-1 (1:1000; Invitrogen, Carlsbad, CA). Alexafluor (Invitrogen) secondary antibodies were used at the following dilutions donkey-anti-mouse 488 (1:500), goat anti-mouse 594 (1:500), donkey anti-rabbit 488 (1:500). In order to quantify inclusion bodies, 6 sections containing the striatum and 6 sections containing the hippocampus per animal were analyzed under an inverted Leica DMIRB fluores-cent microscope using a 40× objective. Images were captured using a Hamamatsu digital C8484 camera using iVision software (BioVision Technologies, Exton, PA). In order to reduce bias, a script, which counted the number of puncta above a certain intensity threshold and then filtered them according to size, was ran in the program iVision (BioVision). Number of inclusions was then averaged.

Colocalization of ubiquilin with htt inclusions was carried out using a Zeiss 510 confocal microscope using 405 (DAPI) 488, and 594 nm laser lines.

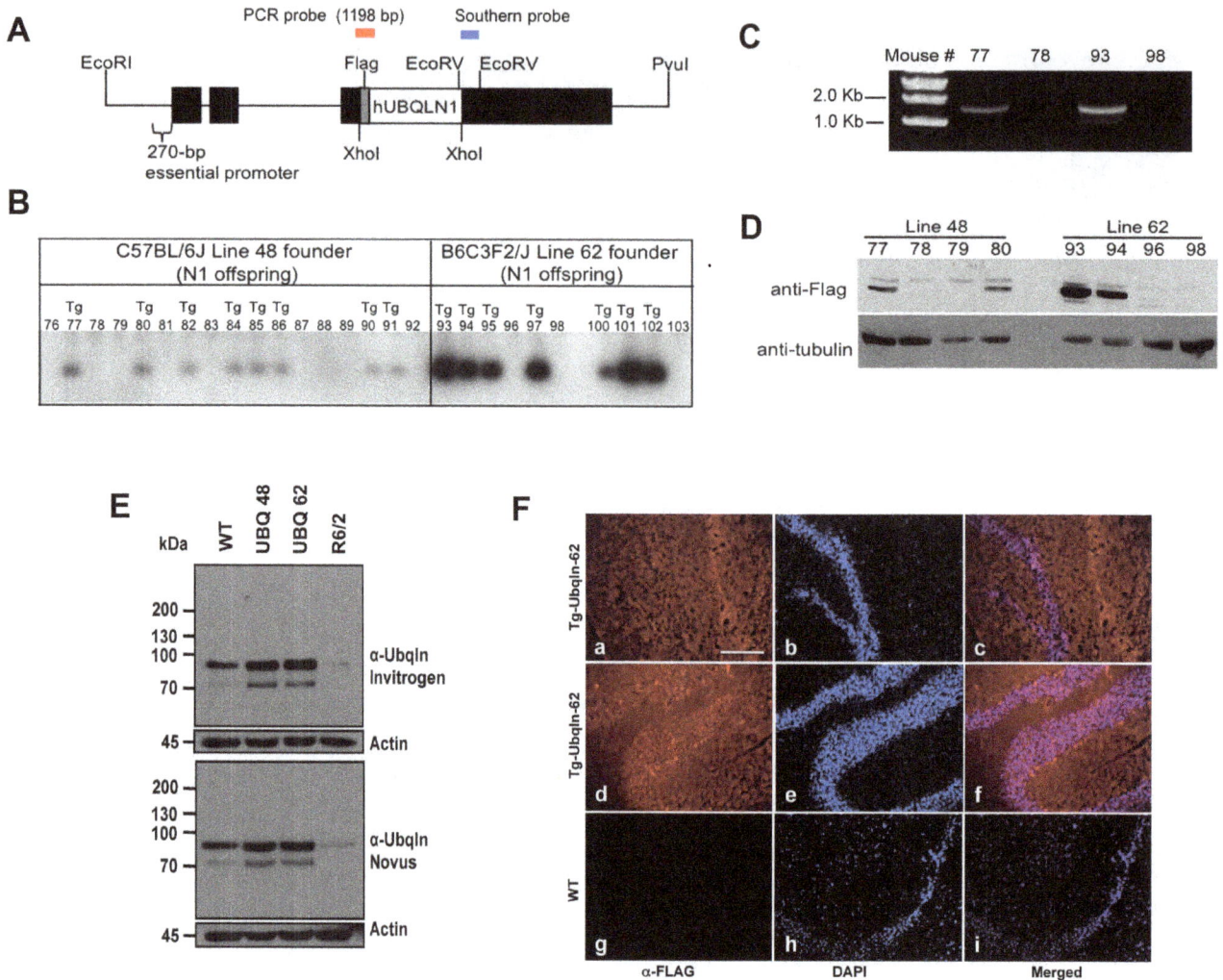

Figure 2. Generation of transgenic mice that overexpress human ubiquilin-1. (A) Schematic of the Thy1.2 expression construct used to generate ubiquilin-1 transgenic mice. Human ubiquilin-1 with an N-terminal FLAG tag was cloned in the appropriate orientation between the XhoI site of the Thy1.2 expression cassette. (B) Southern Blot of the first generation offspring of two founder mice (48 and 62). (C) Validation of a PCR genotyping protocol. Amplification of the transgene was only observed in mice that Southern blotting revealed to be positive. (D) Immunoblots of brain cortical lysates with an anti-FLAG antibody and for tubulin indicated that line 62 offspring express higher levels of FLAG-ubiquilin-1 than line 48. (E) Immunoblots of equal amounts of total brain lysates from 12 month-old WT mouse, 12 month-old Ubqln-1 48 transgenic mouse, 12 month-old Ubqln-1 62 transgenic mouse and end stage 15 week-old R6/2 transgenic mouse. The top panel was probed with a monoclonal anti-ubiquilin antibody (Invitrogen antibody clone 3D5E2) and the lower panel with a different monoclonal anti-ubiquilin antibody (Novus antibody clone 5F5). Note two immunoreactive ubiquilin bands are seen at ~70 kDa and at ~90 kDa, which we presume is a modified form of ubiquilin. Both blots were also probed for actin to ensure equal loading. (F) Cryostat sections of a Ubqln-1 62 transgenic mouse brain (a–f) and WT mouse brain (g–i) showing anti-FLAG antibody staining (Alexa 594, left panels) and corresponding DAPI staining (center panels) and the result of merging the fluorescent and DAPI signals (right hand panels). The brain sections shown are of the hippocampus (a–c and g–i) and cerebellum (d–f). Identical exposure settings were used for the left hand panels.

Statistical Analyses

Students T-Test and Repeated Measures ANOVA were performed using Microsoft Excel and GraphPad Prism.

Results

Ubiquilin levels decline dramatically during late-stages of HD

The R6/2 HD mouse model is used extensively in HD therapeutic trials due to its relatively short life span and ability to recapitulate many of the symptoms and underlying pathology of HD [35]. R6/2 mice express exon 1 of the human HD gene containing approximately 120 CAG repeats [34]. Inclusion bodies can be detected in R6/2 mouse brains as early as 3 weeks of age, which is followed by a rapid decline in behavior, motor and cognitive function, and culminating in premature death at approximately 4 months of age [34,41,42]. Previously we demonstrated that knockdown of ubiquilin-1 expression in Htt-Q74 expressing stable cell lines increases cell death [19]. Moreover, knockdown of ubiquilin expression slows proteasome degradation [21,22] and disrupts autophagy [23,24], the same two pathways that appear to be compromised in HD. We first examined whether ubiquilin proteins are contained in huntingtin aggregates by immunohistochemistry. Similar to previous findings in mouse and human brain [27,43,44], we found ubiquilin staining was colocalized with htt inclusions in brain sections of end-stage

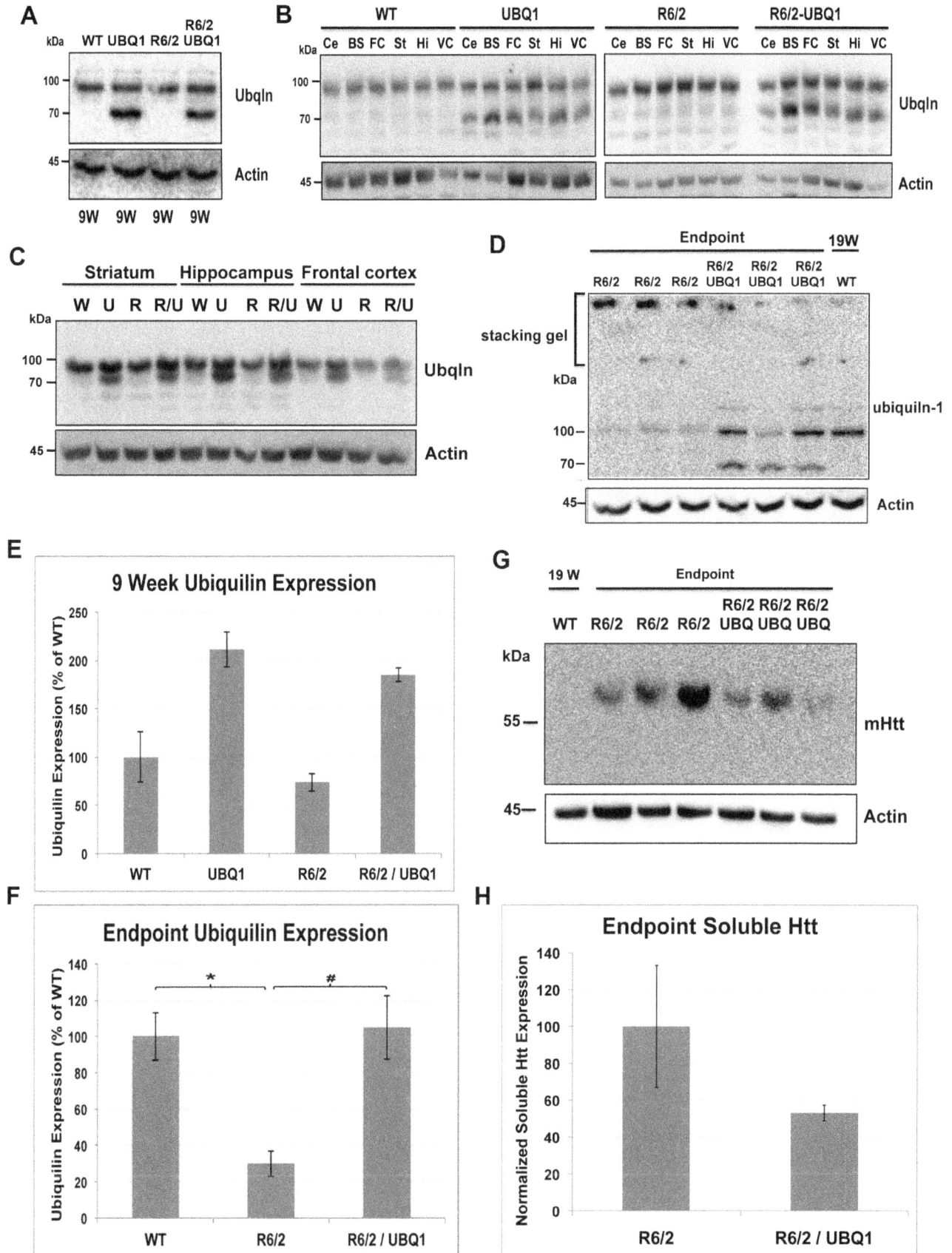

Figure 3. Restoration of ubiquilin levels in animals carrying the R6/2 transgene and accompanying changes in htt protein levels. (A) Immunoblots of lysates made from dissected brain regions of 9 week-old mice with different genotypes obtained after crossing UBQ1 transgenic

mice with R6/2 mice and probed for ubiquilin (upper panel) and for actin (lower panel). Note successful overexpression of ubiquilin in animals carrying the ubiquilin-1 transgene. (B) Immunoblot analysis similar to A, but this time showing ubiquilin levels in different regions of the brain (Ce = cerebellum, BS = Brain stem, FC = frontal cortex, St = striatum, Hi = hippocampus, VC = visual cortex). Note overexpression of ubiquilin in all tissues of transgenic mice carrying the ubiquilin-1 transgene. (C) Immunoblots of equal amounts of protein in lysates made from the striatum (St), hippocampus (Hi) and frontal cortex (FC) of 9 week-old mice from the same cross mentioned above (W = non-transgenic for UBQ1 or R6/6, U = UBQ1 transgenic, R = R6/2 transgenic, R/U = R6/2-UBQ1 double transgenic) and probed for ubiquilin or actin. (D) Immunoblots of brain lysates from end-stage mice showing R6/2-UBQ1 double transgenic mice have higher ubiquilin levels detected in the resolving gel than R6/2 animals, which correlated with decreased ubiquilin that was trapped in aggregates in the stacking gel. Also included is a lysate from a 19-week old WT animal (right lane). (E and F) Quantification of ubiquilin levels in 9-week (E) and end-stage or equivalent time-point animals (F) showing the amount of ubiquilin protein expression in animals with different genotypes used in our study. Note that ubiquilin levels are reduced by 70% in end-stage R6/2 animals compared to wild type age-matched controls ($p = 0.0011$). Furthermore, ubiquilin levels were fully restored in the brains of end-stage R6/2-UBQ1 double transgenic mice compared to R6/2 transgenic mice ($p = 0.0017$). (G) Immunoblot of same lysates shown in D with EM48 antibody showing ubiquilin overexpression in R6/2-UBQ1 double transgenic mice have reduced levels of soluble mutant htt protein accumulation compared to R6/2 animals. Note the WT lysate was loaded on the left lane. (H) Quantification of soluble htt protein shown in panel G.

R6/2 animals (Figure 1A). We next examined whether ubiquilin levels change during disease progression in R6/2 mice. Immunoblotting of whole brain lysates with a monoclonal specific ubiquilin antibody revealed a dramatic 70% decrease in ubiquilin levels at end stage of disease (Figure 1B and 3F). The results suggest a negative correlation between ubiquilin protein accumulation and HD.

Generation and characterization of ubiquilin-1 transgenic mice

In order to test whether increasing ubiquilin expression would be protective in HD, we generated transgenic mice that express N-terminal FLAG-tagged human ubiquilin-1 under the control of the Thy1.2 promoter (Figure 2A), which drives transgene expression specifically in neurons [45–47]. Founders were identified via Southern blotting (Figure 2B) and subsequently confirmed via diagnostic PCR analysis (Figure 2C). Two transgenic lines were further characterized, line 62 and line 48 (Figure 2D). Immunoblots of brain lysates indicated that total ubiquilin levels in line 62 were increased approximately 200% compared to control nontransgenic mice (Figure 2E and 3E), whereas ubiquilin levels were lower in line 48 when the different bands seen in the blots were considered together (Figure 2E). Both lines expressed appropriate size FLAG-tagged human ubiquilin-1 protein that migrated on gels with a mass of ~68 kDa, similar to what we had shown previously [25]. Interestingly, the monoclonal specific ubiquilin antibody reacted with multiple bands in the mouse brain lysates, including the ~68 kDa band and a more prominent band at ~90 kDa, which we presume reflects some post-translation modification of the proteins (Figure 1E and 3A). Further immunoblots of different brain regions, as well as immunohisto-chemistry revealed global overexpression of the ubiquilin-1 transgene in the brain (Figure 2E, F and 3B and C). All subsequent studies were carried out using the higher expressing line 62 mice (henceforth referred to UBQ1 mice).

More detailed characterization of the UBQ1 62 transgenic mice (as well as the 48 line) indicated they were completely normal in terms of appearance, lifespan and according to a battery of behavioral tests (rotarod, grip strength and open field analysis). The mice therefore possessed the appropriate characteristics, overexpression of ubiquilin-1 with no unintended detrimental effects, making them suitable for crossing with mouse models of HD.

Restoration of ubiquilin levels in R6/2 mice brains by transgenic overexpression of ubiquilin-1

Because the R6/2 model is widely used to test therapeutic candidates for HD we crossed our UBQ1 62 transgenic line with R6/2 mice and obtained progeny with appropriate Mendelian transmission of the two transgenes. Comparison of ubiquilin levels in various brain regions of 9-week old animals in each of the four resulting genotypes revealed higher expression in the animals that inherited the ubiquilin-1 transgene, as expected (Figure 3B and C). Importantly, the increase was seen in R6/2-UBQ1 double transgenic mice, indicating successful overexpression of ubiquilin-1 in the presence of the R6/2 transgene (Figure 3A–C). Moreover, immunoblots of brain lysates from end-stage HD-affected R6/2-UBQ1 double transgenic mice revealed ubiquilin levels had been fully restored to the amount that is typically found in age matched wild type animals (Figure 3D and F). An immunoblot with EM48 antibody, which selectively recognizes human mutant huntingtin protein, revealed an ~50% decrease in mutant soluble huntingtin protein in R6/2-UBQ1 double transgenic mice compared to R6/2 animals (Figure 3G and H). However, the changes were not statistically significant at this time-point. There was variability in ubiquilin expression in the mice that inversely correlated with huntingtin protein accumulation. The consequence of increased ubiquilin-1 expression was further evaluated by conducting a battery of tests on the mice, including monitoring effects on survival, body weight, behavioral pheno-types and neuropathology. Changes in all of these parameters are strongly linked with the transmission of the R6/2 transgene [35,48].

Ubiquilin-1 overexpression increased survival of R6/2 mice, but had no effect on motor behavior or body weight

The length of polyglutamine tracts in R6/2 mice has been shown to have a profound effect on phenotype [49]. Moreover, polyglutamine expansions are unstable and have the potential to increase in size upon transmission to progeny [35,50]. In order to ensure that all experimental mice had comparable numbers of repeats the polyglutamine region was sequenced in all R6/2 carriers. The measurements indicated that R6/2 single and R6/2-UBQ1 double transgenic mice had, on average, the same number of CAG repeats (R6/2 (n = 18): Mean = 125, SD = 2, R6/2-UBQ1 (n = 10): Mean = 125, SD = 2).

To eliminate other possible confounds, all of the animals were housed, handled and tested using the recommended guidelines for R6/2 mice [35]. Because of the aggressive nature of male mice, which can profoundly affect the behavior and well being of cage mates, the results presented are those for females only, although similar effects were obtained when the data from both sexes were combined. Ubiquilin-1 overexpression significantly improved survival of R6/2 mice (Figure 4A). Mean survival of R6/2-UBQ1 mice increased by 20% compared to R6/2 mice (R6/2: 119.5 days, n = 14, SEM = 3.16, R6/2-UBQ1: 144 days, n = 7,

Ubiquilin-1 Overexpression Increases the Lifespan and Delays Accumulation of Huntingtin Aggregates...

119

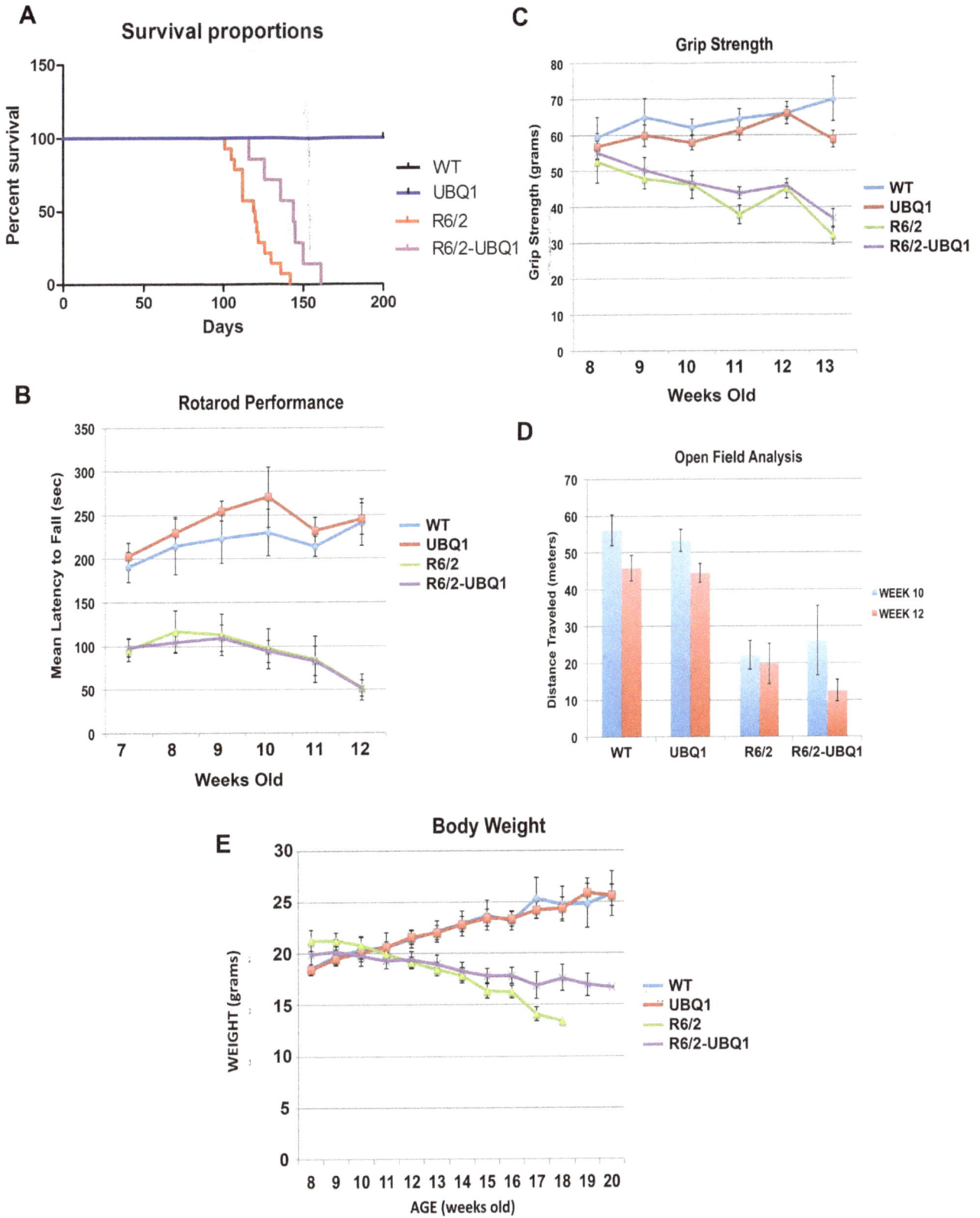

Figure 4. Ubiquilin-1 Overexpression increased survival but had no effect on motor function or loss of body weight. (A) Kaplan-Meier survival curve showing a 20% increase in lifespan in double transgenic mice (R6/2: N = 14, R6/2-UBQ1: N = 7, $p = 0.0032$). (B–D) Ubiquilin-1 overexpression failed to improve rotarod performance (B), grip strength (C) and activity in the open field (D), at any time point tested. (E) There was no significant effect of ubiquilin-1 overexpression on weight loss. For all behavioral experiments: (WT: N = 4, UBQ1: N = 12, R6/2: N = 10, R6/2-UBQ1: N = 7).

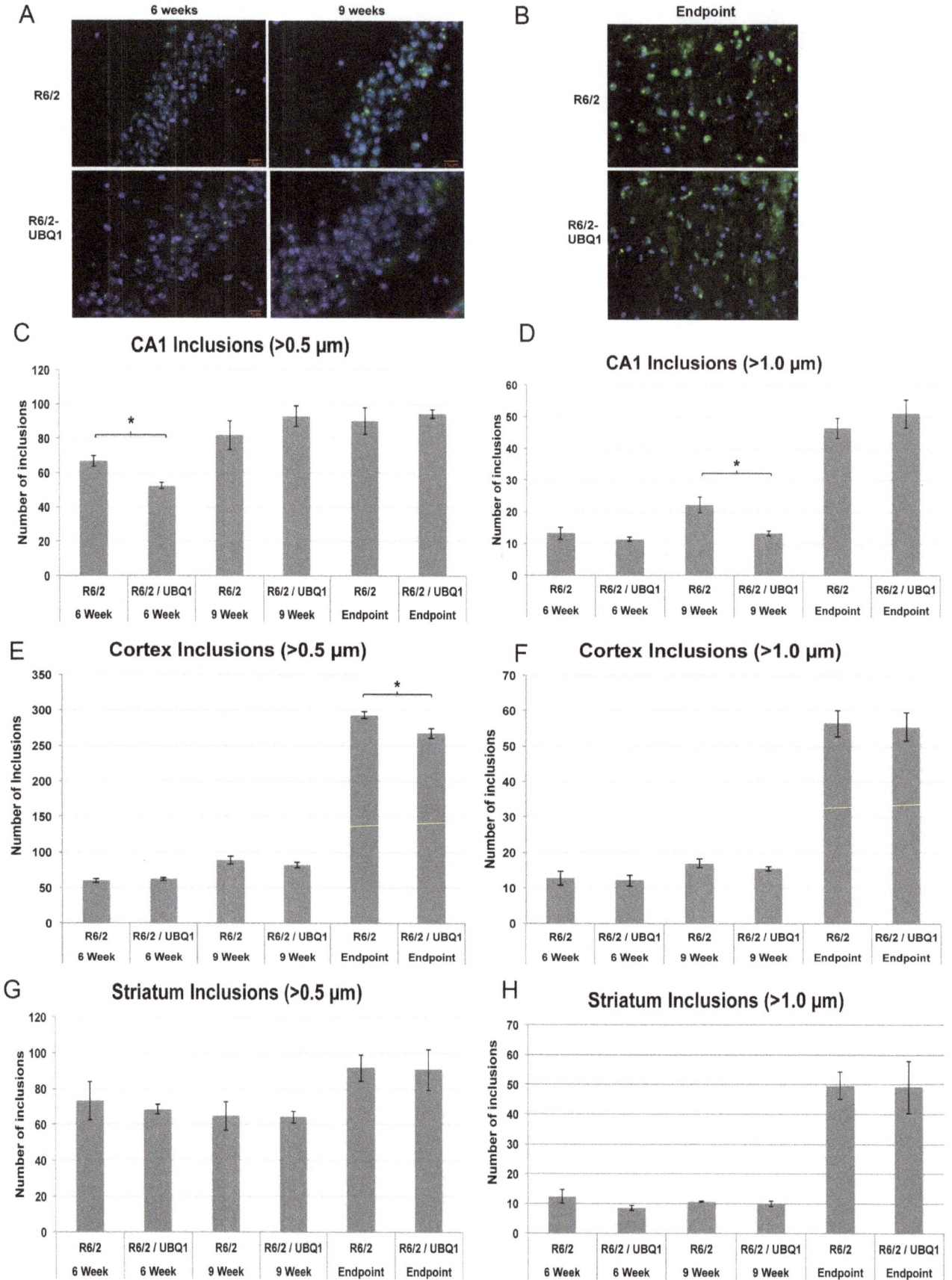

Figure 5. Ubiquilin-1 overexpression modifies aggregate load in the hippocampus and cortex but not the striatum. (A) Representative fluorescence microscopy images of EM48 and DAPI stained cryostat sections of the CA1 region of the hippocampus in R6/2 transgenic and R6/2-UBQ1 double transgenic mouse at 6 weeks, 9 weeks and following end-stage euthanasia. Bar = 15 μm. (B) Similar to A, but showing representative sections from the dentate gyrus in end-stage mice. Bar = 15 μm. (C) Quantification of htt inclusions >0.5 μm in size in the CA1 region of the hippocampus at 6 weeks, 9 weeks and end-stage R6/2 and R6/2-UBQ1 double transgenic mice. The R6/2-UBQ1 double transgenic mice contained 22% fewer inclusions than R6/2 mice at 6 weeks ($p = 0.04$), but not at the other times. (D) Quantification of htt inclusions >1 μm in size in the CA1 region of the hippocampus at 6 weeks, 9 weeks and end-stage R6/2 and R6/2-UBQ1 double transgenic mice. The R6/2-UBQ1 double transgenic mice had 40% fewer inclusions at 9 weeks compared to R6/2 transgenic mice ($p = 0.027$). (E and F) Similar to B and C, but showing htt inclusions in the cortex. R6/2-UBQ1 double transgenic mice had 8.5% fewer inclusions greater than 0.5 μm at the end-stage of disease. (G, H) Similar to E and F, but comparing inclusions in the striatum. There was no difference in the number of inclusions in the striatum between the two genotypes at any time point.

SEM = 5.71; $p = 0.0032$). None of mice lacking the R6/2 transgene died before the longest surviving R6/2 carrier.

Despite this increase in lifespan, R6/2-UBQ1 mice did not exhibit improved motor function. The rotarod test was performed on a weekly basis in order to assess balance and coordination (Figure 4B). Repeated Measures ANOVA revealed a main effect of genotype ($F_{3, 108} = 22.68$, $p < 0.0001$). Bonferroni posttests found significant differences between WT and R6/2 at weeks 7 and from 9 to 12, and between WT and R6/2-UBQ1 mice at all time points. Statistical analysis of the data indicated no significant difference between R6/2 and R6/2-UBQ1 mice at any time point. Therefore, we conclude that ubiquilin-1 overexpression had no effect on rotarod performance.

We observed a similar trend when measuring grip strength (Figure 4C). Repeated Measures ANOVA showed a main effect of age ($F_{5, 150} = 2.48$, $p = 0.0341$), genotype ($F_{3, 150} = 32.92$, $p < 0.0001$), as well as an interaction between age and genotype ($F_{15, 150} = 2.94$, $p = 0.0004$). Bonferroni posttests revealed significant differences between WT and R6/2 from weeks 9 to 13, and between WT and R6/2-UBQ1 mice from weeks 10 to 13. However, no significant differences were observed between R6/2 and R6/2-UBQ1 mice at any time point.

R6/2 mice initially display hyperactivity at 3 weeks of age and later exhibit hypoactivity beginning at 7 weeks, which becomes more pronounced with age [51]. In order to determine whether ubiquilin-1 overexpression had an effect on locomotor activity the open field test was performed initially at 10 weeks, and then repeated at 12 weeks of age (Figure 4D). A main effect of testing session ($F_{1, 23} = 27.71$, $p < 0.0001$), and genotype ($p < 0.0001$), was observed. However, post-hoc tests failed to reveal a difference between R6/2 and R6/2-UBQ1 mice at either time point, indicating that ubiquilin-1 overexpression has no significant effect on locomotor activity. Finally, ubiquilin-1 overexpression also failed to significantly improve body weight at any time point (Figure 4E).

Ubiquilin-1 overexpression delays huntingtin aggregates in the hippocampus, but not the striatum

Previously we demonstrated that ubiquilin-1 overexpression suppresses htt aggregation in cell culture [19]. We therefore explored whether ubiquilin-1 overexpression could reduce or delay the formation of inclusion bodies in the brains of R6/2 mice. We tested several methods to quantify the changes in huntingtin protein aggregation. We found immunohistochemistry was the most reliable and reproducible method to quantify aggregates. iVision software was used to identify inclusions that were greater than 0.5 μm in diameter. We then filtered the inclusions to identify those greater than 1.0 μm in diameter, enabling us to quantify small inclusions (<1.0 μm) from large inclusions (>1.0 μm). We observed a 22% reduction in the number of total inclusions in the CA1 region of the hippocampus of 6 week-old R6/2-UBQ1 mice (N = 3, $p = 0.04$) (Figure 5A and C). This was primarily due to 24%

reduction of small inclusions. This was followed by a 40% reduction in large inclusions in CA1 at 9 weeks of age (N = 3, $p = 0.027$) (Figure 5A and D). Presumably, the reduction in large inclusions at week 9 was due to a smaller pool of inclusions with which to expand at week 6. In the cortex we observed a similar trend during this period and a statistically significant reduction of 0.5 μm diameter inclusions in end-stage animals (Figure 5B, F and G). Ubiquilin-1 overexpression failed to modify htt aggregation at any time point in the striatum (Figure 5G and H).

Overexpression of ubiquilin-1 attenuates ER stress in the hippocampus

Because ubiquilin-1 facilitates ERAD [21,22] and because htt proteins containing polyglutamine expansions have been implicated in disruption of ERAD [12,14] we next investigated whether mice overexpressing ubiquilin-1 had altered ER homeostasis. Interference of ERAD can trigger the unfolded protein response (UPR), which involves the coordinated activation of a series of signaling pathways that function to restore ER homeostasis [52]. Several proteins are activated during UPR, so measuring fluctuations in their levels is used to monitor ER stress [53–55]. We focused on three classical ER stress markers: BiP (immunoglobulin-binding protein or grp78), PDI (protein disulfide isomerase) and CHOP (transcription factor C/EBP homologous protein) [56]. BiP and PDI are two ER chaperones that are activated during UPR to restore protein folding in the ER, whereas activation of CHOP signals execution of the cell death or apoptosis program [55]. Accordingly, we examined if these three proteins were altered by ubiquilin-1 overexpression in the hippocampus of 9 week-old animals. We focused on the hippocampus because it is where we found distinct changes in htt aggregation from ubiquilin overexpression. The immunoblots indicated that BiP, PDI and CHOP were all elevated in R6/2 mice, compared to non-transgenic or ubiquilin-overexpressing mice (Figure 6A and B). More interestingly overexpression of ubiquilin-1 attenuated the increase in each of these stress markers in ubiquilin-1-overexpressing R6/2 mice (Figure 6A and B). Further analysis of hippocampal lysates for huntingtin protein trapped in the stacking gel confirmed EM48 immunoreactivity was reduced in R6/2-UBQ1 mice compared to an R6/2 animal (Figure 6C).

Discussion

Our study provides direct in vivo evidence that increasing ubiquilin-1 expression may be beneficial for HD. The most compelling evidence is that through transgenic overexpression of human ubiquilin-1 protein, mean lifespan of R6/2 mice was increased by 20%. The observed increase in lifespan is amongst the largest improvements in survival in R6/2 therapeutic trials [57]. In fact, most of the large improvements in R6/2 survival have come from drug treatment of animals, and not by transgenic manipulation of proteins, as we have done here. Besides survival,

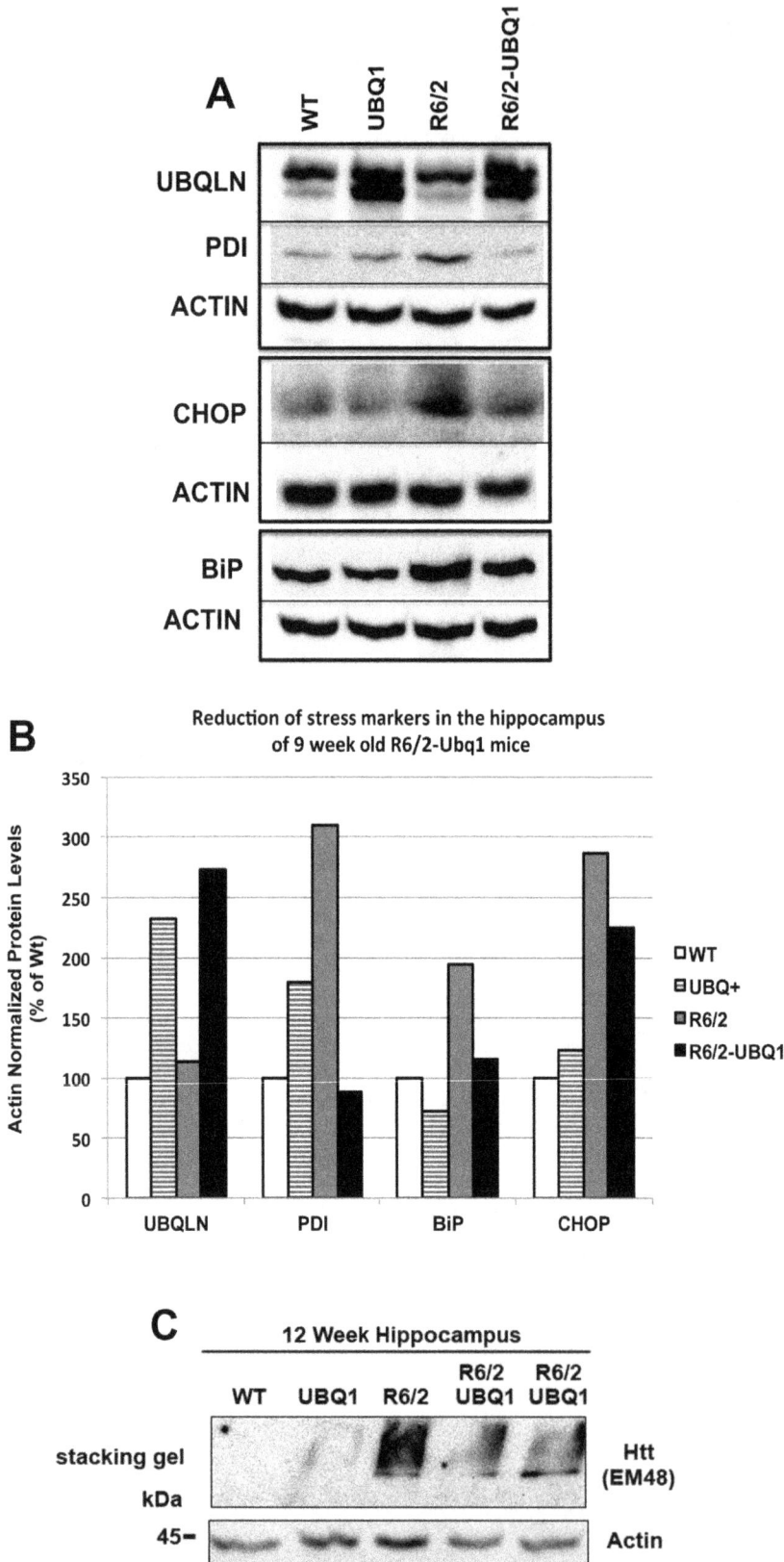

Figure 6. Ubiquilin-1 overexpression attenuates ER stress in R6/2 animals. (A) Immunoblots of equal amounts of brain lysates from 9 week-old female animals from the R6/2 and UBQ1 cross. The blots were probed for PDI, BiP, ubiquilin, and CHOP as well as for actin to monitor protein loading. (B) Quantification of the expression of the different proteins in the four mice after normalization for actin loading. (C) Immunoblot of 12-week hippocampal lysates to detect htt aggregates retained in the stacking gel. The double transgenic mice had an approximately 30% reduction in EM48 immunoreactivity after normalization for actin loading.

ubiquilin-1 overexpression delayed inclusion body formation in the CA1 region of the hippocampus and cortex. Finally, consistent with a functional role of ubiquilin-1 in promoting ERAD, R6/2-UBQ1 double transgenic mice overexpressing ubiquilin-1 had an attenuated ER stress response in the hippocampus compared to an age-matched mouse carrying the R6/2 transgene alone. There are several possible explanations for our findings.

Because the exact mechanisms by which huntingtin proteins containing polyglutamine expansions cause disease is still not known, it is difficult to know if the increase in R6/2 survival from ubiquilin-1 overexpression is related to any specific improvement in a specific pathway(s). However, it is clear that it must represent some improvement in a crucial pathway(s) needed for survival. Two hints suggested by our study is the possible relationship of survival to formation of htt inclusions and/or to an attenuation of ER stress. The reduction in htt aggregates found in the hippocampus and cortex is consistent with our previous findings conducted in cell culture and *C. elegans* showing ubiquilin overexpression reduces htt aggregates and improves a motility defect in nematodes [19]. Moreover, previous studies have shown ubiquilin-1 overexpression increases the turnover of expanded huntingtin protein [20]. Thus a simple explanation for our findings is that R6/2-UBQ1 mice overexpressing ubiquilin-1 have increased turnover of the polyQ-expanded htt protein thereby reducing the amount of mutant htt protein available for aggregation and for inducing toxicity. Consistent with this idea we found ubiquilin-1 overexpression led to a reduction in accumulation of soluble mutant htt protein in the mice (Figure 3E).

A conundrum was why huntingtin inclusions in the striatum were not reduced despite clear evidence of ubiquilin overexpression? Although we do not know the answer we speculate that intrinsic differences in the composition and/or function of neurons in the CA1 region of the hippocampus and cortex compared to the striatum could influence the ability of htt protein to aggregate in one cell type, but not the other. It is interesting that the reduction in inclusions we found occur in regions of the brain that are rich in pyramidal neurons, which suggest that they could be more amenable to rescue by ubiquilin-1 overexpression.

The attenuation of ER stress in R6/2 mice overexpressing ubiquilin-1 is most likely related to facilitation of ERAD by ubiquilin [21,22]. For example, studies have shown that overexpression of ubiquilin-1 enhances degradation of ERAD substrates, whereas knockdown of ubiquilin-1 expression slows degradation [21,22]. In accordance with facilitating ERAD, knockdown of ubiquilin in *C. elegans* induces ER stress [22]. Thus, our results are consistent with ubiquilin-1 overexpression attenuating ER stress. Activation of ER stress is associated with induction of the unfolded protein response to restore ER homeostasis, but if unsuccessful, its persistent activation can trigger cell death [53,54,58,59]. In fact, there is growing evidence that ER stress could be involved in the etiology of many human diseases, particularly neurodegenerative diseases, such as Huntington's disease [52,60]. For example, htt proteins with expanded polyglutamine tracts have been shown to interfere with ERAD resulting in induction of ER stress [12,14]. Furthermore, ER stress markers are increased in different mouse models of HD [61], including the R6/2 model we have used [62]. It remains to be determined whether the increased survival of R6/2-UBQ1 animals is related to alleviation of ER stress and/or whether it is related to the protective effects of ubiquilin in some other pathways.

Despite reducing htt aggregation, attenuating ER stress and increasing mouse survival, we found mice overexpressing ubiquilin-1 had no significant improvement in the deterioration of motor function caused by the R6/2 transgene. Several reasons may account for this apparent discrepancy. The R6/2 mutant huntingtin transgene is ubiquitously expressed [34,63], while expression of our ubiquilin-1 transgene is restricted to neurons. It is possible that degeneration of muscle tissue (or other cell types) may have occluded any potential improvements in neurological function. Another possibility is that the R6/2 model, which expresses high levels of the toxic exon-1 fragment of htt, might have too aggressive and penetrant phenotypes that could mask subtle improvements from ubiquilin-1 overexpression. Mice with more modest behavioral and pathological phenotypes have been generated by expressing full-length htt containing polyQ expansions at more physiological levels, such as the BACHD and YAC128 transgenic models or the knock-in CAG140 and CAG150 models [64–67]. It is therefore possible that ubiquilin-1 overexpression would show some benefit when tested in these less aggressive mouse models of HD. On the other hand, the behavioral impairments in R6/2 mice might be reduced or eliminated by even higher overexpression of ubiquilin-1, either by generating new ubiquilin-1 transgenic mice or by crossing our mice to obtain homozygous transmission of the ubiquilin-1 transgene.

The ubiquilin transgenic mice we generated overexpress approximately 2-fold more ubiquilin protein in total brain lysates. Because the brain lysates are composed of neuronal and non-neuronal cells, and because expression of the Thy1.2 promoter is restricted to neurons, it is likely that ubiquilin-1 overexpression in neurons is even higher. Despite this large increase we have not noticed any overt toxic effects of its overexpression, suggesting any effort to increase ubiquilin overexpression for therapeutic treatment of neurodegenerative disease would likely be safe. Obviously, overexpression in cells other than neurons would have to be tested.

An interesting observation regarding ubiquilin proteins in brain is the presence of a major protein band of approximately 90 kDa that is seen in addition to the normal ~68 kDa band. The ~90 kDa band most likely reflects some post-translation modification of ubiquilin. It will be interesting to determine the nature of this modification, and whether ubiquilin function(s) are altered by the presence or absence of the modification.

In summary, our studies suggest that overexpression of ubiquilin-1 provides some benefit when tested using the aggressive R6/2 model of HD. Similar tests conducted with other less aggressive HD mouse models should reveal whether methods to increase ubiquilin expression or activity would be effective for HD treatment.

Acknowledgments

We thank Dr. Asaf Keller for help with the dissection of different mouse brain regions and Brian Hagan for help with the confocal microscopy.

Author Contributions

Conceived and designed the experiments: NS AEA LC CET TDG DFB MJM. Performed the experiments: NS AEA LC. Analyzed the data: NS AEA CET TDG DFB MJM. Contributed reagents/materials/analysis tools: TDG. Wrote the paper: NS MJM.

References

1. Walker FO (2007) Huntington's disease. Lancet 369: 218–228.
2. Bates G, Harper P, Jones L (2002) Huntington's Disease. 3rd edn. Oxford University Press, Oxford.
3. The Huntington's Disease Collaborative Research Group (1993) A novel gene containing a trinucleotide repeat that is expanded and unstable on Huntington's disease chromosomes. Cell 72: 971–983.

4. Duyao M, Ambrose C, Myers R, Novelletto A, Persichetti F, et al. (1993) Trinucleotide repeat length instability and age of onset in Huntington's disease. Nat Genet 4: 387–392.

5. Wexler NS, Lorimer J, Porter J, Gomez F, Moskowitz C, et al. (2004) Venezuelan kindreds reveal that genetic and environmental factors modulate Huntington's disease age of onset. Proc Natl Acad Sci USA 101: 3498–3503.

6. Finkbeiner S (2011) Huntington's disease. Cold Spring Harbor Laboratory Press.

7. Martindale D, Hackam A, Wieczorek A, Ellerby L, Wellington C, et al. (1998) Length of huntingtin and its polyglutamine tract influences localization and frequency of intracellular aggregates. Nat Genet 18: 150–154.

8. Scherzinger E, Sittler A, Schweiger K, Heiser V, Lurz R, et al. (1999) Self-assembly of polyglutamine-containing huntingtin fragments into amyloid-like fibrils: implications for Huntington's disease pathology. Proc Natl Acad Sci USA 96: 4604–4609.

9. Bennett EJ, Bence NF, Jayakumar R, Kopito RR (2005) Global impairment of the ubiquitin-proteasome system by nuclear or cytoplasmic protein aggregates precedes inclusion body formation. Mol Cell 17: 351–365.

10. Bennett EJ, Shaler TA, Woodman B, Ryu KY, Zaitseva TS, et al. (2007) Global changes to the ubiquitin system in Huntington's disease. Nature 448: 704–708.

11. Hipp MS, Patel CN, Bersuker K, Riley BE, Kaiser SE, et al. (2012) Indirect inhibition of 26S proteasome activity in a cellular model of Huntington's disease. J Cell Biol 196: 573–587.

12. Duennwald ML, Lindquist S (2008) Impaired ERAD and ER stress are early and specific events in polyglutamine toxicity. Genes Dev 22: 3308–3319.

13. Kouroku Y, Fujita E, Jimbo A, Kikuchi T, Yamagata T, et al. (2002) Polyglutamine aggregates stimulate ER stress signals and caspase-12 activation. Hum Mol Genet 11: 1505–1515.

14. Yang H, Liu C, Zhong Y, Luo S, Monteiro MJ, et al. (2010) Huntingtin interacts with the cue domain of gp78 and inhibits gp78 binding to ubiquitin and p97/VCP. PLoS One 5: e8905.

15. Sapp E, Schwarz C, Chase K, Bhide PG, Young AB, et al. (1997) Huntingtin localization in brains of normal and Huntington's disease patients. Ann Neurol 42: 604–612.

16. Ravikumar B, Vacher C, Berger Z, Davies JE, Luo S, et al. (2004) Inhibition of mTOR induces autophagy and reduces toxicity of polyglutamine expansions in fly and mouse models of Huntington disease. Nat Genet 36: 585–595.

17. Sarkar S, Rubinsztein DC (2008) Huntington's disease: degradation of mutant huntingtin by autophagy. FEBS J 275: 4263–4270.

18. Martinez-Vicente M, Talloczy Z, Wong E, Tang G, Koga H, et al. (2010) Cargo recognition failure is responsible for inefficient autophagy in Huntington's disease. Nat Neurosci 13: 567–576.

19. Wang H, Lim PJ, Yin C, Rieckher M, Vogel BE, et al. (2006) Suppression of polyglutamine-induced toxicity in cell and animal models of Huntington's disease by ubiquilin. Hum Mol Genet 15: 1025–1041.

20. Wang H, Monteiro MJ (2007) Ubiquilin interacts and enhances the degradation of expanded-polyglutamine proteins. Biochem Biophys Res Commun 360: 423–427.

21. Kim TY, Kim E, Yoon SK, Yoon JB (2008) Herp enhances ER-associated protein degradation by recruiting ubiquilins. Biochem Biophys Res Commun 369: 741–746.

22. Lim PJ, Danner R, Liang J, Doong H, Harman C, et al. (2009) Ubiquilin and p97/VCP bind erasin, forming a complex involved in ERAD. J Cell Biol 187: 201–217.

23. N'Diaye EN, Kajihara KK, Hsieh I, Morisaki H, Debnath J, et al. (2009) PLIC proteins or ubiquilins regulate autophagy-dependent cell survival during nutrient starvation. EMBO Rep 10: 173–179.

24. Rothenberg C, Srinivasan D, Mah L, Kaushik S, Peterhoff CM, et al. (2010) Ubiquilin functions in autophagy and is degraded by chaperone-mediated autophagy. Hum Mol Genet 19: 3219–3232.

25. Mah AL, Perry G, Smith MA, Monteiro MJ (2000) Identification of ubiquilin, a novel presenilin interactor that increases presenilin protein accumulation. J Cell Biol 151: 847–862.

26. Brettschneider J, Van Deerlin VM, Robinson JL, Kwong L, Lee EB, et al. (2012) Pattern of ubiquilin pathology in ALS and FTLD indicates presence of C9ORF72 hexanucleotide expansion. Acta Neuropathol 123: 825–839.

27. Mori F, Tanji K, Odagiri S, Toyoshima Y, Yoshida M, et al. (2012) Ubiquilin immunoreactivity in cytoplasmic and nuclear inclusions in synucleinopathies, polyglutamine diseases and intranuclear inclusion body disease. Acta Neuropathol 124: 149–151.

28. Satoh J, Tabunoki H, Ishida T, Saito Y, Arima K (2013) Ubiquilin-1 immunoreactivity is concentrated on Hirano bodies and dystrophic neurites in Alzheimer's disease brains. Neuropathol Appl Neurobiol,39: 817–830.

29. Deng HX, Chen W, Hong ST, Boycott KM, Gorrie GH, et al. (2011) Mutations in UBQLN2 cause dominant X-linked juvenile and adult-onset ALS and ALS/dementia. Nature 477: 211–215.

30. Gonzalez-Perez P, Lu Y, Chian RJ, Sapp PC, Tanzi RE, et al. (2012) Association of UBQLN1 mutation with Brown-Vialetto-Van Laere syndrome but not typical ALS. Neurobiol Dis 48: 391–398.

31. Yan J, Ajroud K, Fecto F, Shi Y, Siddique N, et al. (2013) A new mutation in ubiquilin gene family and its effect on protein degradation. Neurology 80: P02.168.

32. Li A, Xie Z, Dong Y, McKay KM, McKee ML, et al. (2007) Isolation and characterization of the Drosophila ubiquilin ortholog dUbqln: in vivo interaction with early-onset Alzheimer disease genes. Hum Mol Genet 16: 2626–2639.

33. Ganguly A, Feldman RM, Guo M (2008) ubiquilin antagonizes presenilin and promotes neurodegeneration in Drosophila. Hum Mol Genet 17: 293–302.

34. Mangiarini L, Sathasivam K, Seller M, Cozens B, Harper A, et al. (1996) Exon 1 of the HD gene with an expanded CAG repeat is sufficient to cause a progressive neurological phenotype in transgenic mice. Cell, 87, 493–506.

35. Hockly E, Woodman B, Mahal A, Lewis CM, Bates G (2003) Standardization and statistical approaches to therapeutic trials in the R6/2 mouse. Brain Res Bull 61: 469–479.

36. Ferrante RJ, Andreassen OA, Dedeoglu A, Ferrante KL, Jenkins BG, et al. (2002) Therapeutic effects of coenzyme Q10 and remacemide in transgenic mouse models of Huntington's disease. J Neurosci 22: 1592–1599.

37. Monteiro MJ, Mical TI (1996) Resolution of kinase activities during the HeLa cell cycle: identification of kinases with cyclic activities. Exp Cell Res 223: 443–451.

38. Xiao J, Monteiro MJ (1994) Identification and characterization of a novel (115 kDa) neurofilament-associated kinase. J Neurosci 14: 1820–1833.

39. Starr R, Xiao J, Monteiro MJ (1995) Production of monoclonal antibodies against neurofilament-associated proteins: demonstration of association with neurofilaments by a coimmunoprecipitation method. J Neurochem 64: 1860–1867.

40. Gutekunst CA, Li SH, Yi H, Mulroy JS, Kuemmerle S, et al. (1999) Nuclear and neuropil aggregates in Huntington's disease: relationship to neuropathology. J Neurosci 19: 2522–2534.

41. Carter RJ, Lione LA, Humby T, Mangiarini L, Mahal A, et al. (1999) Characterization of progressive motor deficits in mice transgenic for the human Huntington's disease mutation. J Neurosci 19: 3248–3257.

42. Gong B, Kielar C, Morton AJ (2012) Temporal separation of aggregation and ubiquitination during early inclusion formation in transgenic mice carrying the Huntington's disease mutation. PLoS One 7: e41450.

43. Doi H, Mitsui K, Kurosawa M, Machida M, Kuroiwa Y, et al. (2004) Identification of ubiquitin-interacting proteins in purified polyglutamine aggregates. FEBS Lett 571: 171–176.

44. Rutherford NJ, Lewis J, Clippinger AK, Thomas MA, Adamson J, et al. (2013) Unbiased Screen Reveals Ubiquilin-1 and -2 Highly Associated with Huntingtin Inclusions. Brain Res 1524: 62–73.

45. Caroni P (1997) Overexpression of growth-associated proteins in the neurons of adult transgenic mice. J Neurosci Methods 71: 3–9.

46. Feng G, Mellor RH, Bernstein M, Keller-Peck C, Nguyen QT, et al. (2000) Imaging neuronal subsets in transgenic mice expressing multiple spectral variants of GFP. Neuron 28: 41–51.

47. Weissman TA, Sanes JR, Lichtman JW, Livet J (2011) Generating and imaging multicolor Brainbow mice. Cold Spring Harb Protoc 2011: 763–769.

48. Menalled L, El-Khodor BF, Patry M, Suarez-Farinas M, Orenstein SJ, et al. (2009) Systematic behavioral evaluation of Huntington's disease transgenic and knock-in mouse models. Neurobiol Dis 35: 319–336.

49. Cummings DM, Alaghband Y, Hickey MA, Joshi PR, Hong SC, et al. (2012) A critical window of CAG repeat-length correlates with phenotype severity in the R6/2 mouse model of Huntington's disease. J Neurophysiol 107: 677–691.

50. Mangiarini L, Sathasivam K, Mahal A, Mott R, Seller M, et al. (1997) Instability of highly expanded CAG repeats in mice transgenic for the Huntington's disease mutation. Nat Genet 15: 197–200.

51. Luesse HG, Schiefer J, Spruenken A, Puls C, Block F, et al. (2001) Evaluation of R6/2 HD transgenic mice for therapeutic studies in Huntington's disease: behavioral testing and impact of diabetes mellitus. Behav Brain Res 126: 185–195.

52. Guerriero CJ, Brodsky JL (2012) The delicate balance between secreted protein folding and endoplasmic reticulum-associated degradation in human physiology. Physiol Rev 92: 537–576.

53. Schroder M, Kaufman RJ (2005) The mammalian unfolded protein response. Annu Rev Biochem 74: 739–789.

54. Ron D, Walter P (2007) Signal integration in the endoplasmic reticulum unfolded protein response. Nat Rev Mol Cell Biol 8: 519–529.

55. Walter P, Ron D (2011) The unfolded protein response: from stress pathway to homeostatic regulation. Science 334: 1081–1086.

56. Lee AS (2001) The glucose-regulated proteins: stress induction and clinical applications. Trends Biochem Sci 26: 504–510.

57. Li JY, Popovic N, Brundin P (2005) The use of the R6 transgenic mouse models of Huntington's disease in attempts to develop novel therapeutic strategies. NeuroRx 2: 447–464.

58. Lin JH, Li H, Yasumura D, Cohen HR, Zhang C, Panning B, et al. (2007) IRE1 signaling affects cell fate during the unfolded protein response. Science 318: 944–949.

59. Tabas I, Ron D (2011) Integrating the mechanisms of apoptosis induced by endoplasmic reticulum stress. Nat Cell Biol 13: 184–190.

60. Malhotra JD, Kaufman RJ (2007) The endoplasmic reticulum and the unfolded protein response. Semin Cell Dev Biol 18: 716–731.

61. Lee H, Noh JY, Oh Y, Kim Y, Chang JW, et al. (2012) IRE1 plays an essential role in ER stress-mediated aggregation of mutant huntingtin via the inhibition of autophagy flux. Hum Mol Genet 21: 101–114.

62. She P, Zhang Z, Marchionini D, Diaz WC, Jetton TJ, et al. (2011) Molecular characterization of skeletal muscle atrophy in the R6/2 mouse model of Huntington's disease. Am J Physiol Endocrinol Metab 301: E49–61.

63. Moffitt H, McPhail GD, Woodman B, Hobbs C, Bates GP (2009) Formation of polyglutamine inclusions in a wide range of non-CNS tissues in the HdhQ150 knock-in mouse model of Huntington's disease. PLoS One 4: e8025.

64. Lin CH, Tallaksen-Greene S, Chien WM, Cearley JA, Jackson WS, et al. (2001) Neurological abnormalities in a knock-in mouse model of Huntington's disease. Hum Mol Genet 10: 137–144.

65. Menalled LB, Sison JD, Dragatsis I, Zeitlin S, Chesselet MF (2003) Time course of early motor and neuropathological anomalies in a knock-in mouse model of Huntington's disease with 140 CAG repeats. J Comp Neurol 465: 11–26.

66. Slow EJ, van Raamsdonk J, Rogers D, Coleman SH, Graham RK, et al. (2003) Selective striatal neuronal loss in a YAC128 mouse model of Huntington disease. Hum Mol Genet 12: 1555–1567.

67. Gray M, Shirasaki DI, Cepeda C, Andre VM, Wilburn B, et al. (2008) Full-length human mutant huntingtin with a stable polyglutamine repeat can elicit progressive and selective neuropathogenesis in BACHD mice. J Neurosci 28: 6182–6195.

13

RanBP9 Overexpression Accelerates Loss of Pre and Postsynaptic Proteins in the APΔE9 Transgenic Mouse Brain

Hongjie Wang[1,⑨], Ruizhi Wang[1,⑨], Shaohua Xu[2], Madepalli K. Lakshmana[1]*

1 Section of Neurobiology, Torrey Pines Institute for Molecular Studies, Port Saint Lucie, Florida, United States of America, 2 Department of Biological Sciences, Florida Institute of Technology, Melbourne, Florida, United States of America

Abstract

There is now compelling evidence that the neurodegenerative process in Alzheimer's disease (AD) begins in synapses. Loss of synaptic proteins and functional synapses in the amyloid precursor protein (APP) transgenic mouse models of AD is well established. However, what is the earliest age at which such loss of synapses occurs, and whether known markers of AD progression accelerate functional deficits is completely unknown. We previously showed that RanBP9 overexpression leads to robustly increased amyloid β peptide (Aβ) generation leading to enhanced amyloid plaque burden in a mouse model of AD. In this study we compared synaptic protein levels among four genotypes of mice, i.e., RanBP9 single transgenic (Ran), APΔE9 double transgenic (Dbl), APΔE9/RanBP9 triple transgenic (Tpl) and wild-type (WT) controls. We found significant reductions in the levels of synaptic proteins in both cortex and hippocampus of 5- and 6-months-old but not 3- or 4-months-old mice. Specifically, at 5-months of age, rab3A was reduced in the triple transgenic mice only in the cortex by 25% ($p < 0.05$) and gap43 levels were reduced only in the hippocampus by 44% ($p < 0.01$) compared to wild-type (WT) controls. Interestingly, RanBP9 overexpression in the Tpl mice reduced gap43 levels by a further 31% ($p < 0.05$) compared to APΔE9 mice. RanBP9 also further decreased the levels of drebrin in the hippocampus by 32% ($p < 0.01$) and chromogranin in the cortex by 24% ($p < 0.05$) compared to APΔE9 mice. At 6-months of age, RanBP9 expression in the cortex led to further reduction of rab3A by 30% ($p < 0.05$) and drebrin by 38% ($p < 0.01$) compared to APΔE9 mice. RanBP9 also increased Aβ oligomers in the cortex at 6 months. Similarly, in the hippocampus, RanBP9 expression further reduced rab3A levels by 36% ($p < 0.01$) and drebrin levels by 33% ($p < 0.01$). Taken together these data suggest that RanBP9 overexpression accelerates loss of synaptic proteins in the mouse brain.

Editor: William Phillips, University of Sydney, Australia

Funding: National Institutes of Health grant numbers: 1R03AG032064-01, M.K. Lakshmana; 1R01AG036859-01, M.K. Lakshmana. The funders had no role in study design, data collection and analysis, decision to publish, or preparation of the manuscript.

* E-mail: mlakshmana@tpims.org

⑨ These authors contributed equally to this work.

Introduction

Alzheimer's disease (AD) is a devastating neurodegenerative disease of elderly that affects more than 35 million people worldwide [1]. AD is characterized by gradual intellectual deterioration and behavioral disturbances throughout the course of the disease. Accumulating data suggest that the progression of AD is more tightly associated with synapse degeneration rather than amyloid plaques or neurofibrillary tangles. For example, substantial evidence indicates that in AD, there is a decrease in the number of synapses, which occurs later than Aβ accumulation and correlates with disease progression [2–4]. Consequently, AD has been suggested to be a form of synaptic plasticity failure [5]. Ultrastructural analysis of autopsied brain tissue from patients with AD within few years after clinical onset revealed progressive synapse loss in the hippocampus, the frontal and inferior parietal cortex and entorhinal cortex [6,7]. Even in cases of mild AD, as much as 55% loss of synapses has been reported within the hippocampus [7]. The role of synaptic proteins, especially their

progressive loss in causing dementia has been the subject of increasing interest ever since the correlation between loss of synapses and AD was first established [8]. Biochemical analysis further showed that both the presynaptic protein synaptophysin [9] and the synaptic membrane and postsynaptic proteins such as synaptobrevin and synaptopodin [10,11] are severely altered in the brains of patients with AD. However, as of today the molecular pathways responsible for either the synapse loss or differential vulnerability is not clear and therefore understanding the cellular and molecular mechanisms responsible for synaptic damage is critical for designing future therapeutic strategies for AD.

Loss of synapses has also been confirmed in several mouse models of AD [12–15]. In a most recent study, a transgenic mouse model with knockin expression of human mutant APP and/or human presenilin showed significant loss of synaptophysin-immunoreactive presynaptic boutons in the CA1-2 region of hippocampus at 10-months of age [16]. In APP/PS1 double transgenic mice (APΔE9), the density of large spines in plaque-free regions of the dentate gyrus is significantly reduced at 12–14 months of age coincident with impairment of cognition [17]. Thus, animal models provide a good opportunity to test the temporal sequence of synaptic protein loss. Despite abundant evidence for loss of synaptic proteins and cognitive impairment, it is not clear precisely when the earliest alterations occur in mouse models of AD. It is also unknown whether known Alzheimer's risk factors accelerate loss of synapses and synaptic proteins. Thus, determining the earliest onset of memory deficits has been one of the main challenges in cognitive studies of AD. This is an important issue because identifying molecules that cause memory deficits depends upon accurately determining when cognitive deficits first appear. Transgenic mouse models provide excellent opportunities to test the effect of risk factors.

RanBP9 is a scaffolding protein that integrates a variety of signals from cell surface receptors to the intracellular targets [18,19]. RanBP9 is known to exist and function in multiprotein complexes [20,21]. We previously demonstrated that RanBP9 forms tripartite protein complex by binding with APP, BACE1 and low-density lipoprotein receptor-related protein (LRP), thereby increases Aβ generation in both transformed cells and primary neurons by enhancing β-secretase-mediated processing of APP at the cost of α-secretase processing [22]. Subsequently we confirmed increased amyloidogenic processing of APP by RanBP9 in vivo, by documenting increased amyloid plaque burden in a mouse model of AD [23]. Because RanBP9 protein levels are increased in the brains of patients with AD [24] as well as APP transgenic mouse models [25,26], increased Aβ levels and associated pathology in AD is at least partly due to RanBP9. In line with this hypothesis, RanBP9 was recently found to be within the clusters of RNA transcript pairs associated with markers of AD progression [27], suggesting that RanBP9 might actually contribute to the pathogenesis of AD. In fact, we recently confirmed an inverse relationship between RanBP9 levels and spinophilin, a marker of spines in the synaptosomes of Alzheimer's brains. RanBP9 overexpression in the APP/PS1 (APΔE9) mouse model was also accompanied by significantly impaired learning and memory skills [28]. These multiple evidences strongly suggest that RanBP9 plays pivotal role in the synaptic damage in AD. However, it is not clear whether RanBP9 overexpression exacerbates synaptic damage and if so whether it anticipates loss of synaptic proteins to earlier ages. Therefore identifying synaptic marker changes that follow cognitive deficits in early AD is critical as the accompanying synaptic changes can be effectively targeted by current treatments. Also detecting synaptic protein changes before the cognitive deficits appear have enormous implications for preventive and prophylactic treatments.

Here to get an overall picture of synaptic damage, we examined the levels of two presynaptic proteins, gap-43 and rab-3a and two postsynaptic proteins, drebrin A and chromogranin B at the earliest ages. Rab3a is a small vesicle protein while chromogranin B is a component of large dense core vesicles, all used as an estimate of synaptic density. GAP-43 is a component of presynaptic membranes while drebrin is a neuron specific major F-actin-binding protein abundantly found in dendritic spines. These four proteins together would represent the synaptic machinery including not only the small and large synaptic vesicles but also the pre and postsynaptic membranes. The levels of these proteins directly reflect number of synapses and since they play crucial role in synaptic plasticity, quantifying their protein levels is very important.

Materials and Methods

Chemicals and Antibodies

Protease inhibitor cocktail (cat # P8340), sodium orthovanadate (cat # S6508) and dithiothreitol (cat # D9779) were purchased from Sigma Aldrich (St. Louis, MO, USA). Microcystin-LR (cat # 475815) was obtained from Calbiochem (La Jolla, CA, USA). Polyclonal chromogranin antibody (cat # ab12242) was purchased from Abcam (Cambridge, MA, USA). Monoclonal anti-drebrin antibody (D029-3) was purchased from MBL international corporation (Woburn, MA, USA). Polyclonal Rab3A antibody (15029-1-AP) was purchased from ProteinTech Group Inc. (Chicago, IL, USA). Rabbit polyclonal antibody against Gap-43 was obtained from Millipore (Temecula, CA, USA). Anti-flag tag antibody (M2; F3165) was purchased from Sigma Aldrich (St. Louis, MO, USA). Polyclonal Aβ oligomer antibody, clone A11 (cat# AHB0052) was obtained from Life Technologies (Grand Island, NY, USA). Mouse monoclonal antibody against beta-actin (cat # A00702) was purchased from Genscript USA Inc. (Piscataway, NJ, USA). Secondary antibodies such as peroxidase-conjugated AffiniPure goat anti-mouse (Code # 115-035-146) and ant-rabbit (code # 111-035-144) IgGs were purchased from Jackson ImmunoResearch Laboratories (West Grove, PA, USA).

Mice

All animal experiments were carried out based on ARRIVE guidelines and in strict accordance with the National Institute of Health's 'Guide for the Care and Use of Animals' and approved by the Torrey Pines Institute's Animal Care and Use Committee (IACUC). Generation of RanBP9 transgenic mice have been described previously [23]. The RanBP9 specific primers used in the polymerase chain reaction (PCR) is as follows. The forward primer is 5′ – gcc acg cat cca ata cca g -3′, and the reverse primer is 5– tgc ctg gat ttt ggt tct c –3′. Positive mice were then backcrossed with native C57Bl/6 mice and the colonies were expanded. RanaBP9 transgenic line 629 was used to breed with B6.Cg-Tg, APPswe, PSEN1ΔE9 (APΔE9) mice for generating triple transgenic mice (APΔE9/RanBP9). We obtained APΔE9 from Jackson Labs (Bar Harbor, Maine, USA). These double transgenic mice express a chimeric mouse/human APP (Mo/HuAPP695swe) and a mutant human presenilin 1 (PS1-ΔE9) both directed to CNS neurons. These APΔE9 transgenic mice were generated by co-injection of APP695swe and PS1-ΔE9 encoding vectors controlled by their own mouse prion protein promoter element. These mice were backcrossed to maintain C57Bl/6 background, expanded and genotyped to confirm the transgene using the following primers. The forward primer is 5′ – gac tga cca ctc gac cag gtt ctg –3′ and the reverse primer is 5 - ctt gta agt tgg att ctc ata tcc g –3′. Only male mice were used for all genotypes. These mice were used to measure synaptic protein levels by standard immunoblots. The number of mice, 6 for the WT and RanBP9 transgenic mice and 8 for the APΔE9 and APΔE9/RanBP9 mice were based on both statistical power analysis and our own previous experience on the same parameters.

Tissue Extraction and Immunoblotting

The mouse brain tissues from four different genotypes, viz., wild-type (WT), RanBP9 single transgenic (Ran), APΔE9 double transgenic (Dbl) and APΔE9/RanBP9 triple transgenic (Tpl), all in C57BL6 background were dissected on ice immediately after euthanasia to obtain cortex and hippocampus tissue. Brain lysates were prepared from 3-, 4-, 5- and 6-months-old male mice from all four genotypes. In brief, we anesthetized the mice with isoflurane, decapitated them immediately and rapidly removed the brain tissues in to 1% NP40 buffer (50 mM Tris-HCl, pH 8.0, 150 mM NaCl, 0.02% sodium azide, 400 nM microcystine-LR, 0.5 mM sodium vanadate and 1% sodium Nonidet P-40) containing complete protease inhibitor cocktail for use with mammalian cell and tissue extracts (Sigma, St. Louis, USA). To extract Aβ oligomers RIPA buffer with SDS was used. Tissue was homogenized using Power Gen 125 (Fisher Scientific, Pittsburgh, USA) and centrifuged at 100,000 g for 1 h in a Beckman ultracentrifuge. Protein concentrations from each sample were measured in duplicates by BCA method (Pierce Biotechnology Inc., Rockford, USA). Before loading on to gels, the lysates were mixed with loading buffer containing dithiothreitol. SDS-PAGE electrophoresis was done exactly as published [22–24]. Briefly, Equal amounts of proteins were loaded into each well and subjected to electrophoresis. To separate Aβ oligomers, NuPAGE gels 4–12% were used. The proteins were then transferred onto PVDF membranes, blocked with 5% milk and incubated overnight with primary antibodies followed by one hour incubation with HRP-conjugated secondary antibodies such as monoclonal mouse anti-Goat IgG light chain or monoclonal mouse anti-Rabbit IgG light chain. The protein signals were detected using Super Signal West Pico Chemiluminescent substrate (Pierce, USA). Quantification of Western blot signals was done using imageJ software.

Immunohistochemistry

Brain sections (16 um) from 6-month-old WT and APΔE9/RanBP9 triple (Tpl) transgenic mice were washed with PBS 1X 3 times each for 5 min. Antigen retrieval was carried out by immersing slides in 10 mM citric acid (pH 6.0) for 10 min at 90–95°C. Sections were washed with PBS 1X for 5 min 3 times, and incubated in blocking solution (10% normal goat serum, 1% BSA, 0.1% Triton X-100 in PBS 1X) for 1 h at room temperature. The sections were incubated overnight with specific antibodies against rab3a, gap43, drebrin and chromogranin in blocking solution (1:200) at 4°C. After washing in PBS 1X for 5 min 3 times, the sections were incubated with Alexa Fluor® 568 goat anti-mouse IgG (Invitrogen) in blocking solution (1:500) at room temperature for 2 h in the dark. Finally, slides were washed with PBS for 5 min 3 times, covered with mounting medium for fluorescence with DAPI (Vector Laboratories) and sealed with nail clear. Sections were visualized in a confocal microscope (Nikon C1Si laser scanning multispectral confocal microscope). Images of frontal cortex and CA1 region of hippocampus were obtained at 20x and processed using Image-Pro Plus (Media Cybernetics) software package. Positive immunoreactive stainings for all four synaptic proteins were observed in the neurons whose nuclei were co-localized with DAPI.

Statistical Analysis

Immunoblot signals for rab3A, gap43, drebrin A and chromogranin B were quantified using publicly available Java-based ImageJ software. Levels of Aβ oligomers in APΔE9 versus APΔE9/RanBP9 mice were analyzed by Student's paired t-test considering two-tailed P value for significance. The protein levels in WT, Ran, APΔE9 and APΔE9/RanBP9 mice were analyzed by one-way analysis of variance (ANOVA) followed by Tukey-Kramer multiple comparison post-hoc test for comparisons among WT, Ran, APΔE9 and APΔE9/RanBP9 mice at different ages for interaction effects using Instat3 software (GraphPad Software, San Diego, CA, USA). The data presented are mean±SEM. The n = 6 for WT and Ran genotypes and n = 8 for the APΔE9 and APΔE9/RanBP9 genotypes. The data were considered significant only if the $p<0.05$, * indicates $p<0.05$, **, $p<0.01$ and ***, $p<0.001$.

Results

RanBP9 Accelerates Loss of Presynaptic Proteins, rab3A and gap43 in the Cortex and the Hippocampus

Although loss of synapses and synaptic proteins is well established in AD brains as well as mouse models of AD, the earliest ages when such anomaly begins to appear and also whether markers of AD progression have any influence on such a loss is unknown. Therefore we quantified the levels of two presynaptic and two post synaptic proteins in the APΔE9 double transgenic mice, and the APΔE9/RanBP9 triple transgenic mice. The generation and characterization of RanBP9 single transgenic and APΔE9/RanBP9 triple transgenic mice from our laboratory have been described [24,28]. In order to restrict the expression of exogenous flag-tagged RanBP9 in the neurons only, we used thy-1 promoter. As there is selective loss of synapses in the cortical and hippocampal brain regions in both AD and in the mouse models of AD, we quantified the synaptic proteins in both the cortex and the hippocampus. The mean densitometric values of rab3A and gap43 were unaltered in both the cortex and the hippocampus of RanBP9 single transgenic, APΔE9 double transgenic and APΔE9/RanBP9 triple transgenic mice at 3 months (Fig. 1A & B) and 4-months (Fig. 2A & B) of age compared to age-matched wild-type (WT) controls. All protein levels were expressed with relative expression levels of actin which was used as loading control (panel 6 in all figures). Flag antibody detected the expression of exogenous flag-tagged RanBP9 only in the RanBP9 single transgenic and APΔE9/RanBP9 triple transgenic mice (panel 5 in all figures).

At 5-months of age, rab3A levels were decreased in the APΔE9 (22%, $p<0.05$) and APΔE9/RanBP9 (25%, $p<0.05$) mice compared to WT controls only in the cortex but not hippocampus (Fig. 3A, panel 1 & 3B). Gap43 levels, on the other hand, decreased in the hippocampus but not the cortex in only APΔE9/RanBP9 mice (44%, $p<0.01$). Compared between APΔE9 and APΔE9/RanBP9 mice, there was 31% ($p<0.05$) further decrease in the levels of gap43 in the hippocampus due to RanBP9 overexpression (Fig. 3A panel 2& 3B,). By 6-months of age, the decreases in the levels of both rab3A and gap43 were more robust in the cortex as well as hippocampus. For rab3A, the decrease was 45% ($p<0.01$) and 35% ($p<0.01$) in the cortex and hippocampus respectively in the APΔE9/RanBP9 triple transgenic mice compared to WT controls, with no significant alterations in the APΔE9 mice (Fig. 4A, panel 1 and 4B). Of note, RanBP9 overexpression in the triple transgenic mice further decreased the levels of rab3A by 30% ($p<0.05$) and 36% ($p<0.01$) in the cortex and hippocampus respectively compared to APΔE9 mice (Fig. 4A, panel 1 and 4B). This suggests that RanBP9 overexpression exacerbates loss of rab3A in both the cortex and the hippocampus. Gap43 levels at 6-months of age were also significantly decreased by 36% ($p<0.01$) in the cortex and by 43% ($p<0.01$) in the hippocampus of APΔE9/RanBP9 triple transgenic mice compared to WT controls (Fig. 4A, panel 2 and 4B), with no significant alterations in the APΔE9 double transgenic mice. Thus, although there is evidence of dynamic changes in the levels of presynaptic proteins starting from 5- and 6-months of ages, the results suggests a consistently decreasing trend for both the proteins.

RanBP9 Accelerates Loss of Postsynaptic Proteins, Drebrin and Chromogranin in the Cortex and the Hippocampus

Having confirmed significant loss of presynaptic proteins, we were interested to study the effect of RanBP9 overexpression in the APΔE9 mice on postsynaptic proteins as well since both pre and

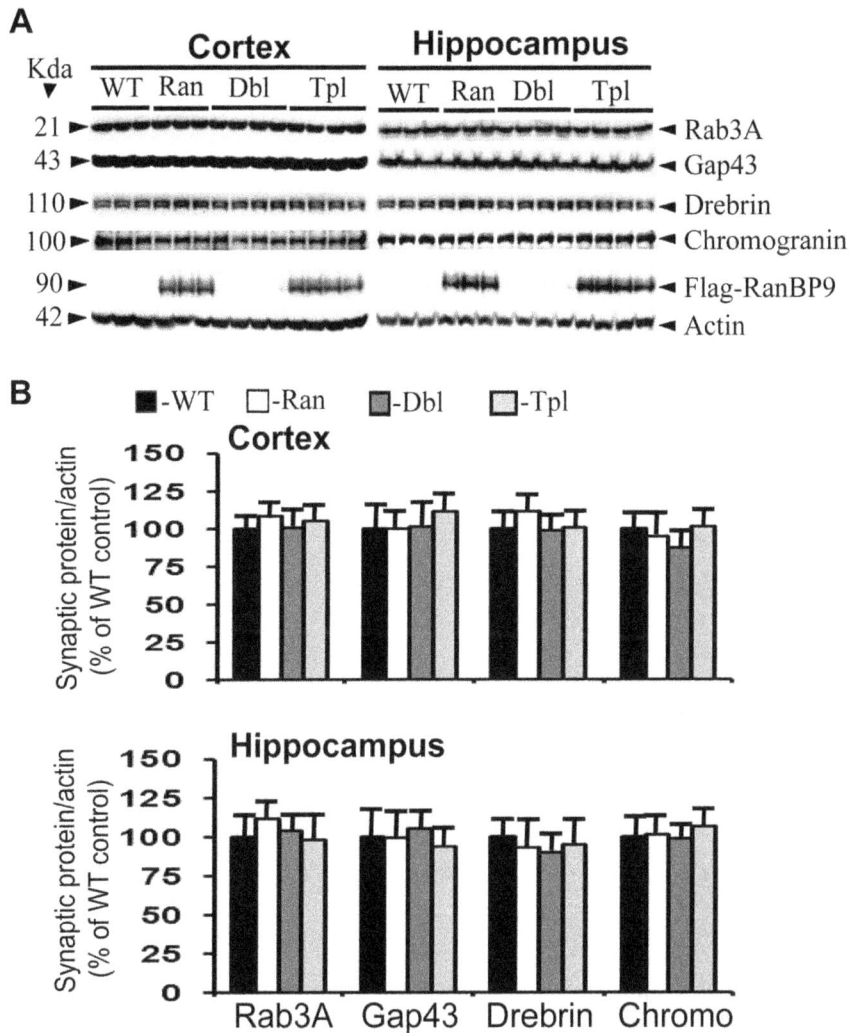

Figure 1. RanBP9 overexpression does not alter synaptic protein levels in the cortex and hippocampus at 3- months of age. A: Shows an immunoblotting analysis of rab3A, gap43, drebrin, chromogranin and the house keeping gene actin in brain samples from cortex and hippocampus. Brain homogenates from RanBP9 transgenic (Ran), APΔE9 double transgenic (Dbl), APΔE9/RanBP9 triple transgenic (Tpl) and age-matched wild-type (WT) control mice at 3-months of age were subjected to SDS-PAGE electrophoresis and probed with their respective antibodies. Flag specific monoclonal antibody detected flag-tagged exogenous RanBP9 in the RanBP9 single transgenic and APΔE9/RanBP99 triple transgenic mice only. Actin was used as loading control. The numbers on the left side indicate the molecular weights of each protein. **B:** Image J quantitation and normalization to actin levels showed no changes in the levels of any of the synaptic proteins at 3 months. The data are mean±SEM, n = 6 for WT and RanBP9 single transgenic, and n = 8 for APΔE9 and APΔE9/RanBP9 genotypes.

postsynaptic terminals are affected in AD. Thus, to ensure overall changes in the whole synaptic structure we examined two postsynaptic proteins, drebrin and chromogranin. Similar to changes in the levels of presynaptic proteins, the levels of drebrin (Fig. 1A, panel 3 and 1B) and chromogranin (Fig. 1A, panel 4 and 2B) were not significantly altered at 3-months of age in either the RanBP9 single transgenic, APΔE9 double transgenic or APΔE9/RanBP9 triple transgenic mice compared to age-matched WT controls in both the hippocampus and the cortex. At 4 months of age also, both drebrin (Fig. 2A, panel 3 and 1B) and chromogranin (Fig. 2A, panel 4 and 2B) were unaffected in any of the genotypes studied in both the cortex and the hippocampus. These results are consistent with those of presynaptic proteins and suggest that even under the condition of overexpression of RanBP9, a known molecular marker of AD progression, it is insufficient to alter any of the synaptic proteins.

However, by 5-months of age significant changes were noted in the levels of both drebrin and chromogranin. Drebrin levels were significantly reduced in the cortex by 33% (p<0.05) in the triple transgenic mice compared to WT controls (Fig. 3A, panel 3 and 3B). Neither the RanBP9 single transgenic nor APΔE9 double transgenic mice showed any alterations in the levels of drebrin in the cortex. In the hippocampus, the reduction was 29% (p<0.01) in the triple transgenic mice relative to WT controls. When the hippocampal drebrin levels were compared between APΔE9 and APΔE9/RanBP9 mice, the reduction was 32% (p<0.01) in the APΔE9/RanBP9 mice implying that RanBP9 overexpression worsens loss of drebrin protein (Fig. 3A, panel 3 and 3B). RanBP9 overexpression similarly worsened the loss of gap43 protein in the hippocampus but not in the cortex. Chromogranin levels, on the other hand were not altered in the hippocampus but showed reduced levels only in the cortex. Normalized levels of chromo-

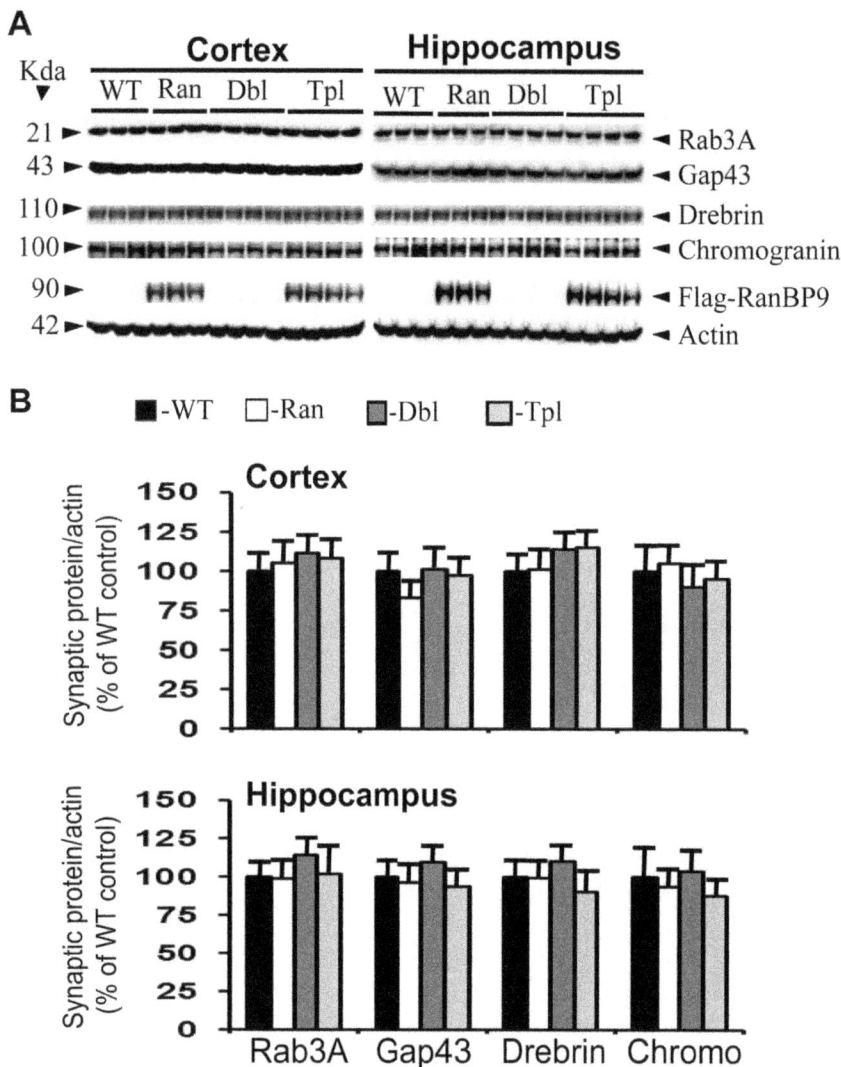

Figure 2. RanBP9 overexpression does not alter synaptic protein levels in the cortex and hippocampus at 4- months of age. A: Shows an immunoblotting analysis of rab3A, gap43, drebrin, chromogranin and the house keeping gene actin in brain samples from cortex and hippocampus. Brain homogenates from RanBP9 transgenic (Ran), APΔE9 double transgenic (Dbl), APΔE9/RanBP9 triple transgenic (Tpl) and age-matched wild-type (WT) control mice at 4-months of age were subjected to SDS-PAGE electrophoresis and probed with their respective antibodies. Flag specific monoclonal antibody detected flag-tagged exogenous RanBP9 in the RanBP9 single transgenic and APΔE9/RanBP99 triple transgenic mice only. Actin was used as loading control. The numbers on the left side indicate the molecular weights of each protein. **B:** Image J quantitation and normalization to actin levels showed no changes in the levels of any of the synaptic proteins at 4 months. The data are mean±SEM, n = 6 for WT and RanBP9 single transgenic, and n = 8 for APΔE9 and APΔE9/RanBP9 genotypes.

granin was reduced by 30% (p<0.05) only in the APΔE9/RanBP9 triple transgenic mice compared to WT controls (Fig. 3A, panel 4 and 3B). A significantly further reduction (24%, p<0.05) was also noted between APΔE9 and APΔE9/RanBP9 genotypes.

Consistent with reductions in the levels of presynaptic proteins, the levels of postsynaptic protein were also worst affected relatively at 6-months of age in both the cortex and the hippocampus. Thus, in the cortex APΔE9/RanBP9 triple transgenic mice but not RanBP9 or APΔE9 mice showed a statistically significant reduction by 58% (p<0.01) in the levels of drebrin when compared to WT controls (Fig. 4A, panel 3 and 3B). RanbP9 overexpression further exacerbated the reduction by 38% (p<0.01) in the triple transgenic mice when compared to APΔE9 double transgenic mice (Fig. 3A, panel 3 and 3B). In the hippocampus the loss of drebrin protein was 40% (p<0.01) in

the APΔE9/RanBP9 mice with no change in the levels either in the APΔE9 or RanBP9 mice (Fig. 3A, panel 3 and 3B) compared to WT controls. RanBP9 exacerbation in the hippocampus was 33% (p<0.01). However, chromogranin levels were reduced in the cortex in both the APΔE9 (44%, p<0.01) and APΔE9/RanBP9 (43%, p<0.01) mice when compared to WT controls. In the hippocampus, chromogranin protein level was reduced (21%, p<0.01) only in the APΔE9/RanBP9 triple transgenic mice (Fig. 3A, panel 4 and 3B). Thus at 6-months of age, both pre and postsynaptic proteins were more severely affected than other ages. This suggests that there is progressive reduction in the levels of synaptic proteins with respect to their age when compared to age-matched WT controls.

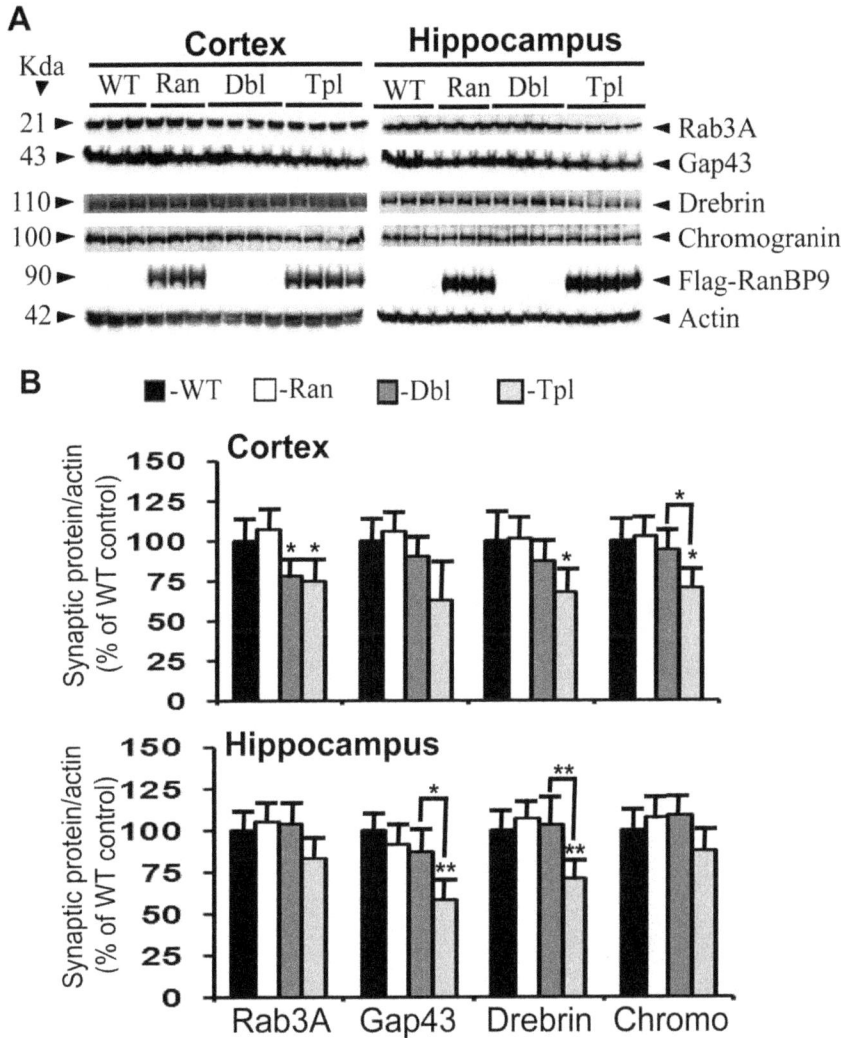

Figure 3. RanBP9 overexpression exacerbates reduction of synaptic protein levels at 5-months of age in the cortex and the hippocampus of APΔE9 mice. A: Brain homogenates were processed and synaptic proteins, flag-tagged RanBP9 and actin were detected as in legend to figure 1. **B:** ImageJ quantitation and normalization to actin levels revealed significant differences. Rab3A levels were reduced in the cortex by 22% and 25% respectively in the APΔE9 and APΔE9/RanBP9 mice compared to WT controls, but no changes were observed in the hippocampus. Gap43 levels were reduced by 44% only in the hippocampus of triple transgenic mice. Drebrin levels were reduced by 33% and 29% respectively in the cortex and the hippocampus. only in the APΔE9/RanBP9 mice compared to WT, but no change in APΔE9 mice versus WT or RanBP9 mice. Chromogranin levels were reduced only in the cortex by 30% in the triple transgenic mice. ANOVA followed by post-hoc Tukey's test revealed significant differences. *, p<0.05, **, p<0.01 in APΔE9/RanBP9 or APΔE9 mice compared to WT mice. The data are mean±SEM, n=6 for WT and RanBP9 mice, and n=8 for APΔE9 and APΔE9/RanBP9 genotypes.

Immunohistochemical Evidence for Loss of Synaptic Proteins in the Cortex and Hippocampus of APΔE9/RanBP9 Mice

To confirm loss of synaptic proteins by another method, we stained brain sections of WT and APΔE9/RanBP9 triple transgenic mice with specific antibodies against rab3a, gap43, drebrin and chromogranin. Because synaptic protein levels were most affected in the triple transgenic mice compared to WT at 6-months of age as revealed by immunoblots, we studied the staining pattern in these two genotypes only at 6-months of age. To assess the role of RanBP9 overexpression, we analyzed synaptic protein immunoreactivity in the CA1 region of the hippocampus and the frontal cortex. Although we did not quantify the fluorescence intensity, an apparent qualitative difference could be seen in the staining intensity in the triple transgenic mice versus WT controls

in both the frontal cortex (Fig. 5) and the CA1 region of the hippocampus (Fig. 6) for all the four synaptic proteins studied. Thus immunohistochemical staining confirmed the immunoblot results and suggests that RanBP9 overexpression significantly reduces both the pre and postsynaptic proteins.

RanBP9 Overexpression in APΔE9 Mice Significantly Reduces Levels of Aβ Oligomers

We previously showed that RanBP9 overexpression in APΔE9 mice reduced both the soluble Aβ monomers and amyloid plaques (24). To understand whether reduced synaptic proteins is due to RanBP9-induced alterations in the levels of Aβ oligomers, we quantified RIPA-soluble Aβ oligomers in the cortex of all four genotypes of mice using oligomer-specific antibody, A11. A11 has proven to recognize only amyloid oligomers of both mouse and

A

B

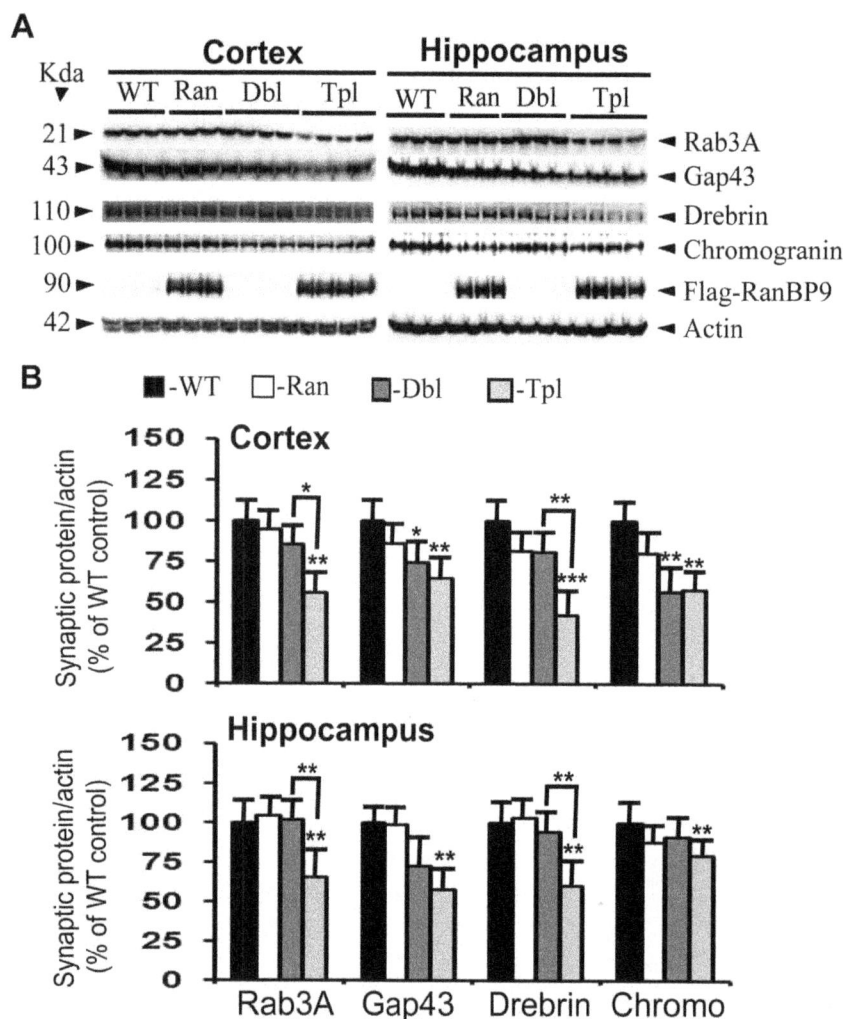

Figure 4. RanBP9 overexpression exacerbates reduction of synaptic protein levels at 6-months of age in the cortex and the hippocampus of APΔE9 mice. A: Brain homogenates were processed and synaptic proteins, flag-tagged RanBP9 and actin were detected as in legend to figure 1. **B:** ImageJ quantitation and normalization to actin levels revealed significant differences. Rab3A levels were further reduced by 30% and 36% in the cortex and hippocampus respectively in the triple transgenic mice compared to double transgenic mice. Similarly, drebrin levels were further reduced by 38% and 33% in the cortex and the hippocampus in the triple transgenic mice compared to double transgenic mice. Gap43 levels were reduced in the cortex by 36% in the triple transgenic mice and by 26% in the double transgenic mice compared to WT controls. In the hippocampus gap43 levels were reduced only in the triple transgenic mice by 43%. In the cortex, chromogranin levels were reduced by 44% in the double and by 44% in the triple transgenic mice, whereas in the hippocampus a 21% reduction was observed only in the triple transgenic mice. ANOVA followed by post-hoc Tukey's test revealed significant differences. *, $p < 0.05$, **, $p < 0.01$, ***, $p < 0.001$ in APΔE9/RanBP9 or APΔE9 mice compared to WT mice as indicated in the figure. The data are mean±SEM, n = 6 for WT and RanBP9 mice, and n = 8 for APΔE9 and APΔE9/RanBP9 genotypes.

human origin but not either amyloidogenic monomers or mature amyloid fibrils. At 3-months of age Aβ oligomers were almost undetectable in the WT and RanBP9 mice while in the APΔE9 and APΔE9/RanBP9 mice a faint band starts to appear suggesting that oligomers have not yet built up to any significant extent. At 4 and 5 months, a clear oligomer band is seen in both the APΔE9 and APΔE9/RanBP9 mice, but quantification revealed no significant differences among them.

At 6 months, oligomer levels were increased by 35% (p<0.001) in the APΔE9/RanBP9 mice compared to APΔE9 mice (Fig. 7A & B). Thus RanBP9 also increases levels of Aβ oligomers which may be responsible for the decreased levels of synaptic proteins.

Discussion

In AD, the degeneration of synapses and neurons occurs in selective regions of the brain, including the frontal and parietal cortices [29,30] and the hippocampus [31], which is the leading cause of cognitive impairment in AD. But the molecular mechanisms responsible for such a loss of synapses and more importantly precisely at what age such a loss of synapses occurs are not yet fully understood. Transgenic mouse models provide great opportunities to address such issues. We previously demonstrated that at 12-months of age, RanBP9 overexpression led to significant reductions in the levels of synaptophysin and PSD95 [23] and learning deficits [25]. In a more recent study [28], we confirmed significant reductions in the levels of spinophilin, a marker of dendritic spines in both the hippocampus and cortex at 12-months

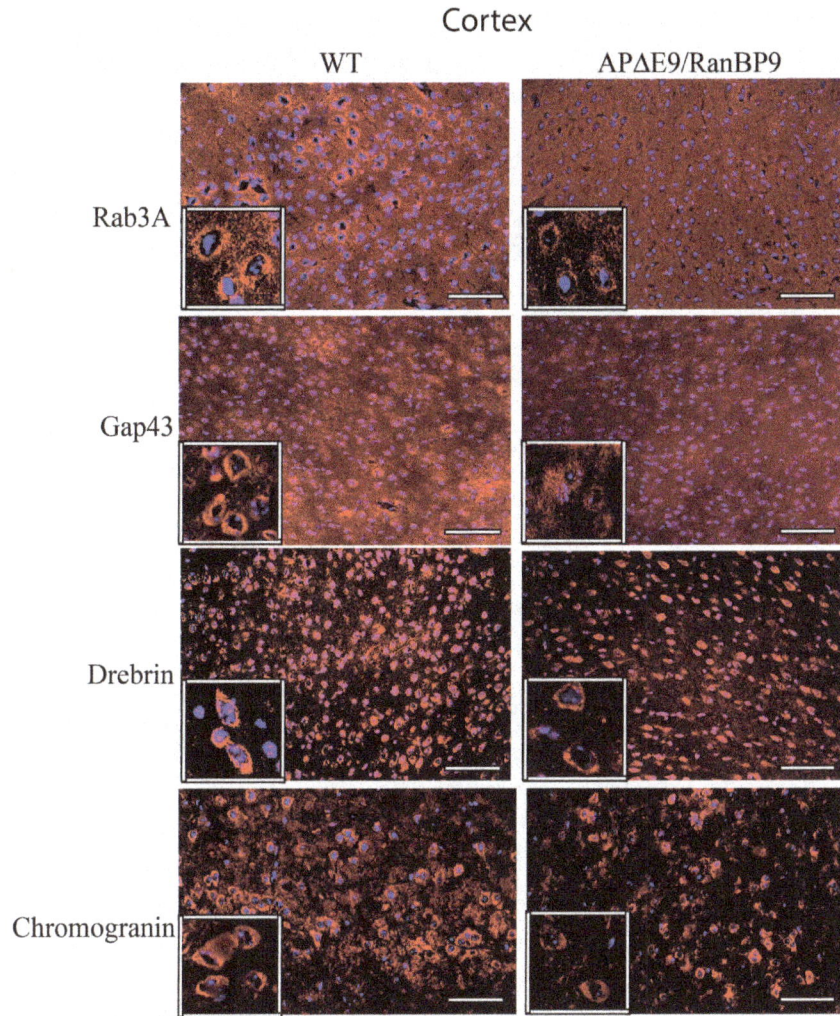

Figure 5. Immunohistochemical evidence for the reduced synaptic proteins at 6-months of age in the frontal cortex. Cortical brain sections from wild-type (WT) and APΔE9/RanBP9 triple transgenic mice were stained with antibodies against rab3a, gap43, drebrin and chromogranin. A qualitative difference is clearly seen with reduced immunoreactive puncta in the triple transgenic mice compared to WT brains for all the four synaptic proteins (red). The neuronal nuclei are stained blue. Scale bar: 100 μm.

of age when RanBP9 was overexpressed in APΔE9 mice. The present study was primarily designed to identify the earliest possible age when loss of other synaptic proteins begins to appear in the APΔE9 mice under RanBP9 overexpression conditions.

The results consistently revealed that synaptic proteins are unaltered at 3- and 4-months of age in both the cortex and the hippocampus. Since we measured both the pre and postsynaptic proteins, the results imply that the synaptic structure as a whole is intact at these ages. At 5-months of age a trend towards significant reductions starts to appear. Interestingly, the reductions were not uniform in both the cortex and the hippocampus at 5-months of age. While rab3A and chromogranin were affected only in the cortex, gap43 was affected only in the hippocampus. Only drebrin protein levels were reduced in both the brains regions studied. Given the functions of synaptic proteins in neurotransmitter vesicle trafficking, docking and fusion to the synaptic membrane, it is not surprising that loss of synaptic proteins in AD is correlated with clinical symptoms [32] and that such a loss of synaptic proteins is brain region-specific [33]. It is conceivable that the reductions in the levels of synaptic proteins simply reflect the loss of synapses

and that AD pathology may not be acting directly on these proteins. While such region specific loss of proteins is known for a long time in AD, the exact reason for such a differential vulnerability is not known. Regional differences in the neuronal activity, energy consumption and expression of specific molecules might play crucial roles in the regional vulnerability. Synaptic proteins play critical role in synaptic plasticity, which is thought to underlie learning and memory. The differential reductions in the levels of synaptic proteins at 5 months of age probably represent highly dynamic and transition state of the changes at this particular age because even by 6-months, the reductions were uniform for all the synaptic proteins in both the cortex and the hippocampus. In contrast to presynaptic markers, the postsynaptic structures have been less studied in both AD brains and the mouse models. As a major F-actin binding protein which is abundantly found in the dendritic spines, inclusion of drebrin in the present study is of major significance. Also, the present results suggest that RanBP9 affects both the pre and postsynaptic components of the synaptic structure as early as 6-months of age.

Hippocampus

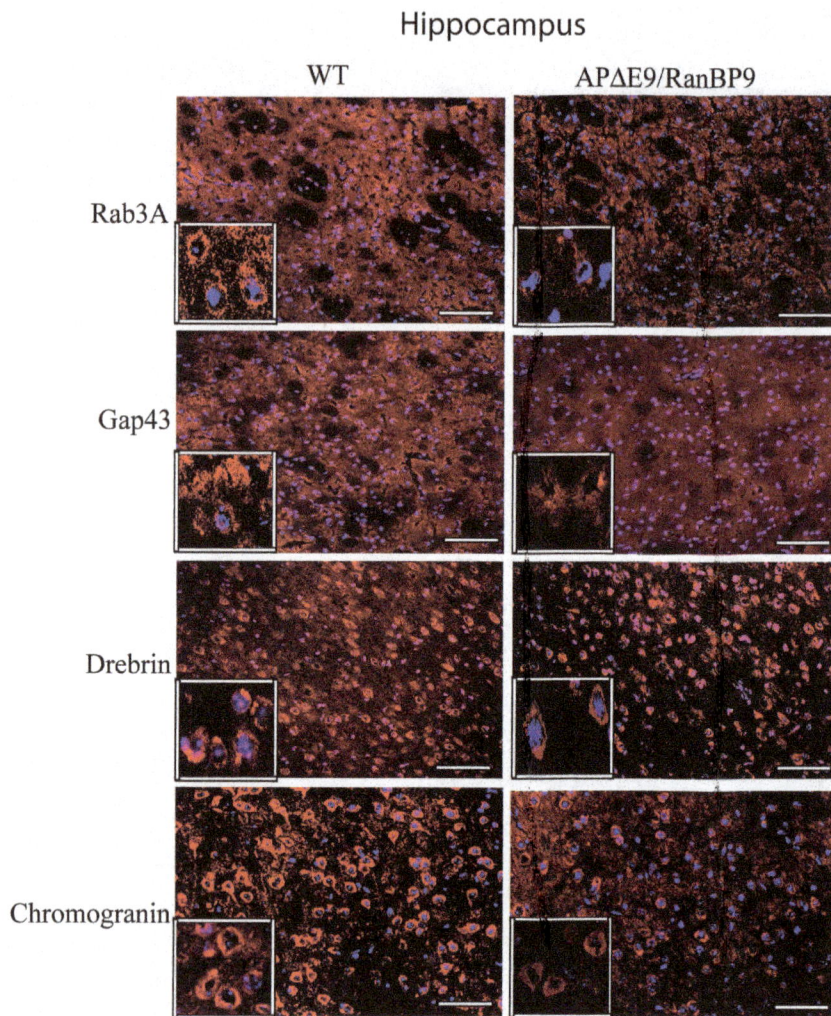

Figure 6. Immunohistochemical evidence for the reduced synaptic proteins at 6-months of age in the CA1 region of the hippocampus. Hippocampal brain sections from wild-type (WT) and APΔE9/RanBP9 triple transgenic mice were stained with antibodies against rab3a, gap43, drebrin and chromogranin. A qualitative difference is clearly seen with reduced immunoreactive puncta in the triple transgenic mice compared to WT brains for all the four synaptic proteins (red). The neuronal nuclei are stained blue. Scale bar: 100 μm.

APΔE9 mice display about 50% reduction in the levels of synaptophysin as estimated by unbiased stereology in the hippocampus of 7 months old mice [34]. In the present study we found 43% reductions in the levels of only chromogranin in the cortex of APΔE9 mice at 6-months of age. Significantly, the expression of RanBP9 was sufficient to induce significant reductions of all four synaptic proteins in both the cortex and the hippocampus. Using another line of double transgenic mice expressing both APP and PS1 mutants similar to APΔE9 mice, Rutten et al., [35] also showed about 33% reduction in synaptophysin levels in the hippocampus. However no data is available for other synaptic markers in this model to our knowledge. Therefore direct comparisons can't be made. Nevertheless, our carefully done study has revealed for the first time that both the pre and postsynaptic proteins can be significantly reduced by overexpressing RanBP9. Because RanBP9 levels are increased in AD brains [24], it is possible that RanBP9 is responsible for the reduced synapses and synaptic proteins in AD. Although at present we have no clue on how RanBP9 levels are increased in AD brains, in future studies we will investigate the contribution of

miRNAs and epigenetic factors which are known to regulate many key genes. Thus RanBP9 accelerates the synaptic protein deficits to earlier ages with more robust reductions with respect to increasing age. Since RanBP9 reduces both pre and postsynaptic proteins, it also indicates that RanBP9 induces gross changes in the synaptic structure.

The present results are consistent with many documented properties of RanBP9. RanBP9 not only increases Aβ and amyloid plaques [22,23], but also induces significant reductions in the levels of sAPPα [22]. Defective APP processing resulting in increased Aβ generation and most importantly decreased sAPPα levels is suggested to play crucial role in reduced synapses. A number of studies have shown that sAPPα exhibits neurotropic properties [36,37]. A more recent study clearly demonstrated that application of sAPPα but not sAPPβ in the conditioned medium significantly restored loss of spines and dendritic branches in neurons cultured from APP−/− mice [38], suggesting that reduced sAPPα levels seen in AD patients might actually be responsible for the loss of synapses. Thus RanBP9-induced reduction in sAPPα levels or increased Aβ oligomer levels observed in the present study is likely

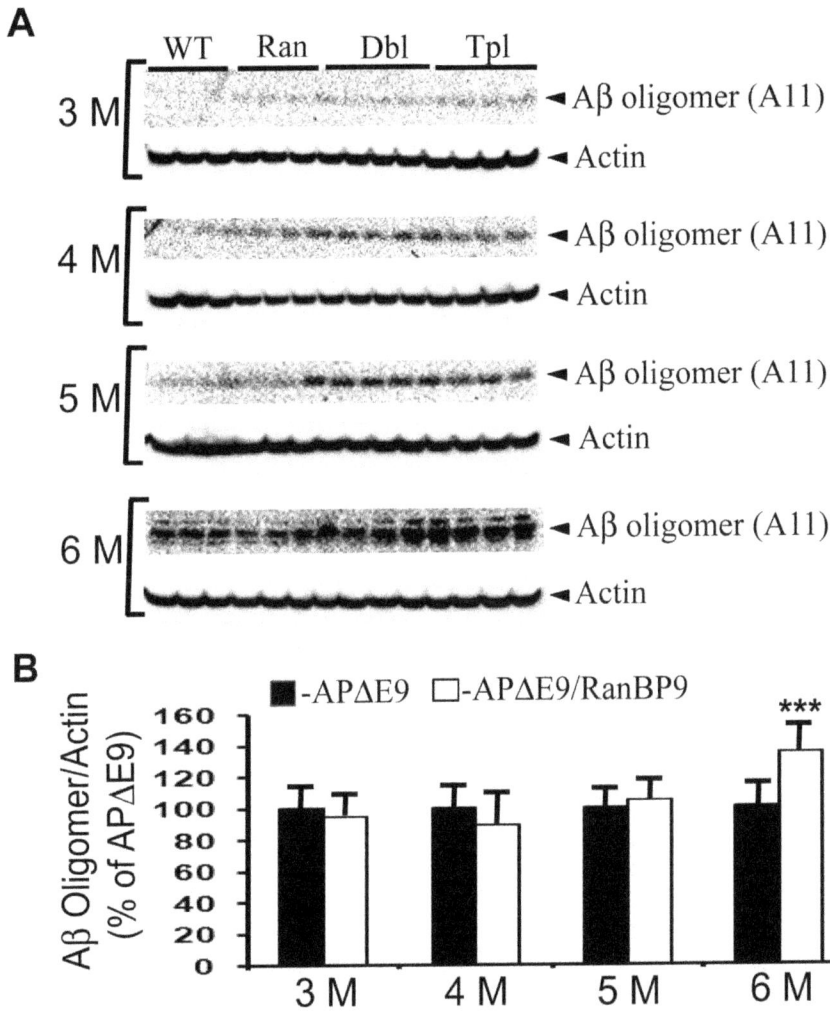

Figure 7. RanBP9 overexpression in APΔE9 mice increases Aβ oligomer levels at 6-months of age in the cortex. A: Proteins extracted from brains of wild-type (WT), RanBP9 single transgenic (Ran), APΔE9 double transgenic (Dbl) and APΔE9/RanBP9 triple transgenic mice (Tpl) of 3, 4, 5 and 6-months of age were homogenized in RIPA buffer and subjected to SDS electrophoresis and subsequently probed with anti-oligomer antibody, A11. **B:** For quantification of oligomers by image j, the data from WT and RanBP9 single transgenic mice were not considered since the levels were insignificant. Protein levels were normalized to actin and data expressed as percentage change. Student's t-test analysis revealed significant increase in oligomer levels only at 6-months of age. Data are ±SEM, n = 3 for each of WT and Ran and 4 for each of APΔE9 and APΔE9/RanBP9 genotypes. ***, p<0.001 in the APΔE9/RanBP9 mice compared to APΔE9 mice.

at least partly responsible for the presently observed loss of synaptic proteins. Such loss of synaptic proteins at 5- and 6-months of age is responsible for the recently observed learning and memory deficits at the same ages in the APΔE9/RanBP9 triple transgenic mice [39]. Moreover, RanBP9 overexpression significantly increased amyloid plaque burden in the APΔE9 mice starting as early as 5-months of age but not at 3- or 4-months of age [39]. Thus there is excellent correlation between the loss of synaptic proteins learning and memory deficits and increased APP metabolism at the same ages. We also recently demonstrated that RanBP9 retards the anterograde transport of mitochondria, consequently leading to reduced synaptic mitochondrial activity [40]. Treatment of primary cortical neurons derived from RanBP9 transgenic mice with Trolox, which inhibits generation of reactive oxygen species (ROS) such as superoxide restored the anterograde transport of mitochondria toward synapses and the corresponding synaptic activity. In fact, direct measurement of mitochondrial activity in the synaptosomes derived from RanBP9 transgenic mice

showed significant reductions [40]. When RanBP9 was overexpressed in the APΔE9 mice, it further exacerbated the deficits in mitochondrial activity, both under basal conditions and under ATP- and KCl-stimulated conditions [28]. Since synaptic terminals have high energy demands, reduced mitochondrial activity can be expected to compromise the viability and survival of synapses. As a result gradual loss of synapses might account for age-dependent loss of synaptic proteins observed in the present study.

The other indirect supporting evidence for RanBP9 to play crucial role at synapses comes from its subcellular localization and protein interactions. In primary neuronal cultures, RanBP9 is present throughout the neuron, especially in the entire network of neurites [23]. In the adult brain also, in addiction to soma, the dendritic processes also show the localization of significant amounts of RanBP9 [23] suggesting that RanBP9 has an important function at the synapses. Additionally, RanBP9 has been shown to interact with L1 receptor [41], integrin LFA-1 receptor [42] and Rho-GTPases [43], all of which are known to play crucial roles in the synaptic plasticity. Given the interaction of RanBP9 with many important proteins it is not surprising that RanBP9 reduces the synaptic proteins. But the most important observation in the present study is that RanBP9 accelerates the loss of synaptic proteins to earlier ages as early as 5-months of age when no abnormalities were found in the APΔE9 mice. Therefore future therapeutic interventions based on RanBP9 and RanBP9

signal transduction pathways may be an excellent approach to bring effective disease modifying therapy for AD.

Author Contributions

Conceived and designed the experiments: MKL. Performed the experiments: HW RW. Analyzed the data: SX. Wrote the paper: MKL.

References

1. Wimo A, Winblad B, Jönsson L (2010) The worldwide societal costs of dementia: Estimates for 2009. Alzheimer's Dement 6; 98–103.
2. Terry RD, Masliah E, Salmon DP, Butters N, DeTeresa R, et al. (1991) Physical basis of cognitive alterations in Alzheimer's disease: synapse loss is the major correlate ofcognitive impairment. Ann Neurol 30; 572–580.
3. Shankar GM, Walsh DM (2009) Alzheimer's disease: synaptic dysfunction and Aβ. Mol Neurodegener 4; 48.
4. Selkoe DJ (2002) Alzheimer's disease is a synaptic failure. Science 298; 789–791.
5. Nistico R, Collingridge GL (2012) The synaptic basis of Alzheimer's disease. Eur J Neurodeg Dis 1; 21–33.
6. Scheff SW, Price DA, Schmitt FA, Mufson EJ (2006) Hippocampal synaptic loss in early Alzheimer's disease and mild cognitive impairment. Neurobiol Aging 27; 1372–1384.
7. Scheff SW, Price DA, Schmitt FA, DeKosky ST, Mufson EJ (2007) Synaptic alterations in CA1 in mild Alzheimer disease and mild cognitive impairment. Neurology 68; 1501–1508.
8. Davies CA, Mann DM, Snupter PQ, Yates PO (1987) A quantitative morphometric analysis of the neuronal and synaptic content of the frontal and temporal cortex in patients with Alzheimer's disease. J Neurol Sci 78; 151–164.
9. Honer WG (2003) Pathology of presynaptic proteins in Alzheimer's disease: more than simple loss of terminals. Neurobiol Aging. 24;1047–1062.
10. Reddy PH, Mani G, Park BS, Jacques J, Murdoch G, et al. (2005) Differential loss of synaptic proteins in Alzheimer's disease: implications for synaptic dysfunction. J Alzheimers Dis 7; 103–117.
11. Ingelsson M, Fukumoto H, Newell KL, Growdon JH, Hedley-Whyte ET, et al. (2004) Early Abeta accumulation and progressive synaptic loss, gliosis, and tangle formation in AD brain. Neurology 62; 925–931.
12. Lanz TA, Carter DB, Merchant KM (2003) Dendritic spine loss in the hippocampus of young PDAPP and Tg2576 mice and its prevention by the ApoE2 genotype. Neurobiol Dis 13; 246–253.
13. Moolman DL, Vitolo OV, Vonsattel JP, Shelanski ML (2004) Dendrite and dendritic spine alterations in Alzheimer models. J Neurocytol 33; 377–387.
14. Tsai J, Grutzendler J, Duff K, Gan WB (2004) Fibrillar amyloid deposition leads to local synaptic abnormalities and breakage of neuronal branches. Nat Neurosci 7; 1181–11833.
15. Spires TL, Meyer-Luehmann M, Stern EA, McLean PJ, Skoch J, et al. (2005) Dendritic spine abnormalities in amyloid precursor protein transgenic mice demonstrated by gene transfer and intravital multiphoton microscopy. J Neurosci 25; 7278–7287.
16. Brasnjevic I, Lardenoije R, Schmitz C, Van Der Kolk N, Dickstein DL, et al. (2013) Region-specific neuron and synapse loss in the hippocampus of APP/PS1 knock-in mice. Transl Neurosci 4; 8–19.
17. Knafo S, Alonso-Nanclares L, Gonzalez-Soriano J, Merino-Serrais P, Fernaud-Espinosa I, et al. (2008) Widespread changes in dendritic spines in a model of Alzheimer's disease. Cereb Cortex 19; 586–592.
18. Murrin LC, Talbot JN (2007) RanBPM, a scaffolding protein in the immune and nervous systems. J Neuroimmune Pharmacol 2; 290–295.
19. Suresh B, Ramakrishna S, Baek KH (2012) Diverse roles of the scaffolding protein RanBPM. Drug Discov Today 17; 379–387.
20. Nishitani H, Hirose E, Uchimura Y, Nakamura M, Umeda M, et al. (2001) Full-sized RanBPM cDNA encodes a protein processing a long stretch of proline and glutamine within the N-terminal region, comprising a large protein complex. Gene 272; 25–33.
21. Zou Y, Lim S, Lee K, Deng X, Friedman E (2003) Serine/Threonine kinase Mirk/Dyrk1B is an inhibitor of epithelial cell migration and is negatively regulated by the Met adaptor Ranbinding protein M. J Biol Chem 278; 49573–49581.
22. Lakshmana MK, Yoon IS, Chen E, Bianchi E, Koo EH, et al. (2009) Novel role of RanBP9 in BACE1 processing of amyloid precursor protein and amyloid beta peptide generation. J Biol Chem 284; 11863–11874.
23. Lakshmana MK, Hayes CD, Bennet SP, Bianchi E, Reddy KM, et al. (2012) Role of RanBP9 on amyloidogenic processing of APP and synaptic protein levels in the mouse brain. FASEB J 26; 2072–2083.
24. Lakshmana MK, Chung JY, Wickramarachchi S, Tak E, Bianchi E, et al. (2010) A fragment of the scaffolding protein RanBP9 is increased in Alzheimer's disease brains and strongly potentiates amyloid-beta peptide generation. FASEB J 24, 119–127.
25. Woo JA, Jung AR, Lakshmana MK, Bedrossian A, Lim Y, et al. (2012) Pivotal role of the RanBP9-cofilin pathway in Abeta-induced apoptosis andneurode-generation. Cell Death Differ 19; 1413–1423.
26. Wang H, Dey D, Carrera I, Minond D, Bianchi E, et al. (2013) COPS5 (Jab1) increases β-site processing of amyloid precursor protein and Aβ generation by stabilizing RanBP9 protein levels. J Biol Chem Aug 7. Epub ahead of print.
27. Arefin AS, Mathieson L, Johnstone D, Berretta R, Moscato P (2012) Unveiling clusters of RNA transcript pairs associated with markers of Alzheimer's disease progression. PLoS One 7, e45535.
28. Palavicini JP, Wang H, Bianchi E, Xu S, Rao JS, et al. (2013) RanBP9 aggravates synaptic damage in the mouse brain and is inversely correlated to spinophilin levels in Alzheimer's brain synaptosomes. Cell Death Dis e667.
29. Davidsson P, Blennow K (1998) Neurochemical dissection of synaptic pathology in Alzheimer's disease, International Psychogeriatrics 10; 11–23.
30. Provoda CJ, Waring MT, Buckley KM (2000) Evidence for a primary endocytic vesicle involved in synaptic vesicle biogenesis. J Biol Chem 275; 7004–7012.
31. Sze CI, Troncoso JC, Kawas C, Mouton P, Price DL, et al. (1997) Loss of the presynaptic vesicle protein synaptophysin in hippocampus correlates with cognitive decline in Alzheimer disease, Journal of Neuropathology and Exp Neurol 56; 933–944.
32. Heffernan JM, Eastwood SL, Nagy Z, Sanders MW, McDonald, etal. (1998) Temporal cortex synaptophysin mRNA is reduced in Alzheimer's disease and is negatively correlated with the severity of dementia. Exp Neurol 150; 235–239.
33. Sze CI, Bi H, Kleinschmidt-DeMasters BK, Filley CM, Martin IJ (2000) Selective regional loss of exocytotic presynaptic vesicle proteins in Alzheimer's disease brains. J Neurol Sci 175; 81–90.
34. Ding Y, Qiao A, Wang Z, Goodwin JS, Lee ES, et al. (2008) Retinoic acid attenuates beta-amyloid deposition and rescues memory deficits in an Alzheimer's disease transgenic mouse model. J Neurosci 28; 11622–11634.
35. Rutten BP, Van der Kolk NM, Schafer S, van Zandvoort MA, Bayer TA, et al. (2005) Age-related loss of synaptophysin immunoreactive presynaptic boutons within the hippocampus of APP751SL, PS1M146L, and APP751SL/PS1M146L transgenic mice. Am J Pathol 167; 161–173.
36. Quast T, Wehner S, Kirfel G, Jaeger K, De Luca M, et al. (2003) sAPP as a regulator of dendritic motility and melanin release in epidermal melanocytes and melanoma cells. FASEB J 17; 1739–1741.
37. Gakhar-Koppole N, Hundeshagen P, Mandl C, Weyer SW, Allinquant B, et al. (2008) Activity requires soluble amyloid precursor protein alpha to promote neurite outgrowth in neural stem cell-derived neurons via activation of the MAPK pathway. Eur J Neurosci 28; 871–882.
38. Tyan S-H, Shih AYJ, Walsh JJ, Maruyama H, Sarsoza F, et al. (2012) Amyloid precursor protein (APP) regulates synaptic structure and function. Mol Cell Neurosc 51; 43–52.
39. Palavicini JP, Wang H, Minond D, Bianchi E, Xu S, et al. (2013) RanBP9 overexpression down-regulates phospho-cofilin, causes early synaptic deficits and impaired learning, and accelerates accumulation of amyloid plaques in the mouse brain. J Alzheimer's. Dis In press.
40. Roh S-E, Woo JA, Lakshmana MK, Uhlar C, Ankala V, et al. (2013) Mitochondrial dysfunction and calcium deregulation by the RanBP9-cofilin pathway. FASEB J Aug. 27, Epub ahead of print.
41. Cheng L, Lemmon S, Lemmon V (2005) RanBPM is an L1-interacting protein that regulates L1- mediated mitogenactivated protein kinase. J Neurochem 94; 1102–1110.
42. Denti S, Sirri A, Cheli A, Rogge L, Innamorati G, et al. (2004) RanBPM is a phosphoprotein that associates with the plasma membrane and interacts with the integrin LFA-1. J Biol Chem 279; 13027–13034.
43. Bowman AL, Catino DH, Strong JC, Randall WR, Kontrogianni-Konstanto-poulos A, et al. (2008) The rhoguanine nucleotide exchange factor domain of obscurin regulates assembly of titin at the Z-disk through interactions with Ran binding protein 9. Mol Biol Cell 19; 3782–3792.

14

Bridging the Species Divide: Transgenic Mice Humanized for Type-I Interferon Response

Daniel Harari[1]*, **Renne Abramovich**[1], **Alla Zozulya**[2,9], **Paul Smith**[2,9], **Sandrine Pouly**[2,9], **Mario Köster**[3], **Hansjörg Hauser**[3], **Gideon Schreiber**[1]*

1 Department of Biological Chemistry, The Weizmann Institute of Science, Rehovot, Israel, 2 MS Platform, Merck-Serono, (a division of Merck KGaA), Geneva, Switzerland, 3 Helmholtz Centre for Infection Research, Dept. Gene Regulation and Differentiation, Braunschweig, Germany

Abstract

We have generated transgenic mice that harbor humanized type I interferon receptors (IFNARs) enabling the study of type I human interferons (Hu-IFN-Is) in mice. These "HyBNAR" (Hybrid IFNAR) mice encode transgenic variants of IFNAR1 and IFNAR2 with the human extracellular domains being fused to transmembrane and cytoplasmic segments of mouse sequence. B16F1 mouse melanoma cells harboring the HyBNAR construct specifically bound Hu-IFN-Is and were rendered sensitive to Hu-IFN-I stimulated anti-proliferation, STAT1 activation and activation of a prototypical IFN-I response gene (MX2). HyBNAR mice were crossed with a transgenic strain expressing the luciferase reporter gene under the control of the IFN-responsive MX2 promoter (MX2-Luciferase). Both the HyBNAR and HyBNAR/MX2-Luciferase mice were responsive to all Hu-IFN-Is tested, inclusive of IFNα2A, IFNβ, and a human superagonist termed YNSα8. The mice displayed dose-dependent pharmacodynamic responses to Hu-IFN-I injection, as assessed by measuring the expression of IFN-responsive genes. Our studies also demonstrated a weak activation of endogenous mouse interferon response, especially after high dose administration of Hu-IFNs. In sharp contrast to data published for humans, our pharmacodynamic readouts demonstrate a very short-lived IFN-I response in mice, which is not enhanced by sub-cutaneous (SC) injections in comparison to other administration routes. With algometric differences between humans and mice taken into account, the HyBNAR mice provides a convenient non-primate pre-clinical model to advance the study of human IFN-Is.

Editor: Roland Lang, Friedrich-Alexander-University Erlangen, Germany

Funding: This study was funded by a research grant awarded to GS from Merck-KGaA. Authors AZ, PS and SP are/were members of Merck-Serono, a division of Merck KGaA. The Weizmann Institute of Science holds patent rights to the interferon superagonists described in this paper. Scientists from the funding agency (AZ,PS,SP) have contributed to this paper with advice of how to enhance pharmacodynamic studies, and thus are listed as co-authors and have additionally contributed to proof-reading this manuscript. Members of the funding agency however have had no role in data collection and analysis, decision to publish, nor have constrained in any way conclusions reached by the senior authors of this study.

Competing Interests: A number of the authors of this manuscript are/were employed by Merck-Serono. Research performed was funded by Merck-KGaA (the parental company of Merck-Serono). Additionally, the authors declare that the disclosed patented technologies relating to IFN superagonists are associated with patent filings: USA: 60/694,810 and PCT/IL2006/000754 (along with downstream national filings). The authors will openly share their reagents described in this paper with academic institutions.

* E-mail: daniel@harari.weizmann.ac.il (DH); gideon.schreiber@weizmann.ac.il (GS)

⑨ These authors contributed equally to this work.

Introduction

Type I Interferons (IFN-Is) comprise a family of sixteen human cytokines, which collectively feed into a signaling network through binding to two-receptor components, IFNAR1 and IFNAR2. Formation of the ternary complex induces a cascade of downstream activities through the reciprocal phosphorylation of Tyk2 and Jak1 that are associated to the receptor. Subsequently, they phosphorylate several tyrosine residues in the membrane-distal intracellular domains of IFNAR1 and IFNAR2, which, in turn, recruit and activate STATs through their phosphorylation (reviewed by [1]). Only a few hundred IFNAR receptors are typically expressed per cell [2,3], their number impacting upon the cell's ability to respond to IFN-I signaling. Perhaps for this reason, IFN-I signaling is particularly sensitive to perturbation by negative feedback loops, dampening the response with time [4–6,2,7–10].

Receptor binding activates a diverse set of overlapping biological outputs such as anti-viral, anti-cancer, and a wide range of innate and adaptive immunity [11–13]. How such diverse outputs can be induced through a single heterodimeric receptor-signaling complex is still an open question. However, it has been shown that differences in receptor-binding affinities of the different IFN-I cytokines, their concentration and duration of activation are important factors. Twelve of the sixteen IFN-Is belong to the IFNα subtype. These, together with all but one of the remaining human homologs IFNε, IFNκ, IFNω all bind the IFNAR1 receptor with low affinity, and when complexed together with IFNAR2 induce the activation of a cascade of anti-viral and pro-immunoregulatory inflammatory effects. Conversely, IFNβ the single high-affinity IFN-I in humans, can exert additional physiological functions such as the induction of growth arrest and apoptosis at low concentrations [3,14], as well as induction of anti-inflammatory activity [15].

The high affinity receptor binding IFNβ induces an anti-proliferative response with a 20–200 decreased EC_{50} in comparison to IFNα [3,16]. Increased receptor-binding affinity is the major determinant differentiating the biological activity of IFNβ from that of the remaining IFN-Is. This is supported by the finding that a variant of human IFNα2 genetically engineered to bind tightly to IFNAR1 converts this "IFNα" into a superagonist with anti-proliferative potency surpassing that of IFNβ [17,18]. This tightly-binding superagonist was generated by exchanging three amino acids receptor-binding interface (HEQ→YNS). To IFN-YNS, we next added the five amino acid carboxyl-tail sequence borrowed from the cytokine IFNα8, which additionally strengthens the binding to the IFNAR2 receptor by 15-fold [3]. Our variant YNS and YNSα8 mutants bind the human IFNARs with higher affinity than any known IFN-I found in nature, and to our knowledge better than any other IFN-I engineered mutant. Their potency in inducing anti-proliferative activity relates directly to their increase in binding affinity. These data all provide a mechanistic explanation to how high affinity IFNβ differs phenotypically to the remaining low affinity members of the IFN-I cytokine family, but what biological properties phenotypically separate the different low affinity IFN-I members still remains a question largely unresolved.

Ultimately, the activity of a particular IFN-I will relate to its amino acid sequence governing its three dimensional space, particularly so at the receptor binding interface for IFNAR1 and IFNAR2. The recently determined ternary structure of human IFN-IFNAR bound complex has revealed a surprising similarity in the interface of cytokine/receptor binding for both high and low affinity cytokines [19], again suggesting that binding affinity rather than structural differences are responsible for differential activation. Although IFN-I signaling is evolutionarily conserved amongst vertebrates [20], genes within this family are subject to strong genetic divergence. The protein sequences of both IFNAR1 and IFNAR2 for humans and mice share only 50–51% sequence identity respectively, with the divergence extending to the regions of cytokine binding. Thus, although sufficient data has emerged to demonstrate that IFN-I signaling in both humans and mice results in similar activation of phenotypic outputs, the fine-tuning dynamics of how the divergent cross-species receptors are being activated by their respective cytokines remains a matter to be resolved.

At a practical level, the cross-species barrier separating humans and mice results in the inability to correctly study human IFN-Is in non-primates. Resultantly, even though there are hundreds of clinical trials taking place today using Hu-IFN-Is or long-life variants of them, these trials have been formulated without the advantage of first testing them directly in the often powerful pre-clinical rodent setting. To overcome this, we generated a transgenic mouse that harbors humanized type I interferon receptors, allowing the cross-species study of human type I interferons (Hu-IFN-Is) in mice. These "HyBNAR" (Hybrid IFNAR) mice encode transgenic interferon receptors, whose extracellular and cytoplasmic segments are encoded by human and mouse sequences respectively. This allows the binding of human IFN-Is to the recombinant receptors but once activated, transduces their signal effectively in a mouse cellular environment. In this paper, we describe the generation of the HyBNAR mice and a mouse cell line harboring the HyBNAR transgenes, and demonstrate to proof-of-concept their ability to induce a sensitive response to Hu-IFN-Is, and provide some insight into suggested dosing regiments, this which differs significantly to the human scenario.

Experimental Procedures

Mouse Stocks, Maintenance and Ethics Statement

Stocks of C57BL/6 mice were purchased from Harlan Laboratories, Israel. Mice were maintained on site in the Weizmann Institute of Science animal facilities. All mouse experiments were performed strictly according to the Weizmann Institute of Science ethics committee guidelines and permissions (IACUC permit numbers 02160412-3 and 012150412-3).

HyBNAR construct design

A RF-cloning strategy was employed to PCR the transmembrane and intracellular domains of mouse IFNAR1 (a gift from M. Rubinstein, the Weizmann Institute) to be inserted in-frame into the corresponding human IFNAR1 cDNA sequence (accession #BC021825, cloned into pCMV-Sport6; Open Biosystems) using the primers AGTGACGCTGTATGTGAGAAAACAAAAC-CAG-GTCAGAATCTTTTATTGTC & CCTGCTGAAAAAC-CTTATACTTGACACAGTTCATTTCTGG-TCAGCAGAG-AAGAGCTGGCTCTGTC). A Similar strategy was used to generate chimeric IFNAR2 but in this case replacing the human 5′ IFNAR2 sequence into a plasmid encoding the full mouse gene (Open Biosystems, Accession #BC071225) primers AG-CAAAACGGACTTAAGAGCTGAGCAGGATG-CTTTTGA-GCCAGAATGCCTTC and CGAAGTAGTTATTCCTACTA-TAGCAGATTCTGATAATCC-TGATTCCTGGCCAGGTG-GAAGGA. Both chimeric transgenes were independently fused to a 512 bp sequence encoding the mouse PGK1 promoter using primers ATTCTACCGGGTAGGGGAGGCGC & AGGTC-GAAAGGCCCGGAGATGAG. The mouse Growth Hormone polyadenylation sequence was then inserted downstream to the termination stop codon of both these transgenes. The two constructs were placed in tandem, generating a double-transgene 7.6 Kb in length (Fig. 1) and were sequence-verified before use.

Cell lines and transgenic mice

The HyBNAR construct was stably co-transfected into mouse B16-F1 melanoma cells (ATCC) along with a carrier plasmid (pEGFP-N1) and subjected to puromycin selection after which resistant clones were further selected by live FACS sorting for human IFNAR2. Transgenic HyBNAR#1 & HyBNAR#2 mouse strains were generated through the ES-cell transfection approach and generation of lines through chimeric mouse intermediates (Regeneron Pharmaceuticals). The remaining HyBNAR mouse strains (inclusive of HyBNAR#11 chosen for further studies) were generated by pronuclear microinjection of KpnI linearized transgene into purebred C57BL/6 embryos and maintained in this purebred background using breeders supplied by Harlan Laboratories (Israel). Transgenic HyBNAR mice were genotyped by TaqMan-based genomic PCR using the external probes CGGGAATTCAGGTCGAAAGG & CATTCTGCACGCTT-CAAAAGC and the internal probe 5′-/56-FAM/GCG CTG TTC/ZEN/TCC TCT TCC TCA TCT C/3IABkFQ/-30′ (IDT). VIK-labeled mouse TERT genomic DNA (Applied Biosystems) was used as a reference gene in order to determine copy-number. MX2-LUC C57BL/6 mice (Pulverer et al., 2010) were kindly provided by Mario Köster (Helmholtz-Zentrum für Infektionsforschung GmbH). Mice were maintained in the Weizmann Institute animal Facilities.

Reagents

Monoclonal anti-human IFNAR1-EC AA3 antibody was a gift from Biogene. Monoclonal anti-human IFNAR2-EC 117.7 antibody was a gift from Daniela Novick. Both of these antibodies do not cross-react with mouse IFNARs and hence specifically stain the transgenic HyBNARs. Antibodies for Western Blot detection of total-STAT1 as well as Phospho-STAT1 (Tyr 701) were purchased from Santa Cruz (sc-C111 & sc-7988-R respectively). Human IFNα2, IFN-YNS and YNSα8 were expressed and

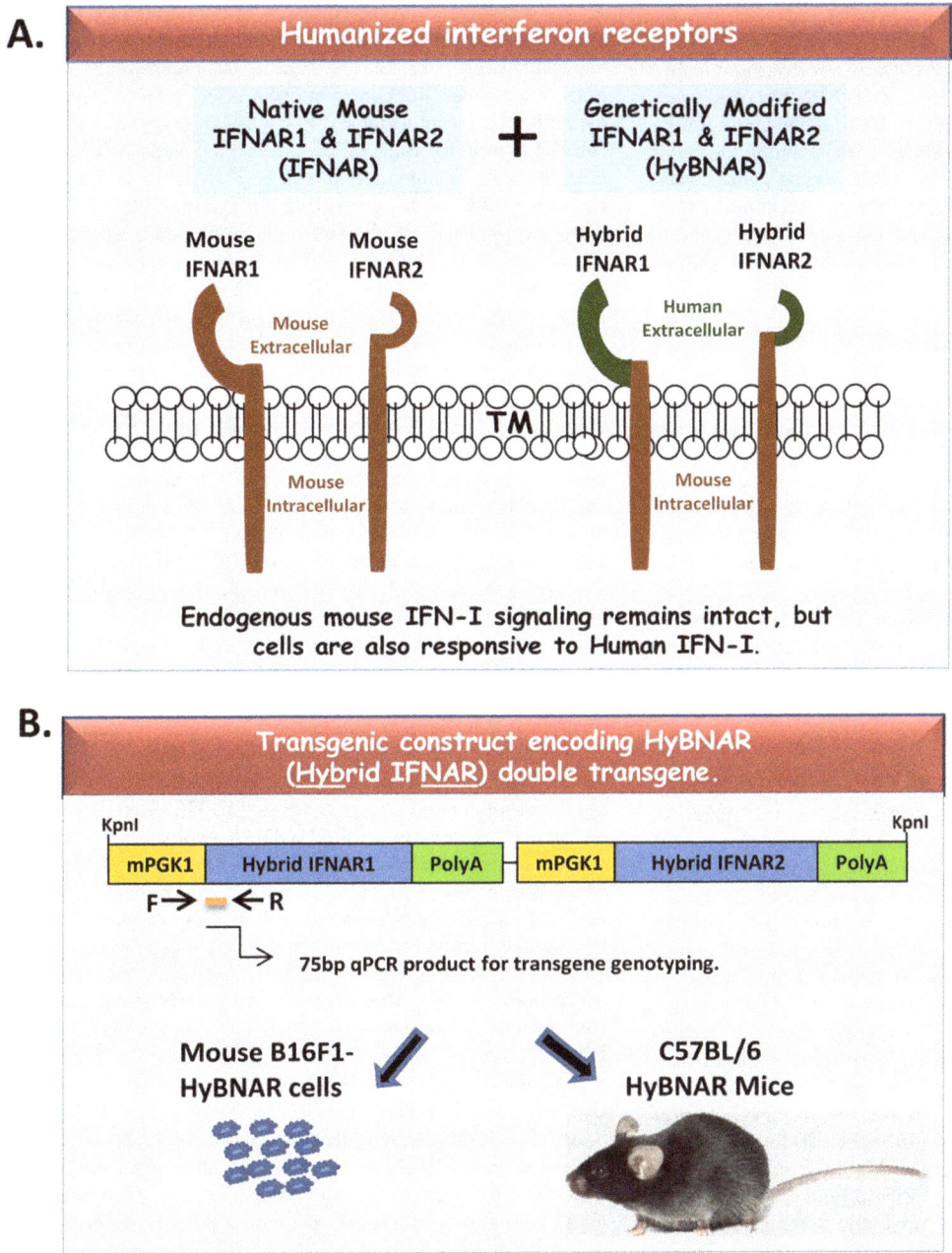

Figure 1. Transgenic HyBNAR Mice and Cell Line. IFN-Is transduce their signals to recipient cells via their binding to and transactivation of a heterodimeric cytoplasmic receptor complex containing IFNAR1 and IFNAR2. But human IFN-Is trans-activate the mouse IFNARs poorly. (A) To overcome this, we have introduced chimeric IFNAR receptors encoding the extracellular domains of IFNAR1 and IFNAR2, fused to their respective transmembrane and intracellular components of mouse origin. By using a regular transgeneisis methodology (IE: not by targeted IFNAR knock-in) we can ensure that regular mouse IFN-I signaling is not disrupted. (B) Diagram of the HyBNAR (Hybrid IFNAR) transgenic construct. Hybrid human and mouse IFNAR1 and IFNAR2 sequences were independently placed under the control of the weak but ubiquitously expressed mouse PGK1 promoter. The two transgenes were then linked together into a single construct. By stable transfection of this construct into the B16F1 mouse melanoma cell line or by pro-nuclear injection into the fertilized embryos of C57BL/6 donors, HyBNAR expressing cells and transgenic mice were generated.

purified as described (3). Mu-IFNβ (recombinant-bacterial) and mammalian-cell-derived human IFNβ were both provided by Merck-Serono, Geneva. D-Luciferin was purchased from Regis Technologies, USA). Tri-Reagent was purchased from MRC-Inc.

Competitive binding assay

IFN-YNS was labeled with 125I by using the chloramine T iodination method as has been described (3). For competition studies, the indicated cold ligands were incubated with a 100-fold molar excess together with the radiolabelled IFN-YNS.

B16F1 cell Bioassays

For the Anti-proliferation study, B16-F1 cell clones were seeded at 5000 cells/well into 96-well plates and grown overnight before addition of indicated concentrations of IFN-Is. The cells were incubated for a further 72 hours before being stained with 5 mg/ml Crystal violet (Sigma) in 70% Ethanol. After staining the plates were rinsed in water and air-dried before adding resuspension buffer (0.1 M NaCitrate, 50% EtOH, pH 4.2) for reading at ODs at 540 nM. Both FACS labeling to quantify surface expression of transgenic IFNAR1 and IFNAR2, and phospho-STAT1 western Blot analysis were performed as described (3).

Quantitative PCR (qPCR) analyses

RNA was extracted from liver and B16F1 cell clones by Tri-Reagent protocol (MRC Inc.). Random-oligonucleotide directed cDNA synthesis was generated from 1 ug RNA per 20 ul reaction (High Capacity cDNA Reverse Transcription kit; Applied Biosystems). Forward and reverse oligonucleotide probes for mouse genes used in this study include MX1; (GGATAATCA-GAGGGATCTGTCTCC & AGGCATTAATAAACCCTGC-TACCT) Trail (TCACCAACGAGATGAAGCAG & TGGAGT-CCCAGAAATCCTCA) and the reference gene HPRT1 (AG-CAGTACAGCCCCAAATG & GGCCTGTATCCAACACT-TCG). DNA was amplified using Power SYBR Green and analyzed using a 96-well 7300 Real-Time PCR System (Applied Biosystems).

Luminescence Studies

For live studies, HyBNAR/MX2-LUC and MX2-LUC mice were injected IP with 100 ul D-Luciferin solution (60 ug/ml, in PBS) 10 minutes prior to luminescence measurements. Mice were lightly anaesthetized with isoflurane and images taken using the IVIS Spectrum optical imaging device (Caliper Life Sciences Inc.) with devoted quantification software (Living Image 4.1) provided by the same supplier. For tissue analyses, IFN-treated mice were perfused with PBS, and tissues were extracted and stored at $-80°C$ before being weighed and homogenized in $10\times$ volume/weight using Reporter Lysis Buffer (RLB); Promega (Cat.# E3971). Immediately prior to luciferase readings, cell extracts were diluted a further 10-fold in RLB and 10 ul samples were placed in 96-well plates. Immediately prior to readings, 50 ul of luciferin buffer (20 mM Tricine, 0.1 mM EDTA, 1.07 mM $(MgCO_3)_4Mg(OH)_2*5H_2O$, 2.67 mM $MgSO_4$, 3.3 mM DTT, 270 μM Coenzyme A, 470 μM luciferin and 530 μM ATP, pH 7.8.) was injected into each well and luminescence was measured using a Modulus Microplate Luminometer (Turner Biosystems).

For luciferase quantification of live animals, Luciferin-treated mice were photographed with IVIS spectrum as described above. For quantification of the live image, a circular region of interest (ROI) of predefined size was chosen and centered upon highest intensity detected (IE: the liver region) for each mouse. Average radiance was then calculated using Living Image version 4.1 software by Caliper Life Sciences.

Results

HyBNAR-transfected B16F1 mouse melanoma cells are humanized for IFN-I signaling

The HyBNAR construct constitutes the human-encoded extracellular domains of both IFNAR1 and IFNAR2, whereas their transmembrane and intracellular domains are encoded by mouse sequence (Fig. 1A). Both transgenes were individually placed under the control of the mouse PGK1 promoter in order to drive constitutive low level expression of the humanized IFNAR1 and IFNAR2 genes (Fig. 1B). The HyBNAR double-transgene was first stably transfected into the murine melanoma B16F1 cell line to test and validate our construct design. Cell surface expression of the HyBNAR receptors for B16F1-HyBNAR cells was detected by live FACS staining using antibodies directed against the extracellular domains of human IFNAR1 or IFNAR2 (Fig. 2A). Comparative staining of human WISH and MDA-MB231 cells (data not shown) suggest that the HyBNAR expression levels are in the same range as IFNAR1 and IFNAR2 expression in human cells. To test for the ability of the B16F1-HyBNAR cells to competitively interact with human IFN-Is, binding of the high affinity IFNα2-YNS mutant to the HyBNAR receptors was evaluated. ^{125}I-YNS bound specifically to B16F1-HyBNAR cells as demonstrated by increased levels of radioactive signal detected in comparison to background levels measured for non-transfected B16F1 cells (Fig. 2B). The strong binding of YNS to human IFNAR1 was displaced by competition binding with 100-fold molar excess of non-labeled YNS itself, or high affinity IFNβ, but not by weaker binding IFNα2A (Fig. 2B). Radiolabelled ^{125}I-YNS could not be displaced by 100-fold excess of mouse IFNβ (Mu-IFNβ), further supporting the species barrier between human and mouse IFN-I binding (Fig. 2B).

To evaluate the biological activity of the B16F1-HyBNAR cells against Hu-IFNs in comparison to B16F1 cells harboring GFP, an anti-proliferative dose response to Hu-IFNα, high affinity Hu-IFN-YNS, and Mu-IFNβ as positive control was conducted. The B16F1-HyBNAR cells demonstrated high sensitivity to the super-agonist ligand IFN-YNS (EC50 2 pM), a value similar to that measured for human cancer cell lines [3]. Hu-IFNα2 also drove antiproliferative activity in the HyBNAR expressing mouse cells, but with a decreased potency of two orders of magnitude (Fig. 3A), which is in line with findings for human cancer cells. In contrast the B16F1-GFP control clone was only sensitive to Mu-IFNβ anti-proliferative response, with minimal effects noted for the human IFN-Is when administered at extremely high doses (Fig. 3B). Thus we demonstrate here that expression of the HyBNAR construct converts this mouse melanoma cell line to being sensitive to Hu-IFN-I.

A major component of IFN-I signaling is via the recruitment and activation of STAT transcription factors [21]. B16F1 and B16F1-HyBNAR cells were treated with increasing doses of either human or mouse IFNβ for 45 minutes and assayed for STAT1 activation by western-blot detection for the phosphorylated STAT1 isoform. Both human and mouse IFNβ activated STAT1 in the B16F1-HyBNAR cells to a similar degree and in a dose-dependent manner (Fig. 3C). However the non-transfected parental B16F1 cells were only responsive to the mouse IFNβ. We did not see cross-activation of STAT1 after Hu-IFN-I stimulation in the parental B16F1 mouse cells, indicating that at least for the maximum concentration of IFN used in this assay (800 pM) the endogenous IFNARs are not being activated by the human cytokine. These data thus support that the activation of STAT1 by human IFNβ is through the HyBNAR receptors, as no activation of parental B16F1 cells was detected when using the human cytokine under these conditions (Fig. 3C).

IFN-I induces the activation of more than a thousand genes [22,23]. The anti-viral genes MX1 and MX2 are particularly well described experimental markers, which are strongly up-regulated after IFN-I exposure [24,25]. B16F1-HyBNAR cells were treated for 4, 12, 24 or 36 hours with 100 pM of either human IFNα2, IFNβ, YNS or Mu-IFNβ and transcript levels of the MX2 gene were measured by quantitative PCR (qPCR). Both mouse and human IFN-Is robustly activated MX2 in the B16F1-HyNBAR

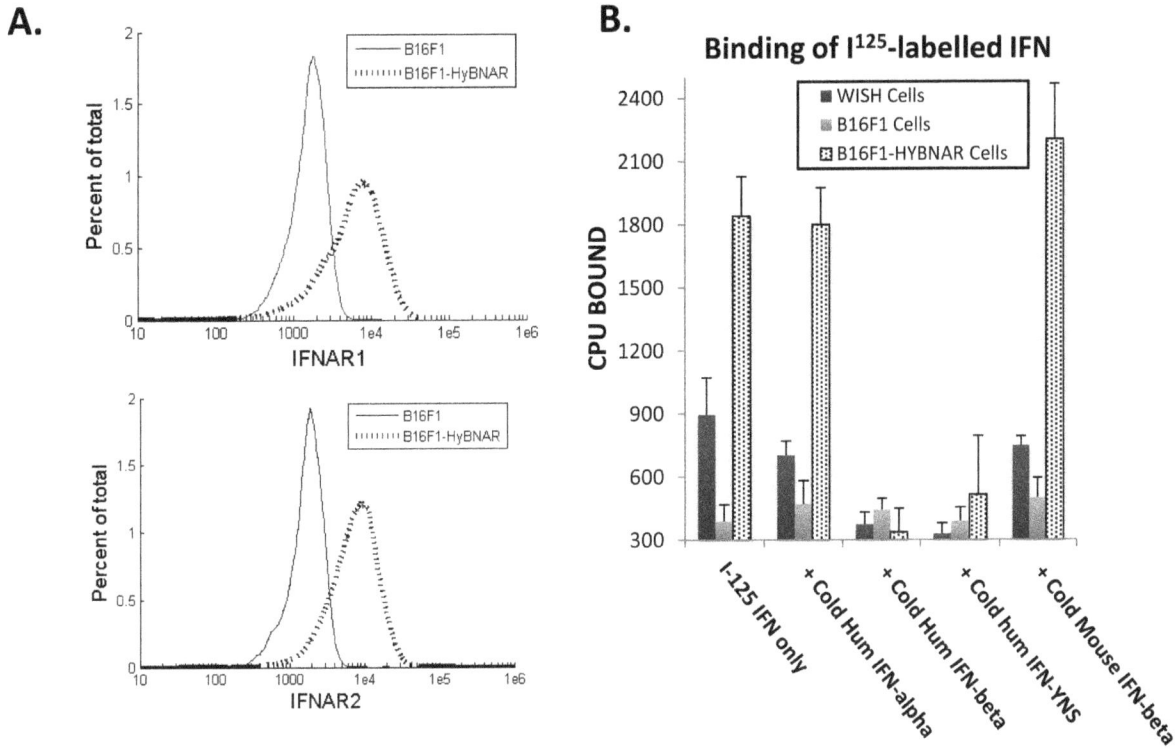

Figure 2. Transgene expression on B16F1-HyBNAR mouse melanoma cells and its binding to human IFN-I. (A) FACS analysis of B16F1-HyBNAR cells using antibodies that specifically bind to the human variants of the extracellular domains of IFNAR1 and IFNAR2. Specificity of binding is demonstrated by the staining of non-transfected mouse B16F1 control cells. (B) YNS, an engineered tight-binding variant of IFNα2 was radiolabelled and incubated with mouse B16F1, B16F1-HyBNAR and human WISH cells. Tightly binding radiolabelled YNS could not be displaced by co-incubation with 100× molar excess of lower affinity Hu-IFNα2a, although high affinity human IFNβ and YNS completely displaced the hot ligand. Importantly, 100× molar excess of mouse IFNβ did not displace binding of hot YNS in this system.

cells (Fig. 3D). After 36 hours of incubation we noticed a down-regulation of IFNα2-induced signal, as also reported previously [17,9]. Conversely, MX2 mRNA was maintained at high level over the time-course of this assay when incubated with high affinity Hu-IFNs, a finding consistent with study of human cells lines. In the parental B16F1 cells, Hu-IFNs did not elicit an activation of MX2 expression (data not shown), again supporting the preferential activation of human IFN-Is via the recombinant HyBNARs.

HyBNAR Mice Respond to human IFN-I

C57BL/6J mice were used for the generation of transgenic HyBNAR strains using either ES cell transfection (HyBNAR strains #1 & #2) or pro-nuclear microinjection (strains HyBNAR #11, #13, and #25). We did not choose to apply a gene knock-in approach as the natural mouse IFNARs are important to maintain endogenous mouse IFN-I signaling (autocrine and paracrine). TaqMan-based genomic qPCR was performed to determine gene copy number for the different strains of HyBNAR mice, which were found to range from approximately one to sixty five copies of the HyBNAR double transgene (Fig. S1). Initial screening for activity of the HyBNAR transgene was performed in transgenic progeny of the different HyBNAR strains by measuring the activation of the IFN-I response gene MX1 from liver extracts taken six hours after injection of Hu-IFNβ. HyBNAR strain #11 which harbors the lowest transgene copy number (Fig. S1), demonstrated sustained activation of IFN-I response markers such as the genes MX1 and USP18 (Fig. 4). The two HyBNAR mouse

strains generated by ES cell electroporation route (HyBNAR#1 and HyBNAR#2) initially exhibited responsiveness to Hu-IFNs, but lost their IFN-I responsiveness after 2–3 generations of backcrossing from their transgenic founders (data not shown) and thus were discontinued. All further studies were conducted with the C57BL/6-HyBNAR#11 strain, which we refer to henceforth as the "HyBNAR mouse".

Tracking IFN-I activity in HyBNAR Mice

A transgenic mouse line that expresses the luciferase reporter gene under the control of the IFN-I responsive MX2 promoter was previously reported, allowing for rapid assessment of the in vivo pharmacodynamic effects of Mu-IFN-I injections [26]. These mice, named MX2-LUC, were obtained and crossed with the HyBNAR mice. Double heterozygous transgenic HyBNAR/MX2-LUC mice were generated and adult mice were injected with 1.0 ug of Hu-IFNα, Hu-IFNβ, Hu-YNSα8, or Mu-IFNβ and analyzed by live imaging for bioluminescence. For control measurements, MX2-LUC mice not harboring the HyBNAR transgenes were similarly treated (Fig. 5). Presence of the HyBNAR transgene rendered the mice sensitive to activation of the MX2 promoter-driven luciferase, with the highest levels of signal emanating from what is likely to be the liver. However, we detected some basal activation of MX2-LUC by Hu-IFN-Is also for mice that are not transgenic for HyBNAR (Fig. 5). This contrasts to our findings for the B16F1 cell lines in which no basal activation of wild type cells by Hu-IFNs was detected unless when administered at extremely high concentrations (Fig. 3B). To

A.

Anti-Proliferative: B16F1-HyBNAR Cells

Legend:
- ○ Hu-IFNα2 (698pM)
- ■ Hu-IFN-YNS (2pM)
- ◇ Mu-IFNβ (92pM)

Y-axis: % MAX (OD 540nM), X-axis: [IFN] pM

B.

Anti-Proliferative: B16F1-GFP Cells

Legend:
- ○ Hu-IFNα2
- ■ Hu-IFN-YNS
- ◇ Mu-IFNβ

Y-axis: % MAX (OD 540nM), X-axis: [IFN] pM

C.

WB: phospho-STAT1

pM IFN 0 0.5 2.0 10 50 200 800

phospho-STAT1

Mouse IFNβ — B16F1-HyBNAR
Human IFNβ

Mouse IFNβ — B16F1-Wild Type
Human IFNβ

STAT1-Total

Mouse IFNβ — B16F1-HyBNAR
Human IFNβ

Mouse IFNβ ND — B16F1-Wild Type
Human IFNβ

D.

B16F1-HyBNAR Cells: MX2 Transcript Levels

Legend:
- ■ Hu-IFNα2
- ▨ Hu-IFNβ
- ▨ Hu-YNS
- ▨ Mu-IFNβ

Y-axis: Fold-Change (Rq), X-axis: Time (Hours) — 4, 12, 24, 36

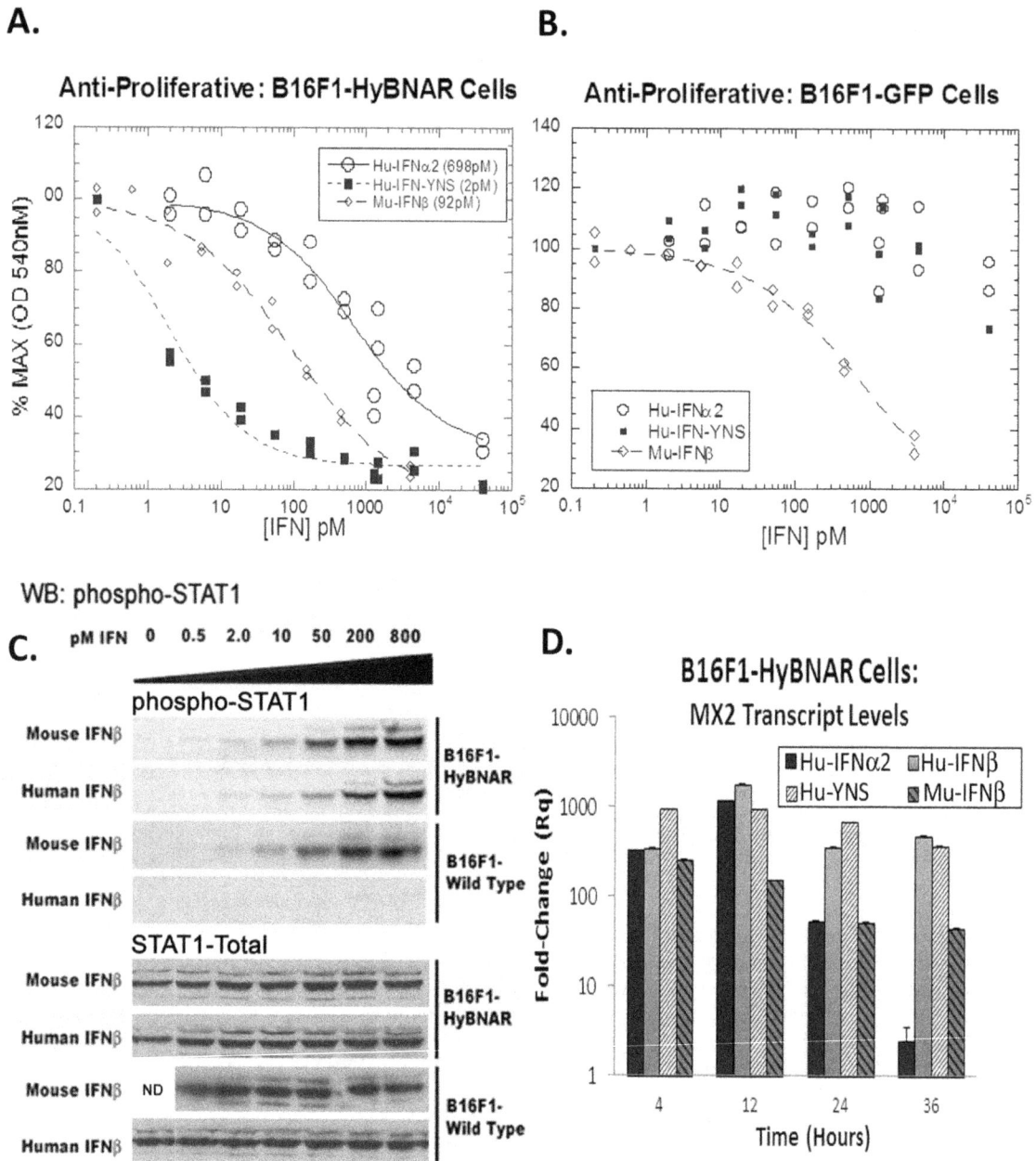

Figure 3. B16F1-HyBNAR mouse melanoma cells are sensitive to human IFN-Is. (A & B) Mouse B16F1 melanoma cells were stably transfected with either the HyBNAR double transgenic construct, or a control GFP vector and tested for their anti-proliferative dose response to Hu-IFNα2, Hu-IFN-YNS and for comparison, mouse IFNβ. (A) HyBNAR transfected B16F1 cells were responsive to both human and mouse IFN-Is (EC50 values shown in parentheses) (B) In the GFP-transfected B16F1 control cells however, the human ligands only promoted a minor loss in proliferation and when administered at very high concentrations. (C) B16F1-HyBNAR and non-transfected control cells were incubated with an increasing dose of both human and mouse IFNβ. After 45 minutes incubation the cells were lysed and analyzed by Western Blot detection for phosphorylated STAT1. Total STAT1 measurements are shown as control. (D). B16F1-HyBNAR cells were treated for the indicated time-points with 100 pM human and mouse IFN-Is. Measurements of the IFN-response gene MX2 were performed by qPCR analysis. Relative fold-change was determined in comparison to untreated cells and normalized using the reference gene HPRT1. Error bars represent standard error of duplicate measurements.

evaluate the difference in responsiveness towards Hu-IFN-Is in HyBNAR versus wild-type mice, we next injected progressively lower doses (1.0, 0.5 and 0.25 ug) of Hu-IFNβ. Injection of this Hu-IFN-I generated very strong up-regulation of the luciferase signal in a dose-responsive manner for the HyBNAR/MX2-LUC mice (Fig. 6). Injections of the same concentrations of Hu-IFNβ into MX2-LUC mice controls showed also a gradual progressive

activation of luciferase signal, however, much lower than obtained in the HyBNAR mice (Fig. 6). We rule out the possibility that contaminants within our Hu-IFN preparations are the cause of the low level of IFN-I activation observed in the non-HyBNAR mice (e.g. residual LPS can induce activation of IFN-I expression in TLR-expressing cells) as the recombinant Hu-IFNβ was derived from mammalian cells (Merck-Serono; human injection quality).

Liver MX1 Transcript Levels

A.

Liver USP18 Transcript Levels

B.

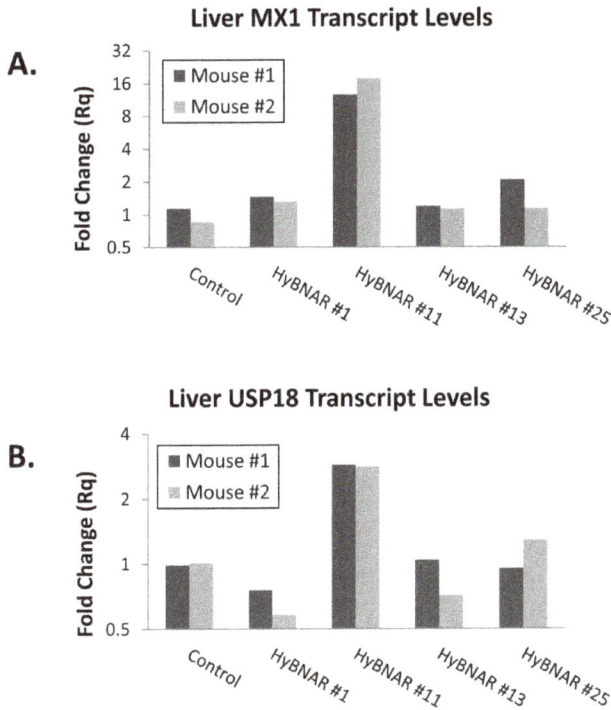

Figure 4. Response of different HyBNAR mouse strains to human IFNβ stimulation. Two representative mice from four independent transgenic HyBNAR strains were injected IV. with 400 ng human IFNβ. Six hours later, livers were collected and RNA was extracted for qPCR measurements. (A) Transcript levels for the IFN-response gene MX1 and (B) USP18. Relative fold-change was determined in comparison to untreated cells and normalized using the reference gene HPRT1.

The HyBNAR/MX2-LUC used in these studies were doubly heterozygous for both HyBNAR and MX2-LUC. We generated homozygote HyBNAR mice (but without the MX2-LUC transgene) in attempt to further enhance Hu-IFN responsiveness. These homozygous mice were injected with increasing concentrations of IFN-YNSα8, after which livers were tested for gene induction. Wild-type C57BL/6J mice were used as baseline controls. MX1 (which is highly sensitive to IFN-I activation) and TRAIL (which requires higher IFN-I concentrations or higher affinity IFN-Is to be activated) served as probes for IFN-I responsive genes. In the HyBNAR mice, injection of low dose (0.1 ug) of YNSα8 resulted in a 140 and 12-fold increase in MX1 and TRAIL transcripts respectively (Fig. 7). At the highest YNSα8 concentration tested (1.6) μg expression levels of both these genes only increased by another two-fold in relation to 0.1 μg injection. These results demonstrate that the HyBNAR transgenic mouse is sensitive to activation by IFN-I, even at low levels of IFN-I concentrations. Consistent with our previous study, low dose (0.1 ug) of YNSα8 injected into wild-type non-transgenic mice elicited almost no induction in MX1 or TRAIL expression. At the highest dose, of 1.6 μg an induction of 110 and 8.5-fold was measured for these two genes (Fig. 7) in the wild-type mice. From these data we infer at least a 20-fold increased EC_{50} for IFN-I activation in the background of the HyBNAR receptors in comparison to wild-type mice. Thus in summary, these results support that the HyBNAR mice are sensitized to Hu-IFN-I signaling. Sub-optimal activation of the mouse IFNARs is likely taking place as well, especially when higher concentrations of Hu-IFNs are being injected.

HyBNAR Mice Respond to Human IFN-I in multiple tissues

Type I interferon receptors are naturally found on all nucleated cells. To test whether Hu-IFN-I are inducing a widespread response in the HyBNAR mice, we examined their activity by indirect luciferase measurement in a panel of different tissues. Transgenic MX2-LUC/HyBNAR mice were injected IP with

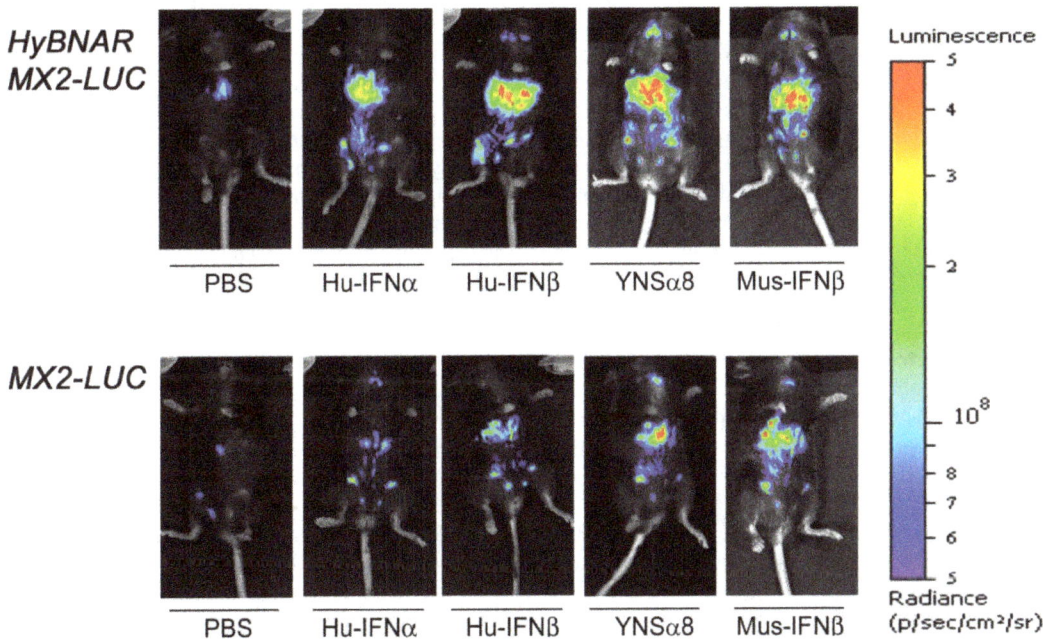

Figure 5. Luciferase signal in HyBNAR/MX2-LUC and MX2-LUC mice in response to different IFN-Is. Mice were injected IP with 1.0 ug of the indicated IFN-Is. At T = 3.0 hours live animals were injected with luciferin, anaesthetized and live luminosity was measured by an image capturing (IVIS spectrum) device. For comparative purposes MX2-LUC mice without the HyBNAR transgene were also analyzed (bottom panel).

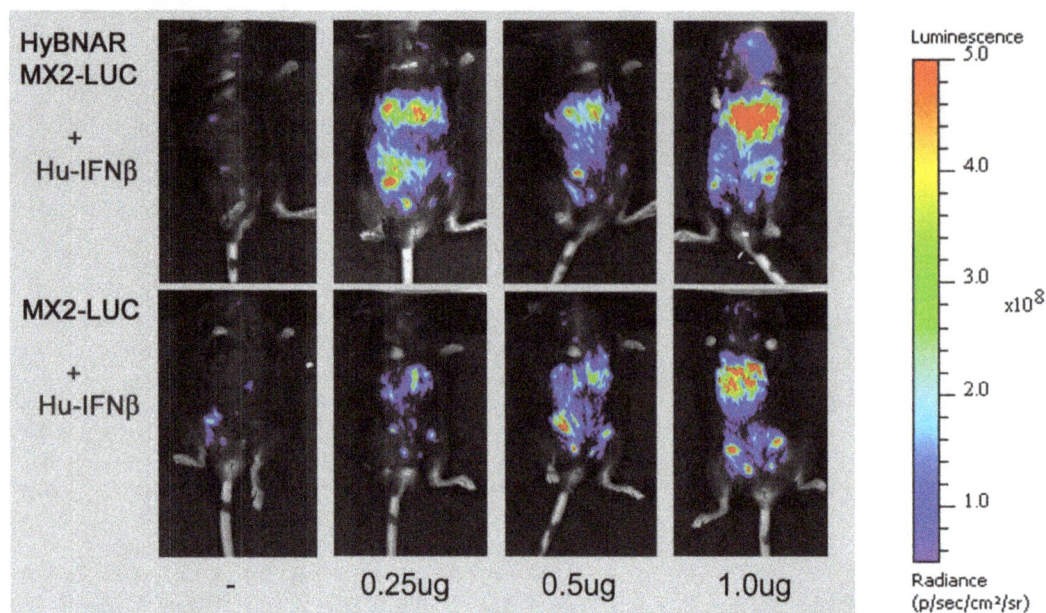

Figure 6. HyBNAR/MX2-LUC mice are both sensitive and responsive to human IFN-Is. Transgenic mice expressing the luciferase reporter gene under the control of the MX2 promoter (MX2-LUC) were interbred with the HyBNAR mice. The HyBNAR/MX2-LUC mice (upper panel) were injected IP with increasing concentrations of human IFNβ. After 6 hours, mice were injected with luciferin, anaesthetized and live luminosity was measured by an image capturing device (IVIS spectrum). For comparative purposes MX2-LUC mice without the HyBNAR transgene were also analyzed (bottom panel).

1 ug of either Hu-IFNα2, Hu-IFNβ or YNSα8, along with PBS and Mu-IFNβ as negative and positive controls respectively. Six hours after injections, the mice were perfused with PBS, a selection of tissues were collected, homogenized and then assayed for luciferase activity. Basal MX2-driven luciferase activity measured in control PBS-injected mice (Fig. 8A) was found to be 10–100-fold higher in the liver than in the other tested tissues. This is in agreement with the *in vivo* luciferase imaging data (Figs. 6 & 7) and with previous studies of the MX2-LUC mice [26]. Interferon injections into the HyBNAR/MX2-LUC mice resulted in a 4 to 25-fold increased activity for the different tissues tested, when internally normalized to their PBS-injection controls (Fig. 8B). Brain tissue exhibited the weakest luciferase signal induction with injection of either human or mouse IFN-Is. As the blood-brain barrier is known to restrict the flow of IFN-Is to the central nervous system [27], we assume this to be the cause of the limited activation that we have measured in the brain in response to injected IFNs. When comparing activation of luciferase signal by the human interferons in relation to control Mu-IFNβ, we noted very little difference for all tissues tested, with exception to the kidney, where the activation of the human cytokines was limited. This data therefore support the widespread effects of HyBNAR transgene in different tissues.

IFN-I induced response in mice is short lived

The serum half-life of IFNβ in humans (pharmacokinetic (PK)) has been reported to range from minutes to a few hours depending on the injection route, with sub-cutaneous injections being the most convenient for patients, and generally showing overall longer half-lives than other routes (reviewed by [28]). Instead of directly measuring the circulating levels of human IFNs in mice, which has been shown to be variable depending on the method of detection used, we used the double transgenic HyBNAR/MX2-LUC mice to follow the pharmacodynamic (PD) activation of the luciferase

signal over time. A relatively modest dose of Hu-IFNβ (200 ng/mouse) was injected by intraperitoneal (IP), sub-cutaneous (SC) or intravenous (IV) route and luciferase levels were measured over time. A time-course of semi-quantitative luminescence data was then measured from the live mice (see experimental procedures). We found that in the HyBNAR mice there was no advantage in performing SC injections over the two other injection routes. If at all, the IFN-I induced luciferase signal was delayed and weaker after SC injection (Fig. 9A). These findings concur with other PD studies that we have performed using singly transgenic HyBNAR mice, where IP injections of IFN-Is provided at least as good as, if not better response than the other two injection routes (data not shown). Importantly, irrespective of the route of administration we found that IFNβ-induced MX2-promoter driven luciferase activation returned to baseline within 24 hours of injection (Fig. 9A). This finding in the mice contrasts to PD measurements in humans; a single dose of IFN injection into humans maintained significant levels of MX transcript expression levels even five days post IFNβ injection [29]. To test the importance of dosage and its effects on IFN-I response over time, we injected IP increasing doses of YNSα8 into HyBNAR/MX2-LUC mice and followed luciferase activation by live imaging. High dose of IFN-I (1.0 ug) conferred a strong increase in the amplitude of activation of luciferase signaling, albeit once again, the signal returned to near baseline levels within 24 hours (Fig. 9B). As expected, MX2-LUC mice that do not express the HyBNAR transgene demonstrate minimal IFN-I induced luminescence (Fig. 9B). These data support that in contrast to humans, there is a very limited timeframe of PD response to single injection of IFN-Is in the mouse, and is consistent with published PK studies suggesting a serum half-life of injected IFN-Is in mice of less than one hour [30,31].

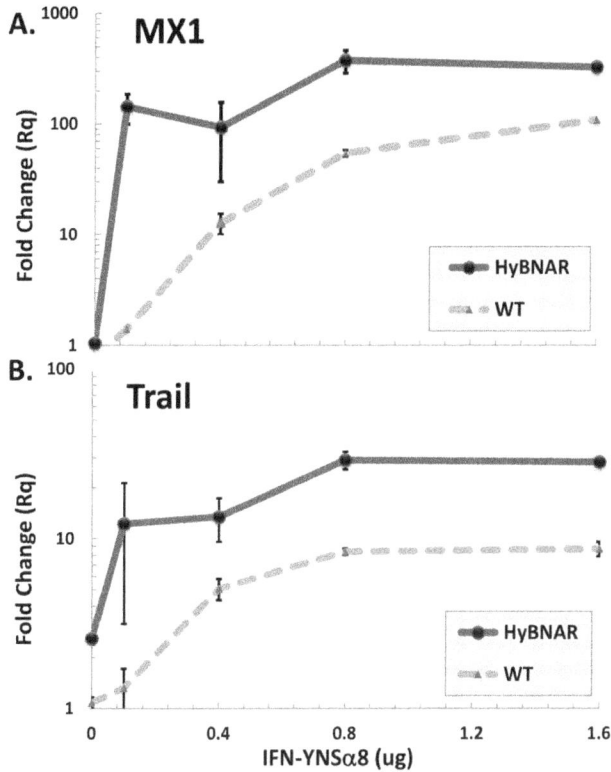

Figure 7. HyBNAR homozygous mice are sensitized to human IFN-I response. Dose response of homozygote HyBNAR after IP injection of increasing doses of the human superagonist IFN-YNSα8. After 6 hours, livers were collected and from the derived RNA/cDNA MX1 (A) and Trail (B) levels were determined by qPCR. Relative fold-change was determined in comparison to untreated wild-type (WT) mice and normalized using the reference gene HPRT1. The average of two mice were used for each injection dosage. The differences in dose response to IFN were significant when comparing control vs. HyBNAR mouse groups (p<0.0005 for both MX1 and Trail expression plots as determined by two-way paired Anova).
doi:10.1371/journal.pone.0084259.g007

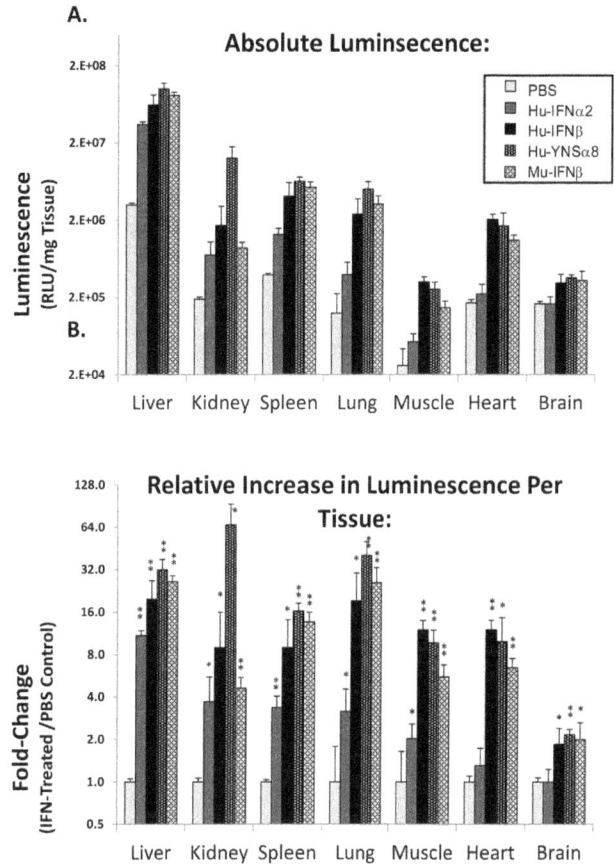

Figure 8. Luciferase activity measurements from tissue homogenates from HyBNAR/MX2-LUC mice. Mice were injected IP with either PBS or 1.0 ug of the indicated type I Interferon. After six hours the mice were perfused with PBS and tissues were collected for homogenization and measured for luciferase activity. (A) Absolute values of luciferase activity standardized per unit wet weight for each tissue. (B) Relative up-regulation of luciferase signal per tissue, in relation to PBS injected controls. Significance values of IFN-treated samples in comparison to PBS controls are shown (one tailed T-Test; *: p≤0.05, **: p≤0.005).

Discussion

Here we describe for the first time the generation of a transgenic mouse model with the capacity to robustly respond to human IFN-Is. The development of human IFN-Is for clinical use has taken place in spite of the lack of a relevant non-primate preclinical model to adequately assess their function. Despite the discovery of type I IFNs more than 50 years ago and their historically verified clinical use, there remains to this day strong interest for expanded use of IFNs as human therapeutics, as demonstrated by their continued listings as drugs to be tested in new clinical trials (Fig. S2). Currently active clinical trials include not only currently approved IFN drugs for the treatment of new disease indications or in new drug cocktail combinations, but also novel long-life IFN formulations or novel combined drug regiments. Our HyBNAR mice can help the acceleration of such studies, providing a pre-clinical platform for rapid screening of Hu-IFNs before moving to clinical trials. To exemplify this point, we are currently testing a novel long-life variant of the YNSα8 superagonist with promising findings. This study has been greatly facilitated by the availability of the HyBNAR mice in which we can study not only the PD of the superagonist in comparison to IFNβ, but also test its efficacy in a multiple sclerosis disease model environment (*manuscript in preparation*).

For reasons not well understood, interspecies IFNAR sequences have diverged to a much greater extent than that for cognate receptors of other signaling systems, as exemplified by the EGFR or Insulin Receptor signaling pathways (Fig. 10). These interspecies comparisons demonstrate that this divergence is not only constrained to the human-mouse species divide, but extends to other organisms as well. As would be expected, a similar rapid interspecies divergence of sequence extends to the IFN-I cytokines as well [20,32]. As a result, human IFN-Is are poor activators of IFNAR signaling in non-primate mammals. By the transgenic co-expression of transgenic humanized IFNAR1 and IFNAR2 in the HyBNAR model we can now study the direct effects of Hu-IFN-Is in mice. Using the B16F1-HyBNAR cells, we first demonstrated that the chimeric IFNARs faithfully and sensitively transduce signals after exposure to human IFN-Is. The HyBNAR mice demonstrate strong activation upon Hu-IFN-Is induction across a variety of tested tissues. Hu-IFN-I response was measured by induction of transcription of IFN-response genes as evaluated by direct qPCR (using HyBNAR homozygotes) and by generation of a luciferase signal using a triple-transgenic model, where the HyBNAR mice expressing chimeric IFNAR1 and IFNAR2 were

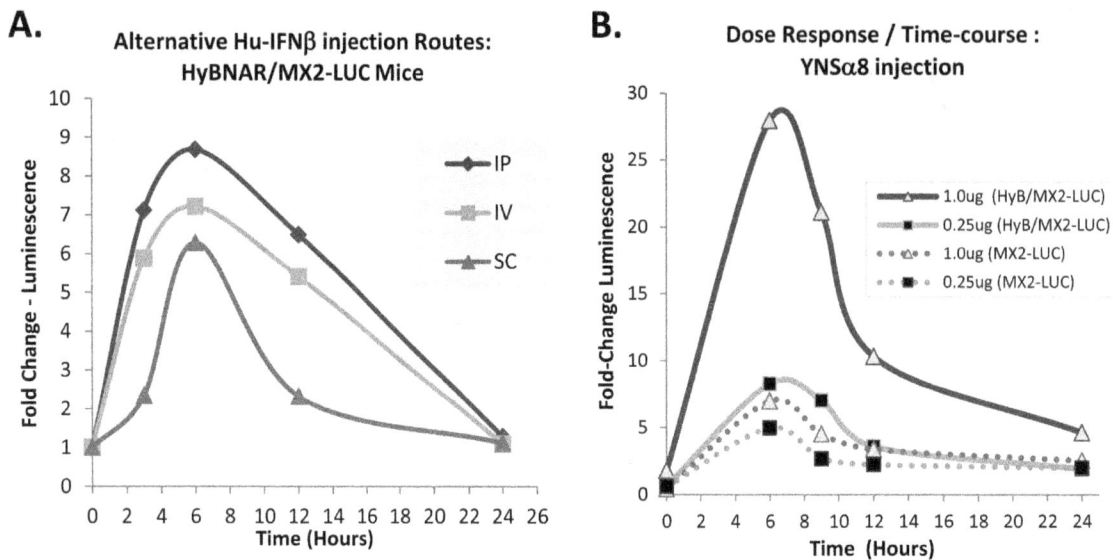

Figure 9. Time and injection regiment affects response to IFN-Is. (A). Response of different injection regiments to 0.2 ug (weight adjusted per 20 g mouse weight) of human IFNβ. Injections were by intraperitoneal (IP), intra venous (IV) or by sub-cutaneous (SC) route. (B). Mice were injected IP with the indicated increasing doses of YNSα8 and luminosity was measured in a time-course from live animal measurements. MX2-LUC mice doubly transgenic for HyBNAR (solid lines) or without presence of HyBNAR (dotted lines) were both measured. The relative luminosity values given are averaged from three animals (A) or two animals (B) per injection group and were measured from a defined region of maximal signal (found in the liver region) for each mouse as described in Figure S2. In both experiments, baseline luminosity measurements were determined from the mice taken immediately prior to IFN induction.

crossed with mice harboring an additional transgene with the MX2 promoter driving luciferase activation (here HyBNAR is heterozygote).

It has been published that the serum half-life of IFN-I in mice ranges from as little as 0.5 hours to as much as 4 hours [30,31,33,34]. The PD data presented in this paper indirectly supports that the PK half-life of IFN-Is in mouse is indeed very short. The resultant effect is that a single-dose injection of IFN-I into the mice provides a pulse of IFN-I activity, but one that is rapidly lost thereafter. This means in real terms, that the pharmacological responses by humans and mice to IFN-I injections are expected to be different, a difference that can be minimized by more frequent injections in mice. Whereas a single dose of IFNβ injection in human induces MX2 activation in blood cells even 120 hours after IFN-I injection [29], in the mouse, we have found that this signal reverts to background levels within 24 hours (Fig. 9). As *in vitro* studies in human cells have shown that some activities of interferon require prolonged time of activation (days of constant IFN exposure [14]), the short half-life in mouse may have significant effects on the outcome of treating disease. In accordance with general considerations of algometric scaling [35], it is not particularly surprising that injected IFN-Is have such a short life-span in mice compared to humans. But the consequence of this difference means that in this respect, the mouse system cannot be simply "humanized", regardless of presence of the HyBNAR transgene, as single dose injection of non-modified IFN-Is in mice is extremely short lived.

When comparing the levels of gene transcript of the wild-type mouse relative to HyBNAR mouse after activation with either IFN-YNSα8 or Hu-IFNβ we noticed a certain amount of transcription activation also in wild type mouse. This is more evident with increasing concentration of IFN-I injected (Figs. 7 & 8). Whereas in comparing B16F1 to B16F1-HyBNAR cells, the presence of chimeric IFNAR receptors sensitizes the cells to human IFN-Is by three orders of magnitude (Fig. 3), we noted only

a 20-fold difference in the EC_{50} of activation of the IFN-response in the HyBNAR mice in comparison to mice without the transgene. Low levels of activation of luciferase signal was also observed in MX2-LUC mice without the HyBNAR transgenes especially after administration of higher levels (1.0 ug) of Hu-IFN-Is. Two possible reasons for the reduced observed species-specificity in animals are: 1) A much higher dose of IFN-Is is typically injected into mice in comparison to humans (typically a 50-fold higher dose in mice after body weight correction [36]), this requirement to inject high doses which may be a consequence to transiently compensate for the rapid clearance of injected IFN-Is in mice. Alternatively, we cannot rule out that endogenous/constitutive Mu-IFNs are priming cells in both wild-type and HyBNAR strains and that even very low activation of the mouse receptor by Hu-IFN leads to a significant effect which is close to saturating levels. This is supported by the fact that at low doses of IFNs the difference is very high (Fig. 7; 0.1 ug values). Despite these potential caveats in our model, it is clear that the presence of the HyBNAR transgene sensitizes mice to Hu-IFN-I signaling, with observed physiological outputs for the transgenic mice expected to emulate the true human effects.

In summary, the generation of the HyBNAR model allows the study of human IFNs in the mouse. This can accelerate *in vivo* study of human IFN-Is as injectable compounds to study added drug effect beyond that of its unaltered endogenous signaling. Moreover, this HyBNAR model can serve as a perfect platform to investigate the relationship between IFNAR receptor levels, IFN-I binding affinity and their potency to the added effects of injected Hu-IFN-Is in cancer and in immune-related diseases. With algometric differences between humans and mice taken into account, the HyBNAR mice provides a convenient non-primate pre-clinical model to advance the study of human IFN-Is.

Percent pairwise sequence identity

IFNAR1	Human	Chimp	Rhesus	Cow	Mouse	Rat	Chicken
Human		99.3	90.7	67.5	50.7	51.3	39.1
Chimp	99.3		91.2	67.6	51	51.1	38.7
Resus	90.7	91.2		64	49.6	50.3	39
Cow	67.5	67.6	64		49.5	50	37.3
Mouse	50.7	51	49.6	49.5		69.2	33.7
Rat	51.3	51.1	50.3	50	69.2		29.9
Chicken	39.1	38.7	39	37.3	33.7	29.9	

IFNAR2	Human	Chimp	Rhesus	Cow	Mouse	Rat	Chicken
Human		99	92.6	59.3	50.4	47	28.1
Chimp	99		92.6	59.4	51.2	47	27.9
Resus	92.6	92.6		57.5	49.1	46	28.5
Cow	59.3	59.4	57.5		43.3	41	28.5
Mouse	50.4	51.2	49.1	43.3		73	26.8
Rat	47	47	46	41	73		28
Chicken	28.1	27.9	28.5	28.5	26.8	28	

EGFR	Human	Chimp	Rhesus	Cow	Mouse	Rat	Chicken
Human		99.7	98.1	88.3	90.6	90.6	78.3
Chimp	99.7		98.1	88.3	90.6	90.6	78.4
Resus	98.1	98.1		89.5	91.3	90.2	80.6
Cow	88.3	88.3	89.5		87	87.5	76.8
Mouse	90.6	90.6	91.3	87		95.9	77.8
Rat	90.6	90.6	90.2	87.5	95.9		77
Chicken	78.3	78.4	80.6	76.8	77.8	77	

INSR	Human	Chimp	Rhesus	Cow	Mouse	Rat	Chicken
Human		99.7	99.3	95.9	95.9	94.6	83.9
Chimp	99.7		99.5	98	95.9	95.1	85.8
Resus	99.3	99.5		96.5	95.3	94.6	83.1
Cow	95.9	98	96.5		94.5	93.6	83.6
Mouse	95.9	95.9	95.3	94.5		98.8	83.2
Rat	94.6	95.1	94.6	93.6	98.8		83
Chicken	83.9	85.8	83.1	83.6	83.2	83	

Figure 10. Interspecies Relatedness for a Selection of Cell Surface Receptors. Interspecies protein sequences derived for IFNAR1, IFNAR2, the epidermal growth factor receptor (EGFR) and the Insulin Receptor (INSR) were each compared by respective pairwise BLASTP alignment. Percent amino acid identity scores are given. Pre-calculated percentage identity scores were extracted from the Homologene Server (NCBI, Release #67). Exceptionally, rat IFNAR2 protein sequence (accession #XP_001073550.1) was not available in the Homologene dataset, and was thus curated manually.

Supporting Information

Figure S1 Genomic Genotyping of HyBNAR mouse strains by qPCR. Genomic Tail DNA was subjected to qPCR by TaqMan-like methodology. Assessment of transgenic signal was performed using external probes specific for the HyBNAR transgene and with an internal fluorescent probe. Background signals (IF: amplification from non-transgenic mice) is shown in green. A reference gene encoding mouse TERT was also amplified as to ascertain approximate transgene copy-number for the different transgenic mouse strains.

Figure S2 Cumulative Number of Clinical Trials for Different IFNs. Newly listed clinical trials using IFNα, IFNβ and (Type II) IFNγ were counted on a year by year basis and cumulative clinical trial numbers over time are plotted. This Data was extracted from ClinicalTrials.gov (a web-based service provided by the U.S. National Institutes of Health). Only the subset of trials listed as "first received" were counted as to avoid possible duplication of same events published over more than one year.

Acknowledgments

We thank the generosity and help from a number of people: To Menachem Rubenstein and Daniela Novick (Weizmann) for providing IFNAR cDNAs and for the 117.7 IFNAR2 MAb. Glenn Baker (Biogen) for the AA3 IFNAR1 MAb, to Robert Weissert and Rob Hoof (Merck-Serono) for early discussions of the HyBNAR project, to Slava Kalchenko and Yuri Kuznetsov for help and advice regarding IVIS Spectrum use, and to Sima Peretz, Anna Tatarin, and other members of the Veterinary Services for help, handling and maintenance of our mice colonies. Mouse and Human IFN**β** were both supplied by Merck-Serono.

Author Contributions

Conceived and designed the experiments: DH SP GS. Performed the experiments: DH RA. Analyzed the data: DH GS. Contributed reagents/materials/analysis tools: MK HH. Wrote the paper: DH AZ PS SP GS.

References

1. Borden EC, Sen GC, Uze G, Silverman RH, Ransohoff RM, et al. (2007) Interferons at age 50: past, current and future impact on biomedicine. Nat Rev Drug Discov 6: 975–990. doi:10.1038/nrd2422.

2. Moraga I, Harari D, Schreiber G, Uzé G, Pellegrini S (2009) Receptor density is key to the alpha2/beta interferon differential activities. Mol Cell Biol 29: 4778–4787. doi:10.1128/MCB.01808-08.

3. Levin D, Harari D, Schreiber G (2011) Stochastic receptor expression determines cell fate upon interferon treatment. Mol Cell Biol 31: 3252–3266. doi:10.1128/MCB.05251-11.

4. Shen X, Hong F, Nguyen VA, Gao B (2000) IL-10 attenuates IFN-alpha-activated STAT1 in the liver: involvement of SOCS2 and SOCS3. FEBS Lett 480: 132–136.

5. Radaeva S, Jaruga B, Kim W-H, Heller T, Liang TJ, et al. (2004) Interferon-gamma inhibits interferon-alpha signalling in hepatic cells: evidence for the involvement of STAT1 induction and hyperexpression of STAT1 in chronic hepatitis C. Biochem J 379: 199–208. doi:10.1042/BJ20031495.

6. Fenner JE, Starr R, Cornish AL, Zhang J-G, Metcalf D, et al. (2006) Suppressor of cytokine signaling 1 regulates the immune response to infection by a unique inhibition of type I interferon activity. Nat Immunol 7: 33–39. doi:10.1038/ni1287.

7. Zheng H, Qian J, Carbone CJ, Leu NA, Baker DP, et al. (2011) Vascular endothelial growth factor-induced elimination of the type 1 interferon receptor is required for efficient angiogenesis. Blood 118: 4003–4006. doi:10.1182/blood-2011-06-359745.

8. Carbone CJ, Zheng H, Bhattacharya S, Lewis JR, Reiter AM, et al. (2012) Protein tyrosine phosphatase 1B is a key regulator of IFNAR1 endocytosis and a target for antiviral therapies. Proc Natl Acad Sci U S A 109: 19226–19231. doi:10.1073/pnas.1211491109.

9. Francois-Newton V, Livingstone M, Payelle-Brogard B, Uzé G, Pellegrini S (2012) USP18 establishes the transcriptional and anti-proliferative interferon α/β differential. Biochem J 446: 509–516. doi:10.1042/BJ20120541.

10. Fuchs SY (2012) Ubiquitination-mediated regulation of interferon responses. Growth Factors Chur Switz 30: 141–148. doi:10.3109/08977194.2012.669382.

11. Müller U, Steinhoff U, Reis LF, Hemmi S, Pavlovic J, et al. (1994) Functional role of type I and type II interferons in antiviral defense. Science 264: 1918–1921.

12. Theofilopoulos AN, Baccala R, Beutler B, Kono DH (2005) Type I interferons (alpha/beta) in immunity and autoimmunity. Annu Rev Immunol 23: 307–336. doi:10.1146/annurev.immunol.23.021704.115843.

13. Piehler J, Thomas C, Garcia KC, Schreiber G (2012) Structural and dynamic determinants of type I interferon receptor assembly and their functional interpretation. Immunol Rev 250: 317–334. doi:10.1111/imr.12001.

14. Apelbaum A, Yarden G, Warszawski S, Harari D, Schreiber G (2013) Type I interferons induce apoptosis by balancing cFLIP and caspase-8 independent of death ligands. Mol Cell Biol 33: 800–814. doi:10.1128/MCB.01430-12.

15. Yong VW, Chabot S, Stuve O, Williams G (1998) Interferon beta in the treatment of multiple sclerosis: mechanisms of action. Neurology 51: 682–689.

16. Lavoie TB, Kalie E, Crisafulli-Cabatu S, Abramovich R, Digioia G, et al. (2011) Binding and activity of all human alpha interferon subtypes. Cytokine 56: 282–289. doi:10.1016/j.cyto.2011.07.019.

17. Kalie E, Jaitin DA, Abramovich R, Schreiber G (2007) An interferon alpha2 mutant optimized by phage display for IFNAR1 binding confers specifically enhanced antitumor activities. J Biol Chem 282: 11602–11611. doi:10.1074/jbc.M610115200.

18. Kalie E, Jaitin DA, Podoplelova Y, Piehler J, Schreiber G (2008) The stability of the ternary interferon-receptor complex rather than the affinity to the individual subunits dictates differential biological activities. J Biol Chem 283: 32925–32936. doi:10.1074/jbc.M806019200.

19. Thomas C, Moraga I, Levin D, Krutzik PO, Podoplelova Y, et al. (2011) Structural linkage between ligand discrimination and receptor activation by type I interferons. Cell 146: 621–632. doi:10.1016/j.cell.2011.06.048.

20. Roberts RM, Liu L, Guo Q, Leaman D, Bixby J (1998) The evolution of the type I interferons. J Interf Cytokine Res Off J Int Soc Interf Cytokine Res 18: 805–816.

21. Darnell JE Jr, Kerr IM, Stark GR (1994) Jak-STAT pathways and transcriptional activation in response to IFNs and other extracellular signaling proteins. Science 264: 1415–1421.

22. Samarajiwa SA, Forster S, Auchettl K, Hertzog PJ (2009) INTERFEROME: the database of interferon regulated genes. Nucleic Acids Res 37: D852–857. doi:10.1093/nar/gkn732.

23. Schoggins JW, Wilson SJ, Panis M, Murphy MY, Jones CT, et al. (2011) A diverse range of gene products are effectors of the type I interferon antiviral response. Nature 472: 481–485. doi:10.1038/nature09907.

24. Staeheli P, Sutcliffe JG (1988) Identification of a second interferon-regulated murine Mx gene. Mol Cell Biol 8: 4524–4528.

25. Staeheli P, Haller O, Boll W, Lindenmann J, Weissmann C (1986) Mx protein: constitutive expression in 3T3 cells transformed with cloned Mx cDNA confers selective resistance to influenza virus. Cell 44: 147–158.

26. Pulverer JE, Rand U, Lienenklaus S, Kugel D, Zietara N, et al. (2010) Temporal and spatial resolution of type I and III interferon responses in vivo. J Virol 84: 8626–8638. doi:10.1128/JVI.00303-10.

27. Kraus J, Oschmann P (2006) The impact of interferon-beta treatment on the blood-brain barrier. Drug Discov Today 11: 755–762. doi:10.1016/j.drudis.2006.06.008.

28. Neuhaus O, Kieseier BC, Hartung H-P (2007) Pharmacokinetics and pharmacodynamics of the interferon-betas, glatiramer acetate, and mitoxantrone in multiple sclerosis. J Neurol Sci 259: 27–37. doi:10.1016/j.jns.2006.05.071.

29. Williams GJ, Witt PL (1998) Comparative study of the pharmacodynamic and pharmacologic effects of Betaseron and AVONEX. J Interf Cytokine Res Off J Int Soc Interf Cytokine Res 18: 967–975.

30. Pepinsky RB, LePage DJ, Gill A, Chakraborty A, Vaidyanathan S, et al. (2001) Improved pharmacokinetic properties of a polyethylene glycol-modified form of interferon-beta-1a with preserved in vitro bioactivity. J Pharmacol Exp Ther 297: 1059–1066.

31. Schlapschy M, Binder U, Börger C, Theobald I, Wachinger K, et al. (2013) PASylation: a biological alternative to PEGylation for extending the plasma half-life of pharmaceutically active proteins. Protein Eng Des Sel PEDS. doi:10.1093/protein/gzt023.

32. Chen J, Baig E, Fish EN (2004) Diversity and relatedness among the type I interferons. J Interf Cytokine Res Off J Int Soc Interf Cytokine Res 24: 687–698. doi:10.1089/jir.2004.24.687.

33. Kagan L, Abraham AK, Harrold JM, Mager DE (2010) Interspecies scaling of receptor-mediated pharmacokinetics and pharmacodynamics of type I interferons. Pharm Res 27: 920–932. doi:10.1007/s11095-010-0098-6.

34. Peleg-Shulman T, Tsubery H, Mironchik M, Fridkin M, Schreiber G, et al. (2004) Reversible PEGylation: a novel technology to release native interferon alpha2 over a prolonged time period. J Med Chem 47: 4897–4904. doi:10.1021/jm0497693.

35. West GB, Brown JH, Enquist BJ (1997) A general model for the origin of allometric scaling laws in biology. Science 276: 122–126.

36. Kalincik T, Spelman T, Trojano M, Duquette P, Izquierdo G, et al. (2013) Persistence on therapy and propensity matched outcome comparison of two subcutaneous interferon Beta 1a dosages for multiple sclerosis. PloS One 8: e63480. doi:10.1371/journal.pone.0063480.

15

Tissue-Specific Expression of Transgenic Secreted ACE in Vasculature Can Restore Normal Kidney Functions, but Not Blood Pressure, of Ace-/- Mice

Saurabh Chattopadhyay[1], Sean P. Kessler[1], Juliana Almada Colucci[1], Michifumi Yamashita[1], Preenie deS Senanayake[2], Ganes C. Sen[1]*

1 Department of Molecular Genetics, Lerner Research Institute, Cleveland Clinic, Cleveland, Ohio, United States of America, **2** Department of Ophthalmic Research, Cole Eye Institute, Cleveland Clinic, Cleveland, Ohio, United States of America

Abstract

Angiotensin-converting enzyme (ACE) regulates normal blood pressure and fluid homeostasis through its action in the renin-angiotensin-system (RAS). Ace-/- mice are smaller in size, have low blood pressure and defective kidney structure and functions. All of these defects are cured by transgenic expression of somatic ACE (sACE) in vascular endothelial cells of Ace-/- mice. sACE is expressed on the surface of vascular endothelial cells and undergoes a natural cleavage secretion process to generate a soluble form in the body fluids. Both the tissue-bound and the soluble forms of ACE are enzymatically active, and generate the vasoactive octapeptide Angiotensin II (Ang II) with equal efficiency. To assess the relative physiological roles of the secreted and the cell-bound forms of ACE, we expressed, in the vascular endothelial cells of Ace-/- mice, the ectodomain of sACE, which corresponded to only the secreted form of ACE. Our results demonstrated that the secreted form of ACE could normalize kidney functions and RAS integrity, growth and development of Ace-/- mice, but not their blood pressure. This study clearly demonstrates that the secreted form of ACE cannot replace the tissue-bound ACE for maintaining normal blood pressure; a suitable balance between the tissue-bound and the soluble forms of ACE is essential for maintaining all physiological functions of ACE.

Editor: Michael Bader, Max-Delbrück Center for Molecular Medicine (MDC), Germany

Funding: This study was supported by the National Institutes of Health grant HL-48258. The funders had no role in study design, data collection and analysis, decision to publish, or preparation of the manuscript.

Competing Interests: The authors have declared that no competing interests exist.

* E-mail: seng@ccf.org

Introduction

The Renin-Angiotensin System (RAS) is a coordinated hormonal cascade that modulates fluid and electrolyte balance as well as blood pressure regulation [1]. In the classical pathway of the RAS, renin, which is secreted in the kidney juxtaglomerular apparatus in response to a wide variety of stimuli, acts on the precursor Angiotensinogen to generate the decapeptide Angiotensin I (Ang I) [2]. Angiotensin-converting enzyme (ACE) plays a central role in the RAS through generation of the octapeptide angiotensin II (Ang II) from its inactive precursor Ang I [3,4]. Ang II induces vasoconstriction, aldosterone release, and other physiologic actions to raise blood pressure [5]. ACE inhibitors block the formation of Ang II and have been used to treat hypertension. ACE exists as two isoforms *viz*, somatic ACE (sACE), the larger isoform which is expressed mainly in the somatic tissues and the smaller isoform, germinal ACE (gACE), which is expressed in sperm cells. Both isoforms are synthesized from a single gene using tissue specific promoters and have distinct and tissue-specific functions in the body [6–8].

Studies with ACE knockout mice revealed additional roles of ACE beyond blood pressure regulation, especially in kidney functions, development and male fertility [9–11]. In addition to these functions, ACE plays roles in fat metabolism, inflammation

and immune responses [12]. Our previous studies using transgenic ACE expression in Ace-/- mice demonstrated tissue- and isoform-specific physiological functions of ACE. The Ts mice, which express transgenic sACE only in the vascular endothelial cells of Ace-/- mice, can restore normal blood pressure [8]. However, gACE is not able to substitute for sACE for the maintenance of normal blood pressure of Ace-/- mice. Expression of gACE, but not sACE, in sperm successfully restores the male fertility of Ace-/- mice [7].

ACE is expressed on the cell surface as type I ectoprotein, with a long ectodomain containing the enzymatic active site, short cytoplasmic domain and a transmembrane domain [13–15]. The ectodomain of ACE can be cleaved by a membrane-bound metalloprotease, to generate a soluble form [16–18]. Although ACE secretase is unknown, its activity has been characterized as a membrane-associated protease. Previous studies have demonstrated that the cytoplasmic domain of ACE interacts with various signaling proteins such as, Calmodulin, Protein Kinase C, Casein Kinase 2 (CK2) and regulate the cleavage-secretion of ACE [19–22]. Moreover, the association of these proteins regulates the phosphorylation of ACE on a specific Ser residue in the cytoplasmic domain [19,22]. Because both the cell-bound and the secreted forms of ACE are enzymatically active, their functional distinction has been of great interest. Cell-bound

ACE, in addition to its enzymatic activity, can act as a signaling protein on the cell surface. Signaling via ACE activates CK2, which phosphorylates the cytoplasmic domain of ACE, suggesting phosphorylation of the cytoplasmic domain of ACE as a potential intracellular event which may contribute to the functions of ACE in the endothelial cells [23].

To investigate the relative roles of secreted and cell-bound ACE in various physiological functions of ACE, we have generated a transgenic mouse model which expressed only the secreted form of sACE, without any cell-bound form. This mutant ACE protein lacks the transmembrane and the cytoplasmic domains and, therefore, is constitutively secreted in the circulation. Our results demonstrated that the secreted ACE could not restore normal blood pressure of Ace-/- mice, although their kidney functions were normal. This study demonstrates that a cell-bound ACE is critical for maintaining normal blood pressure; a secreted ACE cannot substitute for the cell-bound form for this function.

Materials and Methods

Generation of target vector

We used Wt rabbit sACE cDNA to generate esACE, by deletion of the transmembrane and cytoplasmic domain coding region. The esACE cDNA was then used to generate Tie-esACE construct, as described previously [8]. The expression and secretion of ACE was tested by transient transfection of the tie-esACE plasmid in human fibrosarcoma cells (HT1080) followed by Western Blotting of the cell extracts and culture medium. The Tie-esACE-BGHpA transgene was subsequently released from this plasmid by SpeI and AsnI digestion and sent to the University of Cincinnati Transgenic Mouse Core Facility for the generation of transgenic mice utilizing standard techniques.

Establishment of transgenic mice

The tie-esACE transgene was microinjected into the pronuclei of FVB strain zygotes using standard techniques. Adult FVB tie-esACE-BGHpA (TeS) transgenic founder mice (Ace+/+, TeS+/−) were mated with Ace+/− FVB mice to generate Ace+/− TeS+/− mice. Interbreeding between male and female Ace+/−, TeS+/− mice within the same line was performed to generate the Ace-/-, TeS/TeS (TeS). Genotyping of all mice was performed by Southern Blotting as described below. Expression of esACE in the transgenic line was confirmed by Western Blotting of the serum using anti-ACE antibody. All of the mice described in this study, including Wt, KO, Tie-esACE (TeS) and Tie-sACE (Ts), were of the FVB strain (Table 1). All mouse experimental protocols were approved by the Lerner Research Institute's Institutional Animal Care and Use Committee.

Maintenance of control mice

Ace-/- Ts/Ts (Ts) control mice were generated by mating Ace+/− Ts/Ts males with Ace-/- Ts/Ts females as described previously [8]. Ace-/- and Ace+/− control mice were generated by mating Ace+/− male mice with Ace+/− female mice. All of the mice described in this study, including Wt, KO, Tie-esACE (TeS) and Tie-sACE (Ts), were of the FVB strain (Table 1).

Southern Blot hybridization

Southern blot genotyping was performed as described previously utilizing Sac I digestion of genomic tail snip DNA [8]. Heterozygosis or homozygosis of the transgene was determined by normalizing the transgene value to the endogenous mouse *Ace* gene value in the same genomic DNA sample using Imagequant software. The endogenous *Ace* genotype is determined by the presence of a wild-type 6.6 kB SacI genomic fragment or the disrupted 8.4 kB SacI (Ace-/-) genomic fragment [8]. The presence of the tie-esACE transgene is indicated by hybridization of the probe with a 3.7 kB fragment.

ACE enzyme activity assay

The enzymatic activity of ACE was assayed using Hip-His-Leu as a substrate and measuring fluorimetrically the His-Leu liberated at 5 mM of Hip-His-Leu. Serum (1 µl) from retro-orbital eye bleed or tissue homogenates of the transgenic mice were used to measure ACE enzymatic activity. Activity values are reported as µmoles His-Leu per per µl of serum or 25 µg of tissue protein extract liberated from Hip-His-Leu after one-hour incubation at 37°C.

Plasma renin activity

Blood was collected by retro-orbital sinus plexus eye bleed under brief isoflurane anesthesia and plasma renin activity was measured as described before [24]. Briefly, plasma renin activity, defined as the rate of Ang I generation from renin in the sample and excess exogenous substrate provided from nephrectomized rat plasma, was incubated at pH 6.5 (rats) and pH 8.5 (mice) for 90 minutes. Ang I generated in the sample was quantified by radioimmunoassay (DiaSorin Corp, Stillwater, MN).

Serum creatinine measurement

Serum creatinine levels were determined by previously described alkaline picrate method [25]. Standards (Sigma-Aldrich)

Table 1. Transgenic mice strains.

Strain	Transgene	Genotype	ACE isoform	Expression
Wt	None	Ace +/+	Somatic	Vascular endothelial cells
				Proximal tubule cells
				Brain
				Leydig cells
			Germinal	Sperm
KO	None	Ace -/-	None	None
Ts	Tie-sACE	Ace -/-Ts/Ts	Somatic	Vascular endothelial cells
TeS	Tie-esACE	Ace -/- TeS/TeS	Somatic	Vascular endothelial cells (sACE expressed only in serum)

or 20 μl of serum were added into a 96-well microtiter plate. Alkaline picrate solution (10.8 mM picric acid, 29 mM sodium borate, 167 mM NaOH, 1.67% SDS) was added and incubated for 10 min at RT, absorbance was read at 490 nm. After the absorbance measurement, 60% acetic acid was added into all wells and left for 8 min at RT. The absorbance was read at 490 nm again, and subtracted from the first absorbance.

Angiotensin II measurements

For each genotype, blood from four age matched adult mice was pooled to achieve a 1 ml plasma sample. Samples were concentrated on C18 columns (Sep-Pak columns; Waters), evaporated to dryness, and reconstituted in 0.9% NaCl, 0.03% acetic acid and 0.1% BSA. The levels of Ang II were quantified by RIA [26,27]. The results represent the arithmetic mean of assaying a minimum of 5 pools from each genotype ±95% CI for the mean. Renal Ang II was measured from age- and genotype-matched mice, as described previously [28].

Measurement of kidney functions

Age-matched adult mice (Wt, KO, Ts, TeS) were individually placed in a Nalgene metabolic cage supplied with powdered standard chow and water *ad libitum*. The data indicates average daily (24 h) water consumption and urine volume produced for five consecutive days for each of five mice of the same genotype. Urine osmolality was measured for each mouse using the Osmette A (Precision Instruments, Inc., Natick, MA) freezing point osmometer according to the manufacturer's instructions. Triplicate readings were performed on the urine collected as indicated above.

Blood pressure measurement

Blood pressure was measured by two independent methods, by radiotelemetry and tail-cuff plethysmography. For telemetric measurements, the mice were anesthetized with isoflurane (3% in an oxygen stream). The BP transmitter (TA11PA-C10, Data Sciences International, St Paul, MN, USA) was implanted as follows: the catheter was inserted into the left common carotid artery, and the transmitter was positioned in the right flank [29]. The mice received 0.5 mL of 0.25% bupivicaine subcutaneously five minutes before surgery. After a 7 days recovery period each individual mouse cage was placed on the top of a radio-receiver (Model RPC-1) for measurement of BP, heart rate (HR), and spontaneous locomotor activity (SLA). Experimental data was recorded daily during 60 minutes for six consecutive days and analyzed using the Dataquest ART system, version 4.2 (Data Sciences International). A non-invasive computerized RTBP007 tail cuff blood pressure system (Harvard Apparatus, Holliston, MA) was used to obtain systolic blood pressure on conscious mice as described previously [27].

Histology and immunohistochemistry

Kidneys from age-matched adult mice were paraffin embedded, cross-sectioned at 2.5 μm thickness and hematoxylin and eosin stained by the Histology Core (Lerner Research Institute, Cleveland, OH). Immunohistochemistry was performed following deparaffinization as described previously [30]. Slides were incubated in 10 mM sodium citrate pH 6.0 for 30 min at 25°C, and then returned to PBS. The slides were blocked in PBS + 10% horse serum + 0.3% Triton X-100 (blocking buffer) for 2 h at 25°C. The polyclonal goat anti-ACE antibody, diluted 1:1000 in blocking buffer was applied to a slide in a humid chamber for 16 h at 4°C. Following washes in PBS + 0.3% Triton X-100 (PBST), anti-goat-

FITC (Santa Cruz) was applied to polyclonal ACE stained slides at 1:3000 dilution in blocking buffer to each section for 2 h in the dark at 25°C. Following washes in PBST, Vectashield +/− DAPI (Vector Laboratories) diluted 1:1 in PBS was applied. For Ang II staining, kidney slides were incubated with polyclonal rabbit anti-Ang II antibody (generated at Hybridoma Core, Cleveland Clinic, 1:300), using the same protocol as described above. Following the PBS wash, color reaction was carried out using the EnVision + System-HRP (DAB) (Dako) according to the manufacturer's instructions. All stained slides were visualized with a Leica digital fluorescent microscope and processed in Adobe Photoshop software.

Kidney-associated renin

Kidneys from Ace+/− or TeS mice were homogenized in 20 mM Tris HCl, 150 mM NaCl, 0.1% Triton X-100, protease inhibitors and phosphatase inhibitors leaving on ice for 30 min and spun at 14000 rpm at 4°C. The homogenates (50 μg total protein) were separated on 12% SDS-PAGE and Western blotted by anti-renin antibody (1:1000, Abcam). The loading control was confirmed by Western blotting against actin.

Statistics

Data are presented as arithmetic means and variations as 95% confidence interval of the mean. Significance values were obtained by unpaired *t* test, comparing experimental mice to the Wt mice or the KO mice as indicated in the figure legend. The statistical significance has been indicated in the figure legends.

Results

Generation of experimental mice

We have shown previously that the expression of Wt sACE in the vascular endothelial cells, which generates both cell-bound and secreted ACE, can restore normal blood pressure of Ace-/- mice [8]. In the present study, we investigated whether a secreted form of sACE could substitute for cell-bund ACE. We have designed a mutant somatic ACE which was missing the transmembrane and the cytoplasmic domains (esACE) (Figure 1A), and expressed in the vascular endothelial cells of Ace-/- mice. The tissue-specific Tie-esACE targeting vector was constructed from the parental Tie-sACE plasmid, by deletion of the transmembrane and the cytoplasmic domains of sACE (Figure 1B). The Tie-esACE-BGHpA transgene (TeS) was assembled as shown in Fig. 1B and tested *in vitro* by transiently expressing in human cells (HT1080). The esACE protein was expressed in these cells with similar expression as the sACE, with the expected size. This cellular form of esACE is the underglycosylated precursor form of ACE which is not transported to the cell surface; as indicated by two forms (the precursor and the glycosylated cell surface expressing form) for the sACE (Figure 1C, upper panel). As expected, the esACE protein was secreted relatively more to the culture media as compared to the sACE (cleavage-secretion of esACE: 73.1% vs. sACE: 52.3%, Figure 1C, lower panel). The esACE protein was enzymatically active, as tested by ACE activity assay using sACE as a control (data not shown). The SpeI-AseI fragment containing the TeS transgene (Figure 1B) was microinjected into the pronuclei of FVB zygotes and then implanted in the uteri of pseudopregnant mothers. The Tie-esACE (TeS) experimental mice were generated by crossing Ace+/− mice with mice carrying the TeS transgene. Two independent founder lines (8000 and 8400) carrying the TeS transgene were generated and these were separately crossed with Ace+/− mice and the progenies were genotyped by Southern Blot analysis for the presence of the transgene. Both transgenic mice lines transmitted the transgene to their progeny (Figure 1D). The transgenic lines were mated (Ace+/− TeS/TeS males with Ace+/− TeS/TeS females of the same line) to generate the Ace-/- TeS/TeS experimental mice. Serum obtained from retro-orbital eye bleed was used for testing secreted ACE protein; both transgenic TeS lines expressed comparable levels of esACE protein, detected by Western blot or enzymatic activity (Figure 1E and data not shown). Line 8400 was used for all the physiological analyses described in this study and the 8000 line was discontinued.

Physiological properties of the experimental TeS mice

Abnormal fertility.   The TeS mice exhibited normal growth rates, development and restored body weight and size, similar to Wt or Ts mice (data not shown) [8]. This indicates that the secreted sACE can substitute for the cell-bound sACE for maintaining these functions in mice. A striking phenotype of Ace-/- mice is male sterility; only a sperm specific expression of gACE restores the fertility of Ace-/- mice [31]. As expected, male fertility was not corrected in the TeS mice. Moreover, the presence of high levels of secreted ACE in the serum did not adversely affect or reduce the fertility of Ace+/+ or Ace+/− male mice.

Normal kidney structures and functions.   Since esACE was exclusively secreted into the blood flow from vascular endothelial cells, no significant ACE staining was observed in the kidney section of TeS mice by immunohistochemistry (Figure 2). The Ace+/+ mice showed positive ACE staining in

the brush border membrane of the proximal tubular epithelial cells (PT), arterial endothelial cells (A), and peritubular capillaries (C). As expected, Ace-/- kidney did not show any ACE staining. The Ts mice showed positive ACE staining in arterial endothelial cells (A), glomerular endothelial cells (GE), but not in the proximal tubular (PT) cells. No ACE staining was detected in any area of the kidney section tested from TeS mice (Figure 2); confirming the absence of any cell-bound ACE in TeS mice.

Ace-/- mice exhibit hypoplasia of renal cortex; in the renal cross section, the area outside the dotted line is markedly reduced compared to the Ace+/+ mice (Figure 3, left panel). The development of the renal medulla of Ace-/- mice is also impaired: a large renal pelvis appears inside the dotted line in the center because the medulla is largely defective. The gross anatomy of the kidney from TeS mice was indistinguishable from that of Ace+/+ or the Ts mice, and the vessel wall thickness returned to normal (Figure 3, left panel). This was further confirmed by a quantitative analysis of relative cortex thickness from multiple kidney sections of these mice. The results clearly indicate that the TeS mice exhibited normal cortex thickness, which is similar to the Wt or Ts mice (Figure 3, right panel).

To address whether the restored kidney structures correspond to normal kidney functions, we compared their urine concentrating ability by measuring water intake, urine output and urine osmolarity. The overall fluid homeostasis was measured by the amount of urine output and water uptake over a period of five

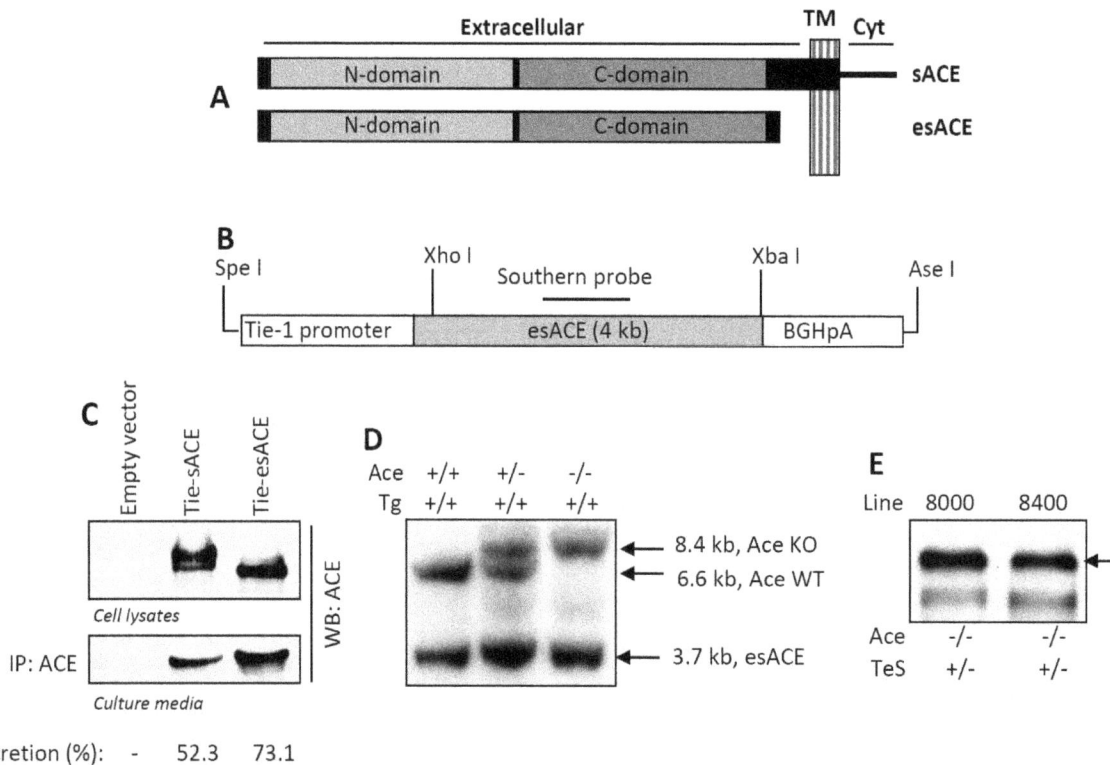

Figure 1. Generation of experimental mice. A Schematic representation of sACE and esACE is shown, different domains are indicated; TM, transmembrane domain, Cyt, cytoplasmic domain. **B** The Tie-1-esACE (rabbit somatic ACE without the transmembrane domain, as shown above)-BGHpA transgene (TeS) and rabbit ACE cDNA Southern Probe were designed. **C** The target vector (Tie-esACE) and the WT sACE (Tie-sACE) were transfected in HT1080 cells, the cell lysates and culture medium were analyzed for ACE by Western Blot by anti-ACE antibody. Secretion of ACE was measured as percentage of total ACE secreted into the culture media, by densitometric analyses. **D** A representative Southern Blot of the transgenic mice containing the Tie-esACE transgene (Tg), using the probe indicated in B, is shown. The genomic DNA, digested with Sac I, yielded a 3.7 kB transgene, a 6.6 kB Ace allele and an 8.4 kB disrupted Ace allele. **E** Serum obtained from retro-orbital eye bleed from TeS transgenic mice lines (8000 and 8400) were analyzed for ACE expression by Western Blot using anti-ACE antibody.

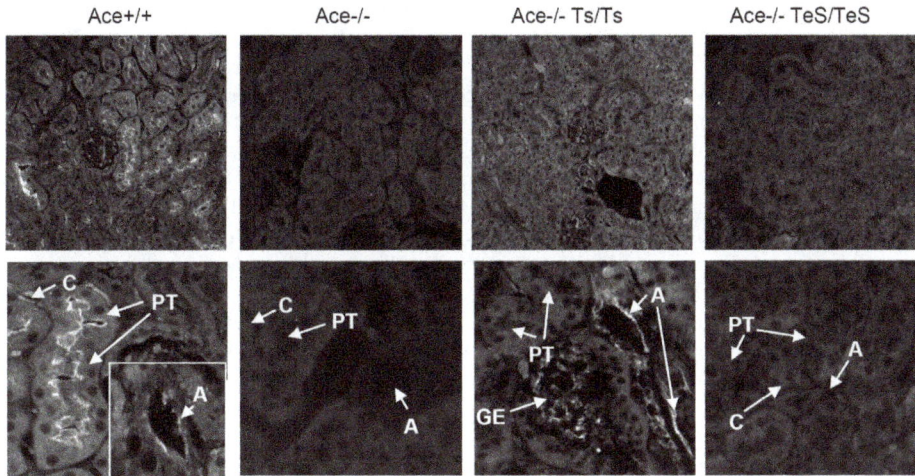

Figure 2. Experimental TeS mice do not express cell-bound ACE in the kidney. Age-matched adult kidneys from Wt (Ace+/+), ACE knockout (Ace-/-), two transgenic (Ace-/- Ts/Ts or Ace-/- TeS/TeS) mice were prepared and stained for ACE expression with anti-ACE polyclonal serum, followed by Alexa 568-conjugated secondary antibody (described in Materials and Methods). Slides were viewed with a Leica fluorescent microscope at 20X magnification. PT: proximal tubular epithelial cells, A: arterial endothelial cells, V: peritubular capillaries, GE: glomerular endothelial cells.

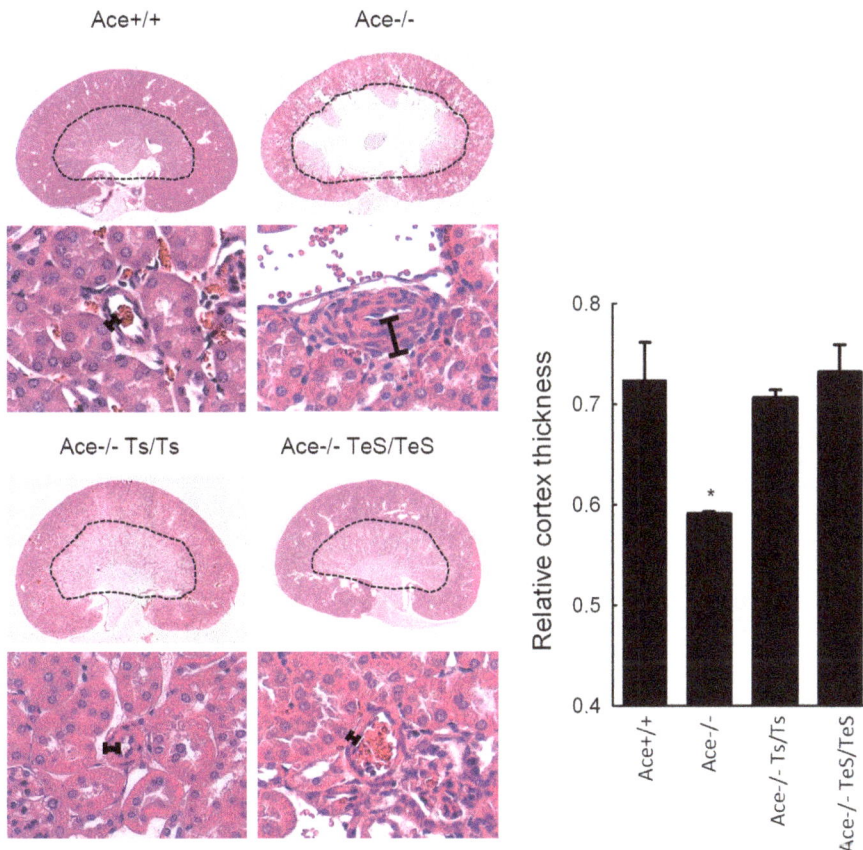

Figure 3. Secreted sACE expressing mice exhibit normal kidney structures. Hematoxylin and eosin-stained tissues from kidneys of age-matched adult mice are shown. All kidneys were photographed at 0.6X (whole cross section) or 20X (tubuli and arteriole) with a digital Leica light microscope. Manually drawn dotted lines indicate the borderline between the cortex and the medulla. The black bars represent wall thickness of arteriole. The lower panels show the vascular structures of the kidney sections. The relative cortex thickness was measured as a ratio between the medulla and the whole kidney section area. Values displayed as mean ± SD (* p<0.01).

days. Consistent with previous findings, the Ace-/- mice exhibited significantly high urine output as well as high water uptake, presumably due to the defect in concentrating urine (Figure 4A, B). However, the average urine output and water uptake in the TeS mice were comparable to that observed in Wt or Ts mice (Figure 4A, B). In addition to this, we measured the urine osmolarity of the mice after overnight water restriction and the results indicate that the TeS mice showed comparable urine osmolarity to that of the Wt or Ts mice (Figure 4C). Therefore, circulating sACE could substitute for cell-bound sACE with respect to the restoration of normal kidney functions. These results, together with our previous studies using the Ts mice, further reinforces the notion that circulating sACE in serum is not only sufficient but can also substitute for the proximal tubular ACE, for maintaining normal kidney phenotypes.

Normal kidney-associated and circulating RAS activity. We tested the integrity of local RAS by measuring kidney-associated renin and Ang II levels of TeS mice. Renin, a critical regulator of local RAS, was analyzed in kidney homogenates. The TeS mice expressed similar levels of kidney-associated renin as compared to Ace+/- mice (Figure 5A). To further investigate whether the normal renin expression correlates with Ang II level, we analyzed kidney-associated Ang II by immunostaining of kidney sections with anti-Ang II antibody. Comparable Ang II staining was observed in Ace+/+ and TeS proximal tubules, specifically in the brush border (Figure 5B). The immunostaining result was further confirmed by an independent assay (RIA) and by comparing with Ace+/- mice. The results indicate the TeS mice showed similar Ang II levels, compared to Ace+/- mice (Figure 5C). As expected, the TeS mice did not show any kidney-associated ACE activity, when compared with Ace+/- mice (Figure 5D). These results indicate that although cell-bound ACE was not present, TeS mice were not defective in kidney associated Ang II generation. To examine the circulating RAS activity, we analyzed ACE, Ang II and renin levels in the circulation. When compared to Ace+/- mice, the TeS mice

expressed similar levels of soluble ACE (Figure 6A) and plasma Ang II (Figure 6B). The unaltered levels of plasma Ang II in the TeS mice was correlated with plasma renin activity, which was similar to the control Ace+/- mice (Figure 6C). No defects in kidney functions were observed for TeS mice even at a later stage of life, compared to the Ace-/- mice, which do not live longer. Serum creatinine level, a marker of kidney functions, was unaltered in TeS mice, when compared with Ace+/+ or Ace+/- control mice; however, the Ace-/- mice showed increased levels of serum creatinine (Figure 6D), indicating defective kidney functions. These results indicate that the absence of cell-bound ACE did not affect the RAS activity, in kidney or in circulation.

Low blood pressure. To examine whether restored renal phenotype, normal levels of circulating ACE, renal and plasma Ang II, plasma renin activity are sufficient to maintain the normal blood pressure of TeS mice, we conducted two independent experiments, computerized tail-cuff plethysmography and radio-telemetry, to measure the systolic and diastolic blood pressures of the mice. The TeS mice exhibited low systolic blood pressure, comparable to that of Ace-/- mice, whereas Wt (Ace+/+) or Ace+/- mice both showed normal blood pressure (Figure 7A). The systolic blood pressure of the TeS mice was 92.3±6.9 mm, compared to 90±2 mm of Ace-/- mice. In contrast, the corresponding numbers for Ace+/+ and Ace+/- mice were 116±4.7 mm and 115±4.6 mm respectively. Although we have successfully used tail-cuff method previously to determine the systolic blood pressure of conscious mice, we understand that this type of measurement impose substantial amounts of thermal and restrain stress that are known to affect blood pressure, heart rate and stress hormones [32]. To address this potential problem, we used wireless radio-telemetry as an additional assay. This technique has the advantage of allowing continuous, direct measurements of blood pressure without the need for restraints. Moreover, the telemetric analysis allowed us to measure diastolic blood pressure, in addition to systolic blood pressure, of the TeS mice. As shown in Figure 7B, the TeS mice exhibited lower

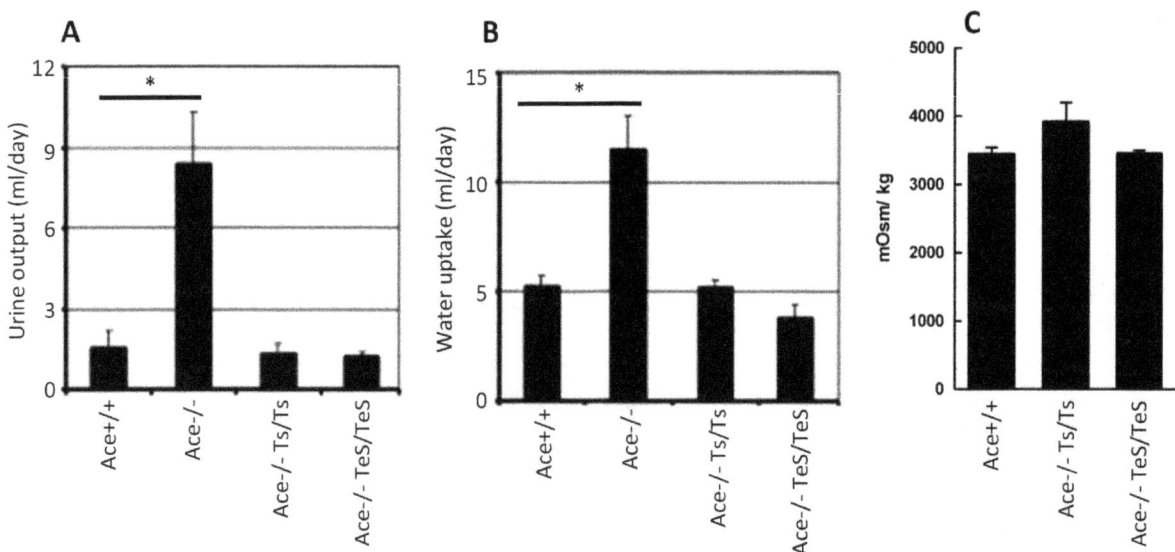

Figure 4. Secreted sACE expressing mice exhibit normal kidney functions. Five, age-matched adult mice (Wt, Ace+/+, KO, Ace-/-, Ts, Ace-/-Ts/Ts, TeS, Ace-/- TeS/TeS) were tested using metabolic cage for a period of five consecutive days. *A* Urine produced over a period of 24 h from WT, KO, Ts and TeS mice is shown. *B* Water uptake (in ml), during the same 24 h period. The data for the control mice (Wt, KO and Ts) were adopted from our previously published paper [8]. *C* Urine osmolality was measured for each mouse using the Osmette A (Precision Instruments, Inc., Natick, MA) freezing point osmometer. Values are displayed as mean ± SD (* $p<0.001$).

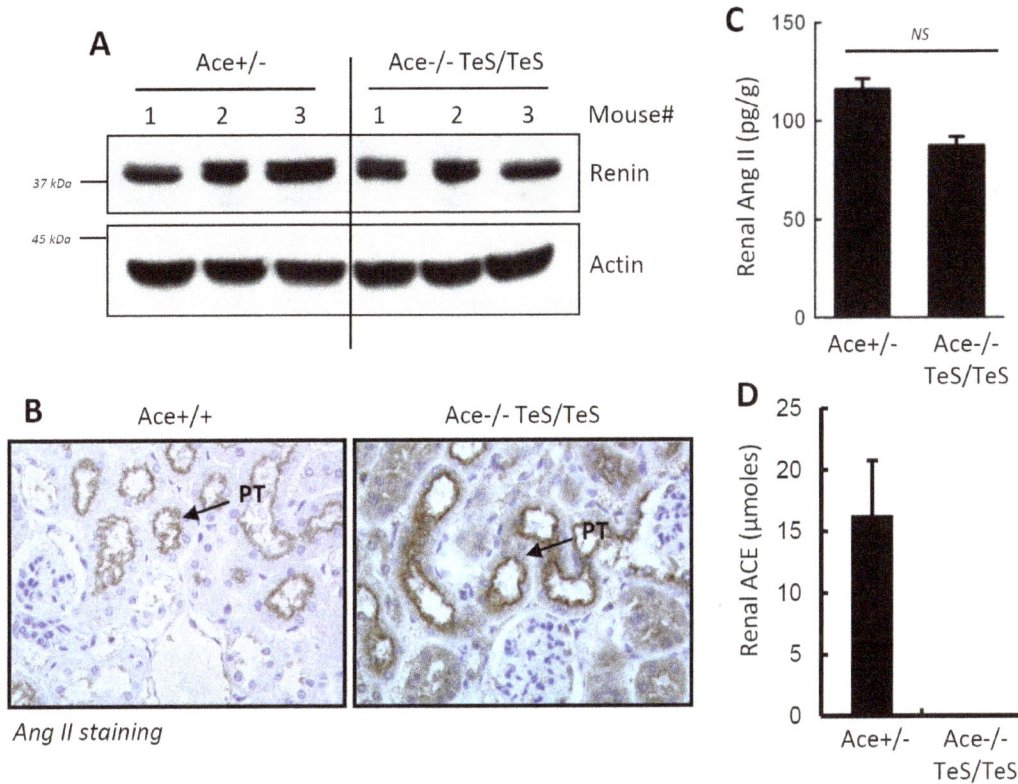

Figure 5. TeS mice exhibit normal kidney-associated RAS activity. *A* The kidney homogenates from Ace+/− and TeS mice were analyzed by SDS-PAGE followed by Western Blot against anti-renin antibody (Abcam). *B* Kidney slides from Wt and experimental TeS mice were stained with anti-AngII antibody followed by EnVision+System-HRP (DAB) detection. Slides were viewed as described in (A); PT: proximal tubular epithelial cells. *C* Renal Ang II levels were assayed as described in the Methods section. Ace +/− and Ace-/- TeS/TeS showed no significant difference in Ang II at the tissue level. *D* Age- and genotype-matched adult mice were used to measure ACE activity in kidney homogenates (per 25 µg of protein), by analyzing the liberated His-Leu using Hip-His-Leu as a substrate. Values are presented as mean values ± SD, from five mice.

systolic and diastolic blood pressures as compared to the Ace+/− mice. From the two independent measurements, we concluded that the TeS mice, which expressed similar levels of circulating ACE and plasma Ang II as the Ace+/− mice, could not maintain normal blood pressure. These results clearly indicate that the secreted form of sACE is not sufficient for maintaining normal blood pressure in mice.

Discussion

Ectodomain shedding of ACE generates an enzymatically active secreted form, whose function is currently unknown. We hypothesized that a suitable balance between the cell-bound and secreted forms of ACE is critical for maintaining the normal physiological functions of the enzyme. To test this hypothesis, we generated transgenic mice expressing a secreted sACE, in the absence of any cell-bound ACE. Our results showed that secreted sACE could not maintain normal blood pressure; however, it was sufficient for restoring normal growth, development, kidney structures and functions, of Ace-/- mice. This strongly indicates the requirement of endothelial cell-bound sACE for maintaining normal blood pressure. This conclusion is supported by our previous results with another transgenic mouse, in which sACE was restrictedly expressed in the kidney proximal tubular cells (Gs mice) [8]. These mice expressed high levels of circulating sACE in the serum. Although the Gs mice exhibited normal kidney development and functions, the high level of circulating sACE in

the serum was unable to restore normal blood pressure. In another study, a transgenic mouse expressing only the N-domain of sACE in the serum has been generated [33]. The circulating truncated sACE expressed in these mice cannot restore normal blood pressure; however, the observed defect could be due to the absence of the C-domain enzymatic site in the protein. In contrast, our study is based on a rational design of a transgenic mouse expressing the entire extracellular domain of sACE, which is the physiological soluble form of ACE in the circulation, by endothelial cells. The cleavage and secretion of ACE is negatively regulated by its cytoplasmic domain. *In vitro* studies using both sACE and gACE proteins show that deletion of cytoplasmic domain leads to elevated levels of soluble ACE [34,35]. The pathological effects of ACE cleavage have been observed in patients with natural mutations of ACE. Studies indicate that mutations of sACE that generate cytoplasmic domain deleted ACE, lead to high levels of circulating ACE, which correlates to disorder in RAS activities [36,37].

Inhibition of ACE enzymatic activity by selective ACE inhibitors blocks the generation of Ang II and thereby acts as a suitable approach for the treatment of high blood pressure. However, it remains unclear whether the Ang II generating function of ACE is the only link between this protein and the maintenance of normal blood pressure. Our study indicates that the level of Ang II is not an accurate indicator of ACE function in blood pressure maintenance. Although, both the TeS mice and the control Ace+/− mice expressed similar levels of renal and plasma

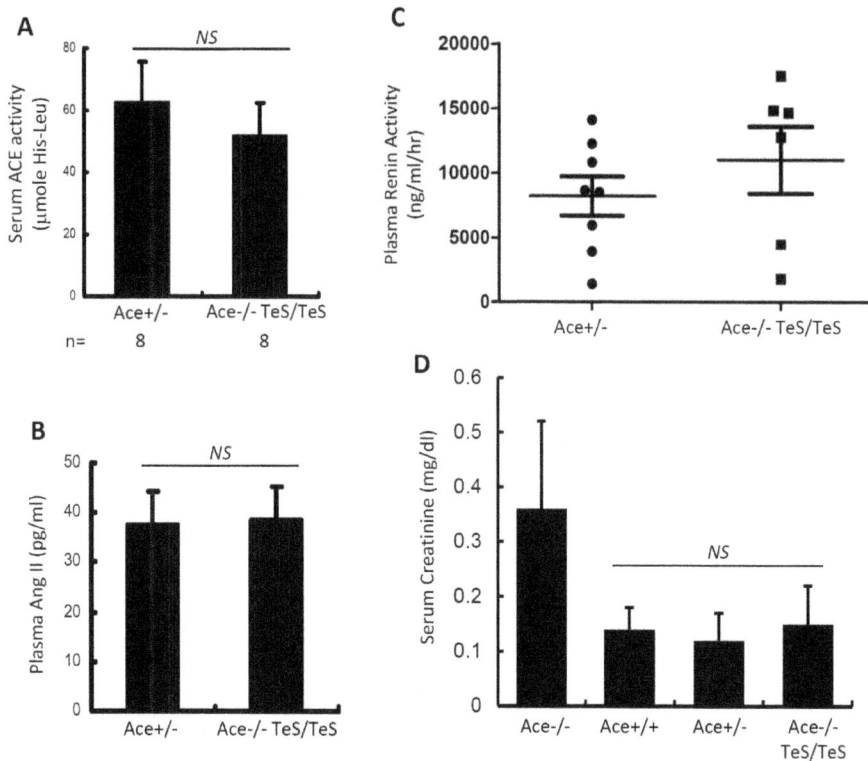

Figure 6. TeS mice exhibit normal circulating RAS activity. *A* Age- and genotype-matched adult mice were used to analyze ACE activity in the serum. The serum (1 μl) was assayed for ACE activity, measured as μmoles of His-Leu liberated from Hip-His-Leu in 1 h at 37°C. Each bar represents an average of multiple mice (number is indicated as n). *B* Age- and genotype-matched adult mice (three for each group) were used to pool plasma (1 ml) and assayed for Ang II levels (described in Materials and Methods). Each data point is an average of two independent measurements (*NS*, non-significant). *C* Age and genotype-matched adult mice (as indicated) were used to pool plasma and assayed for renin activity, as described in Materials and Methods. The difference in renin activity between the two groups of mice is not statistically significant (using paired t-test). *D* Serum collected from Ace-/-, Ace+/+, Ace+/− and TeS mice and used to measure creatinine levels. The difference between the three groups is not statistically significant.

Ang II (Figure 5C, 6B), they could not maintain normal blood pressure (Figure 7). Similar to the TeS mice, the Gs mice, which expressed high levels of plasma Ang II, was also unable to maintain normal blood pressure. Kidney associated RAS activity has also been shown to be involved in blood pressure regulation [38]. Surprisingly, TeS mice exhibited normal levels of kidney Ang II; however, this was not sufficient to restore the normal blood pressure.

So, why is cell-bound ACE essential for blood pressure maintenance? The concept of local vs. systemic RAS may be relevant to this question. The components of the RAS exist in many organs and are postulated to regulate tissue specific Ang II production [39]. Initially considered only a systemic process, the Ang II produced at the tissue level is now accounted for the regulation of the vasculature structure and tone in the microenvironment of each tissue [40]. These local systems depend critically on the presence of tissue-bound ACE and would be, therefore, presumably nonfunctional in the experimental TeS mice because secreted sACE, although capable of generating circulating Ang II in serum, is not able to generate local Ang II in the vascular endothelial bed. However, in addition to normal levels of serum Ang II, TeS mice exhibited kidney Ang II levels comparable to Wt mice. Although, it is not clear at this point, the kidney associated Ang II might be generated by ACE-independent alternative pathways. Multiple studies have described the growing importance of chymase-dependent pathways in generation of Ang

II from Ang I [41]. Ang II, derived from ACE or chymase have also been suggested to have distinct physiological implications. The Ang II levels observed in the TeS mice suggest an alternative hypothesis, that sACE has dual functions, both of which are needed for blood pressure maintenance. One function depends on the traditional enzyme activity ACE, which is responsible for Ang II formation and bradykinin degradation. In the second function, ACE works as a receptor on the surface of endothelial cells and mediates outside-in signaling in them [42,43]. There is strong evidence for the cytoplasmic domain of ACE being capable of activating intracellular signaling [23], a function of ACE that is independent of its enzymatic activity. The ligands that can trigger ACE signaling include ACE inhibitors, which bind to it strongly. Surprisingly, Ang II, a product of the enzyme activity of ACE, can also activate intracellular calcium signaling cascades [44]. Thus, the two alternative models for justifying the need of tissue-bound ACE, for blood pressure maintenance, may be interdependent. Tissue-bound ACE may be needed to produce high levels of Ang II locally, which, in turn, may trigger intracellular signaling by binding to cell-surface ACE. In future, one potential system for testing the relative contributions of the two models will be a genetically engineered mouse expressing, in its endothelial cells, cell-bound ACE that lacks the cytoplasmic tail. Such an ACE mutant will still produce ample quantities of local Ang II in the endothelium but will be incapable of producing intracellular

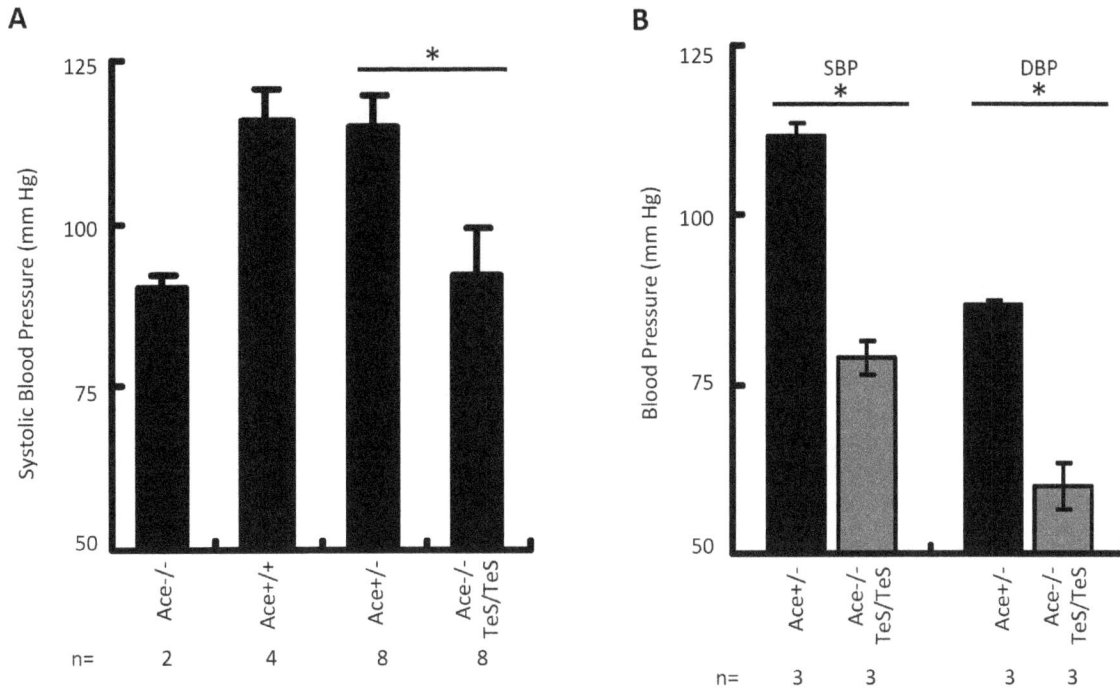

Figure 7. Secreted sACE expressing mice cannot maintain normal blood pressure. *A* The systolic blood pressure of age-matched adult mice (Ace-/-, Ace+/+, Ace+/− and TeS, as indicated) was measured using non-invasive, computerized tail-cuff plethysmography method. The systolic blood pressure (in mm Hg) was calculated using the mean daily blood pressure over a five day reading period (the number of mice is indicated as *n*; * $p<0.001$ Ace+/− vs Ace-/- TeS/TeS). *B* Ace+/− and Ace-/- TeS/TeS (TeS) mice were subjected to radiotelemetric analyses for measurement of both systolic (SBP and diastolic blood pressure (DBP) (described in Experimental Procedures) and a mean blood pressure is represented for each group of mice (n, the number of mice from each group; * $p<0.001$ Ace+/− vs Ace-/- TeS/TeS).

signals. These mutant mice should have low blood pressure, if cytoplasmic-tail signaling is the critical determinant.

Acknowledgments

We thank Indira Sen for the anti-ACE antibody used in the study. We thank Yunhai Chang, Christina Gaughan, DiFernando Vanegas and Pat Kessler for expert technical assistance. We thank the Image Core at Lerner Research Institute for assistance with histology studies, Hypertension core at Wake Forest University for measurement of plasma renin activity.

Author Contributions

Conceived and designed the experiments: SC SPK GCS. Performed the experiments: SC SPK JAC MY PS. Analyzed the data: SC SPK GCS. Contributed reagents/materials/analysis tools: SC SPK GCS. Wrote the paper: SC SPK MY GCS.

References

1. Soffer RL (1981) Angiotensin-converting enzyme. In: Soffer RL, editor. Biochemical Regulation of Blood Pressure. New York: Wiley-Interscience. pp. 123–164.
2. Siragy HM, Awad A, Abadir P, Webb R (2003) The angiotensin II type 1 receptor mediates renal interstitial content of tumor necrosis factor-alpha in diabetic rats. Endocrinology 144: 2229–2233.
3. Sturrock ED, Anthony CS, Danilov SM (2012) Peptidyl-dipeptidase A/ Angiotensin I-converting enzyme. In: Rawlings ND, Salvesen G, editors. Handbook of Proteolytic Enzymes.
4. Bernstein KE, Ong FS, Blackwell WL, Shah KH, Giani JF, et al. (2013) A modern understanding of the traditional and nontraditional biological functions of angiotensin-converting enzyme. Pharmacol Rev 65: 1–46.
5. Navar LG, Inscho EW, Majid SA, Imig JD, Harrison-Bernard LM, et al. (1996) Paracrine regulation of the renal microcirculation. Physiol Rev 76: 425–536.
6. Howard TE, Shai SY, Langford KG, Martin BM, Bernstein KE (1990) Transcription of testicular angiotensin-converting enzyme (ACE) is initiated within the 12th intron of the somatic ACE gene. Mol Cell Biol 10: 4294–4302.
7. Kessler SP, Rowe TM, Gomos JB, Kessler PM, Sen GC (2000) Physiological non-equivalence of the two isoforms of angiotensin- converting enzyme. J Biol Chem 275: 26259–26264.
8. Kessler SP, Senanayake PS, Scheidemantel TS, Gomos JB, Rowe TM, et al. (2003) Maintenance of normal blood pressure and renal functions are independent effects of angiotensin-converting enzyme. J Biol Chem 278: 21105–21112.
9. Krege JH, John SW, Langenbach LL, Hodgin JB, Hagaman JR, et al. (1995) Male-female differences in fertility and blood pressure in ACE-deficient mice. Nature 375: 146–148.
10. Esther CR Jr, Howard TE, Marino EM, Goddard JM, Capecchi MR, et al. (1996) Mice lacking angiotensin-converting enzyme have low blood pressure, renal pathology, and reduced male fertility. Lab Invest 74: 953–965.
11. Cole J, Ertoy D, Bernstein KE (2000) Insights derived from ACE knockout mice. J Renin Angiotensin Aldosterone Syst 1: 137–141.
12. Jayasooriya AP, Mathai ML, Walker LL, Begg DP, Denton DA, et al. (2008) Mice lacking angiotensin-converting enzyme have increased energy expenditure, with reduced fat mass and improved glucose clearance. Proc Natl Acad Sci U S A 105: 6531–6536.
13. Corvol P, Michaud A, Soubrier F, Williams TA (1995) Recent advances in knowledge of the structure and function of the angiotensin I converting enzyme. J Hypertens Suppl 13: S3–10.
14. Riordan JF (2003) Angiotensin-I-converting enzyme and its relatives. Genome Biol 4: 225.
15. Sturrock ED, Natesh R, van Rooyen JM, Acharya KR (2004) Structure of angiotensin I-converting enzyme. Cell Mol Life Sci 61: 2677–2686.
16. Sadhukhan R, Sen GC, Ramchandran R, Sen I (1998) The distal ectodomain of angiotensin-converting enzyme regulates its cleavage-secretion from the cell surface. Proc Natl Acad Sci U S A 95: 138–143.
17. Parkin ET, Turner AJ, Hooper NM (2004) Secretase-mediated cell surface shedding of the angiotensin-converting enzyme. Protein Pept Lett 11: 423–432.
18. Chattopadhyay S, Karan G, Sen I, Sen GC (2008) A small region in the angiotensin-converting enzyme distal ectodomain is required for cleavage-secretion of the protein at the plasma membrane. Biochemistry 47: 8335–8341.
19. Chattopadhyay S, Santhamma KR, Sengupta S, McCue B, Kinter M, et al. (2005) Calmodulin Binds to the Cytoplasmic Domain of Angiotensin-converting

Enzyme and Regulates Its Phosphorylation and Cleavage Secretion. J Biol Chem 280: 33847–33855.

20. Santhamma KR, Sadhukhan R, Kinter M, Chattopadhyay S, McCue B, et al. (2004) Role of tyrosine phosphorylation in the regulation of cleavage secretion of angiotensin-converting enzyme. J Biol Chem 279: 40227–40236.

21. Santhamma KR, Sen I (2000) Specific cellular proteins associate with angiotensin-converting enzyme and regulate its intracellular transport and cleavage-secretion. J Biol Chem 275: 23253–23258.

22. Kohlstedt K, Shoghi F, Muller-Esterl W, Busse R, Fleming I (2002) CK2 phosphorylates the angiotensin-converting enzyme and regulates its retention in the endothelial cell plasma membrane. Circ Res 91: 749–756.

23. Kohlstedt K, Brandes RP, Muller-Esterl W, Busse R, Fleming I (2004) Angiotensin-converting enzyme is involved in outside-in signaling in endothelial cells. Circ Res 94: 60–67.

24. Varagic J, Ahmad S, Voncannon JL, Moniwa N, Brosnihan KB, et al. (2013) Predominance of AT1 Blockade Over Mas-Mediated Angiotensin-(1-7) Mechanisms in the Regulation of Blood Pressure and Renin-Angiotensin System in mRen2.Lewis Rats. Am J Hypertens.

25. Inoshita H, Kim BG, Yamashita M, Choi SH, Tomino Y, et al. (2013) Disruption of smad4 expression in T cells leads to IgA nephropathy-like manifestations. PLoS One 8: e78736.

26. Senanayake PS, Smeby RR, Martins AS, Moriguchi A, Kumagai H, et al. (1998) Adrenal, kidney, and heart angiotensins in female murine Ren-2 transfected hypertensive rats. Peptides 19: 1685–1694.

27. Kessler SP, Senanayake P, Gaughan C, Sen GC (2007) Vascular expression of germinal ACE fails to maintain normal blood pressure in ACE-/- mice. Faseb J 21: 156–166.

28. Kessler SP, Hashimoto S, Senanayake PS, Gaughan C, Sen GC, et al. (2005) Nephron Function in Transgenic Mice with Selective Vascular or Tubular Expression of Angiotensin-Converting Enzyme. J Am Soc Nephrol.

29. Baudrie V, Laude D, Elghozi JL (2007) Optimal frequency ranges for extracting information on cardiovascular autonomic control from the blood pressure and pulse interval spectrograms in mice. Am J Physiol Regul Integr Comp Physiol 292: R904–912.

30. Kessler SP, Gomos JB, Scheidemantel TS, Rowe TM, Smith HL, et al. (2002) The germinal isozyme of angiotensin-converting enzyme can substitute for the somatic isozyme in maintaining normal renal structure and functions. J Biol Chem 277: 4271–4276.

31. Ramaraj P, Kessler SP, Colmenares C, Sen GC (1998) Selective restoration of male fertility in mice lacking angiotensin-converting enzymes by sperm-specific expression of the testicular isozyme. J Clin Invest 102: 371–378.

32. Kurtz TW, Griffin KA, Bidani AK, Davisson RL, Hall JE (2005) Recommendations for blood pressure measurement in humans and experimental animals.

Part 2: Blood pressure measurement in experimental animals: a statement for professionals from the subcommittee of professional and public education of the American Heart Association council on high blood pressure research. Hypertension 45: 299–310.

33. Esther CR, Marino EM, Howard TE, Machaud A, Corvol P, et al. (1997) The critical role of tissue angiotensin-converting enzyme as revealed by gene targeting in mice. J Clin Invest 99: 2375–2385.

34. Wei L, Alhenc-Gelas F, Soubrier F, Michaud A, Corvol P, et al. (1991) Expression and characterization of recombinant human angiotensin I-converting enzyme. Evidence for a C-terminal transmembrane anchor and for a proteolytic processing of the secreted recombinant and plasma enzymes. J Biol Chem 266: 5540–5546.

35. Ehlers MR, Chen YN, Riordan JF (1991) Spontaneous solubilization of membrane-bound human testis angiotensin-converting enzyme expressed in Chinese hamster ovary cells. Proc Natl Acad Sci U S A 88: 1009–1013.

36. Nesterovitch AB, Hogarth KD, Adarichev VA, Vinokour EI, Schwartz DE, et al. (2009) Angiotensin I-converting enzyme mutation (Trp1197Stop) causes a dramatic increase in blood ACE. PLoS One 4: e8282.

37. Persu A, Lambert M, Deinum J, Cossu M, de Visscher N, et al. (2013) A novel splice-site mutation in angiotensin I-converting enzyme (ACE) gene, c.3691+1G>A (IVS25+1G>A), causes a dramatic increase in circulating ACE through deletion of the transmembrane anchor. PLoS One 8: e59537.

38. Crowley SD, Gurley SB, Oliverio MI, Pazmino AK, Griffiths R, et al. (2005) Distinct roles for the kidney and systemic tissues in blood pressure regulation by the renin-angiotensin system. The Journal of clinical investigation 115: 1092–1099.

39. Paul M, Poyan Mehr A, Kreutz R (2006) Physiology of local renin-angiotensin systems. Physiol Rev 86: 747–803.

40. Hsueh WA, Wyne K (2011) Renin-Angiotensin-aldosterone system in diabetes and hypertension. J Clin Hypertens (Greenwich) 13: 224–237.

41. Miyazaki M, Takai S (2006) Tissue angiotensin II generating system by angiotensin-converting enzyme and chymase. Journal of pharmacological sciences 100: 391–397.

42. Kohlstedt K, Busse R, Fleming I (2005) Signaling via the angiotensin-converting enzyme enhances the expression of cyclooxygenase-2 in endothelial cells. Hypertension 45: 126–132.

43. Kohlstedt K, Kellner R, Busse R, Fleming I (2006) Signaling via the angiotensin-converting enzyme results in the phosphorylation of the nonmuscle myosin heavy chain IIA. Mol Pharmacol 69: 19–26.

44. Guimaraes PB, Alvarenga EC, Siqueira PD, Paredes-Gamero EJ, Sabatini RA, et al. (2011) Angiotensin II Binding to Angiotensin I-Converting Enzyme Triggers Calcium Signaling. Hypertension 57: 965–972.

16

Handmade Cloned Transgenic Sheep Rich in Omega-3 Fatty Acids

Peng Zhang[1,6❾], Peng Liu[2❾], Hongwei Dou[2], Lei Chen[1], Longxin Chen[1], Lin Lin[2], Pingping Tan[1], Gabor Vajta[2,3], Jianfeng Gao[4], Yutao Du[2,5*], Runlin Z. Ma[1,6*]

1 State Key Laboratory of Molecular and Developmental Biology, Institute of Genetics and Developmental Biology, Chinese Academy of Sciences, Beijing, China, 2 BGI ARK Biotechnology Co., Ltd, Shenzhen, China, 3 IRIS, Central Queensland University, Rockhampton, Australia, 4 School of Life Sciences, Shihezi University, Shihezi, China, 5 BGI-Shenzhen, Shenzhen, China, 6 Graduate University of the Chinese Academy of Sciences, Beijing, China

Abstract

Technology of somatic cell nuclear transfer (SCNT) has been adapted worldwide to generate transgenic animals, although the traditional procedure relies largely on instrumental micromanipulation. In this study, we used the modified handmade cloning (HMC) established in cattle and pig to produce transgenic sheep with elevated levels of omega-3 (n−3) fatty acids. Codon-optimized nematode *mfat-1* was inserted into a eukaryotic expression vector and was transferred into the genome of primary ovine fibroblast cells from a male Chinese merino sheep. Reverse transcriptase PCR, gas chromatography, and chromosome analyses were performed to select nuclear donor cells capable of converting omega-6 (n−6) into n−3 fatty acids. Blastocysts developed after 7 days of *in vitro* culture were surgically transplanted into the uterus of female ovine recipients of a local sheep breed in Xinjiang. For the HMC, approximately 8.9% (n =925) of reconstructed embryos developed to the blastocyst stage. Four recipients became pregnant after 53 blastocysts were transplanted into 29 naturally cycling females, and a total of 3 live transgenic lambs were produced. Detailed analyses on one of the transgenic lambs revealed a single integration of the modified nematode *mfat-1* gene at sheep chromosome 5. The transgenic sheep expressed functional n−3 fatty acid desaturase, accompanied by more than 2-folds reduction of n−6/n−3 ratio in the muscle ($p<0.01$) and other major organs/tissues ($p<0.05$). To our knowledge, this is the first report of transgenic sheep produced by the HMC. Compared to the traditional SCNT method, HMC showed an equivalent efficiency but proved cheaper and easier in operation.

Editor: Christine Wrenzycki, Justus-Liebig-Universität, Germany

Funding: This study was funded by the research grants from the China Ministry of Agriculture (2009ZX08008-005B) and the Chinese Academy of Sciences (KSZD-EW-005). The funders had no role in study design, data collection and analysis, decision to publish, or preparation of the manuscript.

Competing Interests: PL, HD, LL and YD are currently employed by BGI ARK Biotechnology Co. Ltd. as full-time employees; GV is currently employed by the same company as a part-time consultant. However, none of the above mentioned co-authors has any competing financial interests or potential competing interests. No patent or any other form of intellectual property protection document has been filed as the result of this research.

* E-mail: rlma@genetics.ac.cn (RZM); duyt@genomics.cn (YD)

❾ These authors contributed equally to this work.

Introduction

Sheep is one of the most important domestic animal species for human consumption of meat protein and milk. With new knowledge and understanding that a number of human diseases can be effectively prevented via improved and balanced nutrition, the nutritional value of sheep meat and milk could be further increased by elevated levels of polyunsaturated fatty acids (omega-3 or n−3 PUFAs). Omega-3 is an essential nutrient for human and has been demonstrated to have preventive and therapeutic effects on certain diseases of cardiovascular nature, arthritis, cancer, as well as neuropathic problems [1 4]. Unfortunately, supply of omega-3 in human is totally dependent on dietary intake due to our body's inability to synthesize the essential nutrients. While the exact reason on why human and most other mammals have lost their abilities to synthesize omega-3 during the evolution remains unclear [5], certain primary organisms like *C. elegans* can efficiently covert n−6 into n−3 PUFAs [6–8]. It would be ideal to produce sufficient amount of omega-3 in domestic animals via genetic engineering, which would be much better than capturing and killing of deep-sea fishes for the same essential nutrient. In this regard, transgenic pigs and cattle rich in n−3 fatty acids were generated to explore the possibility [2,9].

Technology of somatic cell nuclear transfer (SCNT) has been adapted worldwide since the successful generation of the sheep "Dolly" [10]. However, the traditional SCNT procedures rely on specialized instruments of micromanipulation, which are expensive and technologically demanding. An alternative technique termed handmade cloning (HMC), established previously [11], has proved to be efficient for animal cloning in several domestic species [12 17]. For the procedure of HMC, in contrast to traditional SCNT, the nucleus of an oocyte is removed manually by slicing off a portion of the zona-free oocyte with a sharp microblade, and a somatic nucleus is introduced by fusion of two enucleated oocytes with the somatic cell under an ordinary light microscope.

In this study, we aimed to establish a reliable and robust HMC procedure to generate transgenic sheep with increased levels of omega-3. We report here that a single copy of nematode *mfat-1* gene was successfully integrated into the sheep *Cep120* genomic locus on chromosome 5. Both of the introduced *mfat-1* and the host gene *Cep120* were functional, and three live births of the transgenic sheep were produced by the HMC. Tissue examination of one of the three new born transgenic lambs showed that the introduced *mfat-1* effectively lowered the n−6/n−3 ratio in the muscle and other major organs/tissues. If the nematode *mfat-1* gene is finally proved to be heritable and functionally stable in the host genome, the transgenic animals would potentially contribute better to human health for evaluated levels of omega-3 in meat and milk.

Results and Discussion

Following the HMC procedure, we successfully generated 3 transgenic sheep that carry and express the nematode *mfat-1* gene. The coding sequence of nematode *fat-1* (1209 bps) was modified for optimal expression in mammalian system (*mfat-1*) prior

integrating to the primary fibroblast cells of the Chinese merino sheep (Figure 1 A, B). Recombinant cells were screened by PCR and RT-qPCR for the presence and level of *mfat-1* mRNA expression (Figure 1C), and the cellular fatty acids composition was assessed by gas chromatography for *mfat-1* function (Figure 1D). In the recombinant clonal cells expressing *mfat-1*, n−6 fatty acids (18:2n−6, 20:4n−6, and 22:4n−6) were successfully converted to the corresponding n−3 (18:3n−3, 20:5n−3, and 22:5n−3), and the ratio of n−6/n−3 were reduced from 8.60:1 to 0.45:1, a change of more than 19-folds of reduction compared to the non-transfected cells (Figure 1D, Table 1). The candidate clones selected (F-1-1, F-6-5, and H-6-6) showed the normal karyotype (data not shown). We eventually selected the clone H-6-6 for the subsequent HMC procedure; partially due to a relatively higher ratio of n−6/n−3 conversion was observed in that clone (Table 1).

For the HMC procedure, approximately 3,740 (68%) out of 5,465 collected ovine oocytes were selected and utilized in the subsequent experiments. According to the established protocol,

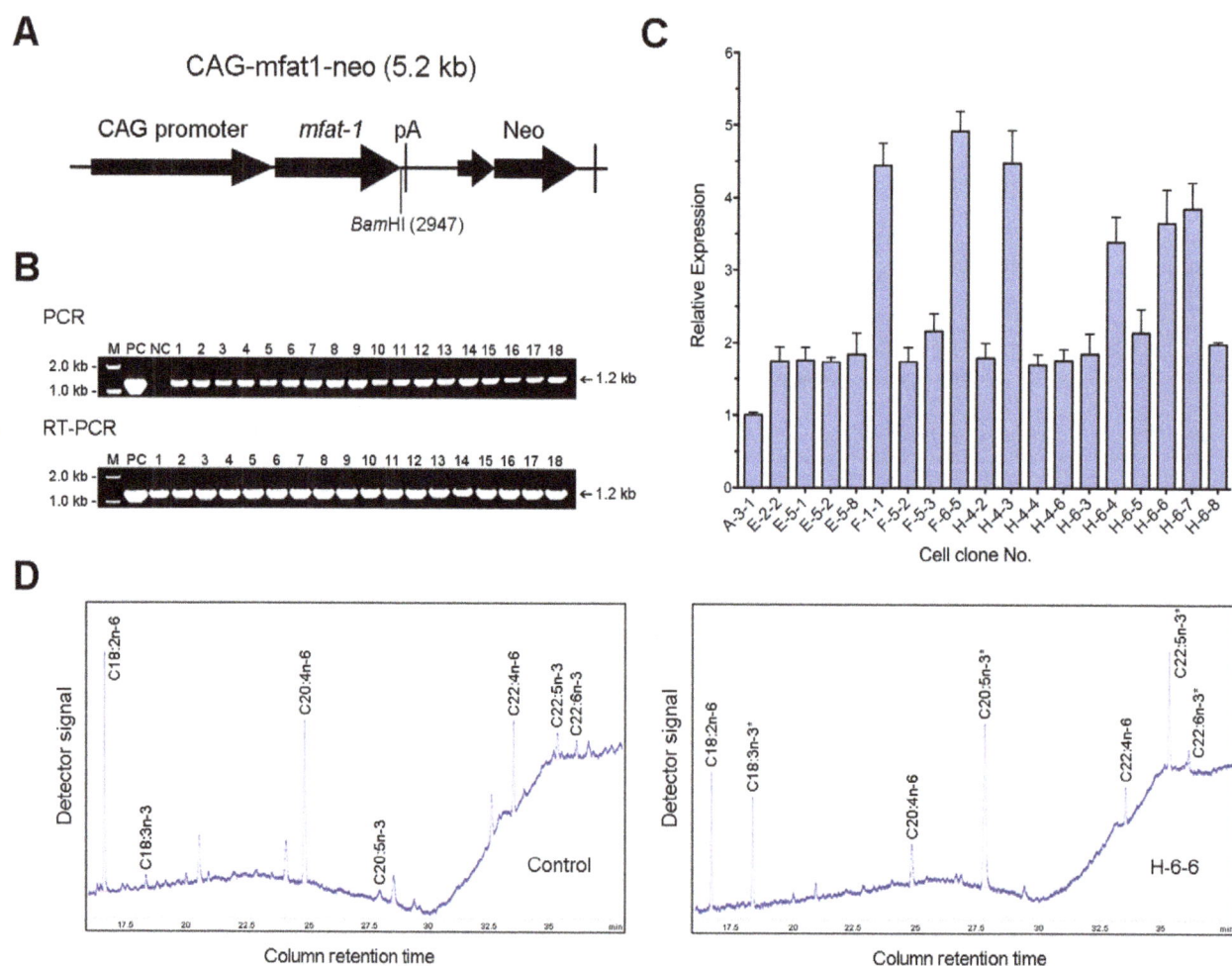

Figure 1. Establishment and analysis of transgenic clonal donor cells. (**A**) Schematic representation of n−3 fatty acid desaturase gene with linearized expression vectors. (**B**) Detection of the *mfat-1* gene in Geneticin-resistant cell clones by PCR and RT-qPCR. *mfat-1* expression vector was used as the template for positive control (PC) and untransfected syngenic cells was used as the negative control (NC). (**C**) Quantitative PCR analysis of *mfat-1* expression in positive cell clones. cDNA representing *mfat-1* was amplified with sequence specific primers. The beta-actin was used as internal control and the expression level observed in the transgenic donor cell was normalized to the value of A-3-1. (**D**) Partial gas chromatograph traces showing the polyunsaturated fatty acid profiles of total cellular lipids from the H-6-6 cells and the control cells. Note the level of n-6 polyunsaturated acids are lower whereas n−3 fatty acids are abundant in the *mfat-1* cells (right) as compared with the control cells (left), in which there is very little n−3 fatty acid.

Table 1. Composition and ratio of n−6 and n−3 PUFA in the syngenic control and the transgenic cells expressing *mfat-1* gene.

Fatty acids	Control Cells	*mfat-1* Cells		
		F-1-1	F-6-5	H-6-6
LA (18:2n−6)	6.35±0.48	4.85±0.03**	5.01±0.13**	4.57±0.19**
ALA (18:3n−3)	0.06±0.05	1.61±0.13**	1.32±0.16**	3.92±0.45**
AA (20:4n−6)	5.90±1.06	4.06±0.50*	5.62±0.14	2.64±0.38**
EPA (20:5n−3)	0.00	6.24±0.06**	4.24±0.06**	10.54±0.18**
ADA (22:4n−6)	2.47±0.39	1.86±0.36	1.83±0.30*	1.73±0.29*
DPA (22:5n−3)	1.11±0.18	3.61±0.16**	1.55±0.18*	4.72±0.26**
DHA (22:6n−3)	0.54±0.12	0.76±0.06*	0.43±0.22	0.71±0.06
Total n−6	14.72±1.90	10.76±0.84*	12.49±0.65	8.95±0.53**
Total n−3	1.71±0.08	12.22±0.35**	7.54±0.55**	19.94±0.65**
n−6/n−3 ratio	8.60±1.50	0.88±0.09**	1.65±0.16**	0.45±0.04**

Fatty acids composition is presented as a percentage of the total cellular lipids from the control cells and *mfat-1* cells. Each value represented the mean ± standard deviation from three cell samples in each group with two independent measurements for each sample.
[a]Statistical analysis using the two tailed student *t*-test. Significant differences between the syngenic control and *mfat-1* cells were marked (*$P<0.05$; **$P<0.01$).

every two enucleated oocytes and a nuclear donor cell were combined by fusion to form one reconstructed embryo for the subsequent *in vitro* development. As a consequence, a total of 925 reconstructed embryos were produced and 82 (~8.9%) of them successfully developed to the blastocyst stage after 7 days of *in vitro* culture (Table 2). A total of 53 blastocysts were surgically transplanted to uterus of 29 naturally cycling female recipients, with 1 to 2 blastocysts per recipient. Four pregnancies were detected and 3 live lambs were born either by caesarean section or natural delivery (Figure 2A and Table 3).

Genotype analysis using 13 ovine microsatellite markers demonstrated that all 3 transgenic sheep shared the same genotype to that of the nuclear donor cells (Table 4). The presence and function of *mfat-1* gene in the transgenic sheep were confirmed by PCR, RT-qPCR, Southern, Northern, and DNA sequencing (Figure 2 B, D, E). All 3 transgenic sheep are males as expected. The efficiency of nuclear transfer varied from 22.2% for the passage 9 to 14.3% for the passage 12 of the donor clonal cells, expressed as the number of live lambs per 100 blastocysts transplanted. The overall efficiency for production of live lambs (5.7%) was close to those reported previously [10,18,19]. The birth weight of the transgenic lambs ranged from 3.86 kg to 5.74 kg. The lambs PP-01 and PP-03 were healthy and behaved normally after the birth. The lamb PP-02, however, died within 48 hours after the birth with the lowest body weight among the three. The dead lamb was utilized as a representative sample to assess the levels of omega-3 in the muscle and other organs/tissues.

Insertion site analysis demonstrated that a single copy of the *mfat-1* was integrated into the sheep genome on chromosome 5, located within the 5th intron of a functional gene *Centrosomal protein of 120 kDa* (*Cep120*, Figure 2C). DNA sequencing of a 1.2 kb PCR fragment across the 3′-end of *mfat-1* and the neighbor host genome showed a 99% sequence homology to exons 5 and 6 of *Cep120* (Figure 2C). The transcriptional orientation of the transgene *mfat-1* was identical to that of the host gene. We found that the insertion

of the *mfat-1* showed no significant interfering effect on the function of *Cep120*, which is involved in centriole assembly and is essential for cell division [20,21]. The remaining 2 transgenic lambs showed no obvious sign of neurogenic or other defects up to the date.

Single-copy insertion of the *mfat-1* to ovine genome was cross-verified by the *Bam*HI restriction enzyme analysis. Southern blot revealed a single target band sized ~8.0 kb in *Bam*HI digested DNA only from the transgenic lambs (Figure 2D). This result was consistent with the previous analysis in that the DNA fragment digested with *Bam*HI was expected to be 7916 bps in length (Figure 2C). The possibility of tandem insertion of multiple *mfat-1* copy was excluded, because the digested *Bam*HI band from the transgenic sheep failed to show a predictive 5.2 kb fragment, a band representing multiple copies of the *mfat-1* with the maximum size of 5.2 kb.

Analyses of fatty acid composition showed that the omega-3 (n−3) peaks in the umbilical cords of all three transgenic lambs were significantly higher than those in the age-matched wild type controls (data not shown). The results was cross-verified by systematic analysis of the fatty acid in major tissues of the lamb PP-02, which showed a two-fold reduction of n−6/n−3 ratio (2.63 to 1.24, $p<0.01$) in muscles compared to the wild-type lamb (Figure 3, Table 5). The lower ratio of n−6/n−3 was also found in other major tissues ($p<0.05$, Table 5). Consistent with these results, the level of *mfat-1* mRNA varied among different tissues, with the highest expression found in the skeletal muscle (Figure 2F). More importantly, the total n−3 fatty acids (a-linolenic acid, eicosapentaenoic acid (EPA), and docosahexaenoic acid (DHA)) constituted approximately 6.2% of total muscle fat in the transgenic lamb, significantly higher than those in the control lambs, which was 3.5% on average ($p<0.01$). All of these indicated that the *mfat-1* gene was functional in the transgenic lambs. Compared to the earlier studies of transgenic animals expressing the n−3 fatty acid desaturase gene [22,23], our results help to provide additional evidence on the feasibility of transgenic animals.

It was not escaped from our attention that, in contrast to the SCNT, the procedure of HMC introduced an obvious heteroplasmy into the transgenic animals, mitochondrial DNAs in particular. While the potential interactions and the long-term consequences of such heteroplasmy remain to be determined, it seems that the developmental reprogramming driven by the enucleated oocytes was not significantly affected. Although certain concern was raised regarding the three origins of mitochondrial DNAs in the developing fetus from the cells of different sheep, no evidence of the negative effect to the transgenic animal is currently available [24]. The two remaining HMC sheep in our study showed normal growth and development up to the date (Figure S1), consistent with those of the HMC pigs [3,15].

Over the past decade the nematode *fat-1* has been successfully introduced into mice [8,25], pigs [2,3] and cattle [9]. The *fat-1* mice produced by pronuclei microinjection have possessed good value for the production of n−3 PUFAs [8,25], particularly for DHA and docosapentaenoic acid (DPA) in transgenic *sFat-1* mice [25]. In the transgenic pig, Lai et al. (2006) reported that there was a fivefold reduction of the n−6/n−3 ratio in tail tissues of *hfat-1* transgenic piglets. The other tissues from transgenic piglets also showed a substantially lower n−6/n−3 ratio [2]. Wu et al. successfully cloned transgenic cow expressing *fat-1* gene by traditional SCNT, with increased levels of n−3 PUFAs in the ear tissue, and reduced ratio of n−6/n−3 in the milk [9]. Our previous results also showed that nematode *mfat-1* could effectively

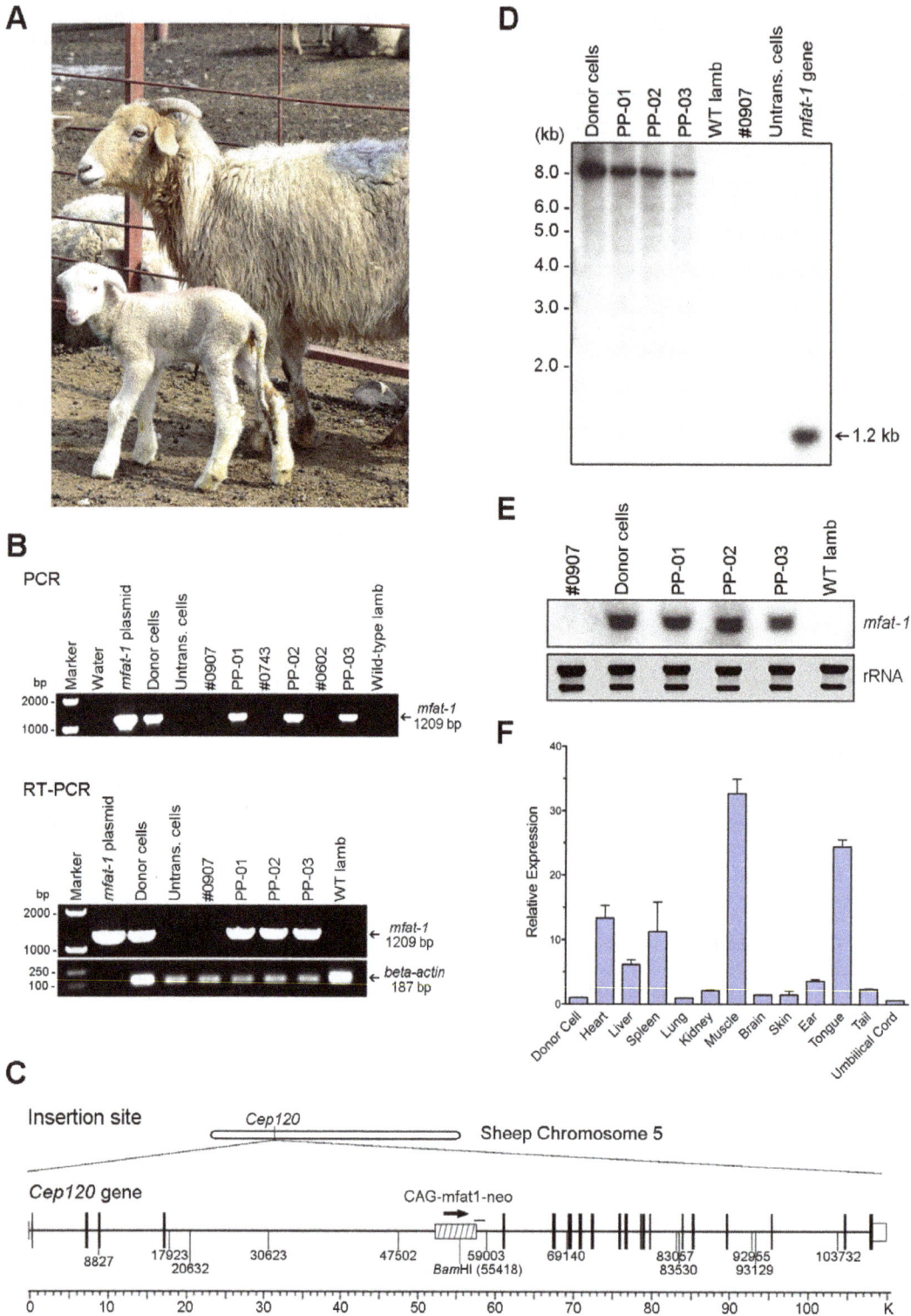

Figure 2. Production of transgenic lambs by handmade cloning. (**A**) The recipient #0907 and the transgenic lamb (PP-01). (**B**) Detection of the *mfat-1* gene in umbilical cord samples of three cloned lambs by PCR and RT-qPCR. (**C**) Insertion site of *mfat-1* vector in the sheep genome. Arrows indicate the *mfat-1* transcriptional direction, which is identical with the endogenous putative sheep *Cep120* gene. PCR fragment obtained in this study are shown by black bars. (**D**) Southern blot using the ^{32}P-labled *mfat-1* specific sequence as a probe to hybridize the genomic DNA from the transgenic donor cells and the cloned lambs. The genomic DNA was digested with *Bam*HI before the gel electrophoresis. (**E**) Northern blot analysis. Total RNAs were loaded on each lane (15 μg per sample) and the coding region of *mfat-1* was used as a probe. Shown below is the gel electrophoresis of rRNA as control. (**F**) Quantitative PCR analysis of *mfat-1* expression in major tissues from the transgenic lambs (PP-02). Compared with the mRNA expression level normalized to the donor cell, the highest level of *mfat-1* expression was observed in transgenic muscle sample.

Table 2. Nuclear transfer efficiencies of the handmade cloning.

Donor cells		No. of reconstructed embryos	No. and (%) of cloned embryos developed to blastocysts	No. of blastocysts transferred	No. of recipients	No. and (%) of pregnant	No. and (%) of lambs born
H-6-6	Passage 8	196	27 (13.8%)	13	6	0	0
	Passage 9	178	12 (6.7%)	9	5	2 (40.0%)	2 (22.2%)
	Passage 10	229	15 (6.6%)	11	6	0	0
	Passage 11	210	18 (8.6%)	13	7	1 (14.3%)	0
	Passage 12	112	10 (8.9%)	7	5	1 (20%)	1 (14.3%)
Total		925	82 (8.9±2.9%)	53	29	4 (13.8%)	3 (5.7%)

lower the $n-6/n-3$ ratio in muscle and other major organs of the transgenic pig [3].

In summary, we used the modified HMC procedure to generate a group of transgenic sheep carrying a functional nematode *mfat-1* gene, and the resulting sheep were rich in omega-3 fatty acids in the muscles and other organ/tissues. Our study demonstrated once again that the alternative HMC is helpful to accelerate the speed of transgenic animal research with cost efficiency and easy in operation. With an increasing demand for dietary intake of omega-3 for a better quality of our life via disease prevention, generation of transgenic animals rich in omega-3 could be an attractive alternative not only for human health, also for environmental conservation.

Materials and Methods

All the chemical reagents used in this research were obtained from Sigma-Aldrich Chemical (St Louis, USA) except for otherwise indicated.

Ethics Statement

This study was carried out following the recommendations in the Guide for the Care and Use of Laboratory Animals of the Chinese Academy of Sciences. Procedures for all of the animal experiments were approved in advance by the Animal Care and Ethics Committee of the Institute of Genetics and Developmental Biology, Chinese Academy of Sciences.

Ovine Primary Fibroblasts Cell Culture

Punch collected ear tissue from a Chinese merino male sheep was washed twice with Ca^{2+}- and Mg^{2+}-free PBS (DPBS), minced with a surgical blade on a 10-cm culture dish (Becton Dickinson, Lincoln Park, USA) and then dissociated with 0.25% (v/v) trypsin-EDTA (GIBCO, New York, USA) containing Dulbecco's modified Eagle's medium (DMEM, GIBCO) at 37°C. Cell pellets after centrifuging were seeded onto a 35 mm culture dish and cultured for 8–10 days in DMEM supplemented with 10% (v/v) fetal bovine serum (FBS, HyClone, Logan, USA) at 37°C and 5% CO_2. The construction of *mfat-1* expression vector CAG-mfat1-neo and the transfection were performed as described previously [3,26]. Cells in each well were trypsinized and seed onto a 10-cm culture dish containing 450 µg/mL Geneticin (Invitrogen, USA) for 24 hrs after the transfection. Drug selection was conducted for 2 weeks with a medium change per 3 days and the Geneticin-resistant colonies were isolated. Each colony was transferred to a 6-well plate for expansion. The resulting clones were frozen at −80°C overnight and stored in liquid nitrogen.

Analysis of *mfat-1* in the Ovine Cell Clones

Geneticin selected cells were analyzed by PCR and RT-PCR as described previously [3]. Quantitative PCR (qPCR) was performed with sequence-specific primer pairs for *mfat-1* and *beta-actin*, respectively. Reagents of Power SYBR® Green PCR Master Mix were used for the PCR amplification in triplicates under the following thermal program: 10 min at 95°C, 40 cycles of 95°C for 15 sec, 60°C for 25 sec, and 72°C for 20 sec.

The *mfat-1* positive cells were cultured in medium supplemented with 10 µM 18:2n−6 and 15 µM 20:4n−6 prior to fatty acid analysis. Cells were collected in a glass methylation tube for fatty acid analysis at 80% confluence. The fatty acid compositions of total cellular lipids were analyzed for both of the transgenic and control clones using the gas chromatography as previously described [27]. Fatty acid methyl esters were quantified using a fully automated 6890 Network GC System (Agilent Technolo-

Table 3. Production of transgenic cloned lambs by HMC.

Lamb name	Donor cell	Recipient no.	Type of delivery (length of gestation)	Birth weight (kg)	Status
PP-01	H-6-6, Passage 9	#0907	Caesarean section (152 day)	5.74	Live (>7 months)
PP-02	H-6-6, Passage 9	#0743	Caesarean section (151 day)	3.86	Dead (2 days)
PP-03	H-6-6, Passage 12	#0602	Naturally (149 day)	5.24	Live (>7 months)

gies, Palo Alto, USA) with an Agilent J&W fused-silica DB-23 capillary column. The injector and detector were maintained at 260°C and 270°C, respectively. The oven program was maintained initially at 180°C for 10 min, then ramped to 200°C at 4°C/min and held for 15 min, finally ramped to 230°C at 10°C/min and maintained for 6 min. Carrier gas-flow rate was maintained at a constant rate of 2.0 mL/min throughput. The peaks were identified by comparison with the internal fatty acid standards, and area percentage for all of resolved peaks was analyzed using the GC ChemStation software (Agilent Technologies, USA).

Procedure Outline for the Handmade Cloning

Sheep ovaries were collected at a local slaughterhouse of Xinjiang Hualing Industry and Trade (Group) Co. Ltd., with the authorization from the business administration for the collection and usage of ovary tissues on the transgenic research. The fresh ovary tissues were transferred to a portable incubator for processing within 120 min and the oocytes were isolated by the ovarian slicing method described previously [28]. Prior to slicing, ovaries were washed in phosphate buffered saline (PBS), and then placed in a Petri dish and covered with Hepes-buffered TCM-199 (GIBCO, USA) containing 60 IU/mL heparin. The compact cumulus-oocytes complexes (COCs) were selected and washed twice in Hepes-buffered TCM-199. Each group of 50 COCs was matured in 400 µL bicarbonate-buffered TCM-199 supplemented with 10% (V:V) fetal bovine serum (FBS), 2 mM L-Glutamine, 0.3 mM sodium pyruvate, 0.1 mM cysteine, 5 µg/mL follicle stimulating hormone, 5 µg/mL luteinizing hormone, 1 µg/mL β-

Table 4. Microsatellite analysis of three transgenic cloned lambs, donor cell, and recipient.

Loci	Lamb PP-01	Lamb PP-02	Lamb PP-03	Donor cell H-6-6	Recipient #0907	Recipient #0743
BMS460	122	122	122	122	122	122
	138	138	138	138	138	122
AE129	145	145	145	145	147	147
	147	147	147	147	149	149
MB009	142	142	142	142	142	142
	142	142	142	142	142	146
ETH3	103	103	103	103	95	103
	103	103	103	103	103	103
TGLA53	132	132	132	132	126	132
	132	132	132	132	126	132
INRA063	172	172	172	172	164	164
	178	178	178	178	172	172
ADCYC	246	246	246	246	246	244
	246	246	246	246	246	244
TGLA126	116	116	116	116	122	116
	116	116	116	116	126	116
MAF209	121	121	121	121	121	109
	127	127	127	127	121	121
TGLA122	146	146	146	146	158	138
	158	158	158	158	158	138
PZ963	253	253	253	253	227	209
	253	253	253	253	227	209
BMS2079	109	109	109	109	109	109
	109	109	109	109	115	109
BMS2104	147	147	147	147	147	147
	147	147	147	147	153	147

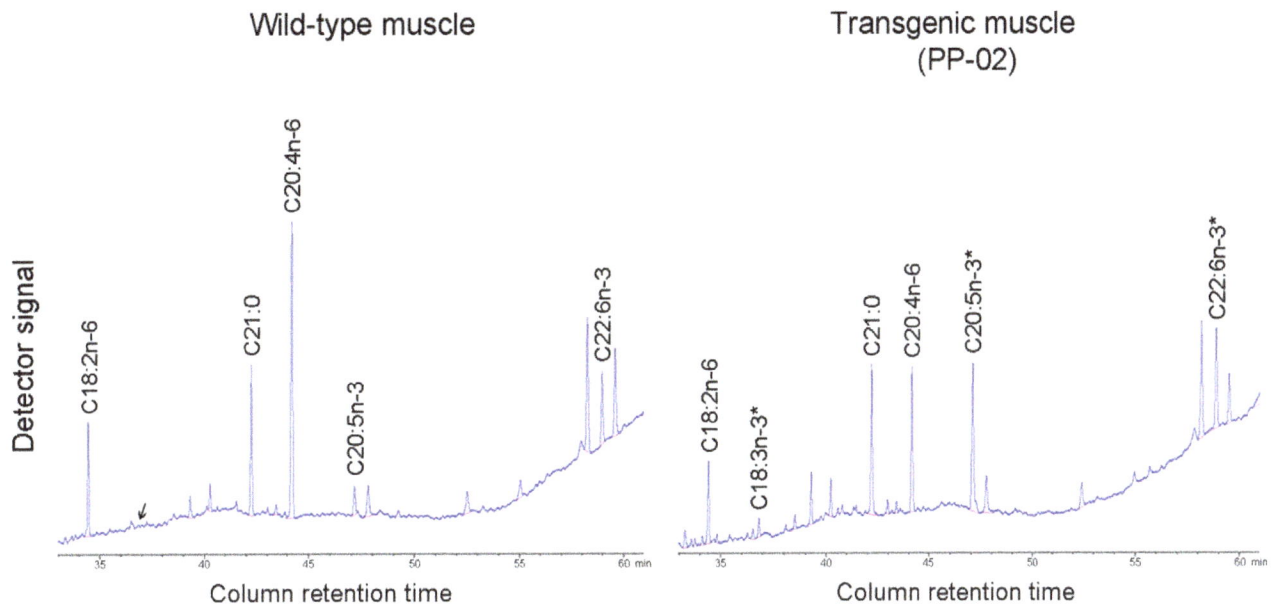

Figure 3. Partial gas chromatograph of fatty acids in muscle sample of *mfat-1* transgenic (PP-02) and the control lamb. Fatty acid methyl esters were quantified using a fully automated 6890 Network GC System with an Agilent J&W fused-silica DB-23 capillary column. The peaks were identified by comparison with the internal fatty acid standards, and area percentage for all of resolved peaks was analyzed using GC ChemStation software. Compared with the wild-type control (left), the level of n−6 polyunsaturated acids in the transgenic muscle (right) are significantly lower, whereas n−3 fatty acids are abundant.

Estradiol in an incubator maintained at 38.6°C with humidified air and 5% CO_2.

The HMC was performed essentially as previously described [13,15], with minor modifications to adapt for the ovine nuclear transferring. Briefly, at 21–22 h after the starting point of maturation, the cumulus cells were removed by repeated pipetting in 1 mg/mL hyaluronidase in Hepes-buffered TCM-199. All the

Table 5. Comparison of n−6/n−3 ratios between wild-type and *mfat-1* transgenic lamb.

Organs or Tissues	n−6/n−3 Ratio[a]	
	Wild-type	Transgenic
Heart	2.28±0.01	0.93±0.02**
Liver	1.56±0.05	0.66±0.04**
Spleen	3.04±0.23	1.50±0.16**
Lung	2.66±0.46	1.40±0.12*
Kidney	2.45±0.05	1.36±0.15**
Brain	0.42±0.01	0.32±0.01**
Ear	2.89±0.02	1.18±0.25**
Tongue	1.66±0.13	1.25±0.02*
Tail	4.60±1.14	2.08±0.63*
Muscle	2.63±0.05	1.24±0.06**

Ratio of n−6/n−3 fatty acid was calculated from n−6 fatty acids [linoleic acid (LA, 18:2n−6) and arachidonic acid (AA, 20:4n−6)] versus n−3 fatty acids [α-linolenic acid (ALA, 18:3n−3), eicosapentaenoic acid (EPA, 20:5n−3), and docosahexaenoic acid (DHA, 22:6n−3)]. Each value represents the mean ± standard deviation from three replicated sample measurements of each tissue.
[a]Statistical analysis using the two tailed student *t*-test. Significant differences between the wild-type and the transgenic lamb samples were marked ($*P<0.05$; $**P<0.01$).

remaining manipulations were performed on a hotplate adjusted to 30°C. Zonae pellucidae were partially digested with 8 mg/mL pronase solution dissolved in T33 (T for Hepes-buffered TCM 199 medium, and the number for the percentage of calf serum supplement), then washed quickly in T2 and T20 drops. The oocytes with softened zonae pellucidae were lined up in T20 drops supplemented with 2.5 µg/mL cytochalasin B (CB), and were enucleated by oriented bisection with an ultra sharp microblade (AB Technology, Pullman, USA) under a stereomicroscope. Approximately 1/3−1/2 cytoplasm close to the polar body was removed manually, and the remaining putative cytoplast were washed twice in T2 drops and collected in a T10 drop. The nuclear donor cells were trypsinized and kept in T2. Fusion was performed in two steps where the second step included the initiation of activation. For the first step, half of the available cytoplasts were transferred into 1 mg/mL of phytohaemagglutinin (PHA; ICN Pharmaceuticals, Australia) dissolved in T0, and then each one was quickly dropped over a single fibroblast. After attachment, cytoplast-fibroblast pairs were equilibrated in a fusion medium (0.3 M mannitol and 0.01% polyvinyl alcohol). Using an alternative current (AC) of 4 V (CF-150/B fusion machine; BLS, Budapest, Hungary), cell pairs were aligned to the wire of a fusion chamber (BTX microslide 0.5 mm fusion chamber, model 450; BTX, SanDiego, USA) with the somatic cells farthest from the wire, then fused with a single direct current (DC) of 1.3 kV/cm for 10 µsec. After the pulse, cytoplast-fibroblast pairs were incubated in T10 drops to examine whether or not the fusion had occurred. Approximately 1 h after the first fusion, each pair was fused with the second cytoplasts and activated simultaneously in activation medium (0.3 M mannitol, 0.1 mM $MgSO4$, 0.1 mM $CaCl_2$ and 0.01% polyvinyl alcohol) with a double DC pulse of 0.9 kV/cm, each pulse for 15 µsec and 1 sec apart. When fusion had been observed in T10 drops, the reconstructed embryos were first incubated in 40 µL Hepes-buffered TCM 199 medium containing 5 µM Ionomycin for 5 min at 38.6°C. After subsequent washing in culture medium 3–5 times, reconstructed embryos were incubated in culture medium containing 2 mM 6-dimethyla-mino-purine (6-DMAP) for 4 h at 38.6°C in 5% CO_2 with maximum humidity, then washed thoroughly with IVC medium for culture.

The reconstructed embryos were *in vitro* cultured in wells of a Nunc 4-well dish in 400 μL SOFaaci medium [29] supplemented with 4 mg/mL BSA and covered with oil. Zona-free embryos produced from the HMC were cultured in modified WOWs system [30,31] at 38.6°C in 5% CO_2 with saturated humidity. Cleavage and blastocyst formation of the reconstructed embryos were monitored during the *in vitro* culture for 7 days.

Embryo Transfer and Pregnancy Diagnosis

The healthy ewes from a local Xinjiang sheep breed aged 12–18 months were selected as the embryo-transfer recipients. The blastocysts with clearly visible inner cell mass were surgically transplanted into the uterine horns of a naturally cycling ewe on 6 or 7 days of standing estrus by a certified veterinarian. Pregnancies were diagnosed by ultrasonography on day 45 after the surgical embryo transfer.

Screening of the Transgenic Sheep

Umbilical cord from each newborn lamb was collected and the genomic DNA was isolated for the genotyping and genetic screening. Presence of *mfat-1* gene in each DNA sample was testified by SP-PCR and Southern blot. The expression of *mfat-1* mRNA was evaluated by RT-PCR and Northern blot.

Comparative genotype analyses were performed for each of the newborn lambs, the recipients, and the donor fibroblasts using a set of microsatellite markers. Thirteen polymorphic microsatellite loci (BMS460, AE129, MB009, ETH3, TGLA53, INRA063, AD-CYC, TGLA126, MAF209, TGLA122, PZ963, BMS2079, BMS2104) located on different ovine chromosomes were amplified by 3-color multiplex PCR and the products were analyzed on a 3130XL Genetic Analyzer (Applied Biosystems) with Gene-Mapper ID Software v3.2 (Applied Biosystems, USA).

For Southern blot analysis, approximately 20 μg of genomic DNA was digested with *Bam*HI for agarose gel electrophoresis. The probes were ^{32}P -labeled PCR fragments of *mfat-1* sequence, generated using the Random Prime Labeling System Redi PrimeTMII (GE Healthcare, Piscataway, USA). Northern blot analysis was performed as described previously [32]. About 15 μg of total RNA were loaded per lane, and the coding region of *mfat-1* was used as a probe.

To identify the insertion site, the genomic DNA flanking inserted *mfat-1* was amplified by a thermal asymmetric interlaced PCR (TAIL-PCR) using Genome Walking Kit (TaKaRa, Japan). The specific primers (CAG4823-1F: 5′-cctcgtgctttacggtatcg-3′;

CAG-4939-2F: 5′-caacctgccatcacgagatt-3′ and CAG5044-3F: 5′-tctcatgctggagttcttcg-3′) were designed with their melting temperatures sufficiently high to ensure a maximum thermal asymmetric priming and to ensure the cloning of genomic DNA flanking the *mfat-1* gene. The tertiary TAIL-PCR products were purified, cloned into pMD18-T (TaKaRa, Japan), and sequenced. The BLAST homology searches and the sequence comparisons with the draft sheep genome assembly OARv2.0 (International Sheep Genomics Consortium) were performed to identify the site of *mfat-1* insertion at the level of nucleotide sequences.

To evaluate the functional activity of desaturase encoded by nematode *mfat-1* gene, the tissues from one of the transgenic lambs (PP-02) was utilized as representative samples for fatty acid analyses. The major organ/tissue samples (heart, liver, spleen, lung, kidney, tongue, ear, muscle, brain, and tail) were collected from both transgenic and age-matched wild-type lamb, and the fatty acid analysis was performed as described previously. Briefly, the oven program of the gas chromatography instrument was maintained initially at 100°C for 5 min, then ramped to 180°C at 6°C/min and held for 10 min, then ramped to 205°C at 2°C/min and held for 10 min, finally ramped to 225°C at 2°C/min and maintained for 7 min. The ratio of n-6/n-3 from each organ/tissue was used for comparison between the transgenic and the wild-type lambs.

Statistical Analysis

Statistical analysis was performed using a two-tailed Student's *t*-test, and p-values <0.05 were considered statistically significant.

Supporting Information

Figure S1 Image of *mfat-1* transgenic lambs of approximately 75 days (PP-01) and 100 days (PP-03) after the birth. The lamb (PP-02) was used for n−3 fatty acid composition analyses of major organ/tissues approximately 3 days after the birth, due to obvious weakness at the time of birth.

Acknowledgments

We gratefully thank Min Han for the gift of *C. elegans* culture stock; Mingjun Liu, Juncheng Huang and Bin Han for the animal surgery facility and technical assistance.

Author Contributions

Conceived and designed the experiments: YD RZM. Performed the experiments: PZ PL HD LC. Analyzed the data: PZ LXC. Contributed reagents/materials/analysis tools: LL PT GV JG. Wrote the paper: RZM.

References

1. Riediger ND, Othman RA, Suh M, Moghadasian MH (2009) A systemic review of the roles of n-3 fatty acids in health and disease. J Am Diet Assoc 109: 668–679.
2. Lai L, Kang JX, Li R, Wang J, Witt WT, et al. (2006) Generation of cloned transgenic pigs rich in omega-3 fatty acids. Nat Biotechnol 24: 435–436.
3. Zhang P, Zhang Y, Dou H, Yin J, Chen Y, et al. (2012) Handmade cloned transgenic piglets expressing the nematode fat-1 gene. Cell Reprogram 14: 258–266.
4. Simopoulos AP (2000) Human requirement for N-3 polyunsaturated fatty acids. Poult Sci 79: 961–970.
5. Kang JX (2005) From fat to fat-1: a tale of omega-3 fatty acids. J Membr Biol 206: 165–172.
6. Spychalla JP, Kinney AJ, Browse J (1997) Identification of an animal omega-3 fatty acid desaturase by heterologous expression in Arabidopsis. Proc Natl Acad Sci U S A 94: 1142–1147.
7. Kang ZB, Ge Y, Chen Z, Cluette-Brown J, Laposata M, et al. (2001) Adenoviral gene transfer of Caenorhabditis elegans n–3 fatty acid desaturase optimizes fatty acid composition in mammalian cells. Proc Natl Acad Sci U S A 98: 4050–4054.
8. Kang JX, Wang J, Wu L, Kang ZB (2004) Transgenic mice: fat-1 mice convert n-6 to n-3 fatty acids. Nature 427: 504.
9. Wu X, Ouyang H, Duan B, Pang D, Zhang L, et al. (2012) Production of cloned transgenic cow expressing omega-3 fatty acids. Transgenic Res 21: 537–543.
10. Wilmut I, Schnieke AE, McWhir J, Kind AJ, Campbell KH (1997) Viable offspring derived from fetal and adult mammalian cells. Nature 385: 810–813.
11. Vajta G, Lewis IM, Hyttel P, Thouas GA, Trounson AO (2001) Somatic cell cloning without micromanipulators. Cloning 3: 89–95.
12. Vajta G, Lewis IM, Trounson AO, Purup S, Maddox-Hyttel P, et al. (2003) Handmade somatic cell cloning in cattle: analysis of factors contributing to high efficiency in vitro. Biol Reprod 68: 571–578.
13. Vajta G, Bartels P, Joubert J, de la Rey M, Treadwell R, et al. (2004) Production of a healthy calf by somatic cell nuclear transfer without micromanipulators and

carbon dioxide incubators using the Handmade Cloning (HMC) and the Submarine Incubation System (SIS). Theriogenology 62: 1465–1472.

14. Kragh PM, Vajta G, Corydon TJ, Purup S, Bolund L, et al. (2004) Production of transgenic porcine blastocysts by hand-made cloning. Reprod Fertil Dev 16: 315–318.

15. Du Y, Kragh PM, Zhang Y, Li J, Schmidt M, et al. (2007) Piglets born from handmade cloning, an innovative cloning method without micromanipulation. Theriogenology 68: 1104–1110.

16. Kragh PM, Nielsen AL, Li J, Du Y, Lin L, et al. (2009) Hemizygous minipigs produced by random gene insertion and handmade cloning express the Alzheimer's disease-causing dominant mutation APPsw. Transgenic Res 18: 545–558.

17. Luo Y, Li J, Liu Y, Lin L, Du Y, et al. (2011) High efficiency of BRCA1 knockout using rAAV-mediated gene targeting: developing a pig model for breast cancer. Transgenic Res 20: 975–988.

18. Schnieke AE, Kind AJ, Ritchie WA, Mycock K, Scott AR, et al. (1997) Human factor IX transgenic sheep produced by transfer of nuclei from transfected fetal fibroblasts. Science 278: 2130–2133.

19. Lee JH, Peters A, Fisher P, Bowles EJ, St John JC, et al. (2010) Generation of mtDNA homoplasmic cloned lambs. Cell Reprogram 12: 347–355.

20. Xie Z, Moy LY, Sanada K, Zhou Y, Buchman JJ, et al. (2007) Cep120 and TACCs control interkinetic nuclear migration and the neural progenitor pool. Neuron 56: 79–93.

21. Mahjoub MR, Xie Z, Stearns T (2010) Cep120 is asymmetrically localized to the daughter centriole and is essential for centriole assembly. J Cell Biol 191: 331–346.

22. Pan D, Zhang L, Zhou Y, Feng C, Long C, et al. (2010) Efficient production of omega-3 fatty acid desaturase (sFat-1)-transgenic pigs by somatic cell nuclear transfer. Sci China Life Sci 53: 517–523.

23. Duan B, Cheng L, Gao Y, Yin FX, Su GH, et al. (2012) Silencing of fat-1 transgene expression in sheep may result from hypermethylation of its driven cytomegalovirus (CMV) promoter. Theriogenology 78: 793–802.

24. Hiendleder S, Zakhartchenko V, Wolf E (2005) Mitochondria and the success of somatic cell nuclear transfer cloning: from nuclear-mitochondrial interactions to mitochondrial complementation and mitochondrial DNA recombination. Reprod Fertil Dev 17: 69–83.

25. Zhu G, Chen H, Wu X, Zhou Y, Lu J, et al. (2008) A modified n-3 fatty acid desaturase gene from Caenorhabditis briggsae produced high proportion of DHA and DPA in transgenic mice. Transgenic Res 17: 717–725.

26. Zhang P, Cao LX, Chen H, Ma RL (2007) Prokaryotic expression of the C. elegans fat-1 gene and preparation of antibody. J Northwest A & F Univ (Nat Sci Ed) 35: 24–28.

27. Kang JX, Wang J (2005) A simplified method for analysis of polyunsaturated fatty acids. BMC Biochem 6: 5.

28. Bhojwani S, Vajta G, Callesen H, Roschlau K, Kuwer A, et al. (2005) Developmental competence of HMC(TM) derived bovine cloned embryos obtained from somatic cell nuclear transfer of adult fibroblasts and granulosa cells. J Reprod Dev 51: 465–475.

29. Holm P, Booth PJ, Schmidt MH, Greve T, Callesen H (1999) High bovine blastocyst development in a static in vitro production system using SOFaa medium supplemented with sodium citrate and myo-inositol with or without serum-proteins. Theriogenology 52: 683–700.

30. Vajta G, Peura TT, Holm P, Paldi A, Greve T, et al. (2000) New method for culture of zona-included or zona-free embryos: the Well of the Well (WOW) system. Mol Reprod Dev 55: 256–264.

31. Feltrin C, Forell F, dos Santos L, Rodrigues JL (2006) In vitro bovine embryo development after nuclear transfer by handmade cloning using a modified wow culture system. Reprod Fertil Dev 18: 126.

32. Chen L, Liu K, Zhao Z, Blair HT, Zhang P, et al. (2012) Identification of Sheep Ovary Genes Potentially Associated with Off-season Reproduction. J Genet Genomics 39: 181–190.

17

Regulation of Endothelial-Specific Transgene Expression by the *LacI* Repressor Protein *In Vivo*

Susan K. Morton[1,2,❧], Daniel J. Chaston[1,2,❧], Brett K. Baillie[1], Caryl E. Hill[1], Klaus I. Matthaei[2]*

1 Blood Vessel Laboratory, The John Curtin School of Medical Research, The Australian National University, Canberra, Australia, 2 Stem Cell & Gene Targeting Laboratory, The John Curtin School of Medical Research, The Australian National University, Canberra, Australia

Abstract

Genetically modified mice have played an important part in elucidating gene function *in vivo*. However, conclusions from transgenic studies may be compromised by complications arising from the site of transgene integration into the genome and, in inducible systems, the non-innocuous nature of inducer molecules. The aim of the present study was to use the vascular system to validate a technique based on the bacterial *lac* operon system, in which transgene expression can be repressed and de-repressed by an innocuous lactose analogue, IPTG. We have modified an endothelium specific promoter (*TIE2*) with synthetic *LacO* sequences and made transgenic mouse lines with this modified promoter driving expression of mutant forms of connexin40 and an independently translated reporter, EGFP. We show that tissue specificity of this modified promoter is retained in the vasculature of transgenic mice in spite of the presence of *LacO* sequences, and that transgene expression is uniform throughout the endothelium of a range of adult systemic and cerebral arteries and arterioles. Moreover, transgene expression can be consistently down-regulated by crossing the transgenic mice with mice expressing an inhibitor protein *LacI^R*, and in one transgenic line, transgene expression could be de-repressed rapidly by the innocuous inducer, IPTG. We conclude that the modified bacterial *lac* operon system can be used successfully to validate transgenic phenotypes through a simple breeding schedule with mice homozygous for the *LacI^R* protein.

Editor: David S. Milstone, Brigham and Women's Hospital, United States of America

Funding: CEH and KIM acknowledge funding support from NH&MRC Project Grant #471421. The funders had no role in study design, data collection and analysis, decision to publish, or preparation of the manuscript.

Competing Interests: The authors have declared that no competing interests exist.

* E-mail: klaus.matthaei@anu.edu.au

❧ These authors contributed equally to this work.

Introduction

For several decades, genetically modified mice have been the premium tool for the study of gene function *in vivo*. However, interpretation of the specific role of candidate genes can often be compromised in transgenic studies by the random nature of the site of transgene insertion into the host DNA [1,2]. This non-specific "position effect" can result from transcriptional regulation imposed on the transgene by the adjacent host sequences, or alternatively, from the loss of function of genes which are accidentally interrupted by insertion of the transgene [1,2].

Attempts to overcome these problems have employed inducible systems, such as the Cre-LoxP or Doxycycline-controlled methods, in which expression of the transgene can be regulated at specific times by an inducer molecule. Thus, phenotypes due to the transgene are isolated from effects due simply to the site of genetic insertion (for reviews see [3,4]). However, both of these systems employ non-innocuous inducers for the switching of gene expression. The inducible CreERT system requires tamoxifen and the Doxycycline system utilises the antibiotic tetracycline, both of which can have off-target effects (reviewed in [4]). Tamoxifen, in particular, as a selective modulator of both membrane bound and intracellular estrogen receptors, is capable of altering vascular tone and blood pressure [5]; effects which are not conducive with uncovering a role for specific genes in cardiovascular function.

With the intention of creating an alternative system which utilises an innocuous inducer, Scrable and colleagues have modified the bacterial *lac* operator system for use in mammalian systems and reported reversible control of the tyrosinase gene and eye pigmentation, as well as the Huntington promoter driving a luciferase reporter, using the lactose analogue, Isopropyl β-D-1-thiogalactopyranoside (IPTG) [6–8]. Their system relied on the introduction into the tyrosinase and Huntington promoters of *Lac* operon (*LacO*) sequences at specific distances spanning the transcription start site (see also [3]); a procedure which did not interfere with promoter function and specificity. However, in the presence of the LacI repressor protein (*LacI^R*), the function of the promoter was repressed due to the binding of the *LacI^R* protein to the *LacO* sequences thereby blocking transcription. Following addition of IPTG, the *LacI^R* protein was modified by the IPTG to undergo a conformational change that weakened its ability to bind to the *LacO* sequences, thus permitting transcription factors to again bind normally and reinstate transcription. In this manner, gene function was controlled in a reversible manner using an agent that had no deleterious effect even after 8 months of continuous administration [6]. Surprisingly, only one report has appeared in the intervening decade, and this investigation, which utilised the haematopoietic system, failed to confirm that the *LacI^R*/*LacO* system could tightly and reversibly control gene expression [9]. However, results of this study were complicated by mosaic transgene expression amongst littermates within the same transgenic line [9]; effects likely due to the use of F2 progeny, in which multiple transgene insertion sites still existed. Furthermore, experiments in which IPTG was unable to reinstate transgene expression *in vivo* employed only a single bolus exposure to IPTG [9].

In the present study, we have re-evaluated the *LacI*^R^/*LacO* system in the vasculature using transgenic mouse strains in which transgene inheritance had stabilised. Our gene of interest was the endothelial gap junctional protein, connexin40 (Cx40), for which identification of a role in the control of vascular function and blood pressure has been complicated by co-ordinate regulation of other connexin family members (e.g. [10]). We have created transgenic mouse lines that express mutant forms of Cx40 and an independently translated reporter, EGFP, selectively in the endothelium, through use of the endothelial specific promoter *TIE2* and its enhancer [11].

We show that addition of *LacO* sequences to the *TIE2* promoter does not interfere with uniform transgene expression in the endothelium of a range of adult systemic and cerebral arteries and arterioles and that repression of promoter activity occurs reliably in the presence of *LacI*^R^, although the capacity for re-activation by daily administration of IPTG appears to vary with transgenic line. Nevertheless, the robust tissue-specific expression of transgenes directed by promoters modified with *LacO* sequences, combined with the ability to significantly repress transgene function through breeding to mice ubiquitously expressing *LacI*^R^, makes this an attractive system to validate vascular phenotypes, ruling out "position effects" due to random transgene insertion.

Materials and Methods

Ethics Statement

All experimental animal procedures were carried out in strict accordance with the recommendations in the Australian code of practice for the care and use of animals for scientific purposes of the National Health and Medical Research Council of Australia. Protocols were approved by the Animal Experimentation Ethics Committee of the Australian National University (JMB.33.07; A2011/72).

Generation of *TIE2LacO*-Cx40-EGFP Constructs

Constructs were created in which Cx40 and Enhanced Green Fluorescent Protein (EGFP) could be expressed separately through the use of an internal ribosome entry site (IRES) which allows the expression of a bicistronic mRNA and production of two independent proteins from the same promoter. This construct was coupled to the endothelial specific promoter, *TIE2*, as described below. The transgenic constructs were composed, in 5′ to 3′ order, of a modified *TIE2* promoter, the coding region of Cx40 with single base mutation, an internal ribosomal entry sequence (IRES), EGFP, human growth hormone poly A tail and a *TIE2* enhancer fragment (Figure 1A). Two different single base mutations encoding amino acids at positions 152 (T152A) and 202 (T202S; [12]) were incorporated into separate constructs.

LacO sites were introduced into the *TIE2* promoter to enable regulation of transgene expression. The *TIE2* enhancer fragment was necessary for endothelial expression of the *TIE2* promoter throughout embryogenesis and adulthood [11].

***TIE2LacO* construct.** The plasmid pHHSDKXK [11], containing the *TIE2* promoter and a minimal enhancer fragment, as well as the plasmid pg50–2.11 containing the full length *TIE2* enhancer, were generous gifts from Professor Tom Sato, The University of Texas, Southwestern Medical Center at Dallas,

A. Cx40Tg

B. Cx40Tg x *LacI*^R^

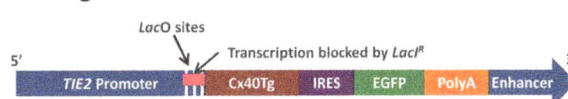

C. Cx40Tg x *LacI*^R^ + IPTG

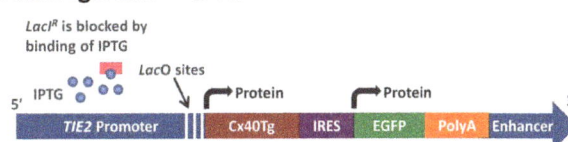

Figure 1. Control of transgene function by the *LacO*/*LacI*^R^ system. (A) Construct used to create Cx40 transgenic mice consists of a modified *TIE2* promoter containing *LacO* transcriptional regulation sites (positions −117, −25 and +67), Cx40 transgene (Cx40Tg), an internal ribosomal entry sequence (IRES), EGFP, PolyA tail and *TIE2* Enhancer element. Proteins expressed from this construct are translated separately due to the presence of the IRES sequence. (B) When bred with *LacI*^R^ mice, expression of the Cx40Tg should be repressed due to binding of the *LacI*^R^ protein (pink) to the *LacO* sites incorporated into the *TIE2* promoter. (C) In the presence of IPTG (blue circles) *LacI*^R^ undergoes a conformational change lessening its affinity for the *LacO* sites and transcription can proceed.

Texas, USA. A 2 kb *TIE2* promoter fragment was recovered from pHHSDKXK by Hind III digestion. This was inserted into pBluescript KS II (pBKSII) by restricting with Hind III and Bam HI, ligating the insert to the vector Hind III site, blunting the vector BamH I and the fragment Hind III sites, and religating (with loss of the BamH I restriction site) to give pBKSII*TIE2*. The correct orientation of the *TIE2* promoter was determined by sequencing.

A BamH I – Nae I fragment was removed from this plasmid and replaced with a synthetic fragment of the *TIE2* promoter now containing *LacO* sites. This synthetic BamH I-Nae I *LacO* fragment, pUC57*TIE2LacO* (synthesised by Genscript, USA) was designed with three 18 bp *LacO* sites (5′-ATT-GTG-AGC-GCT-CAC-AAT-3′), that were introduced to span the transcription start site, with 92 bp intervals, at positions –117, −25 and +67 of the *TIE2* promoter. These locations were determined not to contain known transcription factor binding sites (MatInspector). The resulting construct, *TIE2LacO*, contained the full length *TIE2* promoter with 3 *LacO* sites.

***TIE2LacO*-Cx40-EGFP construct.** Digestion of *TIE2LacO* with Cla I, followed by blunting of the Cla I site, and partial digestion with Bcl I allowed recovery of the 2.2 kb *TIE2LacO* fragment. This was ligated into the Sac II digested, blunted, then BamH I-digested 7.7 kb backbone fragment of pBKSII containing the Cx40-IRES-EGFP construct.

The *TIE2LacO*-Cx40-EGFP plasmid was linearised by digestion with Xho I, blunted, and ligated to the 10 kb *TIE2* enhancer fragment from Nae I and Sal I-digestion of pg50–2.11. The final construct contained the enhancer in the forward orientation after the polyadenylation signal (Figure 1A). Sequencing was performed at each step to confirm the presence, fidelity, and orientation of construct components.

Generation of transgenic mice

TIE2LacO-Cx40Tg-EGFP mice. The final construct was digested with Pvu I and BssH II and gel purified to remove vector sequence prior to use in the generation of transgenic mice. Transgenic C57BL/6 mice were produced by pronuclear injection of the 19 kb construct (TASQ, University of Queensland, Australia). Transgenic founders were identified by PCR.

Selection of transgenic lines. Copy number of the transgene was determined in founder mice (N0) and offspring of the next three generations (N1-N3), using quantitative real time PCR (qPCR) for the reporter gene EGFP relative to a known single copy gene, transferrin. EGFP expression was quantified using Western Blotting. Mice with different transgene copy numbers were bred as separate colonies from N1, N2 and N3 generations to confirm segregation and stabilisation of different copy numbers.

LacIR mice. Mice from this strain were developed to be homozygous for a transgene that expresses the $LacI^R$ inhibitor protein under control of the ubiquitous β-actin promoter [8]. The heterozygous $LacI^R$ strain was a generous gift from Professor Heidi Scrable, University of Virginia, Charlottesville, Virginia, USA. Homozygous mice were identified from heterozygous matings using qPCR of genomic DNA for $LacI^R$. Homozygosity was determined by test breeding with wildtype mice and by quantifying expression of $LacI^R$ protein in offspring by Western Blotting.

TIE2LacO-Cx40Tg-EGFP/LacI^R mice were generated by crossing the *TIE2LacO-Cx40Tg-EGFP* mice with homozygous $LacI^R$ mice (Figure 1B) [8].

Quantitative Real Time PCR (qPCR)

qPCR to determine EGFP Transgene Copy Number. Two standard reference curves were generated. The first was a series of 2-fold dilutions of wildtype mouse genomic DNA, from 8 ng/μl to 0.25 ng/μl, in $T_{10}E_{0.1}$ containing 4 ng/μl tRNA (Sigma) as carrier. The mouse transferrin gene was used as a single copy reference [13] using the primers detailed in Table 1. The second standard used purified plasmid DNA containing the entire *TIE2LacO-Cx40-EGFP* transgene construct. Serial 10-fold dilutions of the plasmid DNA were made from 2.5×10^{-2} ng/μl to 2.5×10^{-7} ng/μl, again using $T_{10}E_{0.1}$ containing 4 ng/μl tRNA as carrier. The primers for EGFP are detailed in Table 1.

Genomic DNA from each mouse to be tested was quantified using the high sensitivity QUBIT assay (Invitrogen) and diluted to 4 ng/μl in $T_{10}E_{0.1}$. Each experimental sample was prepared in triplicate and amplified for EGFP and transferrin. Each qPCR run contained the 6 standard reference samples for transferrin and for EGFP, 15 test samples and a negative control without DNA, all in triplicate.

Each qPCR reaction contained 10μl of 2x Sensimix+SYBR (No ROX) master mix (Bioline), 5μM forward and reverse primer, 5μl of DNA sample and water to a total volume of 20μl. PCR was performed in a Corbett Research Rotor-Gene RG-3000 Real Time Analyser with cycling conditions as follows: 95°C for 10 minutes, followed by 40 cycles of 95°C for 10 s, annealing temp (see Table 1) for 20 s and 72°C for 20 s. Melt curves were performed after each run to ensure the amplification of a single product. The data was analysed using the Rotagene 6.0 software and copy number calculated from the standard curves.

qPCR to determine LacI Zygosity. Genomic DNA from a mouse known to be heterozygous for the $LacI^R$ transgene was quantified using the QUBIT (Invitrogen) and diluted to 16 ng/μl in $T_{10}E_{0.1}$ containing 4 ng/μl tRNA as carrier. A standard curve ranging from 8 ng/μl to 0.25 ng/μl genomic DNA was prepared by 2-fold serial dilutions in triplicate in $T_{10}E_{0.1}$ containing 4 ng/μl tRNA as carrier diluent. Transferrin was again used as a single copy gene reference [13].

DNA samples from test mice were prepared as above in triplicate and amplified with the $LacI^R$ and transferrin primers as detailed in Table 1. The ratio of $LacI^R$ to transferrin for the known heterozygote was normalised to 1.00. Test samples were normalised accordingly and the ratio used as a predictor of zygosity.

Western blot detection of EGFP

Mice were anaesthetised with isoflurane, decapitated, and mesenteric arteries were removed and stripped of fat in ice cold phosphate buffered saline (PBS) prior to being snap frozen in liquid nitrogen. Frozen tissues were pulverised under liquid nitrogen in the presence of 100μl of frozen extraction buffer (37 mM TRIS, 0.5% lithium dodecyl sulphate, 2.5% glycerol, 0.13 mM EDTA, 0.06 mM SERVA blue G250, 0.04 mM phenol red, 50 mM DTT, 1X cOmplete mini Protease Inhibitor Cocktail [Roche], pH 8.5). Frozen extracts were allowed to thaw on ice and the resulting liquid extract was transferred to a plastic sample tube. The remaining sample was washed from the mortar using a further 50μl of sample buffer and recovered into the same plastic sample tube. Extracts were then heated at 70°C for 5 min, vortexed and heated for a further 5 min before being stored at −20°C until analysis.

Proteins present in tissue extracts were resolved by SDS denaturing electrophoresis on NuPAGE 4–12% BIS-TRIS gradient gels using NuPAGE MES buffer according to the manufacturer's recommended protocols (Invitrogen). Proteins were transferred onto PVDF membranes at 180 mA in NuPAGE transfer buffer for 4 h at RT.

Membranes were recovered, and air dried at room temperature for 1 h. Membranes were rehydrated with 100% methanol, followed by 50% methanol in PBS for 5 minutes each.

Table 1. Primer sequences and conditions for quantitative PCR.

Primer	Sequence	Annealing Temp	Amplicon size
Transferrin F	5′-ATCCCAGTGACTACAACGGTTCCA-3′	63°C	189bp
Transferrin R	5′-AGACCACAACATGGTTTGGAGCTT-3′		
EGFP F	5′-TCTATATCATGGCCGACAAGCAGA-3′	65°C	165bp
EGFP R	5′- ACTGGGTGCTCAGGTAGTGGTTGT 3′		
$LacI^R$ F	5′-GGTGTCTCTTATCAGACTGTTTCCA-3′	67°C	206bp
$LacI^R$ R	5′-GCTGCCACAATTTGAGATGGTGCA-3′		

Membranes were then blocked with 2% BSA in PBS for 1 hour at RT. Membranes were washed 6 times with PBST (PBS, 0.05% Tween-20) for 10 minutes each. Primary antibody: mouse anti-GFP monoclonal IgG1 (kindly supplied by Jan Elliot, Research School of Biology, Australian National University); mouse anti-LacI (Millipore), mouse anti-α-smooth muscle actin (αSMA, Sigma), rabbit anti-vonWillebrand Factor (vWBF, Dako); was applied to the blot (1:1000 anti-GFP; 1:2000 anti-LacI; 1:10000 anti-αSMA; 1:2000 anti-vWBF; diluted in 2% BSA in PBST) for 16 hrs at 4°C. Membranes were washed 6 times with PBST for 10 minutes. Secondary antibody (goat anti-mouse HRP conjugate, Millipore, or goat anti-rabbit HRP conjugate, Sigma) was applied to the blot (1:2000 for EGFP; 1:2000 for LacI; 1:10000 for αSMA, 1:2000 for vWBF; diluted in 2% BSA in PBST) for 2 h at RT. Membranes were washed 6 times in PBST, followed by 1 wash in PBS, each for 10 minutes. The signal was developed using Immobilon Western chemiluminescent substrate according to the manufacturer's recommended protocol (Millipore). Image recording was performed on a LAS1000 system (Fujifilm). The EGFP +ve control was from a mouse expressing an EGFP tagged with a nuclear localization signal and the EGFP protein therefore has a slightly higher molecular weight than standard EGFP.

Band signal intensity was measured using ImageQuant TL Software Version 7.0 (GE Healthcare). EGFP and LacI quantification was normalised to either the smooth muscle marker, α-actin, or the endothelial marker, vWBF expression, respectively.

Immunohistochemical detection of EGFP

Mice were anaesthetised with isoflurane, decapitated and kidneys, mesenteric and basilar arteries removed. Kidneys were placed in ice cold PBS and the capsules were removed and the kidneys were cut into 3–4 mm transverse slices. Mesenteric arteries were placed into ice-cold PBS and surrounding fat removed. All tissues were immersion fixed in 2% paraformaldehyde and 0.01% sodium nitrite (100 mM sodium phosphate buffer) for 10 min, then cryprotected in 30% sucrose in PBS overnight. Tissues were finally embedded in Tissue Tek O.C.T (Sakura Fintek USA, Torrance, CA), frozen and cut into 30μm sections using a cryostat.

Cremaster arterioles. Cremaster muscles were removed from mice which were anaesthetised (1 mg/kg medetomidine, 10 mg/kg midazolam, and 0.1 mg/kg Fentanyl, i.p.) and continuously perfused with anaesthetic (0.2 mg/h medetomidine, 0.2 mg/h midazolam, 0.002 mg/h fentanyl) via a jugular vein cannula during the procedure. Isolated cremaster muscles were fixed in 2% paraformaldehyde and cryosections prepared as above. Mice were euthanised by cervical dislocation whilst still deeply anaesthetised.

Tissue sections were preincubated for 30 min in PBS containing 2% bovine serum albumin, 0.2% Triton-X100, and 0.04% sodium azide before being incubated at room temperature for 24 h with rabbit anti-GFP antibodies (Invitrogen) diluted 1:1000 in the pre-incubation solution. Sections were washed three times in PBS and incubated for 2 h with Alexa-488 conjugated donkey anti-rabbit antibodies (Invitrogen) diluted 1:400 with PBS containing 0.02% Triton-X100. Sections were then incubated in PBS containing 0.01% pontamine sky blue for 3 min to shift the green autofluorescence of the vascular internal elastic lamina to longer wavelengths. Finally, sections were washed three times with PBS before being mounted in buffered glycerol.

Confocal microscopy and image analysis

Image series of all stained tissues were taken with a Leica SP2 confocal scanning microscope. For time course studies, the same acquisition settings were used to collect images to enable the degree of staining to be compared at different time points. For quantification of EGFP staining in the kidney, image series were taken using a 20x lens with z plane increments of 1.5 μm. For each image series, an average projection image was generated and three such composite images from each animal were used to quantify EGFP expression.

Quantification of EGFP staining. For each image, areas of endothelial tissue were identified by their location on the inner surface of the auto-fluorescence generated by the internal elastic lamina (red). Using the polygon selection and histogram functions in ImageJ (National Centre for Biotechnology Information), the intensity of EGFP staining (green) was calculated. For each image the EGFP signal within several non-endothelial tissues such as tubules and vascular smooth muscle were also calculated. EGFP expression was quantified as signal to noise ratio minus one using the equation, (Endothelial Cell Intensity/Background Intensity)-1.

IPTG Treatment of TIE2LacO-Cx40Tg-EGFP/LacI^R mice

The TIE2LacO-Cx40Tg-EGFP/LacI^R mice were given IPTG (Inalco, USA; Figure 1C) with doses ranging from 10 to 80 mM in the drinking water for periods of 2 to 14 days. Sucrose (2%) was added to encourage consumption at higher concentrations (40 and 80 mM); an approach commonly used to reduce the bitter taste of Doxycycline [14]. Controls received 2% sucrose in water. Doses as high as 10% sucrose have previously been reported to have no effect on blood glucose, insulin or adiponectin levels, nor on glucose tolerance [15]. Following treatment, mice were euthanised and analysed for EGFP expression by Western Blotting or immunohistochemistry.

Statistical Analysis

Data are presented as means ± SEM where n represents the number of mice. Statistical significance was determined by one-way ANOVA with Bonferonni post-hoc test for multiple comparisons, with $P<0.05$ considered significant.

Results

Transgene Inheritance is unstable in early generations

Transgene copy number was monitored in the T152ATg mouse strain over the first generation (N1) and found to vary from 4 to 22 copies. Subsequent breeding to C57BL/6 wildtype mice showed that this variability continued into the next two generations before stabilizing with breeder selection at 12 copies. The production of equal numbers of transgenic to wildtype offspring in each generation was consistent with a single integration site, suggesting some instability of the multiple transgene copies at the one insertion site.

The T202STg strain was bred for 6 generations to wildtype mice and copy number stabilized at 360 copies. At this stage, equal numbers of transgenic to wildtype offspring were still found in each generation, again consistent with a single integration site.

Insertion of LacO sites in the TIE2 promoter does not compromise endothelial specificity

In order to determine the cellular distribution of the transgene expression, we conducted immunohistochemical studies in the vasculature using antibodies against the reporter EGFP. We did not study Cx40 expression as antibodies against Cx40 were unable to discriminate the Cx40 transgene from wildtype Cx40, which is highly expressed in the vascular endothelium [16].

No staining was detected in arteries or arterioles of wildtype mice (Figure 2A, C, E, G, I). In contrast, EGFP was detected uniformly in the endothelium of arteries and arterioles of several different systemic and cerebral circulations of transgenic mice, while staining was absent from the arterial smooth muscle or adventitia. Vessels studied included mesenteric (Figure 2B) and basilar (Figure 2D) arteries, as well as cremaster muscle arterioles (Figure 2F), renal arteries, arterioles and glomerular capillaries (Figure 2H, J).

Homozygous $LacI^R$ mice can be reliably identified by qPCR

Since repression of transgene expression required interbreeding of transgenic mouse strains ($TIE2LacO$-Cx40Tg-EGFP mice and $LacI^R$ mice), we developed a homozygous $LacI^R$ strain to ensure that all offspring in each generation would express $LacI^R$ and 50% of these mice would also be transgenic for Cx40. Our first aim was therefore to develop a rapid assay to detect homozygous offspring from heterozygous matings and validate the assay by test breeding and Western blot protein assay.

Offspring from matings of heterozygous $LacI^R$ mice were used to provide samples for qPCR using primers and standard curves for both $LacI^R$ and the single copy gene, transferrin. $LacI^R$/transferrin values were calculated for each mouse and zygosity values obtained by normalizing these values to that of a known heterozygote $LacI^R$ mouse, whose zygosity was defined as 1.0. When data from 136 offspring were plotted, a bimodal distribution was found, consistent with a large heterozygous and smaller homozygous population (Figure 3A). From this distribution we defined a homozygote as any animal with a zygosity value ≥ 1.5.

Figure 3B shows that interbreeding of 4 heterozygous mice, defined by zygosity values < 1.5, produced a total of 42 offspring of which 11 were wildtype (zygosity = 0), 21 were heterozygotes (zygosity < 1.5) and 10 were homozygotes (zygosity ≥ 1.5). These numbers accord well with Mendelian segregation which would predict a ratio of 10.5 wildtype: 21 heterozygotes: 10.5 homozygotes from a heterozygous mating producing the same number of offspring.

Subsequent breeding of 5 mice with zygosity values ≥ 1.5 (1.5, 1.8, 1.8 1.9, 2.1) to wildtype mice, each produced 100% $LacI^R$ transgenic mice (41 offspring), with zygosity values < 1.5 (Figure 3B), as expected for a homozygous/wildtype mating.

Western blotting showed that mesenteric arterial samples taken from mice with zygosity values ≥ 1.5 showed higher $LacI^R$ expression than samples from mice with zygosity values < 1.5 (Figure 3C). Quantification of $LacI^R$ bands, relative to the endothelial marker vWBF, showed that $LacI^R$ expression was 2-fold higher in samples from mice with zygosity values of ≥ 1.5 compared to those from mice with zygosity values of < 1.5 (Figure 3D), consistent with assignation of homozygosity and heterozygosity to zygosity values of ≥ 1.5 and < 1.5, respectively.

$LacI^R$ protein significantly reduces transgene expression in vivo

When $TIE2LacO$-Cx40Tg-EGFP mice were interbred with $LacI^R$ mice, EGFP expression in primary mesenteric arteries of both transgenic strains, T152A$LacI^R$ and T202S$LacI^R$, was reduced significantly by 71% and 100%, respectively (Western blots; T152A$LacI^R$: Figure 4A, C; T202S$LacI^R$: Figure 4B, D).

Immunohistochemical analyses of kidney sections of T152A$LacI^R$ and T202S$LacI^R$ mice showed that endothelial specific expression of EGFP was similarly reduced by 65% and 91%, compared to T152ATg and T202STg mice, respectively (T152A$LacI^R$: Figure 5A, C, E, K; T202S$LacI^R$: Figure 5B, D, F, L).

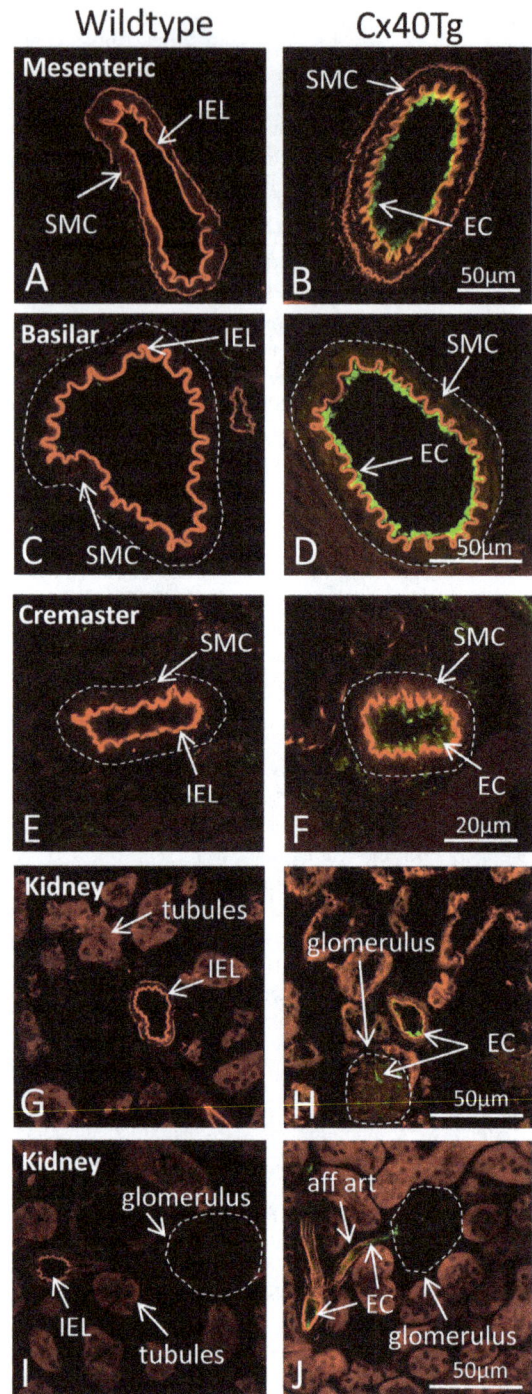

Figure 2. Maintenance of endothelial tissue specificity of the $TIE2$ promoter when modified with $LacO$ sites. EGFP was expressed in the endothelial cells (EC) of primary mesenteric arteries and basilar arteries of transgenic mice (Cx40Tg; B, D), as well as in cremaster muscle arterioles (F) and kidney arteries, arterioles and glomerular capillaries (H, J). EGFP was not detectable in arterial smooth muscle (SMC) or in renal tubules. No EGFP was detected in comparable vessels of wildtype mice (A, C, E, G, I). IEL represents the internal elastic lamina; aff art, afferent arteriole. Examples from Cx40T152ATg (B, J) and Cx40T202STg mice (D, F, H).

Expression of $LacI^R$ protein was not significantly different between mesenteric arterial samples taken from T152A$LacI^R$ and

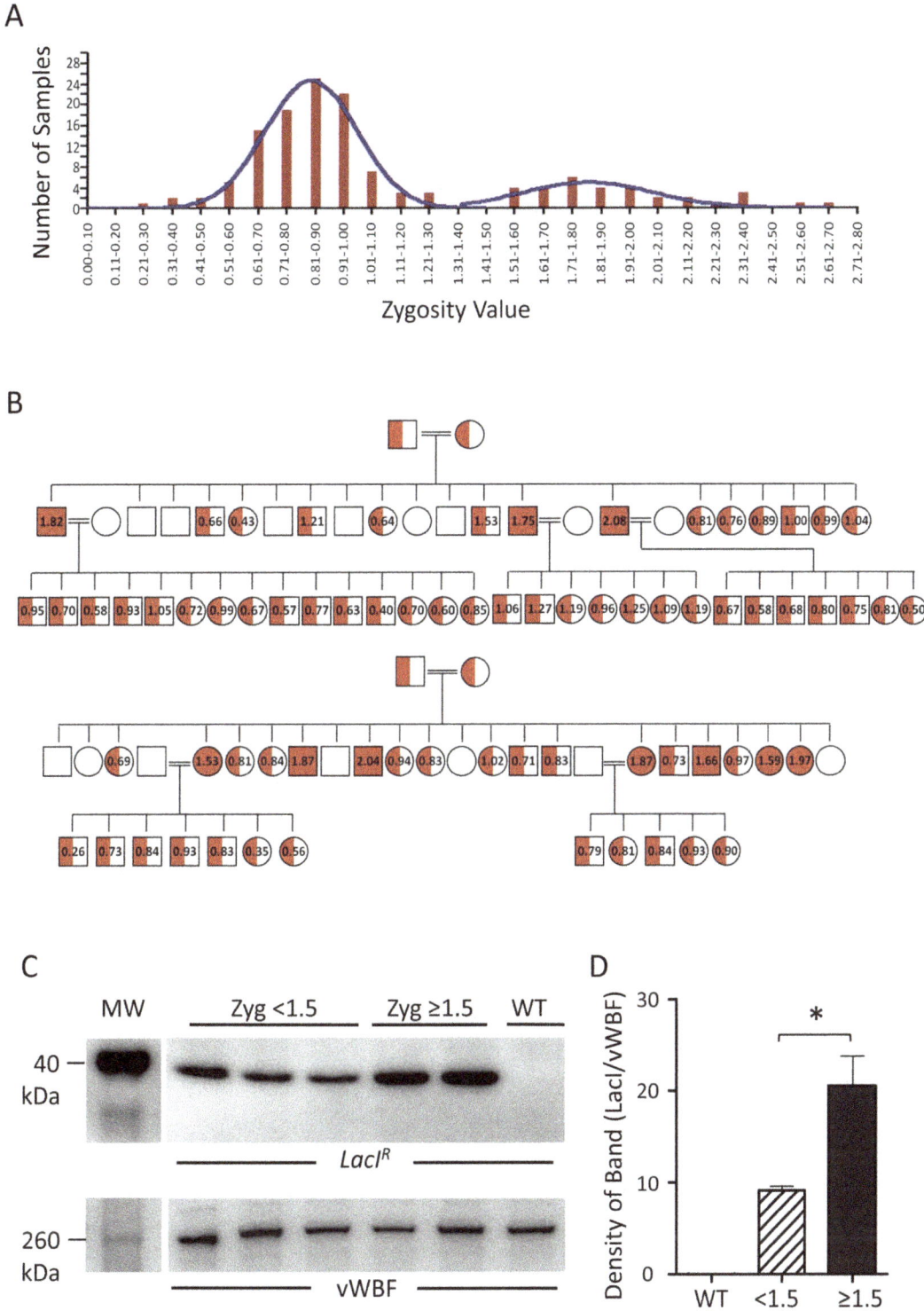

Figure 3. Detection of *LacI^R* homozygosity using quantitative PCR. (A) Binning of zygosity values from heterozygous matings demonstrates two populations indicative of heterozygotes and homozygotes. (B) Interbreeding of predicted heterozygotes produces wildtype, heterozygote and homozygote populations in Mendelian ratios. (C) *LacI^R* expression in predicted homozygotes (zygosity ≥1.5) is greater than in predicted heterozygotes (zygosity <1.5). Representative blots and group data for 3 heterozygote and 5 homozygote samples. WT wildtype. *P<0.05 denotes significant difference.

T202S*LacI^R* mice, when quantified relative to the smooth muscle marker, α-actin (T152A*LacI^R* mice: 3.2±0.5 *LacI^R*/actin, n = 9 mice; T202S*LacI^R* mice: 3.1±0.3 *LacI^R*/actin, n = 6 mice).

Together, these data show that the *LacI^R* protein is expressed to the same extent in the 2 transgenic mouse strains and can reliably and significantly reduce transgene expression *in vivo*.

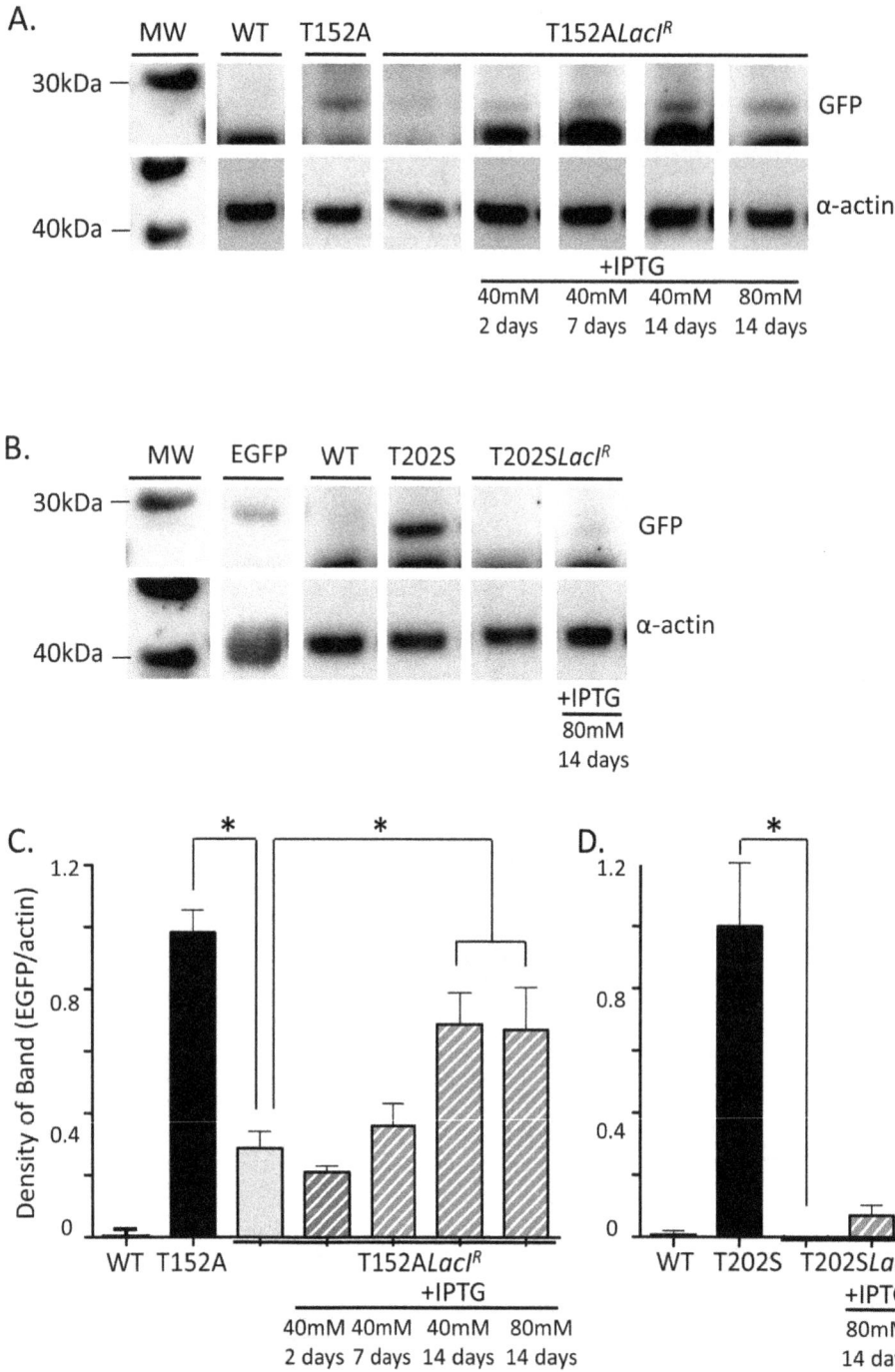

Figure 4. Control of EGFP expression by *Lacl^R* and IPTG: Western blotting. Western blots of mesenteric arterial samples demonstrate repression of EGFP expression by breeding of Cx40 (T152ATg and T202STg) transgenic mice with *Lacl^R* mice (T152A*Lacl^R* mice: A, C; T202S*Lacl^R* mice: B, D). De-repression of EGFP by IPTG is found in T152A*Lacl^R* mice after 14 days (A, C), but not in T202S*Lacl^R* mice (B, D). (C). Group data (left to right): n = 6, 9, 11, 5, 5, 8, 7 mice. (D). Group data (left to right): n = 4, 6, 2, 4 mice. WT, wildtype. Positive control for EGFP is from a mouse expressing EGFP tagged with a nuclear localisation signal and therefore has a slightly higher molecular weight. *P<0.05 denotes significant difference; ANOVA followed by Bonferonni tests for multiple groups.

De-repression of transgene expression can be achieved by administration of IPTG

Preliminary experiments testing low concentrations of IPTG (10 and 20 mM) produced variable induction of transgene expression (data not shown). To encourage consumption of higher concentrations of IPTG (40 and 80 mM), 2% sucrose was added to the

drinking water. Measurement of blood glucose before and after 2% sucrose treatment confirmed that blood glucose levels were not altered by sucrose addition for 7 or 14 days (Before: 7.8±0.4 mM, n=8 mice; 7 day sucrose: 8.1±0.3 mM, n=8 mice; 14 day sucrose: 6.8±0.3 mM, n=8 mice). Addition of either 40 mM or 80 mM IPTG with 2% sucrose to the drinking water did also not

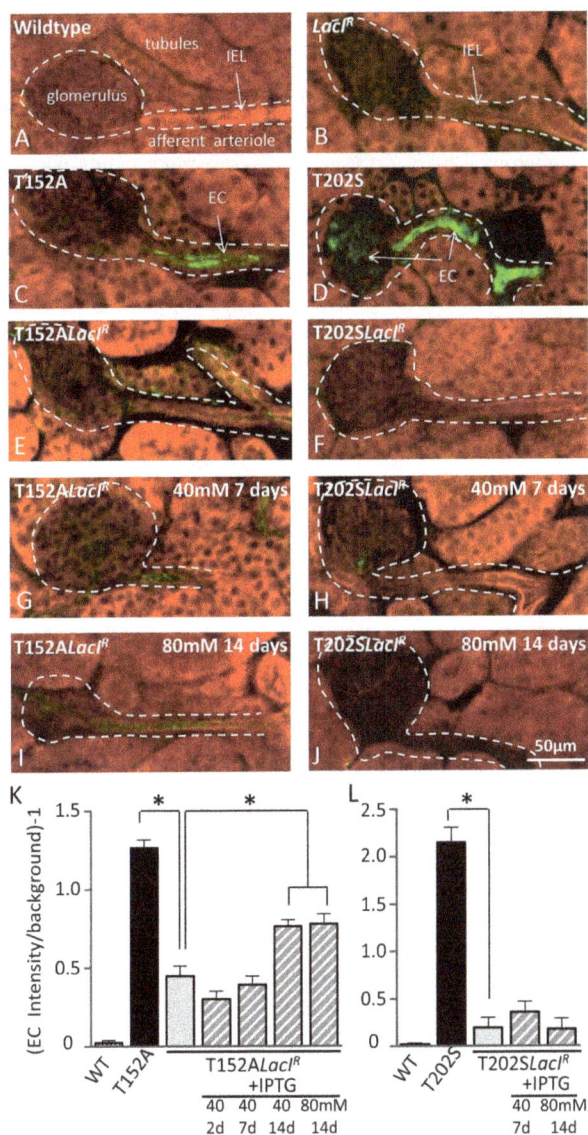

Figure 5. Control of EGFP expression by *LacI*[R] and IPTG: Immunohistochemistry. Immunohistochemical staining for EGFP demonstrates endothelial (EC) staining in T152A (C) and T202S transgenic mice (D) but not in wildtype (A) or *LacI*[R] (B) mice. Interbreeding of transgenic mice with *LacI*[R] mice led to a significant reduction in EGFP staining in both transgenic strains (E, F). De-repression of EGFP by IPTG is found in T152A*LacI*[R] mice after 14 days (I, K), but not in T202S*LacI*[R] mice (J, L). (K) Group data (left to right): n = 7, 12, 7, 4, 5, 7, 5 mice. (L) Group data (left to right): n = 6, 8, 4, 8, 4 mice. *P<0.05 denotes significant difference; ANOVA followed by Bonferonni tests for multiple groups.

significantly higher than in untreated T152A*LacI*[R] mice and had recovered to 70% of that in T152A transgenic mice (Figure 4C). When assayed immunohistochemically in kidney sections, endothelial transgene expression similarly increased over 14 days, with expression after 14 days' treatment significantly increased compared to untreated T152A*LacI*[R]mice and recovery to 60% of that in T152A transgenic mice (Figure 5E, G, K). Doubling the concentration of IPTG to 80 mM did not produce any further increase in transgene expression after 14 days' treatment (Western blot: Figure 4A, C; Immunohistochemistry: Figure 5I, K).

In contrast, administration of 40 mM or 80 mM IPTG to T202S*LacI*[R] mice for 14 days failed to significantly induce transgene expression, assayed either by Western blotting (Figure 4B, D) or immunohistochemistry (Figure 5F, H, J, L). Moreover, the minimally increased transgene expression that was seen in T202S*LacI*[R] mice, not only varied between animals, but also amongst vessels in the same animal (data not shown).

Discussion

The present study has tested a modified bacterial *lac* operon system, developed originally by Scrable and colleagues [6–8], to reversibly control gene expression in the vasculature *in vivo*. We show that this system can be used to reliably repress gene expression selectively in the vascular endothelium, through the use of the modified endothelial specific promoter *TIE2* [11]. This ability thus enables the unambiguous attribution of phenotype to the transgene, eliminating any influence of the site of transgene integration into the genome. Our data also show that transgene expression can be de-repressed rapidly by the innocuous inducer, IPTG, in one transgenic line.

Modification of the promoter with *LacO* sites

Modification of the *TIE2* promoter with three *LacO* sites and generation of transgenic mice did not alter the tissue specificity of this promoter in directing expression of the EGFP reporter. The uniform, endothelial cell specific protein expression for EGFP found here in several different arteries and arterioles of adult mice was entirely consistent with the expression pattern reported in the literature for the *TIE2* promoter with its enhancer element [11]. These data thus extend the findings of Scrable and colleagues in the use of this system, from more generalised promoters [6–8] to a vascular specific promoter, to show that tissue specificity is not compromised by the addition of three *LacO* sequences spanning the transcription start site. However, although our data suggests that a large number of different promoters might be amenable to such a modification, their specificity will have to be carefully verified after introduction of the *LacO* sites.

In contrast to the uniform transgene expression that we found with the *TIE2* promoter, Grespi and colleagues reported variable expression amongst different immune cell populations and different tissues when directed by the *Vav-gene* promoter [9]. While some of this variegation may have resulted from the persistence of multiple transgene insertion sites in F2 progeny, it should also be noted that the *Vav-gene* construct contained viral sequences, both in the intron, as well as the PolyA signal. Mosaic expression might therefore be expected as these sequences would be subjected to silencing through methylation. It is possible that our constructs avoided such modifications since they consisted entirely of mammalian DNA sequences.

Silencing of transgene expression in vivo with *LacI*[R]

Expression of EGFP in both of the transgenic mouse lines established in this study was significantly repressed by interbreed-

significantly alter the average volume of water drunk by each mouse per day (2% sucrose in water: 4.5±0.6 ml, n = 6 mice; 40 mM IPTG+ sucrose: 4.6±0.5 ml, n = 9 mice; 80 mM IPTG+ sucrose: 5.1±0.5 ml, n = 12 mice). Mice given IPTG did not show any abnormalities in regard to coat appearance, hydration, grooming, general activity, or food and water consumption.

Administration of 40 mM IPTG to T152A*LacI*[R]mice over a period of 14 days led to a gradual increase in transgene expression in mesenteric arteries, as measured by Western blotting (Figure 4A, C). After 14 days of IPTG treatment, transgene expression was ing with a mouse line ubiquitously expressing the repressor protein $LacI^R$. In T202S$LacI^R$ mice, expression was fully repressed, while in T152A$LacI^R$ mice, transgene expression was repressed by 70%. Our current studies of cardiovascular function have confirmed that repression in both transgenic lines is extended in a similar manner to the physiological phenotypes resulting from expression of the mutant endothelial Cx40 transgenes. We see altered blood pressure which is significantly reversed by $LacI^R$ in both strains of mice (data under Journal submission). Such repression of transgene expression by $LacI^R$ *in vivo* enables the unequivocal attribution of phenotypes to transgenes, thus eliminating artefacts due to the site of integration into the genome. The development and validation in this study of a simple and reliable method for identifying homozygous transgenic $LacI^R$ mice using qPCR, further facilitates the use of this system by increasing the probability of production of the requisite doubly transgenic mice.

De-repression of transgene expression in vivo using IPTG

Of surprise was the finding that the ability of IPTG to reinstate transgene expression varied between the 2 transgenic mouse strains. Thus, in T152A$LacI^R$ mice, daily administration of IPTG led to a rapid reinstatement of transgene expression to 70% of full activity, while in the T202S$LacI^R$ mice, transgene expression was variable and inconsistent. A correlation thus existed between the extent of repression by the repressor protein $LacI^R$ and the ability of IPTG to reinstate transgene expression. The variation in transgene regulation was unexpected, as the T152A$LacI^R$ and T202S$LacI^R$ transgenic mice were produced using the same promoter construct with the three LacO sites located in the same positions. Indeed, the only differences in the constructs were the two, single nucleotide changes within the connexin gene sequence: a change unlikely to affect the ability of a promoter to regulate transgene expression.

The variability in de-repression *in vivo* of the T152A and T202S transgenes suggests that the activity of the *TIE2Lac*O promoter may have been affected by the site of transgene integration. However, this is unlikely since the use of the *TIE2* enhancer sequences, in conjunction with the *TIE2* promoter, has been shown to ensure consistent and endothelial cell specific activity [11,17,18], as we confirm here. Differences in the availability of the $LacI^R$ protein to the LacO sequences also seem unlikely, since $LacI^R$ protein expression was identical in mesenteric arteries of the T152A$LacI^R$ and T202S$LacI^R$ transgenic mice. Finally, variability in access and activity of IPTG in the target endothelial cells is unlikely, as IPTG was replenished daily over the 14 day exposure period and previous studies have shown that IPTG can remain active after crossing both the blood-brain barrier and the placenta [8].

In spite of the similarity in the 2 transgenic constructs, a major difference that did exist between the T152A and T202S transgenic mouse strains was transgene copy number. Interestingly, the high copy number strain (T202S) was the one in which transgene

expression could be completely repressed by $LacI^R$ but not de-repressed by IPTG. Since it is known that high copy number transgenic mice have more highly compacted heterochromatin (see for e.g.[19,20]) and that the $LacI^R$ protein forms tetramers between a number of different LacO sites [21], it is possible that the numerous T202S transgene copies are more tightly packed and more compressed by $LacI^R$, than the less numerous T152A transgene copies. It would then follow that this conformation could not be de-repressed by IPTG. Indeed if correct, this phenomenon could be an advantage for future studies where different founders could be chosen based on copy number, allowing a gene to be totally repressed in high copy number founders, and de-repressed in low copy number founders. However, more founder lines with varying copy numbers are required to confirm this association.

Conclusions

Using a vascular cell type specific promoter, we have shown that the repressor protein $LacI^R$ of the bacterial *lac* operon system, when modified for use in mammalian cells, can reliably repress transgene expression *in vivo*, thus enabling the attribution of phenotype to the transgene, rather than to the insertion site. Moreover, the introduction into the promoter of LacO sites does not compromise tissue specificity, making this an attractive system to validate transgenic phenotypes. While rapid de-repression of transgene expression by the innocuous inducer, IPTG, has been reported previously for both the tyrosinase and Huntington promoters [6–8], in the present study this only occurred reliably in one of our transgenic lines, even though the constructs were essentially identical. Should studies in which de-repression be of interest, for example, to study gene expression in adulthood without developmental complications, the development of several founders may be required to identify a line in which this feature of the system is possible. In this connection, the inverse correlation between the degree of repression by $LacI^R$ and the ability of IPTG to reinstate expression, a feature also described in a previous study [9], may be a useful index to rapidly identify transgenic lines in which de-repression by IPTG will be possible.

Acknowledgments

We gratefully acknowledge Professor Heidi Scrable, University of Virginia, Charlottesville USA, for advice about placement of the LacO sites and technical advice on IPTG administration as well as for providing the $LacI^R$ repressor mice. We thank Ms Helen Taylor for expert technical support and Dr T. Hilton Grayson for assistance with construct development.

Author Contributions

Conceived and designed the experiments: KIM CEH. Performed the experiments: SKM DJC BKB. Analyzed the data: SKM DJC BKB CEH KIM. Wrote the paper: KIM CEH. Approved the final manuscript: SKM DJC BKB CEH KIM.

References

1. Feng G, Mellor RH, Bernstein M, Keller-Peck C, Nguyen QT, et al. (2000) Imaging neuronal subsets in transgenic mice expressing multiple spectral variants of GFP. Neuron 28: 41–51.

2. Palmiter RD, Brinster RL (1986) Germ-line transformation of mice. Annu Rev Genet 20: 465–499.

3. Matthaei KI (2004) Caveats of Gene Targeted and Transgenic mice. In: Lanza R, Gearhart J, Hogan B, Melton DW, Pederson R, et al., editors. Handbook of Stem Cells: Elsevier, Academic Press. 589–598.

4. Matthaei KI (2007) Genetically manipulated mice: a powerful tool with unsuspected caveats. J Physiol 582: 481–488.

5. Meyer MR, Prossnitz ER, Barton M (2011) The G protein-coupled estrogen receptor GPER/GPR30 as a regulator of cardiovascular function. Vascul Pharmacol 55: 17–25.

6. Cronin CA, Gluba W, Scrable H (2001) The lac operator-repressor system is functional in the mouse. Genes Dev 15: 1506–1517.

7. Cronin CA, Ryan AB, Talley EM, Scrable H (2003) Tyrosinase expression during neuroblast divisions affects later pathfinding by retinal ganglion cells. J Neurosci 23: 11692–11697.

8. Ryan A, Scrable H (2004) Visualization of the dynamics of gene expression in the living mouse. Mol Imaging 3: 33–42.

9. Grespi F, Ottina E, Yannoutsos N, Geley S, Villunger A (2011) Generation and evaluation of an IPTG-regulated version of Vav-gene promoter for mouse transgenesis. PLoS One 6: e18051.

10. Simon AM, McWhorter AR (2003) Decreased intercellular dye-transfer and downregulation of non-ablated connexins in aortic endothelium deficient in connexin37 or connexin40. J Cell Sci 116: 2223–2236.

11. Schlaeger TM, Bartunkova S, Lawitts JA, Teichmann G, Risau W, et al. (1997) Uniform vascular-endothelial-cell-specific gene expression in both embryonic and adult transgenic mice. Proc Natl Acad Sci U S A 94: 3058–3063.

12. Chaston DJ, Baillie BK, Grayson TH, Courjaret RJ, Heisler JM, et al. (2013) Polymorphism in endothelial connexin40 enhances sensitivity to intraluminal pressure and increases arterial stiffness. Arterioscler Thromb Vasc Biol 33: 962–970.

13. Nakamasu K, Kawamoto T, Yoshida E, Noshiro M, Matsuda Y, et al. (2001) Structure and promoter analysis of the mouse membrane-bound transferrin-like protein (MTf) gene. Eur J Biochem 268: 1468–1476.

14. Bachmanov AA, Reed DR, Beauchamp GK, Tordoff MG (2002) Food intake, water intake, and drinking spout side preference of 28 mouse strains. Behav Genet 32: 435–443.

15. Jürgens H, Haass W, Castaneda TR, Schurmann A, Koebnick C, et al. (2005) Consuming fructose-sweetened beverages increases body adiposity in mice. Obes Res 13: 1146–1156.

16. Hill CE, Phillips JK, Sandow SL (2001) Heterogeneous control of blood flow amongst different vascular beds. Med Res Rev 21: 1–60.

17. Constien R, Forde A, Liliensiek B, Grone HJ, Nawroth P, et al. (2001) Characterization of a novel EGFP reporter mouse to monitor Cre recombination as demonstrated by a Tie2 Cre mouse line. Genesis 30: 36–44.

18. Kisanuki YY, Hammer RE, Miyazaki J, Williams SC, Richardson JA, et al. (2001) Tie2-Cre transgenic mice: a new model for endothelial cell-lineage analysis in vivo. Dev Biol 230: 230–242.

19. Garrick D, Fiering S, Martin DI, Whitelaw E (1998) Repeat-induced gene silencing in mammals. Nat Genet 18: 56–59.

20. Henikoff S (1998) Conspiracy of silence among repeated transgenes. Bioessays 20: 532–535.

21. Scrable H (2002) Say when: reversible control of gene expression in the mouse by lac. Semin Cell Dev Biol 13: 109–119.

18

Incorporation of a Horizontally Transferred Gene into an Operon during Cnidarian Evolution

Catherine E. Dana[1,2], **Kristine M. Glauber**[1,2], **Titus A. Chan**[1,2], **Diane M. Bridge**[3], **Robert E. Steele**[1,2]*

1 Department of Biological Chemistry, University of California Irvine, Irvine, California, United States of America, 2 Developmental Biology Center, University of California Irvine, Irvine, California, United States of America, 3 Department of Biology, Elizabethtown College, Elizabethtown, Pennsylvania, United States of America

Abstract

Genome sequencing has revealed examples of horizontally transferred genes, but we still know little about how such genes are incorporated into their host genomes. We have previously reported the identification of a gene (*flp*) that appears to have entered the *Hydra* genome through horizontal transfer. Here we provide additional evidence in support of our original hypothesis that the transfer was from a unicellular organism, and we show that the transfer occurred in an ancestor of two medusozoan cnidarian species. In addition we show that the gene is part of a bicistronic operon in the *Hydra* genome. These findings identify a new animal phylum in which trans-spliced leader addition has led to the formation of operons, and define the requirements for evolution of an operon in *Hydra*. The identification of operons in *Hydra* also provides a tool that can be exploited in the construction of transgenic *Hydra* strains.

Editor: Ahmed Moustafa, American University in Cairo, Egypt

Funding: The work described in this manuscript was supported by grant 1R24GM080537-01A1 to R.E.S. from the National Institute of General Medical Sciences and grant 1RO1AG037965-01 to D.M.B. from the National Institute of Aging. The funders had no role in study design, data collection and analysis, decision to publish, or preparation of the manuscript.

Competing Interests: The authors have declared that no competing interests exist.

* E-mail: resteele@uci.edu

Introduction

Horizontal transfer of genes is widely accepted as a significant feature of genome evolution in prokaryotes [1], and likely in unicellular eukaryotes as well [2]. However, it has been much more difficult to build convincing cases for horizontal gene transfer (HGT) into animal genomes. The amount of HGT in animals is expected to be much less than in unicellular organisms. This expectation is due to the absence in animals of the facile routes for DNA uptake seen in prokaryotes and the necessity of targeting the germ line, which is segregated in most metazoans, in order for a horizontally transferred gene to be propagated [3].

The most commonly used evidence for HGT is anomalous phylogenetic distribution of the gene being considered. Whether such a distribution constitutes strong evidence for HGT depends on the number of genomes being assessed. A relatively small number of animal genomes have been sequenced, making models invoking gene loss [4] viable contenders for explaining anomalous phylogenetic distributions in animals. Ultimately one makes the argument for HGT based on parsimony – with a single horizontal transfer event being considered more parsimonious than multiple secondary losses of a gene. Without information on how difficult it is for a gene to be lost versus the difficulty of being horizontally transferred into a given animal genome, it is not known whether this assumption is correct.

The genome of the cnidarian *Hydra* presents a potentially fertile hunting ground for identifying horizontal gene transfers into an animal genome. In the adult *Hydra* polyp, all cells are separated from the environment by no more than a single cell layer [5]. Thus all cells are readily exposed to exogenous sources of DNA (e.g. bacteria and unicellular eukaryotes). *Hydra* propagates primarily by asexual

budding and its germ line is not segregated [6,7,8], features that greatly increase the potential for a horizontally transferred gene spreading within the population. Finally, *Hydra* mRNAs undergo trans-spliced leader addition [9], which gives the animal the potential for having operons [10]. Operons provide an opportunity for a gene entering the genome by a horizontal route to "piggy-back" onto an existing gene and thus to be expressed without need for its own promoter. This potential for immediate incorporation into the genetic circuitry of the animal makes it possible for the gene to come under selection for a function quickly upon entering the genome.

Sequencing of the *Hydra* genome has led to the identification of putative horizontal gene transfers from bacteria [11]. Habetha and Bosch [12] have reported the presence of a peroxidase gene in *Hydra* that may have entered by horizontal gene transfer from a plant. We identified a *Hydra* gene, called *flp*, whose only homologues at the time of its discovery were in the genome of the parabasalid protist *Trichomonas vaginalis* [13]. The function of the *flp* gene is unknown, although its expression has been shown to respond to iron levels in *T. vaginalis* [14], and the amino acid composition of the 14.8 kDa *flp* protein suggests that it might be a metal-binding protein [13]. We report here additional findings regarding the *flp* gene that provide insight into its evolutionary history and its incorporation into the *Hydra* genome. In addition, we demonstrate that *Hydra* has operons in its genome, and that the requirements for forming an operon in *Hydra* are surprisingly simple.

Results

Phylogenetic distribution of the *flp* gene

Since our original report of a *Hydra* homologue of the *Trichomonas vaginalis flp* genes [13], *in silico* screens of genome sequences and ESTs have revealed *flp* genes in several additional organisms (Fig. 1). These include a marine bacterium (*Lentisphaera araneosa*), a human gut bacterium (*Akkermansia muciniphila*), a glaucophyte (*Glaucocystis nostochinearum*), a euglenid (*Euglena gracilis*), a pelagophyte (*Aureococcus anophagefferens*), a red alga (*Porphyridium purpureum*), an additional parabasilid protist (*Tritrichomonas foetus*), termite gut symbionts, and a metagenome from the Sargasso Sea. The gene is also present in the cnidarian *Clytia hemisphaerica*. Like *Hydra*, *Clytia* is a member of the subclass Hydroidolina in the cnidarian subphylum Medusozoa [15]. The *flp* gene is absent from the genomes of the anthozoan cnidarians *Nematostella vectensis* [16] and *Acropora digitifera* [17]. The divergence of Anthozoa and Medusozoa occurred at the base of the cnidarian radiation [18,19]. The *flp* gene is also absent from the genomes of the choanoflagellate *Monosiga brevicollis* [20], *Capsaspora owczarzaki* (an opisthokont that diverged between fungi and choanoflagellates) (GenBank Accession Number ACFS00000000), the sponge *Amphimedon queenslandica* [21], the placozoan *Trichoplax adhaerens* [22], the ctenophore *Mnemiopsis leidyi* (Joseph Ryan, personal communication), and all publicly available bilaterian animal, plant, and fungal genome sequence and EST datasets. Fig. 2 shows the evolutionary relationships of the opisthokonts for which we have presence/absence information regarding the *flp* gene.

From a phylogenetic analysis of *flp* protein sequences (Fig. 3), we conclude that the three *flp* genes in *Trichomonas vaginalis* are paralogues, as are the two *flp* genes in *Euglena*. This analysis also shows that the *Clytia* and *Hydra flp* genes are vertically descended from a common ancestor within Medusozoa.

Genomic organization of the *Hydra flp* gene

The *Hydra flp* gene was originally identified from ESTs [13]. Subsequently we amplified the coding sequence of the *flp* gene from *Hydra* genomic DNA and found that it lacks introns, as do all three copies of the *flp* gene in *Trichomonas*. Using the genome assembly from the *Hydra* genome project [11], we identified the region containing the *flp* gene. Immediately upstream of the *flp* gene is a gene encoding the 140 kDa subunit of replication factor C (RFC140) (Fig. 4A). From mapping of ESTs onto the genome sequence and blasting with RFC140 sequences from other metazoans, we determined the coding sequence (GenBank Accession Number FJ154842) and exon/intron structure of the *Hydra* RFC140 gene. From EST sequences, we found that the RFC140 mRNA, like the *flp* mRNA [13], is trans-spliced. RFC140 ESTs containing spliced leaders SL-B1, SL-B3, SL-C, and SL-F3 [9,11] were identified. The stop codon of RFC140 is only 417 nucleotides upstream of the trans-splicing acceptor dinucleotide of the *flp* gene (Fig. 4B). This 417 nucleotide sequence must include the RFC140 3′ UTR (between 63 and 83 nucleotides in length if the predicted polyadenylation signal is correct and polyadenylation begins 10–30 nucleotides downstream from this signal). Thus the distance between the end of the RFC140 gene and the beginning of the *flp* gene is less than 417 nucleotides. This short distance and the fact that the *Hydra flp* mRNA undergoes trans-spliced leader addition [13], suggested that the *flp* gene is in an operon with the RFC140 gene. Upstream of the putative trans-splicing acceptor in the *flp* gene is a T-rich (U-rich in RNA) sequence (doubly underlined in Fig. 4B). Huang et al. [23] have

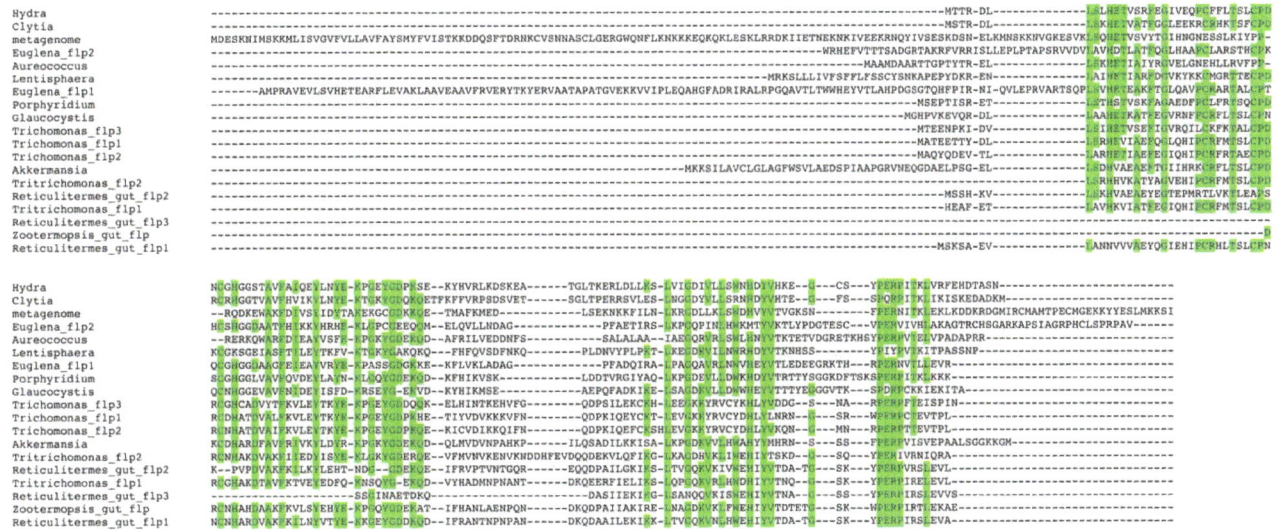

Figure 1. Alignment of *flp* amino acid sequences. Sequences were aligned using MUSCLE [50]. Amino acids highlighted in green are conserved in a majority of the sequences. Sequence sources are as follows: meta, translated from GenBank accession number AACY020544127 (DNA from the Sargasso Sea); Euglena_1, translated from *Euglena gracilis* EST with accession number EC679450; Euglena_2, translated from *Euglena gracilis* EST with accession number EC678321; Glaucocystis, translated from *Glaucocystis nostochinearum* EST with accession number EC122554; Tf_1, translated from *Tritrichomononas foetus* EST with accession number CX156355; Tf_2, translated from *Tritrichomononas foetus* EST with accession number CX157959; Reticulitermes_gut_1, translated from *Reticulitermes flavipes* symbiont ESTs with accession numbers FL643370 and FL643898; Reticulitermes_gut_2, translated from *Reticulitermes flavipes* symbiont ESTs with accession numbers FL637453 and FL638405; Reticulitermes_gut_3, translated from *Reticulitermes flavipes* symbiont EST with accession number GO904605; Tv_1, from *Trichomonas vaginalis* protein XP_001316516; Tv_2, from *Trichomonas vaginalis* protein XP_001322305; Tv_3, from *Trichomonas vaginalis* protein XP_001324076; Clytia, translated from a *Clytia haemispherica* EST with accession number FP933787. Zootermopsis_gut, translated from *Zootermopsis* symbiont EST with accession number EG751663; Aureococcus, from *Aureococcus anophegefferens* protein EGB07434; Lentisphaera, from *Lentisphaera araneosa* protein ZP_01876528; Porphyridium, translated from *Porphyridium purpureum* EST with accession number HS847715; Akkermansia, from *Akkermansia muciniphila* protein YP_001878433. Neither of the two *Euglena* ESTs or the two *Tritrichomonas foetus* ESTs contained a start codon preceded by an in frame stop codon. Thus it is unclear if the amino terminal sequences shown here are correct for these species. The Reticulitermes_gut_3 and Zootermopsis_gut ESTs encode only a portion of the protein.

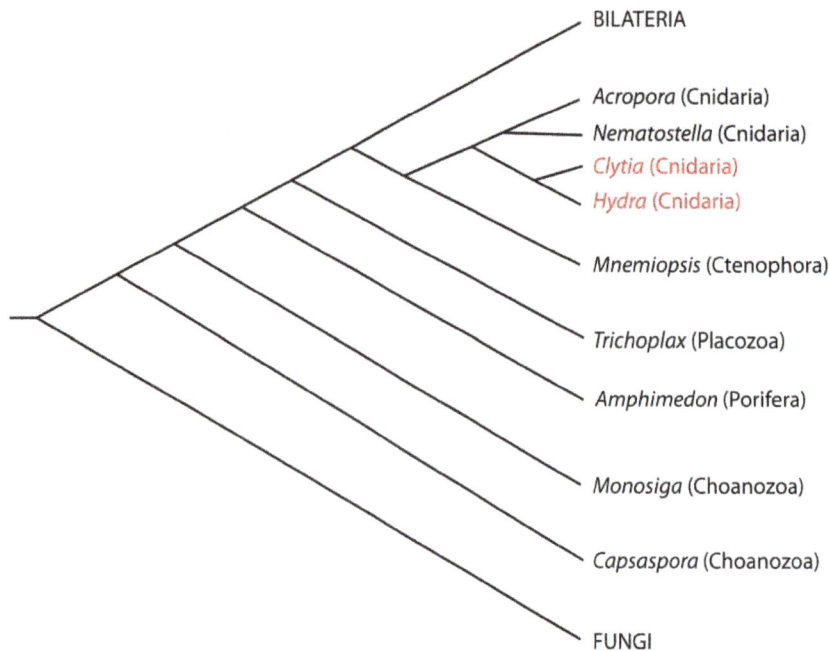

Figure 2. Distribution of the *flp* gene among opisthokonts. Evolutionary relationships of opisthokonts are shown, with the two medusozoans that contain a *flp* gene indicated in red. The placement of Ctenophora among opisthokonts is still unresolved. We have chosen the position proposed by Philippe et al. [55]. The position of *Capsaspora* is as proposed by Shalchian-Tabrizi et al. [56].

shown that a U-rich sequence upstream of a trans-spliced gene is essential for splicing in operons in *C. elegans*. Candidate operons have been reported in other metazoans that undergo trans-spliced leader addition [24,25,26], including *Hydra* [11], but none have been experimentally verified. Using multiple approaches, we tested the hypothesis that the *Hydra* RFC140 and *flp* genes are in an operon.

The RFC140 and *flp* genes are in an operon

To test whether RFC140 and *flp* are in an operon, we constructed a plasmid in which a single promoter is located upstream of two fluorescent protein genes, with these genes separated by the RFC140/*flp* intergenic sequence. In order to construct the plasmid, we needed a fluorescent protein gene in addition to the GFP gene previously used in *Hydra* and contained in the plasmid hoTG [27,28]. Due to *Hydra*'s strongly A+T-biased codon usage [29], we had a version of the DsRed2 red fluorescent protein gene synthesized using *Hydra* codon and codon pair preferences. The sequence of the gene was generated using computationally optimized DNA assembly (CODA) [30] and commercially synthesized by CODA Genomics (now Verdezyne).

Using the GFP gene from hoTG, the promoter, 3′ UTR, and polyadenylation site from a *Hydra* cytoplasmic actin gene [29], the synthetic DsRed2 gene, and the intergenic region between the RFC140 and *flp* genes, we constructed an operon plasmid with the GFP gene in the upstream position and the DsRed2 gene in the downstream position (Fig. 5A). The resulting plasmid (pHyVec7, GenBank accession number EF539830) was introduced into *Hydra* using particle bombardment [27]. Cells in the bombarded animals that expressed GFP were always positive for DsRed2 expression (Fig. 5B, C). To further confirm our finding of co-expression, we generated a stably transgenic line with transgene expression in the ectoderm. This line co-expresses GFP and DsRed2 (Fig. 6). To rule out the possibility that a promoter is

present in the RFC140/*flp* intergenic region and is responsible for expression of *flp*, we made a plasmid in which the intergenic region is located upstream of the GFP gene. *Hydra* bombarded with this plasmid, mixed with a separate plasmid expressing DsRed2 under control of the actin promoter, contained cells that were red but not green (Fig. 7A, B). Thus the intergenic region does not contain a promoter.

To confirm that trans-spliced leader addition to the downstream transcript was occurring in this artificial operon, we carried out RT-PCR on RNA from the transgenic line with a primer for one of the spliced leaders (SL-B1) known to be used on the *flp* RNA [9] and a primer for the DsRed2 coding sequence. The sequence of the resulting PCR product confirmed that the DsRed2 mRNA was trans-spliced at the same site as *flp* mRNA (Fig. 8).

While trans-splicing of a bicistronic mRNA to produce two mRNAs is the most straightforward explanation for co-expression of the GFP and DsRed2 genes, alternative explanations include internal ribosome entry or translational read-through of a bicistronic mRNA [31]. The later explanation seems unlikely as there are nine in-frame stop codons between the GFP stop codon and the DsRed2 start codon. If the bulk of DsRed2 expression is due to internal ribosome entry, prevention of trans-splicing should not affect DsRed2 expression. To test whether expression of DsRed2 is dependent on trans-splicing, we mutated the AG dinucleotide that is the acceptor for trans-splicing (see Fig. 3B) to AA. We also mutated the three AG dinucleotides between the trans-splicing acceptor and the DsRed2 start codon to AA in order to prevent trans-splicing from occurring at alternative sites. Particle bombardment of *Hydra* with the mutated construct yielded cells expressing GFP but not DsRed2 (Fig. 7C, D). Thus blocking of trans-splicing prevents expression of the DsRed2 gene but has no effect on GFP expression. From this result, we conclude that *flp* gene expression requires trans-splicing and that expression of the RFC140 gene does not require trans-splicing of *flp*.

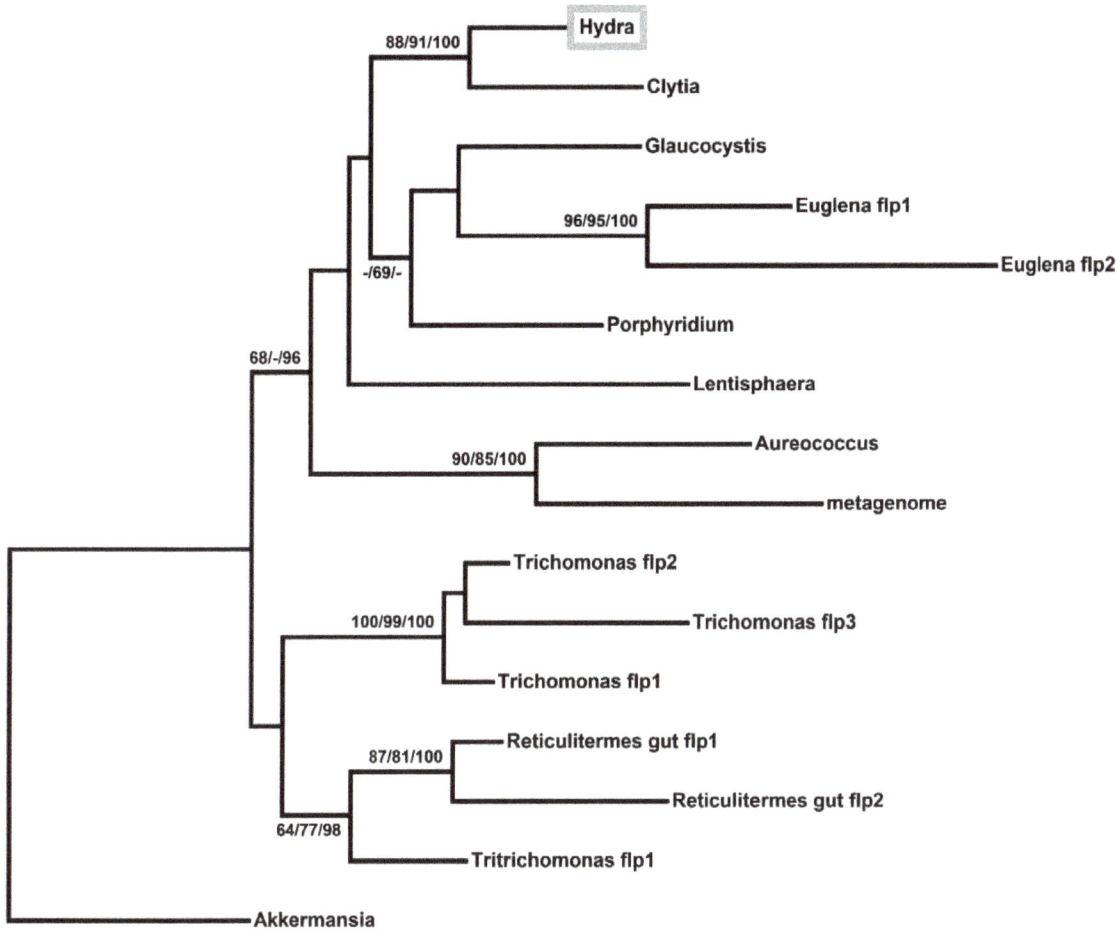

Figure 3. Phylogenetic relationships among *flp* proteins. Relationships are shown as an unrooted maximum likelihood phylogram. The numbers at the nodes indicate maximum likelihood bootstrap support values/maximum parsimony bootstrap support values/Bayesian posterior probabilities. A dash indicates a value less than 60. Nodes without numbers have bootstrap support and Bayesian posterior probabilities of less than 60.

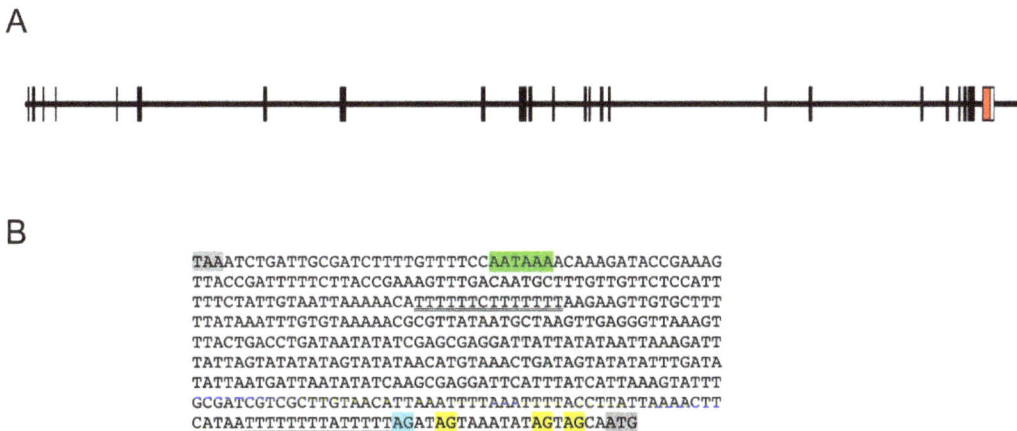

A

B

Figure 4. RFC140/*flp* exon-intron organization (A) and intergenic region sequence (B). The RFC140 gene exons are shown in black. The single *flp* exon and 3′ UTR are shown in red and white respectively. The coding sequence of the RFC140 gene has been deposited in GenBank under accession number FJ154842. Annotations in Panel B are as follows: gray, RFC140 stop codon and *flp* start codon; green, putative polyadenylation signal; double-underline, T-rich sequence; single-underline, polypyrimidine tract associated with the trans-splicing acceptor dinucleotide; blue, trans-splicing acceptor dinucleotide; yellow, additional AG dinucleotides changed to AA in mutant plasmid.

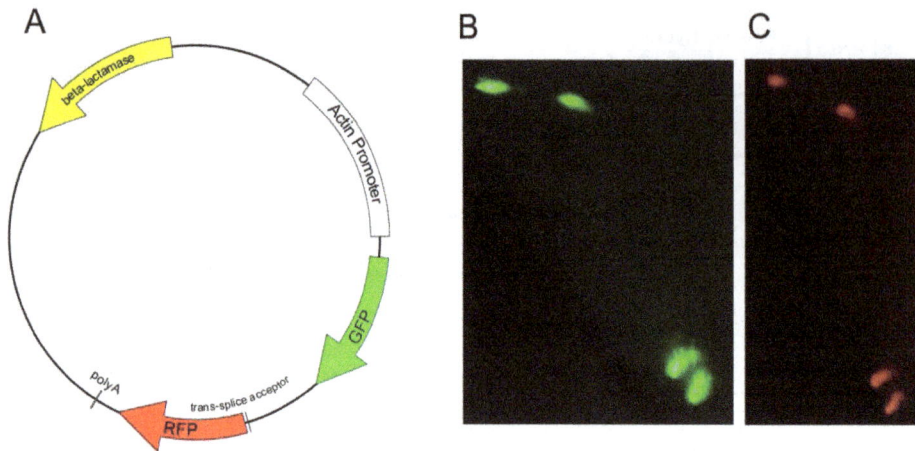

Figure 5. Structure of the operon plasmid and expression of the plasmid following particle bombardment. Panel A shows the structure of pHyVec7, the artificial operon plasmid. Panels B and C show expression of GFP (B) and DsRed2 (C) in four ectodermal epithelial cells in the body column of a live polyp that was bombarded with gold particles coated with the pHyVec7 DNA.

Discussion

How frequently horizontal transfer of genes into the genomes of metazoans occurs, what taxa the genes come from, and the routes they take to get into the target genome are questions about which we have very little information. All sequenced metazoan genomes contain some genes that show anomalous phylogenetic distributions, consistent with horizontal transfer. Absence of introns in such genes is taken as additional evidence that they were horizontally transferred, either from a prokaryote or by retrotransposition. These properties, while expected of a horizontally transferred gene, cannot be taken as proof of such transfer. Secondary loss is always a possibility, and is a particularly viable hypothesis when one is examining the genome of an organism from a part of the metazoan tree that is not well-sampled [4]. The strongest case for horizontal transfer of a gene into a metazoan would be if: (1) the gene has an anomalous phylogenetic distribution; (2) the gene lacks introns; (3) the gene has no homologues in a substantial number of taxa that diverged immediately before and after the organism being considered; (4) there is a reasonable hypothesis for how the gene entered the germ line. The final criterion is the most difficult to satisfy for most metazoans, since the germ line is usually segregated early during embryonic development and a route for an exogenous gene to reach it is not easily imagined.

The *Hydra flp* gene provides a particularly compelling case for horizontal gene transfer in a metazoan. The gene has not been identified by genome or EST sequencing in any opisthokonts other than *Hydra* and *Clytia*. This includes genome sequences from metazoan phyla that diverged immediately before and after Cnidaria. The *Hydra flp* gene lacks introns, a feature that is consistent with either retrotransposition or acquisition from an organism that lacks introns. The known bacterial *flp* genes and the *flp* genes in *Trichomonas vaginalis* lack introns. In *Hydra*, the germ line remains unsegregated throughout the life of the animal. The germ cells in the adult polyp arise from the interstitial cell lineage, which contains multipotent stem cells that give rise to some classes of somatic cells (e.g. nerve cells and nematocytes) in addition to germ cells [32]. These stem cells arise during embryogenesis [33], divide continuously in the adult, and are transmitted to the progeny produced asexually by budding. An adult *Hydra* polyp contains about 3000 such multipotent stem cells [34], which have a cell cycle of about 24 hours [35]. A horizontally transferred gene that is incorporated into a single such multipotent stem cell could be present both in gametes and in the stem cells of asexually-produced progeny. An interstitial cell lineage similar to that seen in *Hydra* is present in at least some other hydrozoans but not in other cnidarian classes [36]. *Clytia* and *Hydra* share this feature [37], and thus we assume that their last common ancestor had this feature as well. Phagocytosis by multipotent stem cells in *Hydra* has not been

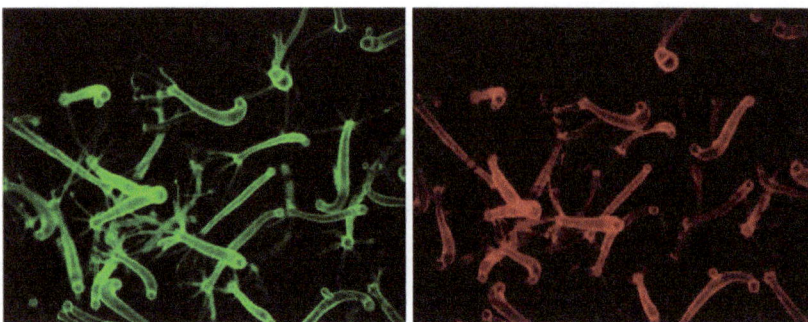

Figure 6. Stably transgenic *Hydra* produced with pHyVec7. Panels show expression of GFP (left) and DsRed2 (right) in a stably transgenic line produced by injecting an embryo with pHyVec7. In this line, only the ectodermal epithelial cell lineage is transgenic.

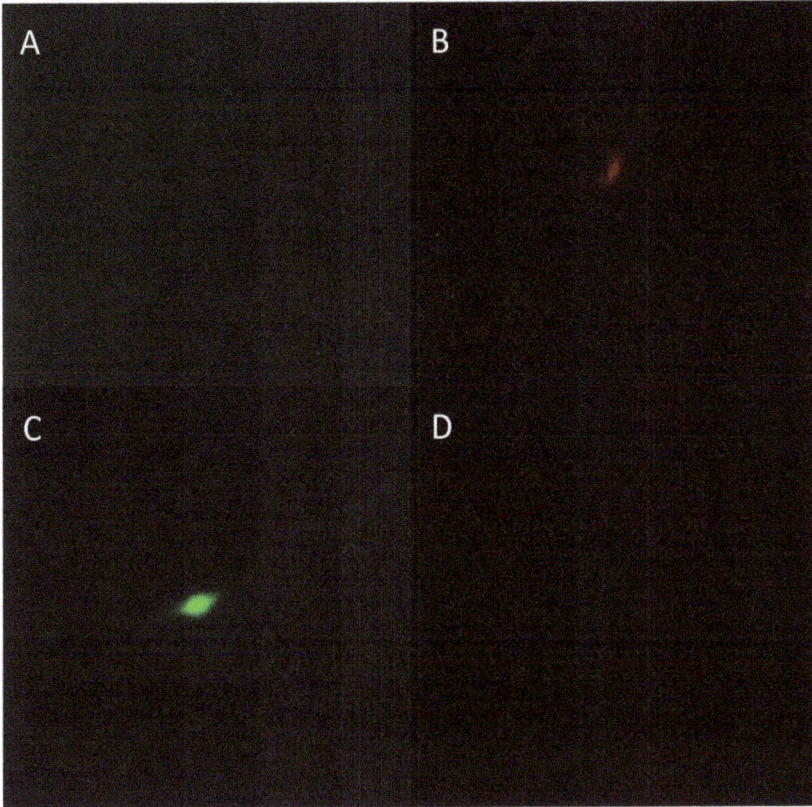

Figure 7. Tests of the intergenic region for promoter activity and of an operon construct with a mutated trans-splicing acceptor dinucleotide. The intergenic region was placed upstream of the green fluorescent protein gene and the resulting plasmid was introduced into adult *Hydra* polyps by particle bombardment together with a plasmid in which DsRed2 expression is driven by an actin promoter. Panels A and B show an ectodermal epithelial cell in the body column of a bombarded polyp that expresses DsRed2 from the control plasmid (B) but not green fluorescent protein from the plasmid with the intergenic region (A). Panels C and D show an ectodermal epithelial cell in the body column of a polyp that was bombarded with the operon construct in which the trans-splicing acceptor dinucleotide was mutated. The cell expresses the upstream green fluorescent protein gene (C) but not the downstream DsRed2 gene (D).

reported, but the interstitial cell that becomes the oocyte is phagocytic [38,39], and the phagocytic ability of various cell types in the medusozoan ancestor of *Hydra* and *Clytia* is obviously not known. Bacteria have been found between and within cells in *Hydra* [40,41,42]

The scenario we envision is that a *flp* gene entered a medusozoan genome from the genome of a unicellular organism ingested or associated with the host and that it entered a cell in the lineage that gives rise to the germ line (i.e. the interstitial cell lineage). We cannot, however, rule out the possibility that the *flp* gene was acquired from a unicellular organism by a virus, which then carried it into a cnidarian host. Virus-like particles have been reported in *Hydra* [43]. The simplest version of events has the gene

landing immediately downstream of the RFC140 gene. This would allow immediate expression of the gene under control of the RFC140 gene promoter. This scenario requires that the host carried out trans-splicing, which the ancestor of *Clytia* and *Hydra* did [11,44]. Support for this scenario will be provided if the *flp* gene is in an operon with the RFC140 gene in other medusozoans.

The unicellular organisms in which a *flp* gene has been identified are not taxonomically close to each other (except for *T. vaginalis* and *T. foetus*) and the phylogenetic tree for *flp* (Fig. 3) does not provide information on which clade of unicellular organisms was the likely source of the *flp* gene in cnidarians. It is possible that the *flp* gene has moved horizontally among unicellular organisms as well, complicating the understanding of its evolutionary history.

```
genome                              ...TTTTTTTTATTTTTAGATAGTAAATATAGTAGCAATG
SL-B1        ACGGAAAAAAACACATACTGAAACTTTTTAGTCCCTGTGTAATAAGATAGTAAATATAGTAGCAATG
SL-B1 RT-PCR          CACATACTGAAACTTTTTAGTCCCTGTGTAATAAGATAGTAAATATAGTAGCAATG
```

Figure 8. RNA from the pHyVec7 transgene is trans-spliced. RNA from the operon transgenic line was reverse-transcribed to produce first strand cDNA. The cDNA was amplified with primers for the spliced leader SL-B1 and the DsRed2 coding sequence. The top line shows sequence between the RFC140 and *flp* genes extending from the polypyrimidine tract associated with the trans-splicing AG acceptor dinucleotide (doubly underlined) to the start codon of the *flp* gene (shaded in gray). The second line shows the sequence from a previously identified *flp* EST containing the SL-B1 spliced leader sequence [13]; the spliced leader sequence is highlighted in yellow, and the *flp* start codon is shaded. The third line shows sequence from the RT-PCR product obtained from the pHyVec7 transgenic line using SL-B1 and DsRed2 primers. The sequence corresponding to the SL-B1 primer is highlighted in green, and the DsRed2 start codon is shaded.

While our data are consistent with the *flp* gene entering the cnidarian lineage after the divergence of Anthozoa and Medusozoa, a caveat associated with this conclusion is that only two anthozoan genome sequences have been published [16,17]. It is possible that the *flp* gene was present in the stem cnidarian and that the anthozoans whose genomes have been sequenced have secondarily lost the gene. As more cnidarian genome sequences become available, it should be possible to determine with some precision when the gene entered the cnidarian lineage and whether any secondary loss has occurred.

Our results with an artificial operon in *Hydra* are in contrast to findings in the nematode *Brugia malayi*, in which synthetic operons have also been tested. In *Brugia*, a synthetic operon was constructed in which the upstream gene encoded firefly luciferase and the downstream gene encoded *Renilla* luciferase [45]. Both genes were expressed, but surprisingly the downstream gene was not trans-spliced. This appears to be due to the requirement for a sequence motif in an intron of the downstream gene for proper trans-splicing [46,47]. Our results show that trans-splicing of the downstream gene in the *Hydra* RFC-*flp* operon does not depend on any sequences in or 3′ to the *flp* gene, since the *flp* gene is absent and the 3′ UTR in the transgene construct is from a *Hydra* cytoplasmic actin gene. By obtaining successful trans-splicing of the bicistronic RNA produced by our construct, we have demonstrated that everything necessary to produce a functional operon is contained in the intergenic region. This result suggests that evolution of a bicistronic operon in *Hydra* is a relatively simple process, not requiring introns in either of the two genes, nor trans-splicing of the upstream gene (the cytoplasmic actin gene that was the source of the promoter in our operon construct does not undergo trans-splicing). Additional facilitation is provided by the A+T-richness of the *Hydra* genome, which should aid in the generation of a polypyrimidine tract (using T) near the splice acceptor dinucleotide and a T-rich sequence in the intergenic region which may be required for trans-splicing.

The ability to construct artificial operons in *Hydra* has practical applications. Creation of transgenic *Hydra* expressing genes whose proteins cannot be fused to a fluorescent protein is problematic. Transgenic animals are initially mosaic [28]. Without a fluorescent marker, it is impossible to identify transgenic tissue and thus to track the formation of fully transgenic animals by asexual propagation. The presence of operons in *Hydra* offers a solution to this problem. By placing the gene of interest upstream of a fluorescent protein gene in a bicistronic operon, expression of the fluorescent protein gene serves as a proxy for expression of the gene of interest. To allow easy cloning of genes into the upstream position of the operon, the plasmid vector pHyVec11 was constructed (GenBank accession number EU183365). In pHyVec11 a linker replaces the GFP gene in pHyVec7. The linker contains an NheI site, a HpaI site, and a BamHI site. This allows insertion of genes engineered with an NheI site upstream of the start codon and a blunt end (compatible with HpaI) or a BamHI-compatible site after the stop codon. The resulting construct will express the gene of interest in unfused form and DsRed2. Thus transgenic animals that express DsRed2 will also express the gene of interest. We have used pHyVec11 to express several genes in transgenic *Hydra* (Steele et al., unpublished observations), indicating that it functions as predicted.

Materials and Methods

Hydra strains and culture

Hydra magnipapillata strain 105 and *Hydra vulgaris* strain AEP were cultured using standard methods, including feeding with *Artemia*

nauplii [48]. *H. magnipapillata* strain 105 is the strain that was used for genome sequencing [11]. *H. vulgaris* AEP is the strain used for making transgenic *Hydra* [28]. The phylogenetic relationship between these two strains, both members of the Vulgaris clade, is described in Martínez et al. [49].

Synthetic DsRed2 gene

A DsRed2 gene with *Hydra* codon and codon pair preferences was synthesized commercially by CODA Genomics (now Verdezyne) using the method of Larsen et al. [30] and a dataset of codons and codon pairs generated from several hundred *Hydra* genes. The sequence of the gene has been deposited in GenBank under accession number EF451141.

Expression Plasmid Constructions

Expression plasmids used in this study were originally derived from hoTG [27,28], a plasmid that contains ~1.5 kb of a *Hydra* actin gene promoter, a green fluorescent protein gene, and the 3′ UTR and polyadenylation site from the same actin gene as the promoter. The pHyVec1 plasmid was constructed by excising the expression cassette from hoTG at NsiI sites in the actin promoter and the actin 3′ flanking sequence, blunting these sites, and cloning the resulting fragment into pBluescript II SK+ blunted at the KpnI and SacI sites of the multiple cloning site. The actin promoter fragment in pHyVec1 has had ~500 bp deleted from the 5′ end of the promoter present in hoTG. The actin 3′ flanking sequence in pHyVec1 has had ~400 bp deleted from the 3′ flanking sequence present in hoTG. The pHyVec5 plasmid was constructed by replacing the GFP gene in pHyVec1 with the synthetic DsRed2 gene described above. The pHyVec7 plasmid was constructed using a combination of PCR and standard recombinant DNA methods. When needed, appropriate restriction sites were added to DNA fragments by PCR. The intergenic sequence between the *flp* and RFC140 genes was isolated by amplification from *H. magnipapillata* strain 105 genomic DNA. The intergenic fragment was inserted between the actin promoter and the DsRed2 gene in pHyVec5. The GFP gene was then inserted between the actin promoter and the intergenic segment to yield the completed operon. The sequence of pHyVec7 has been deposited in GenBank under accession number EF539830.

DNA particle bombardment

Introduction of transgenes into adult *Hydra* polyps by particle bombardment [27] was carried out using a Bio-Rad Biolistic PDS-1000/He Particle Delivery System as follows. Approximately 200 polyps of the 105 strain of *Hydra magnipapillata* that had been starved for 24–48 hours were placed in a 35 mm plastic dish and the culture medium removed. By use of a disposable plastic inoculating loop, the polyps were pushed into a pile in the center of the dish. Gold particles (1 micron; Bio-Rad) were washed as suggested by the supplier and stored at 60 mg/ml in 50% glycerol at −20°C. Coating of gold particles with DNA and preparation of the macrocarriers were carried out immediately before use as described in the Bio-Rad user manual. The Particle Delivery System was evacuated using house vacuum (approximately 25 inches of mercury), and 650 psi rupture disks were used. Each batch of polyps was bombarded five times. Between bombardments, the polyps were gently washed by addition of a small volume of *Hydra* medium to the dish. After washing, the medium was removed and the polyps were again concentrated in the center of the dish using a plastic inoculating loop. Following the last bombardment, the polyps were transferred to a 100 mm Petri dish containing *Hydra* medium and placed at 18°C. On the following day, the polyps were transferred to fresh *Hydra* medium. Two to four days following bombardment, live animals were photographed in *Hydra* medium in a six well tissue culture plate using either an Olympus IX inverted microscope or an Olympus SZX-ILLD2-100 stereomicroscope. Because the gold particle containing the DNA must hit the nucleus in order for the transgene to be expressed [27], many polyps show no expression and those that do have only one to a few expressing cells.

Generation of Transgenic *Hydra*

Transgenic *Hydra* lines were generated essentially as described by Wittlieb et al. [28]. Polyps that had initiated egg production were collected from mass cultures of *Hydra vulgaris* strain AEP and placed in a Petri dish with polyps of the same strain that had testes. A single blastomere of embryos at the 2–8 cell stage was injected with approximately 0.2 nanoliters of plasmid DNA (0.5 μg/μl) using a Narishige IM-300 Microinjector. Plasmid DNA for injection was prepared using a Qiagen EndoFree Plasmid Giga Kit. The DNA was resuspended in filter-sterilized Milli-Q water and stored at −20°C. Injections were carried out using needles pulled on a Model P-87 Flaming/Brown Micropipette Puller (Sutter Instruments) from filament-containing capillary tubing (Sutter Instruments, catalog number BF100-50-10). Injections were done in *Hydra* medium and injected embryos were kept individually in microtiter plate wells in *Hydra* medium. The plates were initially incubated for two weeks in the dark at 18°C. The plates were then kept at room temperature (18–21°C) on a 17 hour light, 7 hour dark cycle until the embryos hatched. Newly hatched polyps were fed *Artemia* nauplii. Transgenic polyps were cultured until animals were obtained in which all the cells in the lineage that contained the transgene were transgenic. Live transgenic animals were photographed in *Hydra* medium in a six well tissue culture plate using an Olympus SZX-ILLD2-100 stereomicroscope.

Mutagenesis

Generation of a construct in which the *flp* gene trans-splicing acceptor dinucleotide was mutated was carried out as follows. The intergenic region between the *flp* and RFC140 genes contains a BsaBI cleavage site nine nucleotides downstream from the RFC140 stop codon. A BstAPI cleavage site is present seven nucleotides downstream of the *flp* ATG start codon. A version of the 436 bp segment between these two sites in which the trans-splicing AG acceptor dinucleotide and the three AG dinucleotides immediately downstream of the splice acceptor were replaced by AA (see Fig. 3B) was synthesized commercially by Blue Heron Biotechnologies. The wild-type intergenic segment in pHyVec7 was replaced by the mutated segment to produce the plasmid pHyVec7-mut. The construct was confirmed by sequencing.

RT-PCR

Total RNA for RT-PCR was isolated from five polyps of the transgenic line using a Quick-RNA MicroPrep kit (Zymo Research). Oligo dT-primed first strand cDNA synthesis was carried out with reagents from the GeneRacer Core Kit (Invitrogen) and reverse transcriptase from the SuperScript III RT Module (Invitrogen). Amplification of spliced leader-containing DsRed2 cDNA was carried out with AccuPrime *Pfx* polymerase (Invitrogen) and the following primers: SL-B1,

CACATACTGAAACTTTTTAGTCCC; DsRed2, TGTGGTG-ATAAAATATCCCACGC. The resulting amplification product was purified using a DNA Clean & Concentrator Kit (Zymo Research) and ligated into the pCR-Blunt plasmid using a Zero Blunt PCR Cloning Kit (Invitrogen). The ligation mixture was used to transform One Shot TOP10 chemically competent *E. coli* cells (Invitrogen), and transformed cells were selected on LB plates containing kanamycin. The recombinant plasmid was sequenced using M13 forward and reverse primers (Eton Bioscience, Inc.).

Bioinformatics and Phylogenetic Analyses

Sequences related to *flp* were identified using blastp and tblastn to query the protein, EST, whole-genome shotgun read, and environmental samples databases at NCBI. The protein sequence alignment in Fig. 1 was carried out using the CLUSTALW2 server at the EMBL European Bioinformatics Institute (http://www.ebi.ac.uk/Tools/msa/clustalw2/).

For phylogenetic analyses, amino acid sequences were aligned using Multiple Sequence Alignment by Log-Expectation (MUS-CLE) [50] on the server at the EMBL European Bioinformatics Institute (http://www.ebi.ac.uk/Tools/msa/muscle/). The portion of the alignment used for analyses began at the start of the *Hydra* sequence and ended at the last point with sequence present for all proteins. *Reticulitermes* gut 3, *Zootermopsis* gut, and *Tritrichomonas foetus* flp2 sequences were not used in analyses, because they are incomplete at the amino terminal ends.

ProtTest [51], on the server at the University of Vigo (http://darwin.uvigo.es/software/prottest_server.html), was used to select among models of protein evolution. The best fitting model of those evaluated, based on AICc and BIC frameworks, was WAG+G. This model was used for the maximum likelihood and Bayesian analyses. The maximum likelihood analysis was performed using the ATGC Bioinformatics Platform PhyML 3.0 server [52] (http://www.atgc-montpellier.fr/phyml/). Bootstrap values were based on 1000 replicates. Bayesian analysis was performed using MrBayes 3.1.2 [53], with four Markov Chain Monte Carlo chains run for 2,000,000 generations and sampled every 100 generations, and with the first 500,000 generations discarded as burn-in. Parsimony analysis was conducted using PAUP* 4.10 [54], with 1000 bootstrap replicates, each involving a heuristic search with 10 random addition replicates and TBR branch swapping. Amino acid substitutions were weighted using the PAUP protpars matrix.

Acknowledgments

We thank Dr. Thomas Bosch for providing the hoTG plasmid. We thank Dr. Tom Schilling for providing injection needles. We are especially grateful to Drs. Kyoko Yokomori and Ken Cho for use of their fluorescence microscopes.

Author Contributions

Conceived and designed the experiments: CED KMG DMB RES. Performed the experiments: CED KMG TAC DMB RES. Analyzed the data: CED KMG DMB RES. Contributed reagents/materials/analysis tools: DMB. Wrote the paper: CED KMG DMB RES.

References

1. Koonin EV, Wolf YI (2008) Genomics of bacteria and archaea: the emerging dynamic view of the prokaryotic world. Nucl Acids Res 36: 6688–6719.
2. Andersson JO (2009) Gene transfer and diversification of microbial eukaryotes. Annu Rev Microbiol 63: 177–193.
3. Andersson JO (2005) Lateral gene transfer in eukaryotes. Cell Mol Life Sci 62: 1182–1197.
4. Technau U, Rudd S, Maxwell P, Gordon PM, Saina M, et al. (2005) Maintenance of ancestral complexity and non-metazoan genes in two basal cnidarians. Trends Genet 21: 633–639.
5. Campbell RD, Bode HR (1983) Terminology for morphology and cell types. In: Lenhoff HM, ed. Hydra: Research Methods. New York: Plenum Press. pp 5–14.

6. Bosch TCG, David CN (1987) Stem cells of *Hydra magnipapillata* can differentiate into somatic cells and germ line cells. Dev Biol 121: 182–191.

7. Tannreuther GW (1909) Observations on the germ cells of hydra. Biol Bull 16: 205–209.

8. Tardent P (1974) Gametogenesis in the genus Hydra. Amer Zool 14: 447–456.

9. Stover NA, Steele RE (2001) Trans-spliced leader addition to mRNAs in a cnidarian. Proc Natl Acad Sci USA 98: 5693–5698.

10. Blumenthal T (2004) Operons in eukaryotes. Brief Funct Genomics Proteomics 3: 199–211.

11. Chapman JA, Kirkness EF, Simakov O, Hampson SE, Mitros T, et al. (2010) The dynamic genome of *Hydra*. Nature 464: 592–596.

12. Habetha M, Bosch TC (2005) Symbiotic Hydra express a plant-like peroxidase gene during oogenesis. J Exp Biol 208: 2157–2165.

13. Steele RE, Hampson SE, Stover NA, Kibler DF, Bode HR (2004) Probable horizontal transfer of a gene between a protist and a cnidarian. Curr Biol 14: R298–R299.

14. Crouch MV, Alderete JF (2001) Trichomonas vaginalis has two fibronectin-like iron-regulated genes. Arch Med Res 32: 102–107.

15. Cartwright P, Evans NM, Dunn CW, Marques AC, Miglietta MP, et al. (2008) Phylogenetics of Hydroidolina (Hydrozoa: Cnidaria). J Mar Biol Assoc 88: 1663.

16. Putnam NH, Srivastava M, Hellsten U, Dirks B, Chapman J, et al. (2007) Sea anemone genome reveals ancestral eumetazoan gene repertoire and genomic organization. Science 317: 86–94.

17. Shinzato C, Shoguchi E, Kawashima T, Hamada M, Hisata K, et al. (2011) Using the Acropora digitifera genome to understand coral responses to environmental change. Nature 476: 320–323.

18. Bridge D, Cunningham CW, DeSalle R, Buss LW (1995) Class-level relationships in the phylum Cnidaria: molecular and morphological evidence. Mol Biol Evol 12: 679–689.

19. Bridge D, Cunningham CW, Schierwater B, DeSalle R, Buss LW (1992) Class-level relationships in the phylum Cnidaria: evidence from mitochondrial genome structure. Proc Natl Acad Sci U S A 89: 8750–8753.

20. King N, Westbrook MJ, Young SL, Kuo A, Abedin M, et al. (2008) The genome of the choanoflagellate *Monosiga brevicollis* and the origin of metazoans. Nature 451: 783–788.

21. Srivastava M, Simakov O, Chapman J, Fahey B, Gauthier ME, et al. (2010) The Amphimedon queenslandica genome and the evolution of animal complexity. Nature 466: 720–726.

22. Srivastava M, Begovic E, Chapman J, Putnam NH, Hellsten U, et al. (2008) The *Trichoplax* genome and the nature of placozoans. Nature 454: 955–960.

23. Huang T, Kuersten S, Deshpande AM, Spieth J, MacMorris M, et al. (2001) Intercistronic region required for polycistronic pre-mRNA processing in Caenorhabditis elegans. Mol Cell Biol 21: 1111–1120.

24. Davis RE, Hodgson S (1997) Gene linkage and steady state RNAs suggest *trans*-splicing may be associated with a polycistronic transcript in *Schistosoma mansoni*. Mol Biochem Parasitol 89: 25–39.

25. Satou Y, Hamaguchi M, Takeuchi K, Hastings KE, Satoh N (2006) Genomic overview of mRNA 5′-leader trans-splicing in the ascidian Ciona intestinalis. Nucl Acids Res 34: 3378–3388.

26. Satou Y, Mineta K, Ogasawara M, Sasakura Y, Shoguchi E, et al. (2008) Improved genome assembly and evidence-based global gene model set for the chordate Ciona intestinalis: new insight into intron and operon populations. Genome Biol 9: R152.

27. Böttger A, Alexandrova O, Cikala M, Schade M, Herold M, et al. (2002) GFP expression in Hydra: lessons from the particle gun. Dev Genes Evol 212: 302–305.

28. Wittlieb J, Khalturin K, Lohmann JU, Anton-Erxleben F, Bosch TCG (2006) Transgenic *Hydra* allow in vivo tracking of individual stem cells during morphogenesis. Proc Natl Acad Sci U S A 103: 6208–6211.

29. Fisher DA, Bode HR (1989) Nucleotide sequence of an actin-encoding gene from Hydra attenuata: structural characteristics and evolutionary implications. Gene 84: 55–64.

30. Larsen LSZ, Wassman CD, Hatfield GW, Lathrop RH (2008) Computationally optimized DNA assembly of synthetic genes. Int J Bioinform Res Appl 4: 324–336.

31. Jackson RJ (2005) Alternative mechanisms of initiating translation of mammalian mRNAs. Biochem Soc Trans 33: 1231–1241.

32. Bode HR (1996) The interstitial cell lineage of hydra: a stem cell system that arose early in evolution. J Cell Sci 109: 1155–1164.

33. Martin VJ, Littlefield CL, Archer WE, Bode HR (1997) Embryogenesis in hydra. Biol Bull 192: 345–363.

34. David CN, Gierer A (1974) Cell cycle kinetics and development of Hydra attenuata. III. Nerve and nematocyte differentiation. J Cell Sci 16: 359–375.

35. Campbell RD, David CN (1974) Cell cycle kinetics and development of Hydra attenuata. II. Interstitial cells. J Cell Sci 16: 349–358.

36. Frank U, Plickert G, Müller WA (2009) Cnidarian interstitial cells: the dawn of stem cell research. In: Rinkevich B, Matranga V, eds. Stem Cells in Marine Organisms. Heidelberg: Springer.

37. Denker E, Manuel M, Leclere L, Le Guyader H, Rabet N (2008) Ordered progression of nematogenesis from stem cells through differentiation stages in the tentacle bulb of *Clytia hemisphaerica* (Hydrozoa, Cnidaria). Dev Biol 315: 99–113.

38. Honegger TG, Zurrer D, Tardent P (1989) Oogenesis in Hydra-carnea - A new model based on light and electron microscopic analyses of oocyte and nurse cell differentiation. Tissue Cell 21: 381–393.

39. Zihler J (1972) Zur Gametogenese und Befruchtungsbiologie von Hydra. Wilhelm Roux' Arch fur Entwicklungsmechander Org 169: 239–267.

40. Davis LE, Haynes JF (1968) An ultrastructural examination of the mesoglea of Hydra. Z Zellforsch Mikrosk Anat 92: 149–158.

41. Fraune S, Bosch TC (2007) Long-term maintenance of species-specific bacterial microbiota in the basal metazoan *Hydra*. Proc Natl Acad Sci U S A 104: 13146–13151.

42. Hufnagel LA, Myhal ML (1977) Observations on a spirochaete symbiotic in *Hydra*. Trans Amer Micros Soc 96: 406–411.

43. Bonnefoy AM, Kolenkine X, Vago MC (1972) Pathologie des invértebratés-particules d'allure virale chez les Hydres. Compte Rendus Acadmie Science Paris, Serial D 275: 2163–2165.

44. Derelle R, Momose T, Manuel M, Da Silva C, Wincker P, et al. (2010) Convergent origins and rapid evolution of spliced leader trans-splicing in metazoa: insights from the ctenophora and hydrozoa. RNA 16: 696–707.

45. Liu C, Oliveira A, Chauhan C, Ghedin E, Unnasch TR (2010) Functional analysis of putative operons in Brugia malayi. Int J Parasitol 40: 63–71.

46. Higazi TB, Unnasch TR (2004) Intron encoded sequences necessary for trans splicing in transiently transfected Brugia malayi. Mol Biochem Parasitol 137: 181–184.

47. Liu C, de Oliveira A, Higazi TB, Ghedin E, DePasse J, et al. (2007) Sequences necessary for trans-splicing in transiently transfected Brugia malayi. Mol Biochem Parasitol 156: 62–73.

48. Lenhoff HM, Brown RD (1970) Mass culture of *Hydra*: an improved method and its application to other aquatic invertebrates. Lab Anim 4: 139–154.

49. Martínez DE, Iñiguez AR, Percell KM, Willner JB, Signorovitch J, et al. (2010) Phylogeny and biogeography of *Hydra* (Cnidaria: Hydridae) using mitochondrial and nuclear DNA sequences. Mol Phylogenet Evol 57: 403–410.

50. Edgar RC (2004) MUSCLE: multiple sequence alignment with high accuracy and high throughput. Nucl Acids Res 32: 1792–1797.

51. Abascal F, Zardoya R, Posada D (2005) ProtTest: selection of best-fit models of protein evolution. Bioinformatics 21: 2104–2105.

52. Guindon S, Dufayard JF, Lefort V, Anisimova M, Hordijk W, et al. (2010) New algorithms and methods to estimate maximum-likelihood phylogenies: assessing the performance of PhyML 3.0. Systematic Biol 59: 307–321.

53. Ronquist F, Huelsenbeck JP (2003) MrBayes 3: Bayesian phylogenetic inference under mixed models. Bioinformatics 19: 1572–1574.

54. Swofford DL (2003) PAUP*. Phylogenetic Analysis Using Parsimony (*and Other Methods). Version 4 ed. SunderlandMassachusetts: Sinauer Associates.

55. Philippe H, Derelle R, Lopez P, Pick K, Borchiellini C, et al. (2009) Phylogenomics revives traditional views on deep animal relationships. Curr Biol 19: 706–712.

56. Shalchian-Tabrizi K, Minge MA, Espelund M, Orr R, Ruden T, et al. (2008) Multigene phylogeny of choanozoa and the origin of animals. PLoS One 3: e2098.

19

Induction of Body Weight Loss through RNAi-Knockdown of APOBEC1 Gene Expression in Transgenic Rabbits

Geneviève Jolivet[1]*, Sandrine Braud[2], Bruno DaSilva[1], Bruno Passet[4], Erwana Harscoët[1], Céline Viglietta[1], Thomas Gautier[3], Laurent Lagrost[3], Nathalie Daniel-Carlier[1], Louis-Marie Houdebine[1], Itzik Harosh[2]*

1 INRA UMR1198, Biologie du Développement et Reproduction, Jouy en Josas, France, 2 ObeTherapy Biotechnology, Evry, France, 3 INSERM UMR866, Université de Bourgogne, Dijon, France, 4 INRA UMR1313, Génétique Animale et Biologie Intégrative, Jouy-en-Josas, France

Abstract

In the search of new strategies to fight against obesity, we targeted a gene pathway involved in energy uptake. We have thus investigated the *APOB* mRNA editing protein (*APOBEC1*) gene pathway that is involved in fat absorption in the intestine. The *APOB* gene encodes two proteins, APOB100 and APOB48, via the editing of a single nucleotide in the *APOB* mRNA by the APOBEC1 enzyme. The APOB48 protein is mandatory for the synthesis of chylomicrons by intestinal cells to transport dietary lipids and cholesterol. We produced transgenic rabbits expressing permanently and ubiquitously a small hairpin RNA targeting the rabbit *APOBEC1* mRNA. These rabbits exhibited a moderately but significantly reduced level of *APOBEC1* gene expression in the intestine, a reduced level of editing of the *APOB* mRNA, a reduced level of synthesis of chylomicrons after a food challenge, a reduced total mass of body lipids and finally presented a sustained lean phenotype without any obvious physiological disorder. Interestingly, no compensatory mechanism opposed to the phenotype. These lean transgenic rabbits were crossed with transgenic rabbits expressing in the intestine the human *APOBEC1* gene. Double transgenic animals did not present any lean phenotype, thus proving that the intestinal expression of the human *APOBEC1* transgene was able to counterbalance the reduction of the rabbit *APOBEC1* gene expression. Thus, a moderate reduction of the APOBEC1 dependent editing induces a lean phenotype at least in the rabbit species. This suggests that the *APOBEC1* gene might be a novel target for obesity treatment.

Editor: Hervé Guillou, INRA, France

Funding: This study was funded by Agence Nationale de la Recherche (ANR-06-RIB-FATSTOP) to ObeTherapy as leader of the program and to INRA. Sandrine Braud and Itzik Harosh are employees of and shareholders in ObeTherapy Biotechnology. ObeTherapy Biotechnology provided support in the form of salaries for authors SB and IH, had a direct role in the study design, and analysis, decision to publish, and preparation of the manuscript.

Competing Interests: Sandrine Braud and Itzik Harosh are employees of and shareholders in ObeTherapy Biotechnology. There are no products in development or marketed products to declare.

* Email: genevieve.jolivet@jouy.inra.fr (GJ); harosh@obetherapy.com(IH)

Introduction

Obesity is becoming a major problem all over the world spreading like global epidemic with a higher prevalence in the USA [1]. Overweight and obesity are important risk factors for diabetes and cardiovascular disease. Several hundreds of genes are involved in obesity and the estimation is that one quarter of our genome is involved in weight management and energy metabolism [2,3]. In the search of new targets for obesity, we have investigated the *APOB* mRNA editing protein (*APOBEC1*) gene pathway that is involved in fat absorption in the intestine.

The *APOB* gene encodes two proteins, APOB100 and APOB48, via the editing of a single nucleotide in the mRNA by a specialized enzyme, the *APOB* mRNA editing protein (APOBEC1). This enzyme, a catalytic deaminase expressed in human and rabbit in the intestine but not in the liver, is part of a complex that deaminates a cytidine residue to an uridine one in

the intestine *APOB* mRNA (at position 6666 in the human and 6529 in the rabbit) thus generating a STOP codon; it results in the production of the shorter polypeptide designated APOB48 [4] [5] [6]. APOB48 is essential for chylomicron formation, secretion and transport of dietary cholesterol and triglyceride from the intestine [7,8]. Besides, in the liver, where the editing protein is not expressed, and editing does not occur, the unaltered mRNA gives rise to APOB100 that is an integral part of VLDL and LDL.

With the aim to show that *APOB* mRNA editing is a target mechanism for fighting against obesity, we searched to modulate APOBEC1 enzymatic activity *in vivo* in the rabbit species by modulating *APOBEC1* gene expression through transgenesis. Rabbits have the same lipid metabolism as human [9] as opposed to mice that express *APOBEC1* gene both in the liver and intestine [10], do not have CETP and have higher level of HDL and lower level of LDL, that altogether makes mice a less suitable model to study lipid metabolism than rabbits. Thus, we generated transgenic rabbits by knocking down the endogenous *APOBEC1* gene using RNA interference strategy and expressing permanently a small hairpin RNA (shRNA) targeting specifically the rabbit *APOBEC1* mRNA. We generated also transgenic rabbits expressing the human *APOBEC1* gene, and double transgenic animals by inter-crossing these two models. We observed interesting differences in the phenotypes of these rabbits, especially as regard to their body weight and total lipid content. Finally, our results suggest that APOBEC1 could be considered as a potential target for metabolic disorder treatment.

Results

Production of transgenic rabbits

We aimed to produce transgenic animals expressing a shRNA targeting the rabbit *APOBEC1* mRNA in order to knock down the expression of this gene. A construct encompassing a shRNA expressing gene (rbapobec1-shRNA, Figure 1) was therefore introduced by microinjection in the pronuclei of fertilized unicellular rabbit embryos. The sequence of the shRNA targeting the rabbit *APOBEC1* mRNA was chosen among a set of sequences designed by using the OligoWalk tool [11] after assessment of its high efficiency by using an *in vitro* test as previously described [12] (Figure S1).

Twenty-five rabbits were born after microinjection of the rbapobec1-shRNA construct in pronuclei of unicellular rabbit embryos. The screening of newborn rabbits led us to identify 5 (20%) rbapobec1-shRNA transgenic founders. Transgenic lines were successfully established from 3 (shL21, shL23, shL27) of these founders by breeding each one with a wild type animal of the facility. One copy of integrated transgene was integrated in each line. The efficiency of transgenesis and germline transmission was similar to what is currently observed in our rabbit transgenesis facility and led us to suppose that the transgenes were not deleterious for the survival of the rabbits.

rbapobec1-shRNA transgene expression

The rbapobec1-shRNA transgene was expected to produce a shRNA able to knock down the expression of the rabbit *APOBEC1* gene that is known to be specifically expressed in the

intestine [13]. The transgene expression was measured in scrapped duodenum cells. Within each line, the expression of the rbapobec1-shRNA transgene was stable over generations and not significantly different in males and females (Figure 2). Note that in shL21 line, the shRNA transgene expression was the highest compared to lines shL23 and shL27. The line shL23 was not further studied.

Expression of the rabbit *APOBEC1* gene

As presented in Figure 3, the level of the rabbit *APOBEC1* gene expression was moderately (2 to 3 times) but significantly reduced in both males and females in the rbapobec1-shRNA lines shL21 and shL27. This suggests that the shRNA produced by the rbapobec1-shRNA transgene targeted the rabbit *APOBEC1* gene probably through a RNA interference mechanism.

Unfortunately, no antibody was available to detect by Western blot the rabbit APOBEC1 protein in intestinal cell extracts. Thus we are unable to confirm that the level of rabbit APOBEC1 enzyme was lower in rbapobec1-shRNA transgenic animals than in wild type ones.

Indirect quantitative estimation of the level of APOB mRNA editing in intestinal cells

In numerous mammals, it has been already reported that the APOBEC1 induced *APOB* mRNA editing introduces a STOP codon in the *APOB* mRNA [14]. In the rabbit species, this phenomenon is responsible for the conversion of a C residue in a U one at the 2177th codon of the rabbit *APOB* mRNA [15]. We have attempted to quantify the level of editing in the various transgenic lines and in wild type animals to test whether the reduction of *APOBEC1* gene expression could modify the *APOB* mRNA editing.

This was achieved by analyzing the chromatograms of the sequence of DNA fragments encompassing the edited nucleotide and produced in each animal by PCR using reverse transcribed intestinal RNAs as template and the LapoB48F/LapoB48R set of primers (Figure 4A and Table S1). Editing was responsible for the

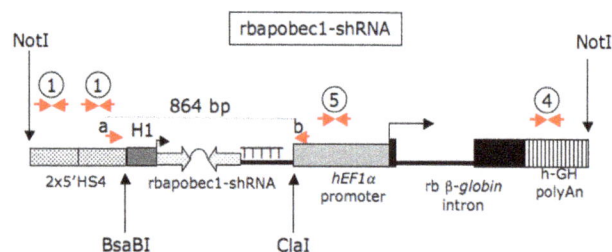

Figure 1. Structure of the rbapobec1-shRNA construct. The rbapobec1-shRNA construct encompassed the H1-rbapobec1-shRNA gene that expressed the shRNA under the activity of the H1 promoter. A gene expression insulator element (two copies of the chicken ß-GLOBIN gene fragment 5'HS4) and a transcription unit composed of the *hEF1alpha* – promoter, the rabbit ß-GLOBIN second exon and intron, and the human *GH* gene polyadenylation signal were expected to protect the shRNA expression from transcriptional extinction that occurs frequently in transgenesis. Transgenic animals were detected by PCR using sets of primers 1, 4, and 5 (Table S1). Moreover, we checked that after PCR amplification using the a/b set, a 864 bp long fragment with the expected sequence was amplified.

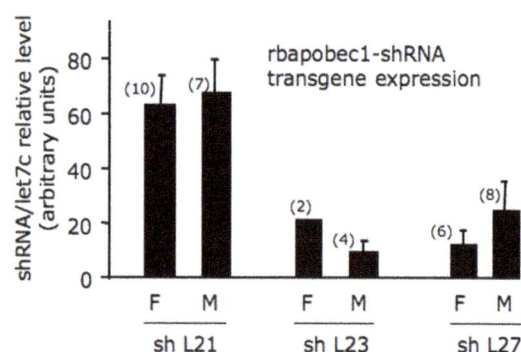

Figure 2. rbapobec1-shRNA transgene expression in rabbit intestine. The amount of shRNA targeting the rabbit *APOBEC1* mRNA was measured in RNAs prepared from duodenum cells as described in "materials and methods" section in 3 rbapobec1-shRNA lines (shL21, shL23 and shL27). Values are given in females (F) and males (M) after normalization to the level of Let7c miRNA determined simultaneously as reference gene in each sample. The number of animals in each group is indicated in brackets. Values are given with the standard error of the mean (sem). All shRNA expressing lines harbored one copy of the rbApobec1-shRNA transgene. Note that in shL21 line, the shRNA transgene expression was the hig hest compared with lines shL23 and shL27.

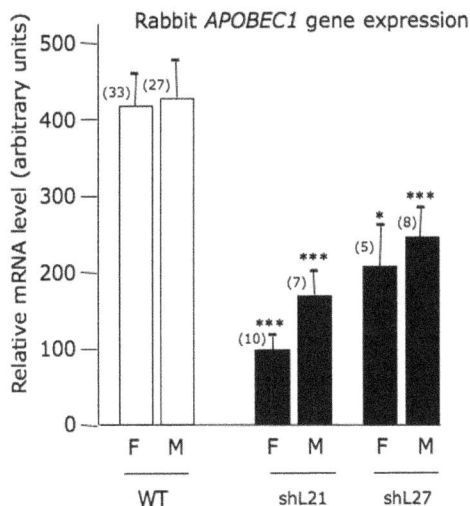

Figure 3. Expression of the rabbit *APOBEC1* gene in wild type and rbapobec1-shRNA transgenic rabbits. The amount of rabbit *APOBEC1* mRNA was measured in RNAs prepared from duodenum cells as described in "materials and methods" section in wild type animals (WT) and in two rbapobec1-shRNA lines (shL21, shL27). Values are given in females (F) and males (M) after normalization to the level of expression of three reference genes (*RPLT9, YHWAZ, HPRT*) determined simultaneously in each sample. The number of animals in each group is indicated in brackets. Values are given with the standard error of the mean (sem). Comparisons were made with control animals of the same sex (*** = $p < 0.001$; * = $p < 0.05$).

modification of the "C" nucleotide in a "T" one at the expected position in the amplified product. We postulated that after amplification, the yield of amplified fragment with a "T" residue was similar to the yield of edited mRNA. We deduced the later from the height of the peaks of each sequence chromatogram (Figures 4B and 4C).

Typical chromatograms are presented in Figure S2 showing that in wild type rabbits, editing occurred in the intestine and not in the liver. More than 95% of *APOB* mRNA was edited in the intestine in wild type animals (Figure 5). Interestingly, the level of editing was clearly lower in rbapobec1-shRNA expressing lines (shL21 and shL27, Figure S2 and Figure 5), with a significant reduction of the number of STOP/edited codon (UAA) encompassing *APOB* mRNA and a concomitant significant increase of the number of non-edited *APOB* mRNA. This led us to propose that the alteration of the level of intestinal editing in shRNA expressing animals was consecutive to the probable reduction of the level of APOBEC1 enzyme in this tissue.

APOB48 amount in plasma

The APOB48 protein is produced by the translation of the *APOBEC1* dependent edited *APOB* mRNA. Since the *APO-BEC1* gene expression and the APOBEC1 dependent editing differed in wild type and rbapobec1-shRNA expressing rabbits, it was expected that the plasma level of APOB48 protein also differed in these rabbits.

No efficient antibody was available to detect the rabbit APOB48 protein by Western blot in intestinal extracts. However, we attempted to assay the concentration of APOB48 in the plasma of rabbits using an ELISA specific for the rabbit APOB48 [16]. Firstly, we assayed APOB48 in all plasma samples collected when animals were sacrificed. Surprisingly, all values were similar to the background level of the ELISA. In Kinoshita's paper, it was

reported that the plasma level of APOB48 was enhanced in rabbits fed for at least 8 days with a cholesterol- and triglyceride-enriched regimen. Thus, we decided to feed wild type rabbits and rbapobec1-shRNA expressing rabbits with a soybean oil (8%) and cholesterol enriched (0.2%) regimen [17]. As shown in Figure 6, the plasma level of APOB48 was significantly detected in all wild type animals after feeding for 9 days with the high fat regimen. Besides, the plasma level of APOB48 was not detected in any rbapobec1-shRNA transgenic animal on the four that have been tested. We propose that the undetectable level of plasma APOB48 in plasma samples of most rbapobec1-shRNA transgenic animals was the consequence of the reduction of *APOBEC1* gene expression in the intestine and of the modification of the *APOB* mRNA editing.

APOBEC1-mediated changes in plasma lipid levels and lipoprotein distribution

We hypothesized that chylomicron formation and secretion were impaired in rbapobec1-shRNA transgenic rabbits as a consequence of the reduction of intestinal *APOB* mRNA editing, leading to modifications of the transport of dietary cholesterol and triglyceride from the intestine. With the aim to assess the extent of this phenomenon, we analyzed the concentration of cholesterol and triglycerides in the various lipoproteic fractions of the plasma. Cholesterol (total, free and esterified) and triglycerides were assayed in rabbits fed with a normal diet. The daily food intake was not different in transgenic and wild type animals. Plasma samples were collected after 20 hours fasting, and 4 hours after re-feeding. As shown in Figure 7, after 20 hours fasting, and in all classes of lipoproteins, the concentration of lipids was not different in wild type and in transgenic rabbits (comparison of white bars in WT and shL21 rabbits in each lipid fraction). Besides, after re-feeding, the expected increase of triglycerides and cholesterol in the chylomicron + VLDL fraction was significantly reduced in rbapobec1-shRNA transgenic animals (line shL21, comparison of starved and fed rabbits in each category). Indeed, after feeding, the levels of cholesterol and triglycerides increased clearly in the chylomicron + VLDL fraction in wild type animals only, and not significantly in transgenic animals. As regard to the other classes of lipoproteins (LDL and VLDL) and after re-feeding, there were no significant differences between wild type and shRNA transgenic animals.

The ordinary diet of the rabbit is devoid of cholesterol and poor in lipids (around 2% instead of 8% in the high fat diet). To further investigate the lipoprotein distribution in the rbapobec1-shRNA transgenic rabbits and their ability to respond to a high fat diet challenge, animals were fed with a diet enriched with triglycerides and cholesterol. In this experiment, as in the alimentary challenge performed with the normal diet, the daily food intake was not different in transgenic and wild type animals. Triglycerides and cholesterol were assayed in plasma samples collected after 8 days feeding with the enriched diet, after a further 20 hours fasting, and 4 hours after re-feeding with the enriched diet. The pattern of triglycerides concentration in the plasma differed clearly in wild type and transgenic animals (Figure 8). Indeed, the plasma concentration of triglycerides was not enhanced after high fat feeding in transgenic animals as it was in wild type rabbits. More precisely, the chylomicrons + VLDL fraction was not enhanced by the diet challenge in transgenic rabbits. Taken altogether, these data support the hypothesis of an inability of the transgenic animals to produce rapidly large amounts of chylomicrons + VLDL in response to the food supply. The food intake being similar in all animals, this suggests a lower lipid absorption in transgenic animals than in wild type ones.

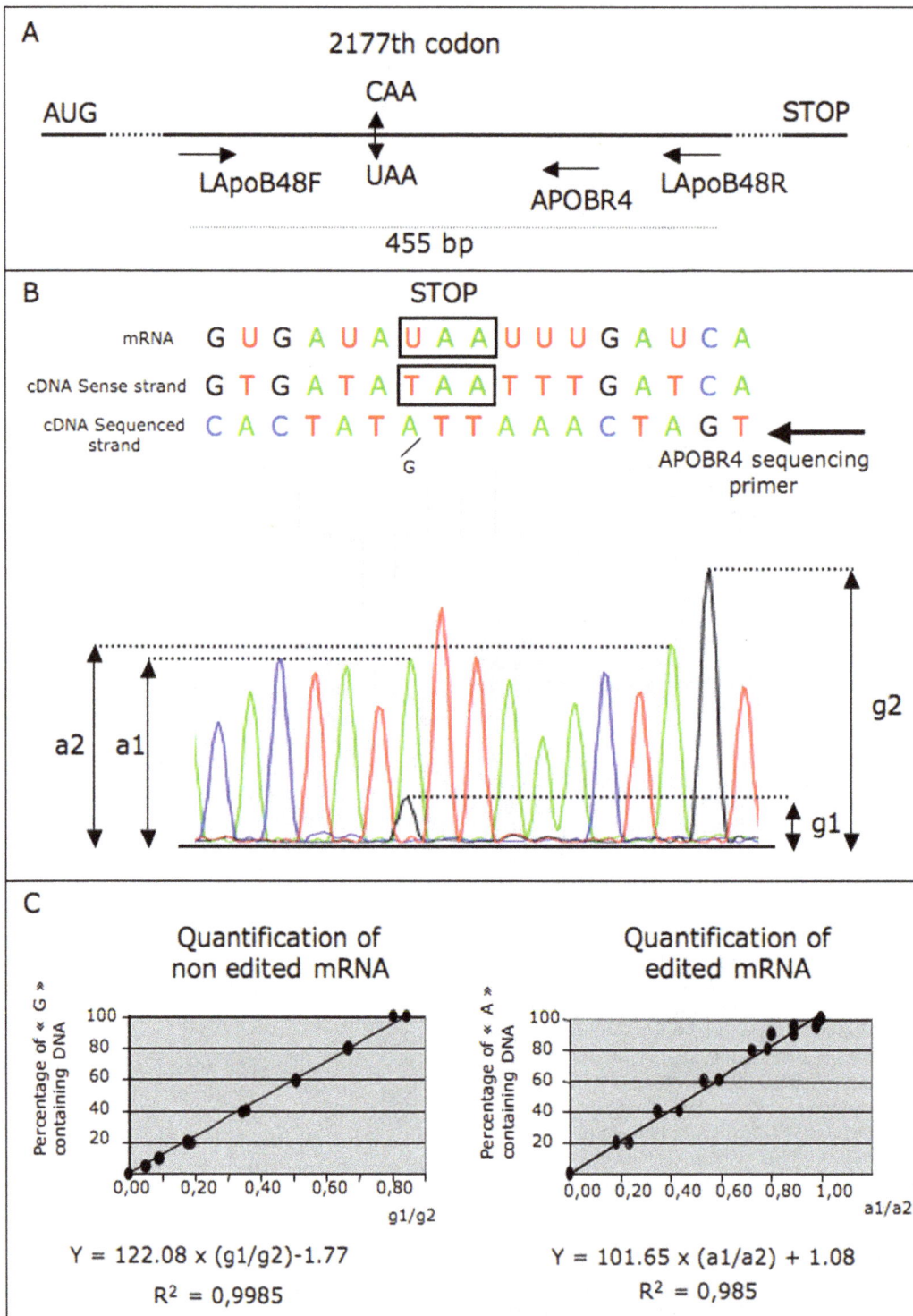

Figure 4. Indirect estimation of the level of "CAA" to "UAA" editing. A: schematic representation of the rabbit *APOB* mRNA from the AUG translation initiation codon until the STOP codon. At the 2177th codon, the "C" residue is edited in a "U" residue. Using reverse transcribed RNA as template, the LApob48F/LApoB48R set of primers amplifies a 455 bp long amplicon encompassing the 2177th codon. When using the APOBR4 primer as sequencing primer, the chromatogram shows the antisense sequence. **B:** detail of a characteristic chromatogram showing how the heights of the peaks were measured at the level of the 2177th codon. Here, the "A" residue was the major one (a1), and the "G" the minor one (g1). Consequently, a large majority of DNA strands in this mixture encompassed the edited TAA (STOP) codon at position 2177. (a2) and (g2) are measured as references. **C:** standard equations obtained by plotting the a1/a2 and g1/g2 ratios against the amount of "A" or "G" containing DNA 455 bp fragment in the sequenced sample. Amounts are given as percentage of "A" or "G" containing DNA.

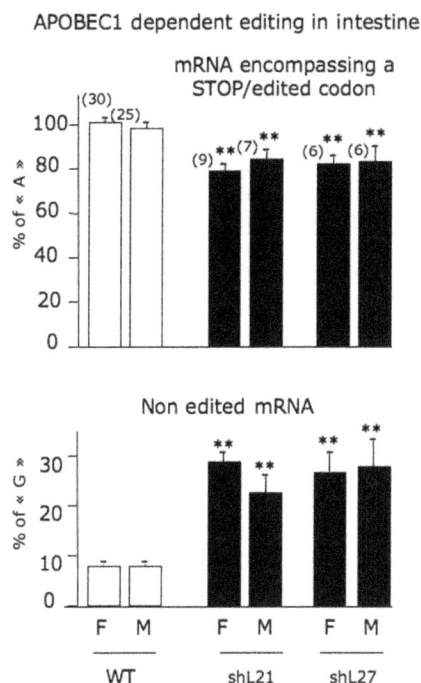

APOBEC1 dependent editing in intestine

Figure 5. Indirect estimation of editing in wild type and rbapobec1-shRNA transgenic rabbits. APOBEC1 dependent editing was measured in the intestine of wild type and transgenic animals. Values were deduced from sequence chromatograms of a PCR fragment encompassing the edited codon as described in "material and methods" section and in Figure 4. The amount of DNA with a "A" residue was representative of the amount of APOB mRNA with a 2177th STOP/edited codon; the amount of DNA with a "G" residue was representative of the amount of full length APOB mRNA. The number of studied animals in each group is indicated in brackets. Mean values are given as percentages with the standard error of the mean (sem). Comparisons were made with control animals (** = p<0.001).

Interestingly, in the high fat diet animals, and not in the normal diet ones, the concentration of triglycerides and cholesterol in the HDL fraction was obviously reduced in transgenic animals compared to wild type animals. This could be related to the low rate of synthesis of chylomicrons by the intestine in transgenic animals, since chylomicrons and their remnants contribute significantly to the production of HDL.

Storage of total body lipids

Since the production of chylomicrons was impaired in transgenic rabbit, one could expect that the uptake of lipids from the diet would be reduced leading to a decreased storage of lipids. To assess this hypothesis, the total mass of body fat was estimated using TOBEC analysis at around 12–16 weeks after birth. The total mass of fat was always the lowest in rbapobec1-shRNA transgenic animals (Figure 9), and the highest in wild type animals. We propose that the reduced body mass of lipids in rbapobec1-shRNA transgenic animals was the result of the reduced uptake of diet lipids consecutive from a low production of chylomicrons and low absorption of fatty acids.

Growth curves in transgenic and wild type rabbits

Our main objective was to study whether modifications of APOBEC1 gene expression in the intestine induced a lean phenotype in the rabbit species. Thus, all animals were weighed weekly from birth during 12–18 weeks. All transgenic litters were

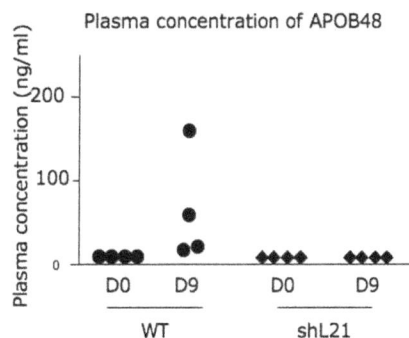

Figure 6. Plasma concentration of APOB48 in rabbits challenged by a high fat/high cholesterol regimen. Plasma concentration of APOB48 was assayed by a specific ELISA kit. Four wild type rabbits and four transgenic rabbits expressing the rbapobec1-shRNA transgene were fed ad libitum with a high fat/high cholesterol regimen for 9 days. Blood samples were collected before the high fat/high cholesterol regimen (D0) and 9 days after the starting of the regimen (D9). Each point indicates the plasma concentration of APOB48 (in ng/ml) in one animal.

obtained by breeding a transgenic male with a wild type female. Newborns being thus nourished by wild type mothers, this eliminated any possible incidence of the transgenic milk on growth.

At birth, the weight of newborns was not significantly different whatever animals were transgenic or not. However, after three weeks and for the whole length of the experimentation, transgenic rabbits expressing the shRNA targeting the rabbit APOBEC1 gene (shL21 and shL27) were always the lightest animals (by 10% to 20%) as shown within each litter (Figure 9). Thus, this led us to conclude that the rbapobec1-shRNA transgene expression induced actually a lean phenotype in the rabbit species. The lean phenotype could result from the low production of chylomicrons + VLDL possibly leading to a reduced uptake of diet lipids and a reduced absorption of energy deriving from fatty acids. However, additional experiments should be performed to confirm this hypothesis, and specifically to study whether the energy expenditure was affected in a different manner in transgenic and wild type animals.

Rescue of the normal phenotype in double transgenic rabbits expressing both the rbapobec1-shRNA and the human APOBEC1 gene

In order to eliminate the possibility that the lean phenotype was not consecutive to the reduction of rabbit APOBEC1 gene expression but was due to any other phenomenon induced by the rbapobec1-shRNA transgene, we decided to produce double transgenic rabbits expressing simultaneously the rbapobec1-shRNA transgene and the human APOBEC1 gene.

We first produced transgenic rabbits expressing the human APOBEC1 gene in the intestine through the tissue specific activity of the rat IFABP gene promoter [18] added in the construct (Figure 10A). Fifty-four rabbits were born after microinjection of the NotI insert, giving 4 (7.4%) rIFABP-hAPOBEC1 transgenic founders. Transgenic lines were successfully established from 2 (L01 and L02) of these founders, harboring respectively 2 and 6 copies of the transgene. A small number of double transgenic animals expressing both the human APOBEC1 gene and the shRNA targeting the rabbit APOBEC1 mRNA were produced by breeding rIFABP-hAPOBEC1 (L01 or L02) and rbapobec1-shRNA transgenic lines (shL21 or shL27). The analysis of

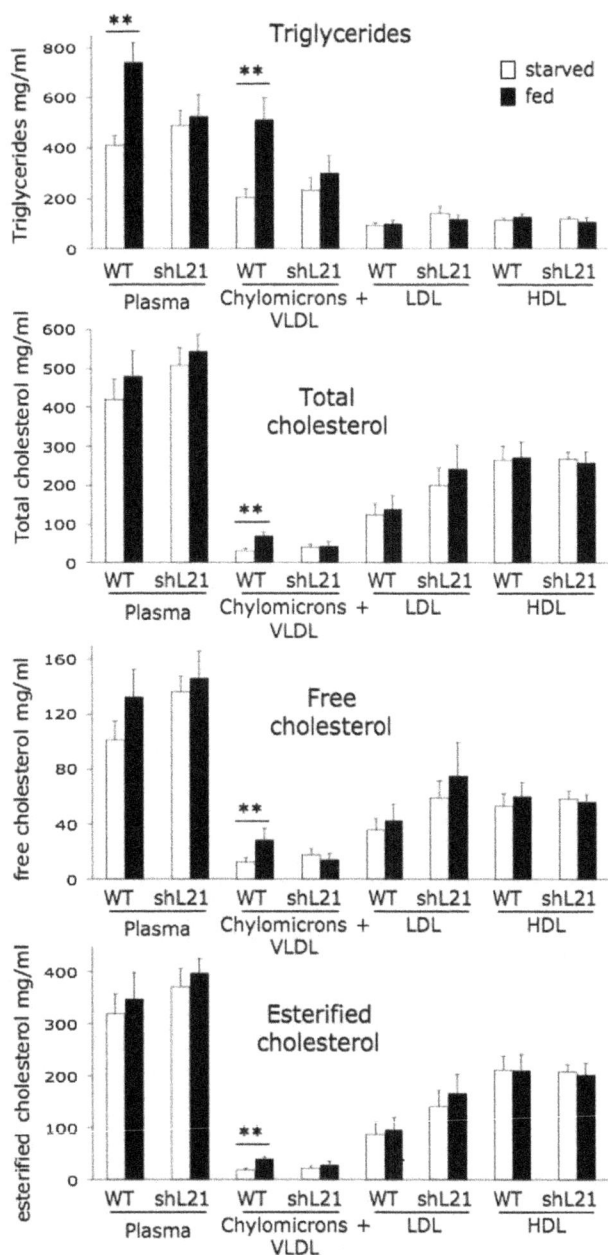

Figure 7. Plasma concentrations of triglycerides and cholesterol (total, free and esterified) in rabbits fed with a normal diet. Triglycerides and cholesterol were assayed in the plasma, and in three lipoproteic compartments separated by ultracentrifugation. Blood samples were collected in rabbits (6 wild-type and 7 transgenic rabbits from line shL21) fed with a normal diet and starved for 20 hours (white bars) and 4 hours after re-feeding with the normal diet (black bars). Values are given in mg/ml, with the standard error of the mean. Comparisons were made between starved and fed animals within each group (** = p<0.001).

transgenic lines L01, L02 and double transgenic rabbits is presented in Figures 10 and 11.

Both transgenic lines L01 and L02 expressed the human *APOBEC1* gene stably over generations, with similar levels in males and females (Figure 10B, left graph). In double transgenic lines, the level of the human *APOBEC1* gene expression was not significantly different from that in lines L01 and L02, which proves

that the shRNA produced by the rbapobec1-shRNA transgene did not alter the expression of the human *APOBEC1* transgene. The presence of the human APOBEC1 enzyme was confirmed in the intestine by western blot assay in line L02 (Figure S3) with the expected 27 kD molecular weight. Interestingly, in the transgenic rIFABP-hapobec1 line L02, an unexpected leaking expression of the rIFABP-hapobec1 transgene was detected in the liver but with a 50 times lower level than in the intestine.

As expected, in double transgenic rabbits, the rbapobec1-shRNA was significantly expressed in the intestine (Figure 10B middle graph).

The level of rabbit *APOBEC1* gene expression was similar in transgenic rabbits expressing the human *APOBEC1* gene and in wild type animals (Figure 10B, right graph). Besides, it was lower in double transgenic rabbits, as we had previously observed in rabbits from lines shL21 and shL27. Thus, as we already suggested, the shRNA targeted the expression of the intestinal rabbit *APOBEC1* gene probably through a RNA interference mechanism, without altering that of the human *APOBEC1* gene.

In transgenic animals expressing the human *APOBEC1* gene, the level of editing was at around 95% of the maximum, as it was previously determined in wild type animals (Figure 10C). This was surprising since we were expecting for an increase consecutive to the additional human APOBEC1 enzyme. Though, the human APOBEC1 enzyme was actually efficient in *APOB* mRNA editing in the rabbit as the rabbit APOBEC1 enzyme. Indeed, editing was observed in the liver of some transgenic rIFABP-hapobec1 animals (Figure 10C, line L02, middle panel) harboring a leaking expression of the human *APOBEC1* transgene in the liver, when editing is never observed in liver in wild type rabbits. Thus, the lack of any modification in *APOB* mRNA editing in transgenic animals over-expressing the APOBEC1 enzyme was not due to the inefficacy of the enzyme but probably the consequence of the saturation of the mechanism of editing.

In double transgenic animals, the level of editing was similar to that of wild type animals, despite the reduced expression of the rabbit *APOBEC1* gene in the intestine. This proves that the human APOBEC1 enzyme expressed in the intestine by the transgene was able to counterbalance the default of rabbit APOBEC1 enzyme due to the shRNA targeting the rabbit *APOBEC1* mRNA.

Interestingly, the plasma level of APOB48 was highly enhanced in the human *APOBEC1* transgenic rabbits L02 by the high fat/high cholesterol diet challenge (Figure 10C, right graph). Since editing was not modified in the intestine of these animals, it is likely that the high plasma concentration of APOB48 originated from the liver, where a significant editing of the *APOB* mRNA was measured consecutively to the leaking expression of the human *APOBEC1* transgene.

The plasma lipid levels and lipoprotein distributions were assayed in human *APOBEC1* transgenic rabbits (L02, Figure S4) submitted to the high fat/high cholesterol diet and starvation/feeding challenge. Surprisingly, the concentrations of triglycerides in the plasma and also in the chylomicrons + VLDL fraction were not enhanced by the diet, by opposition to what we were expecting for in these rabbits characterized by a high level of circulating APOB48. Clearly, in these animals, the high circulating APOB48 did not contribute to a high synthesis of chylomicrons. Other differences were further detected throughout the starvation/feeding challenge. These could be consecutive to the leaking expression of the human *APOBEC1* gene in the liver, which induced the liver editing of the *APOB* mRNA and thus the reduction of the hepatic synthesis of APOB100 protein.

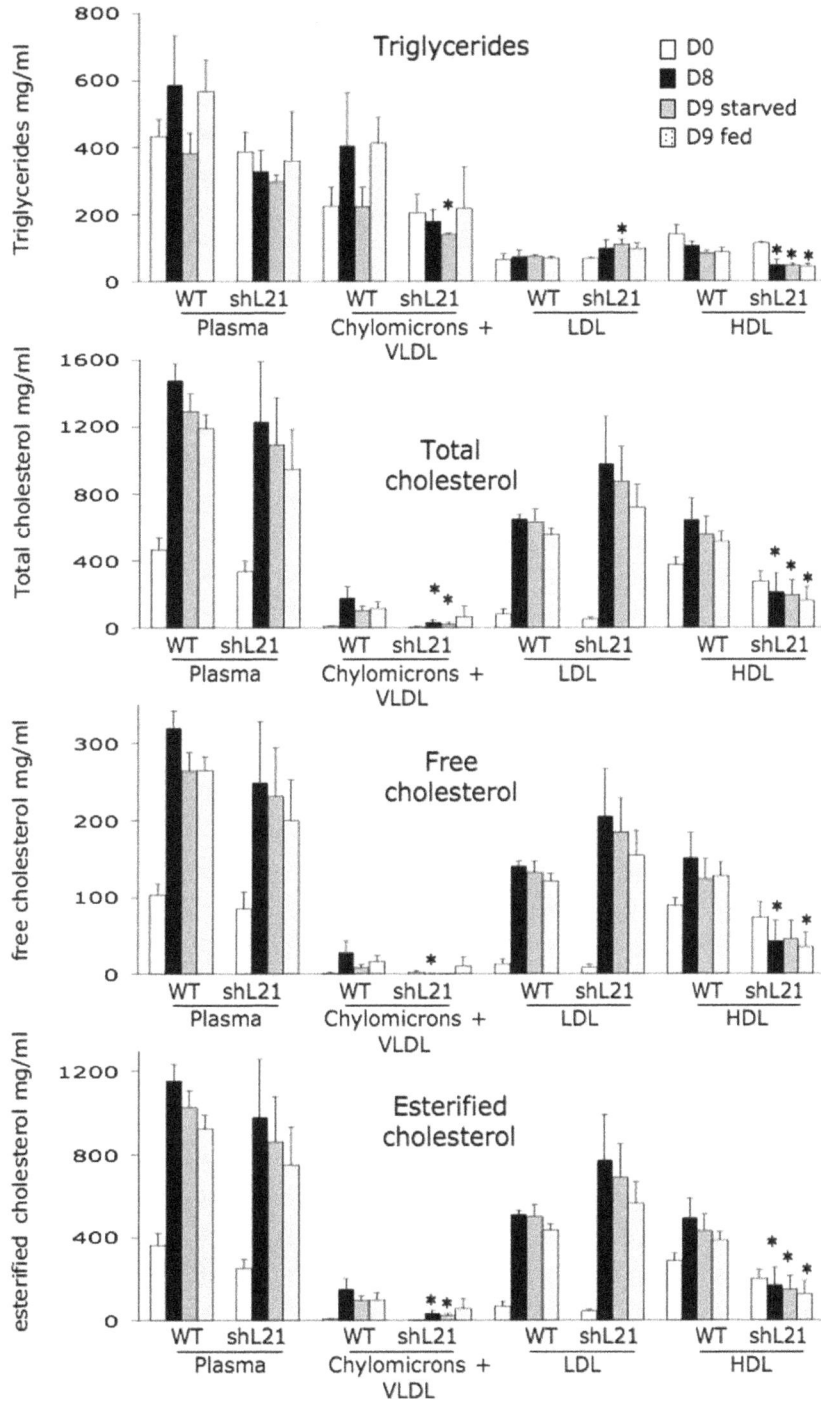

Figure 8. Plasma concentration of triglycerides and cholesterol in rabbits fed with a high fat/high cholesterol regimen. Rabbits (4 wild type, and 3 transgenic rabbits from line shL21) were fed for 8 days with a high fat/high cholesterol diet. Plasma samples were collected before the diet (D0, white bars), after feeding for 8 day with the diet (D8, black bars), after 20 hours starvation (D9 starved, grey bars) and 4 hours after re-feeding with the high fat diet (D9 fed, dotted bars). Triglycerides and cholesterol were assayed as in Figure 7. Values are given in mg/ml, with the standard error of the mean. Comparisons were made between transgenic and wild type animals for each day of the challenge (* = p<0.05).

The total mass of body lipids and growth curves were determined from a series of litters including newborns of each genotype (wild type, rbapobec1-shRNA, rIFABP-APOBEC1, and double transgenic animals, Figure 11). The transgenic animals expressing the human *APOBEC1* gene gained weight and possessed a total lipid mass as the wild type animals. This was

not surprising since in these transgenic animals, the APOB mRNA editing and the production of chylomicrons were similar to those determined in wild type rabbits. A small number of animals of rIFABP-hapobec1 transgenic lines L01 and L02 were weighed for a longer time (Figure S5), in order to detect possible long-term modifications consecutive to limited but sustained modifications of

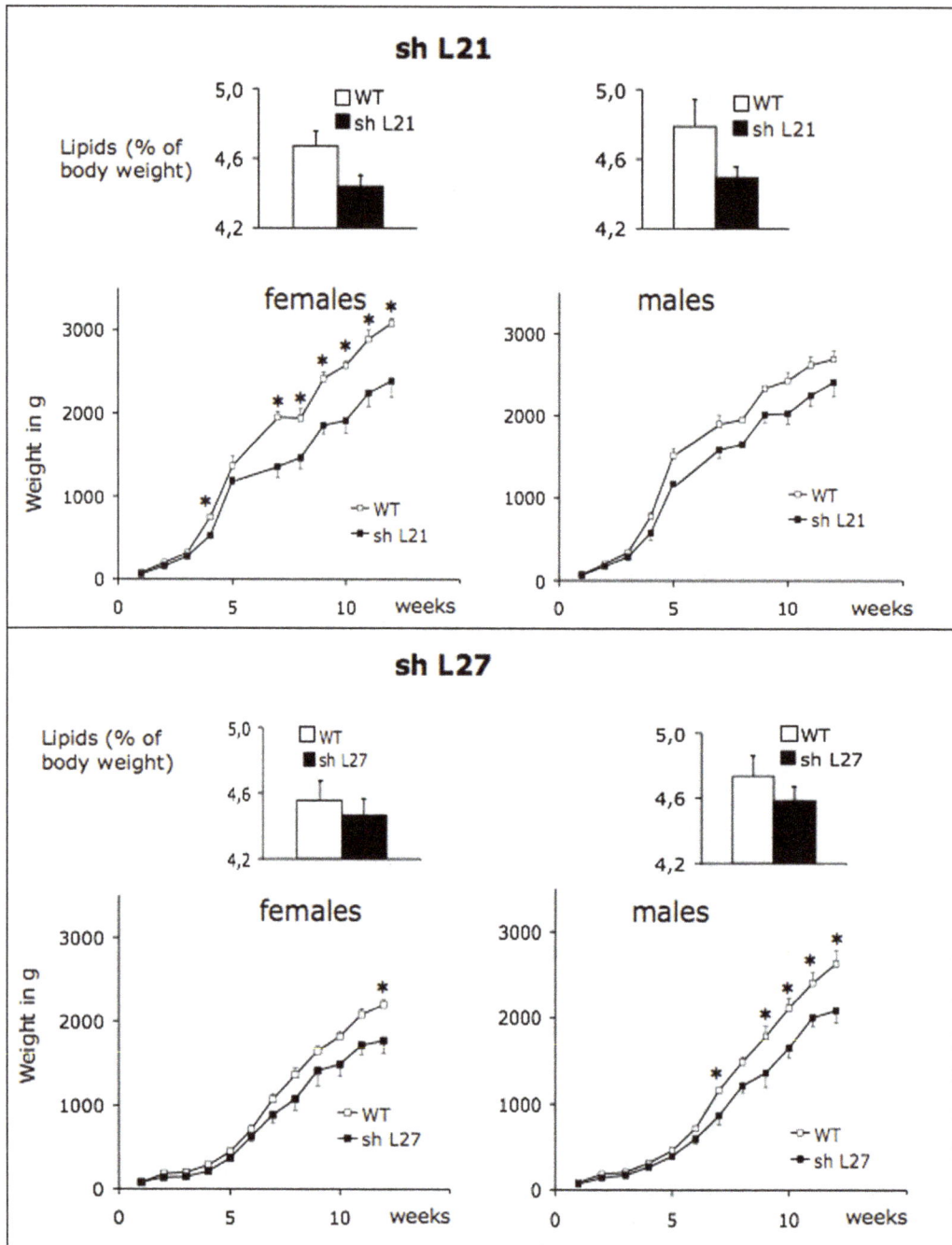

Figure 9. Total content of body lipids and growth curves of wild type and transgenic rabbits from lines shL21 and shL27. The total content of body lipids and growth curves were established on the same rabbits. All rabbits (mothers during pregnancy and lactation and their litters after weaning) were fed with the normal diet. Wild type mothers nourished all newborns (transgenic or wild type ones). The total content of body lipids, expressed as the percentage of the body weight, was measured in transgenic (shL21 and shL27, black bars) and wild type (white bars) rabbits at around 12–16 weeks after birth. Three animals at least were considered for each point. Values are means +/− sem. Note that the percentage was always the lowest in shRNA expressing animals, and the highest in wild type animals. Growth curves were established by weighing weekly each rabbit from 3–5 weeks to 12–18 weeks after birth. Males and females are shown in separate graphs. * = p<0.05 comparison of shRNA expressing animals and wild type ones.

the level of editing that we might have not been able to detect earlier. However, the weight of transgenic animals was not different from that of wild type ones, showing that even in older animals, the long term-expression of human APOBEC1 enzyme induced no significant over-weight gain.

More interestingly, in spite of the small number of animals of each genotype in the litters, the double transgenic animals were clearly heavier than the shRNA expressing animals and their total mass of body lipids was similar to that of wild type animals. This shows once more that the presence of the human APOBEC1

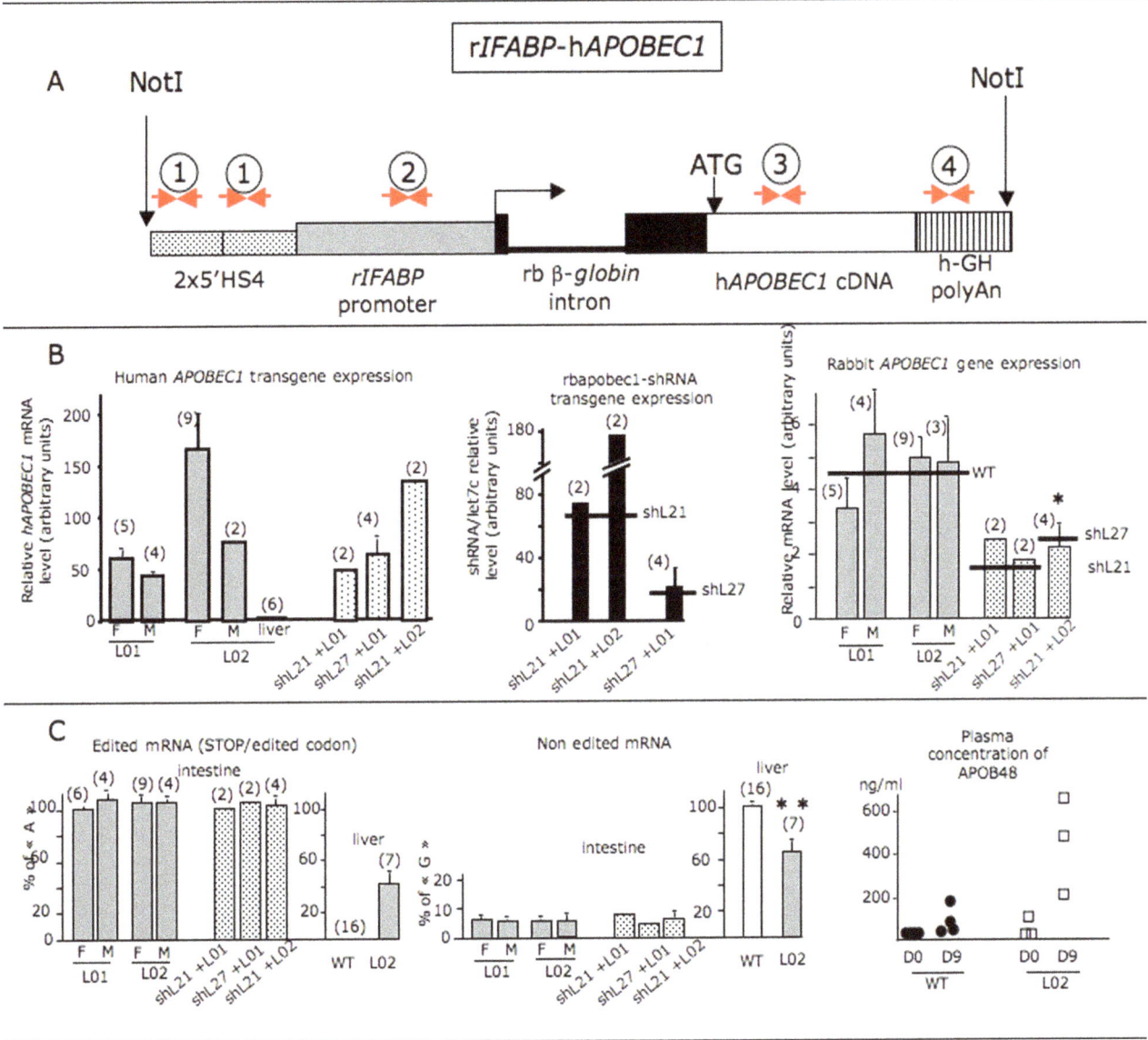

Figure 10. Analysis of rIFABP-hapobec1 transgenic and double transgenic rabbits. 10A: Structure of the recombinant gene to express the human APOBEC1 cDNA in the intestine of transgenic rabbits. The rIFABP-hAPOBEC1 construct encompassed two copies of the chicken ß-GLOBIN gene fragment 5'HS4 (gene expression insulator element, dotted box), the promoter of the rat intestinal fatty acid binding protein gene (rIFABP; grey box), the rabbit (rb) ß-GLOBIN second intron (black boxes and thick line), the human APOBEC1 cDNA (white box) produced by PCR amplification from reverse transcribed RNA of HT29 cells (derived from a human colon tumor that have retained the ability to express the APOBEC1 gene), and the human growth hormone polyadenylation sequences (box with vertical bars). The horizontal black arrow points the position of the transcription start site. ATG = translation initiation site of hAPOBEC1 cDNA. Numbers and small horizontal arrows represent the sets of primers. All studied transgenic animals were PCR positive for the sets 1–4. **10 B: gene expression.** The levels of human APOBEC1 mRNA (left panel) and shRNA (middle panel) were measured in RNAs prepared from duodenum cells in two transgenic lines (L01 and L02) and in double transgenic animals (shL21+L01; shL21+L02; shL27+L01). Values are given in females (F) and males (M) after normalization to the level of reference gene expression determined simultaneously in each sample: Let7c miRNA in the case of shRNA, and RPL19, YHWAZ, HPRT in the case of human APOBEC1. In double transgenic animals, males and females were not distinguished, considering the small number of animals in these groups. The number of animals in each group is indicated in brackets. Values are given with the standard error of the mean (sem). The mean level of shRNA in shL21 and shL27 as presented in Figure 2 is indicated with a horizontal line. The level of expression of the human APOBEC1 transgene measured in the liver is given in L02. In L01, this level was not significantly detected. The level of rabbit APOBEC1 mRNA (right panel) was measured in intestinal RNAs as described in Figure 3. The level found in wild type rabbits and in lines shL21 and shL27 is indicated with a horizontal line. **10 C: APOBEC1 dependent editing in intestine and liver and Plasma concentration of APOB48 in rabbits expressing the human APOBEC1 gene.** The estimation of editing was made as described in the legend of Figure 5. Plasma concentration of APOB48 was performed as described in legend of Figure 6 in 3 transgenic rabbits from line L02. Wild type animals are the same than those in Figure 6.

enzyme was able to counterbalance the effect of the shRNA targeting the rabbit APOBEC1 gene. Taken altogether, our results suggest strongly that the lean phenotype observed in rbapobec1-

shRNA transgenic rabbit was the consequence of the reduced level of APOBEC1 gene expression.

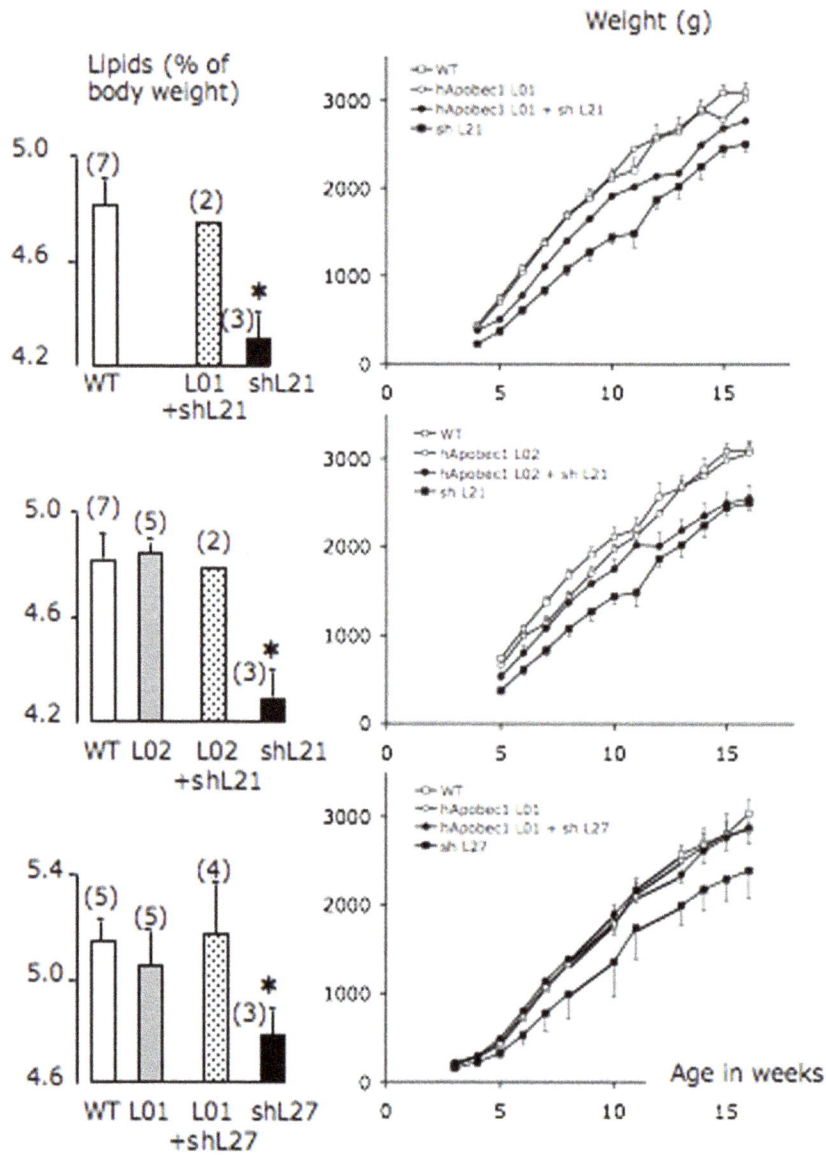

Figure 11. Total mass of body lipids and growth curves in double transgenic rabbits. Double transgenic animals (shL21+L01; shL21+L02; shL27+L01) were produced by breeding rIFABP-hAPOBEC1 (L01 or L02) and rbapobec1-shRNA transgenic lines (shL21 or shL27). In these litters, the total mass of lipids was significantly lower in shL21 or shL27 transgenic animals than in animals from all other groups (* = p<0.05). Numbers in brackets indicate the number of animals in each group. Growth curves were established by weighing weekly each rabbit from 3–5 weeks to 12–18 weeks after birth. Males and females are shown in separate graphs. * = p<0.05 comparison of shRNA expressing animals and wild type ones.

Discussion

A great number of genes are devoted to the storage of energy, and it is reasonable to propose that evolution has selected organisms able to survive in scarce conditions thanks to efficient mechanisms of energy storage. Limiting energy uptake and storage is probably a valuable strategy to fight against obesity. Thus, our approach consisted of looking for critical genes in people with lean phenotype. If a monogenic slimness disease resulting from a deficiency of fat absorption can be found, the implicated gene likely plays a critical role in the disease and is a potential target for new anti-obesity drugs. When this gene is not compensated by other mechanisms, it is therefore a powerful target for obesity treatment.

Three human genetic diseases have been described with very similar lean phenotypes: abetalipoproteinemia, hypobetalipoproteinemia, and chylomicron retention disease also known as Anderson's disease [19]. The genes involved in the first two diseases, abetalipoproteinemia and hypobetalipoproteinemia, have now been identified, but it is not yet the case in the Anderson's disease [20,21]. All three diseases are characterized by a severe reduction or total absence of APOB48 protein in intestinal cells and plasma and of chylomicrons production. This led us to investigate further the possibility of fighting against obesity through regulating APOB48 production. APOB48 resulting exclusively from the translation of the APOBEC1 dependent edited *APOB* mRNA, we decided to target the expression of the *APOBEC1* gene.

The phenotype of mice harboring a complete invalidation of the *APOBEC1* gene has been already reported by a series of laboratories [22–25]. As expected, the editing of the *APOB* mRNA was suppressed, and no APOB48 was produced in these mice. It was observed that intestinal fat absorption was less efficient in $APOBEC1^{-/-}$ mice containing only APOB100 than in wild type mice but it was not totally abolished. Probably, APOB100 could replace to some extent APOB48 in chylomicron formation and finally the plasma lipoprotein cholesterol and triglycerides profiles were not different in knock out and wild type mice [7] [8]. In contrast, in the human, APOB100 is not able to form chylomicrons and carry lipids from intestine to liver [26]. Clearly, the metabolism of lipids differs between species, and thus we decided to target the *APOBEC1* gene in another species than the mouse and closer to the human as regard to the metabolism of lipids, in order to investigate whether the *APOBEC1* dependent editing could be a valuable target for fighting against obesity through modulating the lipid uptake.

RNA interference is a natural cellular process mediated by small double strand RNA that induces knockdown of gene expression through mRNA targeting. Here, we produced transgenic rabbits expressing permanently a small interfering RNA (siRNA) targeting the rabbit intestinal *APOBEC1* gene. This was achieved through the introduction in the rabbit genome of a DNA construct expressing a small hairpin RNA by using a strategy that we had followed in a previous study [12]. This strategy had the advantage to provoke the sustained production of the siRNA, and a moderate but significant and permanent decrease of the rabbit *APOBEC1* gene expression. Our objective was to observe long term and prolonged effects of the gene knockdown that is totally different to what can be observed after a total invalidation of the gene. To validate our findings, we produced transgenic rabbits expressing the human *APOBEC1* gene in the intestine, with the aim to rescue the knockdown induced by the RNA interference mechanism.

The Figure 12 presents a model that could explain how targeting the *APOBEC1* gene induces a lean phenotype in the rabbit species. In wild type rabbits (Figure 12 A), the *APOBEC1* gene expressed in the intestine only is responsible for the editing of the *APOB* mRNA. The APOB48 protein thus produced exclusively by the intestine is processed to synthesize chylomicrons that are responsible for the lipid uptake from the diet. In the liver, the APOB mRNA is translated in the APOB100 protein that is involved in the synthesis of VLDL and LDL. In transgenic rabbits expressing the rbapobec1-shRNA (Figure 12B), the level of *APOBEC1* gene expression and the level of editing are significantly reduced in the intestine. The ability to synthesize chylomicrons in response to a diet challenge is reduced. The productions of LDL and VLDL are not significantly modified; but the production of HDL is modified since HDL is processed from chylomicrons and remnants that are reduced. The lean phenotype is thus consecutive to the reduced lipids uptake by the enterocytes through the low synthesis of chylomicrons after each food challenge.

The absence of obese phenotype in transgenic rabbits expressing the human APOBEC1 enzyme (Figure 12C) in the intestine is amazing. Indeed, we were expecting for a long-term gain of weight in these animals since the human *APOBEC1* transgene should have enhanced the level of editing of the *APOB* mRNA. It has been already published that in the rabbit species, around 90% of the *APOB* mRNA was edited in the intestine [14]. Accordingly, our results have shown that more than 95% of *APOB* mRNA were edited. However, in the present study, the expression of the human *APOBEC1* transgene was not able to enhance the level of editing, and consequently, animals did not elicit any obese phenotype. One

explanation is that the editing was already at its maximum in wild type animals, and that the over-expression of APOBEC1 was inefficient. It has been reported that the editing results from the activity of the APOB mRNA editing complex (the editosome), a multicomponent protein complex with enzymatic and regulatory activities [27]. Thus, we suggest that the editing was limited by the availability or activity of other components of the editing complex.

More surprisingly, in transgenic rabbits expressing the human *APOBEC1* gene, the plasma level of triglycerides in the chylomicrons + VLDL fraction was not enhanced after a fat-rich diet for 8 days (Figure S4) as it was in wild type rabbits. Yet, we were expecting for an enhanced production of chylomicrons, since the plasma level of APOB48 was clearly enhanced in these animals (Figure 10C). Notably, the excess of APOB48 protein was produced by the liver, in response to the leaking expression of the human *APOBEC1* transgene. Thus, it suggests that the APOB48 protein originating from the liver was not processed into chylomicrons, and that only the APOB48 produced by the intestinal cell can form chylomicrons. Finally, this study shows that in spite of a significant high expression of the human *APOBEC1* gene in the intestine, no obese phenotype can be observed in the rabbit species.

In double transgenic rabbits (Figure 12D), the expression of the human *APOBEC1* gene in the intestine counterbalances the shRNA induced knockdown of the rabbit *APOBEC1* gene expression. Consequently, the level of editing is similar to that in wild type animals, as the level of synthesized chylomicrons. The rescue of the normal phenotype in the double transgenic rabbits is a solid argument to demonstrate the relation between *APOBEC1* gene expression and the lean phenotype.

In conclusion, this study presents for the first time evidences that targeting the *APOBEC1* gene is a valuable strategy to induce a lean phenotype in the rabbit. Importantly, it has to be further confirmed that the lean phenotype is the consequence of a moderate modification of the lipid uptake through chylomicrons. Remarkably, animals did not suffer from any disease, and their breeding capacity was not apparently affected for more than two years that is a long experimental duration in this species. However, a series of additional experiments should be performed to investigate the impact of targeting the APOBEC1 gene expression on health, by specifically studying the impact of alterations of intestinal absorption of lipids and other nutrients. Finally, the success of this strategy lies probably in the fact that it concerns a gene pathway without compensatory mechanisms that affects the lipids uptake of the diet. Moreover, it suggests that looking for new genes associated to lean phenotype is probably a valuable tool to highlight novel targets for obesity treatment.

Materials and Methods

Animals

Californian rabbits (GD24 strain) were bred at the UCEA rabbit facility (Unité Commune d'Expérimentation Animale, Jouy-en-Josas, France). All experiments were performed with the approval of the local committee for animal experimentation (COMité d'ETHique appliqué à l'Expérimentation Animale (COMETHEA), Jouy-en-Josas, accreditation number 12/017). All researchers working directly with the animals possessed an animal experimentation license delivered by the French veterinary services.

All rabbits were weighed each week from birth until week 12–18, few before puberty that occurs at around 20–24 weeks in this species. Breeders were nourished with a normal diet. Since the growth rate depends on the number of newborns in each litter,

A : WILD TYPE RABBIT

B : rbapobec1-shRNA transgenic rabbits

C : Human APOBEC1 transgenic rabbits

D : Double transgenic rabbits

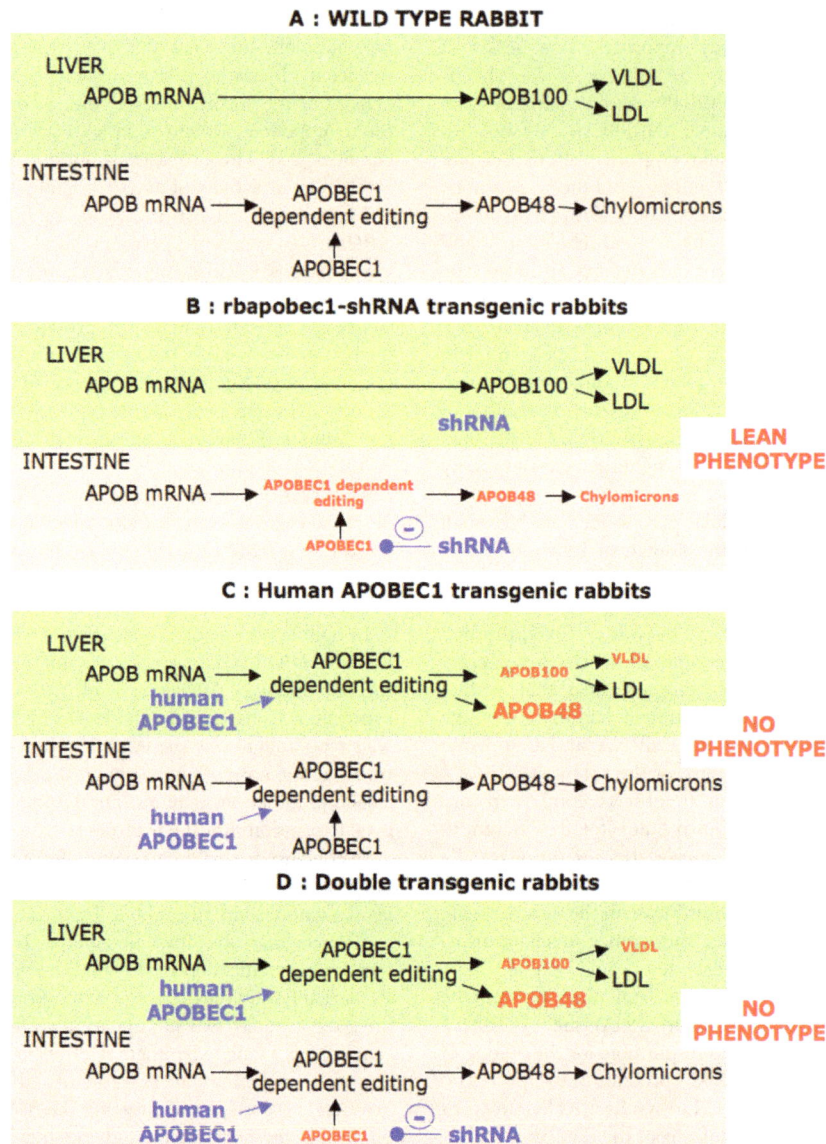

Figure 12. Intestinal and liver regulation of APOB48, APOB100, and chylomicron production. Four schematic representations are given, to simulate the regulation of *APOB* mRNA editing and the consequence upon the phenotype in wild type and transgenic rabbits. The models depict the situation after a diet challenge with normal of high fat/high cholesterol diet. Blue characters are used for the transgene expression (shRNA targeting the rabbit *APOBEC1* mRNA, and human *APOBEC1* gene); red characters indicate the measured parameters with significant modifications; the size of the letters is related to the level of production.

care was taken to compare rabbits issued from litters encompassing approximately the same number of newborns. Animals were currently weaned at around 7 weeks. After weaning, animals were fed with the normal diet except otherwise mentioned.

At around 18 weeks after birth, animals were starved for 24 hours, blood samples were collected on EDTA to prevent for coagulation, and food was immediately provided. Four hours after feeding, animals were sacrificed; blood and tissue samples were collected. Blood samples were centrifuged (10 minutes, 3000 g), then plasma and tissues were frozen at −80°C until used.

Construction of recombinant genes

The rbapobec1-shRNA-producing gene (Figure 1) encompassed two inverted repeats of the rbapobec1-shRNA and a stretch of five T residues as gene terminator. The shRNA

transcription unit was constructed from synthetic DNA fragments (Eurofins, Ebersberg, Germany). This H1-rbapobec1-shRNA gene was then inserted into the pM10 vector [12] at the enzymatic BsaBI-ClaI restriction sites as presented in Figure 1. The final construct used for microinjection encompassed the DNA fragment included between the two NotI restriction sites.

The human *hAPOBEC1* construct (rIFABP-*hAPOBEC1*, Figure 10) encompassed the human *APOBEC1* cDNA from 3 nt upstream of the site of initiation of translation (ATG) to 3 nt downstream the STOP codon linked to the rabbit (rb) ß-*globin* second intron. Transcription was driven by the promoter of the rat intestinal fatty acid binding protein gene (r*IFABP*) spanning from nucleotides −1150 to +51 as regard to the *IFABP* gene transcription start site. A tandem of the chicken ß-*globin* gene fragment 5'HS4 was added as insulator. The termination signal was brought by the terminator from the human growth hormone gene (h-GH polyAn).

The sequences of all plasmids encompassing these constructs are available upon request.

Generation of transgenic rabbits

The inserts to be used for microinjections were released from plasmids by NotI digestion, separated on 1% agarose gel in 1x TBE, purified using the Qiaquick gel extraction kit (Qiagen, Courtaboeuf, France) and then EluTipD filtration (Schleicher & Schuell, Mantes la Ville, France). The resulting DNA preparations were microinjected into Californian rabbit embryo pronuclei at a concentration of 2 ng/µl. The transgenic rabbits were identified using PCR performed on ear clip DNA extracts. Four sets of primers were designed to cover the integrity of each integrated construct: sets 1, 2, 3 and 4 for the *hAPOBEC1* expressing construct, and sets 1, 4, 5 and a/b for the shRNA expressing construct (Figures 1 and 10, table S1). Sets 1, 2, 3, 4, and 5 were used in real time PCR with the fast SYBR Green master mix (Applied Biosystems). In parallel, a set of primers (cas1, cas2, table S1) amplifying a non-coding region upstream of a control gene (the rabbit ß-*CASEIN* gene) was used as control amplification. All sets of primers were designed by the Primer Express software (Applied Biosystem) and all amplicons were 100 base-pairs long.

The number of copies of integrated gene was deduced from real time PCR amplifications by the $2^{(\Delta (\Delta Ct))}$ method. The sets of primers 1, 2, 3, 4 and 1, 4, 5 were used as transgene specific probes for the rIFABP-hapobec1 and rbapobec1-shRNA constructs respectively. We used the set of primers cas1/cas2 located on the rabbit ß-*CASEIN* gene as reporter probe to normalize to a two copies endogenous gene. A reference rabbit genome was produced by mixing one copy of transgene per copy of genome and used as standard. The number of copies of integrated transgene was similar for sets 1, 2, 3, 4 in the case of the rIFABP-hapobec1 transgene and for sets 1, 4, 5 in the case of the rbapobec1-shRNA transgene, thus suggesting that transgenes were intact.

shRNA assay

The concentration of shRNA produced by the shRNA constructs in transgenic rabbit tissues was estimated by RT-qPCR [28] [12]. Briefly, 5 µg of total RNA, prepared as previously described [29], were polyadenylated according to Ambion's protocol (PolyA Polymerase, Ambion, Applied Biosystems, France). The polyadenylated RNAs were reverse transcribed (High Capacity cDNA Archive kit, Applied Biosystems) using as reverse primer a polyT adapter encompassing a series of twelve "T" residues and a universal primer (Table S1). Quantification was achieved by SYBR Green quantitative PCR (Applied Biosystems) using a set of primers composed of the universal primer corresponding to the 5′ end of the polyT-adapter and a primer specific to the shRNA sequence, resulting in the amplification of a 65 bp long fragment.

The concentration of shRNA in tissue samples was estimated after normalization by the concentration of Let7c miRNA determined by the same method in each sample. It was thus given by the formula $2^{(CtLet7c-CtsiRNA)}$. A set of samples was chosen as calibrators and was assayed in all compared runs. Care was taken to consider Ct values within the linear amplification zone.

Quantification of human and rabbit *APOBEC1* gene expression

Total RNAs were extracted from tissues as previously described [29]. Reverse Transcription (RT) was performed on 1 µg of total RNA using the High Capacity cDNA Archive kit (Applied Biosystems) and the random primer mix included in the kit.

Quantification was achieved using SYBR Green quantitative PCR (fast SYBR Green master mix, Applied Biosystems) with dilutions of the RT reactions and sets of primers designed by the Primer Express software (Applied Biosystem). Whenever possible, primers were chosen on separate exons in order to avoid contaminant DNA amplification, and all amplicons were 100 base-pairs long (Table S1). The sequence of primers was chosen in order to avoid cross reactivity between human and rabbit *APOBEC1* gene measurements. Moreover, for all samples, a RT minus reaction was performed with all RT components except the reverse transcriptase enzyme, and assayed as a complete RT reaction to ensure that no amplification was due to contaminant DNA.

Three normalizing genes (*RPL19, YHWAZ, HPRT*) were tested on all samples for their stable expression in the studied tissues. The GeNorm program included in Biogazelle QBasePlus software (Biogazelle NV, Ghent, Belgium) was used to analyze the data. In order to correct for inter-run fluctuations, a set of samples was chosen as calibrators and was assayed in all compared runs. Care was taken to consider Ct values within the linear amplification zone. Gene expression was considered as significant when Ct values obtained using 2–5 ng of cDNA in each q-PCR reaction were lower than 34, and when one single DNA fragment with the expected size was amplified as template in each q-PCR reaction.

Characterization of the intestinal human APOBEC1 protein

Human APOBEC1 was characterized in intestinal protein extracts by Western blotting. Scrapped intestinal mucosa cells were homogenized with a Dounce homogenizer in RIPA buffer (50 mM Tris-HCl, pH 7.4; 1% IGEPAL; 0.5% Na-deoxycholate; 0.1% SDS; 150 mM NaCl; 2 mM EDTA; 50 mM NaF; 0.2 mM sodium orthovanadate) with protease inhibitors (Complete Protease Inhibitor Cocktail, Roche; 1 mM PMSF; 1 mM Benzamidine) extemporaneously added. After incubation on ice for 30 minutes and centrifugation at 10 000 g for 10 minutes at 4°C, the supernatant was collected and frozen in aliquots at −80°C. Protein concentration was determined by Bradford assay (BioRad, France) using BSA as standard.

Proteins were separated on a 16% acrylamide gel electrophoresis then transferred on Hybond-P membrane. The human APOBEC1 enzyme was detected after incubation with a rabbit anti-APOBEC1 antibody (Sigma, SAB2100132), a goat anti-rabbit IgG peroxidase conjugate (Sigma, A-0545) and the immunofluorescence ECL 2 detection kit (Pierce).

Indirect quantification of the proportion of edited ApoB mRNA

Reverse transcribed (RT) RNAs were obtained as for the measurement of rb*APOBEC1* gene expression. A 455 bp long DNA fragment was amplified from RT RNAs using a set of primers specific of the rabbit *APOB* cDNA (LApoB48F/ LApoB48R, Table S1). The *APOBEC1* edited 2177th codon of the *APOB* cDNA was included in this amplified DNA fragment (Figure 4A). Editing was responsible for the modification of the "CAA" codon in a "UAA" one (a TAA codon in the amplified product). The amplified fragment was purified by MSB Spin PCRapace (Stratec, Eurobio, France), then sequenced (Eurofins, MWG) using the APOBR4 oligonucleotide as sequencing primer. Using this primer, the antisense strand was sequenced. We measured the yield of 455 bp fragments with a "T" residue in the mixture by analyzing the chromatogram of the sequence of each amplified 455 bp fragment. As shown in Figure 4B, the height of the "G" (g1) and "A" (a1) peaks in the chromatogram was compared to that of the "G" (g2) and "A" (a2) peaks chosen in the vicinity to normalize for sequencing efficiency (in the antisense sequenced strand, "G" and "A" sequenced residues corresponded to the "C" and "T" residue of the edited 2177th codon). A standard curve was performed by sequencing a definite amount of the 455 bp DNA fragment containing a mixture of varying proportions of the two types of DNA strands elsewhere purified, which sequence encompassed the "C" or "T" residue at the 2177th codon (Figure 4C). Two linear equations were deduced by plotting the g1/g2 or a1/a2 ratio against the amount of "C" or "T" encompassing DNA in the mixture. These equations were then used to determine the percentage of "C" or "T" encompassing DNA in each RT RNA mixture produced from the various studied samples. Thus, we indirectly measured the yield of edited mRNA.

Plasma APOB48 levels

Plasma APOB48 levels were assayed using an ELISA [16] as indicated by the manufacturer (Shibayagi, X-Celtis GmBH, Germany). To enhance the level of APOB48 in the plasma, rabbits were fed ad libitum for 9 days with a high fat/high cholesterol regimen containing 0,2% cholesterol and 8% soybean oil [17].

Plasma cholesterol and lipid levels

The major classes of lipoproteins were isolated from plasma samples by sequential ultracentrifugation ensuring the separation of chylomicrons + VLDL, LDL and HDL [30]. Cholesterol (total, esterified, and free) and triglycerides were further determined in each fraction using commercially available enzymatic kits.

Total body content of lipids

The total content of lipids was deduced from the measure of total body electrical conductivity (TOBEC) as previously described [31] with modifications brought by L Lamothe and C Bannelier for using an EM-SCAN SA-3000-type chamber. Animals were not anesthetized during measurements. Measurements were made at around 12-16 weeks after birth, and rabbits were immediately weighed. The content of lipids was deduced from the E-value given by the TOBEC and the weight of the rabbit using a prediction equation as follows: total lipid content (% of live weight, LW) = $3.33843 + 0.00248 \times LW - 0.00196 \times E$ with LW = live weight and E = TOBEC measurement.

Supporting Information

Figure S1 *In vitro* **assessment of the efficiency of shRNA expressing constructs.** The OligoWalk web server generated a list of small hairpin RNA candidate sequences ranked by the probability of being efficient to knock down the targeted gene expression. Four sequences (named "a", "b", "c", and "d") were chosen within this list (their probability of being efficient ranged from 88% to 95.5%), and were tested *in vitro* using a cell transfection assay. These sequences targeted the 3'UTR region of the rabbit *APOBEC1* transcript. The transfections were carried out in CHO.K1 cells (ATCC number CCL-61) using ExGen500 (Euromedex, Souffelweyersheim, France), according to the manufacturer's protocol. The test aimed to measure the efficacy of rbapobec1-shRNA constructs to target the rabbit *APOBEC1* gene expression, in order to select an efficient one that will be used

to produce transgenic rabbits. Four constructs harboured the "a", or "b" or "c" or "d" shRNA sequence. A fifth construct harboured two H1-shRNA genes, one with the "a" and the other with the "d" sequence (see rbapobec1-shRNA diagrams). In the absence of rabbit intestinal cell cultures expressing the *APOBEC1* gene, we designed a chimeric target gene encompassing the *luciferase* gene fused to the targeted sequence of the rabbit *APOBEC1* gene (upper diagram). The 3'UTR of the rabbit *APOBEC1* gene (from nucleotide 756 to 905 respectively to the ATG translation initiation codon) was added at the 3' position of the *luciferase* gene. Degradation of the 3'UTR region in the target construct by shRNAs was expected to prevent translation of the *luciferase* cistron. Thus, a quantification of shRNA-induced knockdown could be achieved by measuring *luciferase* activity in the transfected cells. The reliability of this method was previously established <Hung, 2006 #30>, showing that it is possible to fuse short target sequences (such as the rabbit *APOBEC1* gene 3'UTR sequences) in the UTR of a reporter gene in order to establish a quantitative reporter-based shRNA validation system. Each shRNA-expressing constructs (0.75 μg/P35 dish) was transfected with the target *luciferase* construct (0.75 μg/P35 dish) or the control empty vector pM10, and the *β-galactosidase* vector pCH110 (Pharmacia, 1 μg/P35 dish) to correct for transfection efficiency. Luciferase and ß-galactosidase activities were measured 48 h after transfection. The results are given as percentages of luciferse activity in cells transfected by the target *Luciferase* vector and the empty pM10 vector. All luciferase values were normalized to ß-galactosidase activities. The graph is representative of two independent experiments.

Figure S2 **Chromatograms of sequence of DNA amplified from RT-RNA of intestine and liver in wild type or shRNA expressing transgenic animal.** The product of amplification of RT-mRNA encompassing the 2177th codon was sequenced using the APOBR4 oligonucleotide as sequencing primer. By using this oligonucleotide, the antisense strand was sequenced. Three typical chromatograms are reported showing the amplitude of A, C, G and T peaks. The sequence of the sense and the antisense strands are written below. The edited 2177th codon is boxed. The lines and small letters indicate how was measured the height of the peaks. Editing converts the "G" residue of the antisense sequence in a "A" residue". In the liver in wild type animals (**upper panel**), the APOB mRNA was not edited. A "G" residue was detected at the position of the 2177th codon and no "A" residue was possible to be detected. It was considered that all DNA strands in the amplified sample harbored a CAA codon. In the intestine in wild type animals (**middle panel**), the codon was edited. Most strands harboured a "A" residue in place of the non edited "G" residue. However, a small proportion of strands harboured the "G" residue. The **lower panel** shows that in intestine of transgenic animals expressing the shRNA (sh L21), the height of the "A" peak was reduced and that of the "G" peak was enhanced compared to the chromatogram in intestine of wild type animals. The sample was a mixture of DNA fragments harbouring the CAA or the TAA sequence. The measure of a1, a2, g1 and g2 ensured the determination of the proportion of "G" and "A" containing fragments in the mixture.

Figure S3 **Western blot detection of the human APO-BEC1 enzyme in intestinal cell extracts in L02 transgenic rabbits.** Intestinal cell extracts (100 μg of protein in each sample) prepared from a wild type rabbit (WT) and a L02 transgenic rabbit expressing the human APOBEC1 enzyme were fractionated on

SDS-PAGE (16%). The human APOBEC1 enzyme was detected by Western blotting using the APOBEC1 antibody (1/1000 dilution). A similar amount of spleen extract was assayed on the same gel as negative control. One specific band (labelled by an arrow) was seen in the L02 transgenic animal at the expected migration rate according to the size of the human protein (27 kD). No band was possible to be detected in the wild type extract or in the spleen extract.

Figure S4 Plasma triglycerides and cholesterol concentrations in transgenic rabbits expressing the human *APOBEC1* gene fed with a high fat/high cholesterol diet. The experiments and symbols are similar to those described in the legend of Figure 8. Rabbits (4 wild type, and 3 transgenic rabbits from line L02) were fed for 8 days with a high fat/high cholesterol diet. Plasma samples were collected before the diet (D0, white bars), after feeding for 8 day with the diet (D8, black bars), after 20 hours starvation (D9 starved, grey bars) and 4 hours after re-feeding with the high fat diet (D9 fed, dotted bars). Triglycerides and cholesterol were assayed as in Figure 7. Values are given in mg/ml, with the standard error of the mean. Comparisons were made between transgenic and wild type animals for each day of the challenge (* = p<0.05).

Figure S5 Long-term recording of weight curves of rIFABP-hApobec1 transgenic rabbits. A series of animals (8 wild type; 3 from line L01; 5 from line L02) were weighed for up to 40 weeks after birth. Clearly, the long-term expression of the human APOBEC1 transgene did not induce an excess weigh. Values are means +/− sem.

Acknowledgments

We would like to thank Sonia Prince for the contribution to the production of transgenic rabbits; Michel Baratte, Gilbert Boyer, Jean-Pierre Albert and Gwendoline Morin (Unité Commune d'Expérimentation animale, INRA, Jouy-en-Josas, France) for their excellent care of the animals; Mathieu Leroux-Coyau (Biologie du Développement et Reproduction, INRA, Jouy-en-Josas) and Stéphanie Lemaire-Ewing for their excellent technical assistance.

Author Contributions

Conceived and designed the experiments: GJ LMH SB IH. Performed the experiments: GJ SB BP BDS EH CV TG. Analyzed the data: GJ TG LL IH. Contributed reagents/materials/analysis tools: BDS BP EH NDC. Contributed to the writing of the manuscript: GJ IH LL.

References

1. Mathus-Vliegen EMH (2012) Prevalence, Pathophysiology, Health Consequences and Treatment Options of Obesity in the Elderly: A Guideline. Obes Facts 460–483.
2. González Jiménez E (2011) Genes and obesity: a cause and effect relationship. Endocrinol Nutr 58: 492–496.
3. Reed DR, Lawler MP, Tordoff MG (2008) Reduced body weight is a common effect of gene knockout in mice. BMC Genetics 9:doi:10.1186/1471-2156-1189-1184.
4. Chan L (1995) Apolipoprotein B messenger RNA editing: An update. Biochimie 77: 75–78.
5. Teng B, Burant CF, Davidson NO (1993) Molecular cloning of an apolipoprotein B messenger RNA editing protein. Science 260: 1816–1819.
6. Davidson NO, Shelness GS (2000) APOLIPOPROTEIN B: mRNA Editing, Lipoprotein Assembly, and Presecretory Degradation. Annu Rev Nutr 20: 169–193.
7. Kendrick JS, Chan L, Higgins JA (2001) Superior role of apolipoprotein B48 over apolipoprotein B100 in chylomicron assembly and fat absorption: an investigation of apobec-1 knock-out and wild-type mice. Biochem. J. 356: 821–827.
8. Lo C-M, Nordskog BK, Nauli AM, Zheng S, vonLehmden SB, et al. (2008) Why does the gut choose apolipoprotein B48 but not B100 for chylomicron formation? Am J Physiol Gastrointest Liver Physiol 294:G344–G352.
9. Zhang X-J, Chinkes DL, Aarsland A, Herndon DN, Wolfe RR (2008) Lipid Metabolism in Diet-Induced Obese Rabbits Is Similar to That of Obese Humans. J. Nutr. 138: 515–518.
10. Nakamuta M, Oka K, Krushkal J, Kobayashi K, Yamamoto M, et al. (1995) Alternative mRNA splicing and differential promoter utilization determine tissue-specific expression of the apolipoprotein B mRNA-editing protein (Apobec1) gene in mice. J Biol Chem 270: 13042–13056.
11. Lu ZJ, Mathews DH (2008) OligoWalk: an online siRNA design tool utilizing hybridization thermodynamics. Nucl. Acids Res. 36:W104–108.
12. Daniel-Carlier N, Sawafta A, Passet B, Thépot D, Leroux-Coyau M, et al. (2013) Viral infection resistance conferred on mice by siRNA transgenesis. Transgenic Res 22: 489–500.
13. Yamanaka S, Poskay KS, Balestra ME, Zeng G-Q, Innerarity TL (1994) Cloning and mutagenesis odf the rabbit ApoB mRNA editing protein. A zinc motif is essential for catalytic activity, and noncatalytic auxiliary factor(s) of the editing complex are widely distributed. J Biol Chem 262: 21725–21734.
14. Greeve J, Altkemper I, Dieterich J-H, Greten H, Windler E (1993) Apolipoprotein B mRNA editing in 12 different mammalian species: hepatic expression is reflected in low concentrations of apoB-containing plasma lipoproteins. J Lipid Res 34: 1367–1383.
15. Greeve J, Jona VK, Chowdhury NR, Horwitz MS, Chowdhury JR (1996) Hepatic gene transfer of the catalytic subunit of the apolipoprotein B mRNA editing enzyme results in a reduction of plasma LDL levels in normal and Watanabe heritable hyperlipidemic rabbits. J Lipid Res 37: 2001–2017.
16. Kinoshita M, Matsushima T, Mashimoto Y, Kojima M, Kigure M, et al. (2010) Determination of Immuno-Reactive Rabbit Apolipoprotein B-48 in Serum by ELISA. Exp Anim 59: 459–467.
17. Picone O, Laigre P, Fortun-Lamothe L, Archilla C, Peynot N, et al. (2011) Hyperlipidic hypercholesterolemic diet in prepubertal rabbits affects gene expression in the embryo, restricts fetal growth and increases offspring susceptibility to obesity. Theriogenology 75: 287–299.
18. Yamamoto T, Yamamoto A, Watanabe M, Matsuo T, Yamazaki N, et al. (2009) Classification of FABP isoforms and tissues based on quantitative evaluation of transcript levels of these isoforms in various rat tissues. Biotechnol Lett 31: 1695–1701.
19. Dannoura A, Berriot-Varoqueaux N, Amati P, Abadie V, Verthier N, et al. (1999) Anderson's disease: exclusion of apolipoprotein and intracellular lipid transport genes. Arterioscl Throm Vasc 19: 2494–2508.
20. Okada T, Miyashita M, Fukuhara J, Sugitani M, Ueno T, et al. (2011) Anderson's disease/chylomicron retention disease in a Japanese patient with uniparental disomy 7 and a normal SAR1B gene protein coding sequence. Orphanet J Rare Dis 6: 78.
21. Georges A, Bonneau J, Bonnefont-Rousselot D, Champigneulle J, Rabes J, et al. (2011) Molecular analysis and intestinal expression of SAR1 genes and proteins in Anderson's disease (Chylomicron retention disease). Orphanet J Rare Dis 6: 1.
22. Nakamuta M, Chang BH-J, Zsigmond E, Kobayashi K, Lei H, et al. (1996) Complete Phenotypic Characterization of apobec-1 Knockout Mice with a Wild-type Genetic Background and a Human Apolipoprotein B Transgenic Background, and Restoration of Apolipoprotein B mRNA Editing by Somatic Gene Transfer of Apobec-1. J Biol Chem 271: 25981–25988.
23. Hirano K-I, Young SG, Farese RV Jr, Ng J, Sande E, et al. (1996) Targeted Disruption of the Mouse apobec-1 Gene Abolishes Apolipoprotein B mRNA Editing and Eliminates Apolipoprotein B48. J Biol Chem 271: 9887–9890.
24. Morrison JR, Paszty C, Stevens ME, Hughes SD, Forte T, et al. (1996) Apolipoprotein B RNA editing enzyme-deficient mice are viable despite alterations in lipoprotein metabolism. P Natl Acad Sci USA 93: 7154–7159.
25. Blanc V, Xie Y, Luo J, Kennedy S, Davidson N (2012) Intestine-specific expression of Apobec-1 rescues apolipoprotein B RNA editing and alters chylomicron production in Apobec1 −/− mice. J Lipid Res 53: 2643–2655.
26. Anant S, Davidson NO (2001) Molecular mechanisms of apolipoprotein B mRNA editing. Curr Opin Lipidol 12: 159–165.
27. Anant S, Davidson NO (2002) Identification and Regulation of Protein Components of the Apolipoprotein B mRNA Editing Enzyme: A Complex Event. Trends Cardiovas Med 12: 311–317.
28. Shi R, Chiang V (2005) Facile means for quantifying microRNA expression by real-time PCR. Biotechniques 39: 519–525.
29. Chomczynski P, Sacchi N (1987) Single step method of RNA isolation by acid guanidium thiocyanate-phenol-chloroform extraction. Anal. Biochem. 162: 156–159.
30. Hatch F (1968) Practical methods for plasma lipoprotein analysis. Adv Lipid Res 6: 1–68.
31. Fortun-Lamothe L, Lamboley-Gaüzere B, Carole B (2002) Prediction of body composition in rabbit females using total body electrical conductivity (TOBEC). Livest Prod Sci 78: 133–142.

20

Assessment of Fetal Cell Chimerism in Transgenic Pig Lines Generated by *Sleeping Beauty* Transposition

Wiebke Garrels[1]¤, **Stephanie Holler**[1], **Ulrike Taylor**[1], **Doris Herrmann**[1], **Heiner Niemann**[1], **Zoltan Ivics**[2], **Wilfried A. Kues**[1]*

1 Institut für Nutztiergenetik, Friedrich-Loeffler-Institut, Mariensee, Germany, **2** Paul-Ehrlich-Institute, Langen, Germany

Abstract

Human cells migrate between mother and fetus during pregnancy and persist in the respective host for long-term after birth. Fetal microchimerism occurs also in twins sharing a common placenta or chorion. Whether microchimerism occurs in multiparous mammals such as the domestic pig, where fetuses have separate placentas and chorions, is not well understood. Here, we assessed cell chimerism in litters of wild-type sows inseminated with semen of transposon transgenic boars. Segregation of three independent monomeric transposons ensured an excess of transgenic over non-transgenic offspring in every litter. Transgenic siblings (n = 35) showed robust ubiquitous expression of the reporter transposon encoding a fluorescent protein, and provided an unique resource to assess a potential cell trafficking to non-transgenic littermates (n = 7) or mothers (n = 4). Sensitive flow cytometry, fluorescence microscopy, and real-time PCR provided no evidence for microchimerism in porcine littermates, or piglets and their mothers in both blood and solid organs. These data indicate that the epitheliochorial structure of the porcine placenta effectively prevents cellular exchange during gestation.

Editor: Xiuchun (Cindy) Tian, University of Connecticut, United States of America

Funding: Financial support was provided by the Deutsche Forschungsgemeinschaft (DFG) via KU-1586/2-1 and SPP1313 is acknowledged. The funders had no role in study design, data collection and analysis, decision to publish, or preparation of the manuscript.

Competing Interests: The authors have declared that no competing interests exist.

* E-mail: wilfried.kues@fli.bund.de

¤ Current address: Medical School Hannover, Institute of Laboratory Animal Sciences, Hannover, Germany

Introduction

Microchimerism refers to the presence of a small ratio of foreign cells (<1:100) within the tissues of a host organism. Microchimerism can occur after iatrogenic interventions, such as transplantation or transfusion, or naturally between twins, or mother and fetus. In utero, cell transfer between littermates or materno-fetal cell exchange can have important consequences for the function of the immune system, health of the offspring and for tissue compatibility [1–3]. Microchimerism is found in women [4,5], and in their progeny by genotyping and detection of transferred tumor cells [6,7]. The human hemochorial placenta, in which maternal blood is in direct contact with the chorionic trophoblast, facilitates trafficking of cells and/or macromolecules, such as antibodies [7,8].

Most of the domestic species possess an epitheliochorial placenta [9], which consists of maternal and fetal epithelia and is thought to be much tighter than the hemochorial placenta. Nevertheless, cell trafficking has been described in these species [10]. In cattle, fetal microchimerism was observed in monochorionic twins [11–13]. Fetal microchimerism has also been reported in goats [14]. However, cell transfer in multiparous, multichorionic species like the domestic pig has only rarely been studied. A recent report suggested microchimerism in porcine littermates under specific experimental situations [15], in which xenogenic human cord blood derived cells had been injected into the peritoneum of porcine fetuses around day 40 of gestation. Evidence for long-term maintenance of human cells in the treated animals, but also in untreated littermates was reported [15].

The domestic pig is an important model for biomedical studies and preclinical cell therapy approaches [16–19]. Fetal chimerism could be biologically relevant in the pig, specifically for immunology, since the consequences of fetal microchimerism might affect immunity, immune surveillance and tissue repair [8,20–26]. In recipient cows, which were pregnant with transgenic embryos, transgene-specific DNA sequences were found in blood of 50% of the surrogate mothers [27]. In another study, enhanced green fluorescent protein (EGFP) was detected in maternal parts of the placenta [28]. These findings suggest that transplacental leakage of fetal DNA and proteins into the maternal circulation can occur in cattle despite the epitheliochorial placenta. Whether the leakage of fetal DNA correlated with transfer of cells and long-term persistence of fetal cells is not yet known.

Here, we assessed potential inter-fetal and feto-maternal cell transfer in lines of transgenic pigs expressing a Venus fluorophore reporter [29]. Venus is a derivative of the enhanced yellow fluorescent protein (EYFP), and is compatible with live imaging. The *Venus* construct was integrated into the porcine genome by *Sleeping Beauty* transposition and was found to be ubiquitously expressed [29–31]. Offspring from two founder boars, each carrying three monomeric reporter transposons in their genome were used for this study [29,32–34]. The faithful expression of Venus in a ubiquitous manner is an excellent tool for the identification of even few transgenic cells in circulation and solid organs of non-transgenic littermates or wild-type animals.

Table 1. Assessment of potential chimerism in non-transgenic littermates and sows.

Animal ID	Status	FACS (Venus pos. leukocytes)	Tissue sections (Venus pos. cells)	Real time PCR (Ct)
#506	Non-tg littermate	<1: 200 000	<1: 50 000	40.1±1.1
#512	Non-tg littermate	<1: 240 000	<1: 50 000	∞
#522	Non-tg littermate	<1: 180 000	<1: 50 000	36.2±0.9
#524	Non-tg littermate	<1: 200 000	<1: 50 000	38.5±1.0
#527	Non-tg littermate	<1: 140 000	<1: 50 000	37.2±1.9
#531	Non-tg littermate	<1: 120 000	<1: 50 000	38.0±0.4
#533	Non-tg littermate	<1: 100 000	<1: 50 000	36.2±0.9
#515	Tg littermate	All (99.97%)	All	<25
#517	Tg littermate	All (99.90%)	All	<25
Sow #404[§]	Mother of #506, #531, #533	<1: 300 000	<1: 50 000	41.6±0.9
Sow #408	Mother of #512, #515, #517	<1: 120 000	<1: 50 000	40.1±1.8
Sow #412	Mother of #522, #524	<1: 50 000	<1: 50 000	38.4±1.2
Sow #420	Mother of #527	n.d.	n.d.	∞
Sow 1	Unrelated control	<1: 200 000	<1: 50 000	38.3±1.9
Sow 2	Unrelated control	<1: 200 000	n.d.	∞
Sow 3	Unrelated control	<1: 200 000	n.d.	n.d.

ID, unique identification number; Ct, threshhold cycle in a quantitative real time PCR;
∞, no threshold cycle reached during PCR;
n.d., not done;
[§], sow with two "transgenic" pregnancies.

Materials and Methods

Ethics Statement

Animals were maintained and handled according to German and international laws regulating animal welfare and genetically modified organisms (GMO), and all experiments were approved by an external animal welfare committee at the Niedersächsisches Landesamt für Verbraucherschutz und Lebensmittelsicherheit (LAVES) in Oldenburg, Germany (AZ 33.9-42502-04-09/1718).

Collection of Ejaculated Sperm and Artificial Insemination

Sperm-rich fractions were collected from boars using a dummy and by gloved – hand technique [31,32]. The semen samples were extended with Androhep (1:1) and transported to the laboratory at 37°C. Sperm concentration was determined by NucleoCounter SP-100 system (ChemoMetec, Denmark). Inseminations of wild-type sows at oestrus were performed according to standard procedure.

Fluorescence Microscopy and Macroscopic Excitation of Venus Fluorochrome

For fluorescence microscopy, images were obtained by an Olympus BX 60 (Olympus, Hamburg, Germany) fluorescence microscope equipped with a 12-bit digital camera (Olympus DP 71). For specific excitation of live *Venus*-transgenic piglets and pigs, a blue floodlight LED (40 W; eurolite Germany, Germany) and an electronic camera (Canon Powershot) equipped with a yellow emission filter was used.

Identification of Venus Positive Cells by Flow Cytometry

Flow cytometry analysis of leukocytes was performed using a FACScan (BD Bioscience, Heidelberg, Germany) equipped with an argon laser (488 nm, 15 mW) [31,32]. Samples were diluted to 0.5×10^6 cells/ml and measured in triplicates acquiring 50 000–

300 000 cells per sample. Membrane impaired cells were excluded from analysis by staining with propidium iodide (20 µM).

Fluorescence Histology

Tissue samples (~5 mm^3) were snap frozen in liquid nitrogen and then fixed in 3.6% formaldehyde for 12–24 hours. After washing in phosphate-buffered saline, the samples were soaked in 30% sucrose for 24 hours, then frozen in tissue-tec (Hertenstein, Germany), and 10 µm sections were prepared in a cryostat (Micom Laborgeräte, Walldorf, Germany) and dried on glass slides. The sections were mounted with Vecta-shield (H-100) and viewed under a fluorescence microscope.

Genotyping of Offspring

Genomic DNA was isolated with the proteinase K method, digested with NcoI, separated on a 0.6% agarose gel by electrophoresis and blotted on a polyvinylidene fluoride (PVDF) membrane. Then a transgene specific probe labelled with digoxigenin was used for hybridization as described [29,31,32].

Real Time-PCR

Genomic DNA was isolated from leukocytes with the proteinase K method. The PCR mix in each well included 15 µl of 2x Power SYBR Green PCR Master Mix (Applied Biosystems), 11.4 µl dH$_2$O, 1.6 µl each of the primer pairs (5 µM), 2 µl of DNA (25 ng) in a total volume of 30 µl. Primer and PCR characteristics are summarized in Table S1. The PCR program included denaturation and activation of the Taq polymerase for 10 min at 95°C followed by 45 cycles of 95°C for 15 s and the appropriate annealing temperature given in Table S1 for 1 min. Then a dissociation curve of the product was assessed (ABI 7500 Fast Real-Time System, Applied Biosystems).

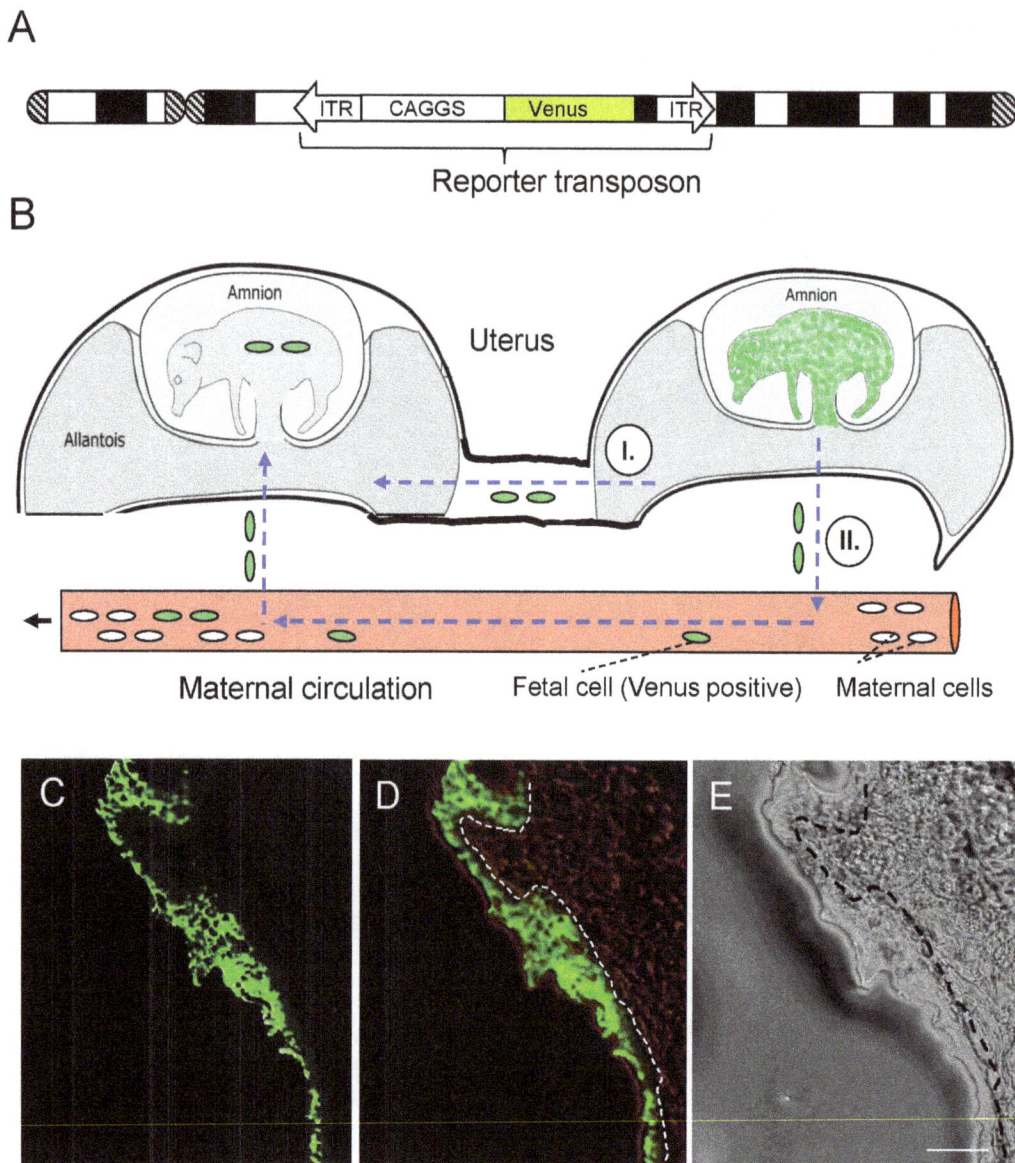

Figure 1. Reporter transposon and potential routes of inter-fetal and feto-maternal cell trafficking in the pig. A) Schematic depiction of a monomeric reporter transposon in chromosomal context. The SB transposon consists of *CAGGS* promoter, *Venus* cDNA and polyadenylation sequence flanked by SB inverted terminal repeats (ITRs). Drawing not at scale. B) Fetal cells might traffic directly between fetuses (I.) or via the maternal circulation (II.). In both cases the non-transgenic fetuses should carry Venus-expressing cells. In case (II.), the mother should also show cell chimerism. For simplification only two fetuses are depicted and only the transgenic fetus is shown in green, however, cells of amnion and allantois are also Venus positive. C) Specific Venus fluorescence in embryonic allantois at day 25 of gestation. Cryosection of a *CAGGS-Venus* transgenic implantation in a wild-type sow. D) Overlay of fluorescence and brightfield views. E) Corresponding brightfield view. A dotted line indicates the border between embryonic and maternal tissue. Bar = 10 μm.

Results

Two transgenic founders, each carrying three monomeric integrations of the *cytomegalovirus enhancer, chicken beta actin* hybrid promoter (*CAGGS*)-*Venus* transposon, were used for collection of semen, and six wild-type sows were artificially inseminated. This resulted in 5 pregnancies that went to term and yielded a total of 44 piglets, of which 35 were transgenic and 9 were non-transgenic. Seven of the non-transgenic littermates were analysed in detail for signs of fetal chimerism (Tab. 1). An additional pregnancy of a wild-type sow carrying *GAGGS-Venus* transgenic embryos was interrupted at day 25 of gestation. Out of 12 fetuses, 10 were

transgenic and expressed the Venus reporter in fetal tissue, amnion and allantois (Fig. 1). The ratio of transgenic to non-transgenic genotypes fitted with the independent inheritance of three monomeric transposon copies according to Mendelian rules as shown recently [32].

In principle, either direct cell trafficking between fetuses or indirectly via circulation of the mother is conceivable (Fig. 1). First, the robust and ubiquitous expression of Venus reporter in transgenic offspring was confirmed. All transgenic animals showed homogenous expression of the Venus protein, which could be detected already by direct fluorescence of pigs in the barn, but also by FACS measurements, cryosections, and with molecular

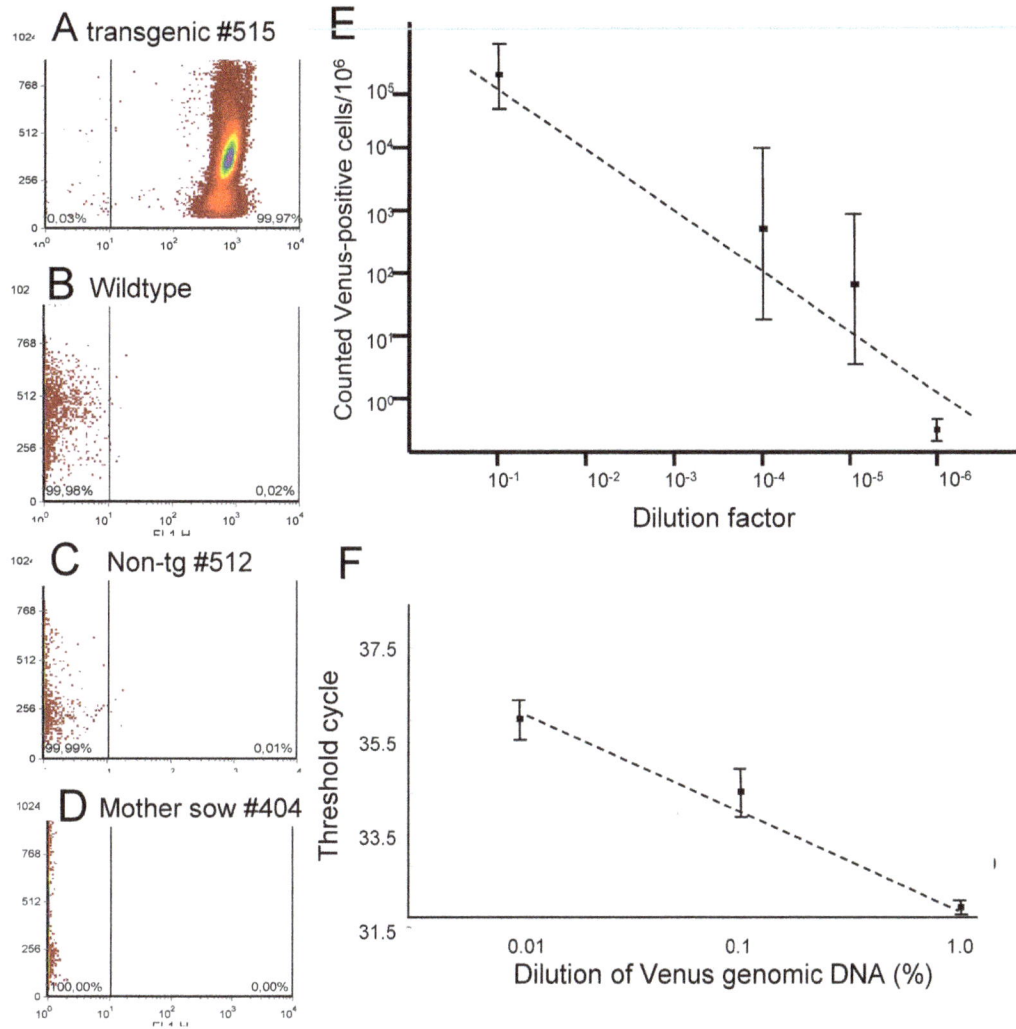

Figure 2. Flow cytometric measurement of porcine leucocytes for Venus-positive cells. A) Venus transposon transgenic pig (#515), B) Wild-type pig, C) Non-transgenic littermate, D) Sow #404 (delivered two litters of transposon piglets), E) Determination of the detection limit of flow cytometry. Leukocytes from a wild-type animal were spiked with decreasing amounts of Venus-positive leukocytes (n = 3 technical and biological replicates for each dilution). The dotted line indicates the theoretically expected cell counts. A detection limit of 1 Venus cell in 100 000 wild type cells was determined. F) RT-PCR of a dilution series of *Venus* gDNA in wildtype DNA.

methods [32,33]. Terminally differentiated gametes of these animals were also Venus positive [32]. A macroscopic inspection of inner organs revealed tissue-specific expression levels of Venus [32].

The 7 vital non-transgenic offspring (Tab. 1) were of particular interest for the assessment of fetal chimerism. Since all non-transgenic offspring were obtained from litters with a surplus of transgenic littermates, potential microchimerism should be reflected by a certain ratio of transgenic cells in the non-transgenic animals and/or in their wild-type mothers (Fig. 1).

To investigate the potential scenarios for cell chimerism, blood samples were collected from pregnant sows at two different time points of gestation (4 weeks and 12 weeks) and after they gave birth, as well as from the non-transgenic piglets (Tab. 1), and were analysed by flow cytometry, PCR and fluorescence microscopy. In parallel, blood and tissue samples from two transgenic animals were analysed. The samples from transgenic animals were processed separately to prevent any cross-contamination during handling.

First, we determined the expression of Venus in leukocytes of transgenic and non-transgenic littermates. In leukocytes isolated from transgenic animals, all cells exhibited prominent Venus fluorescence which was about two log orders higher than in wild-type cells (Fig. 2). In artificially spiked leukocyte preparations (transgenic : non-transgenic cells = $1:10^1$ to $1:10^6$) the Venus-positive cells could unequivocally be identified with a detection limit of 1 Venus-positive cell per 10^5 wild-type leukocytes (Fig. 2). Flow cytometric results were further substantiated by re-analysis via fluorescence microscopy (Fig. S1). In leukocytes from non-transgenic animals (n = 7) Venus-positive cells were never found, even by analysing 50 000–300 000 cells per measurement (2–3 replicates per animal).

In addition, leukocytes from three mothers (animal #404 had been pregnant twice and delivered two transgenic litters) and three unrelated wild-type sows (negative controls) were analysed by flow cytometry. To scan for rare events, a minimum of 50 000 cells and a maximum of 300 000 cells were counted. The negative control samples, but also the samples from the mothers did not reveal any

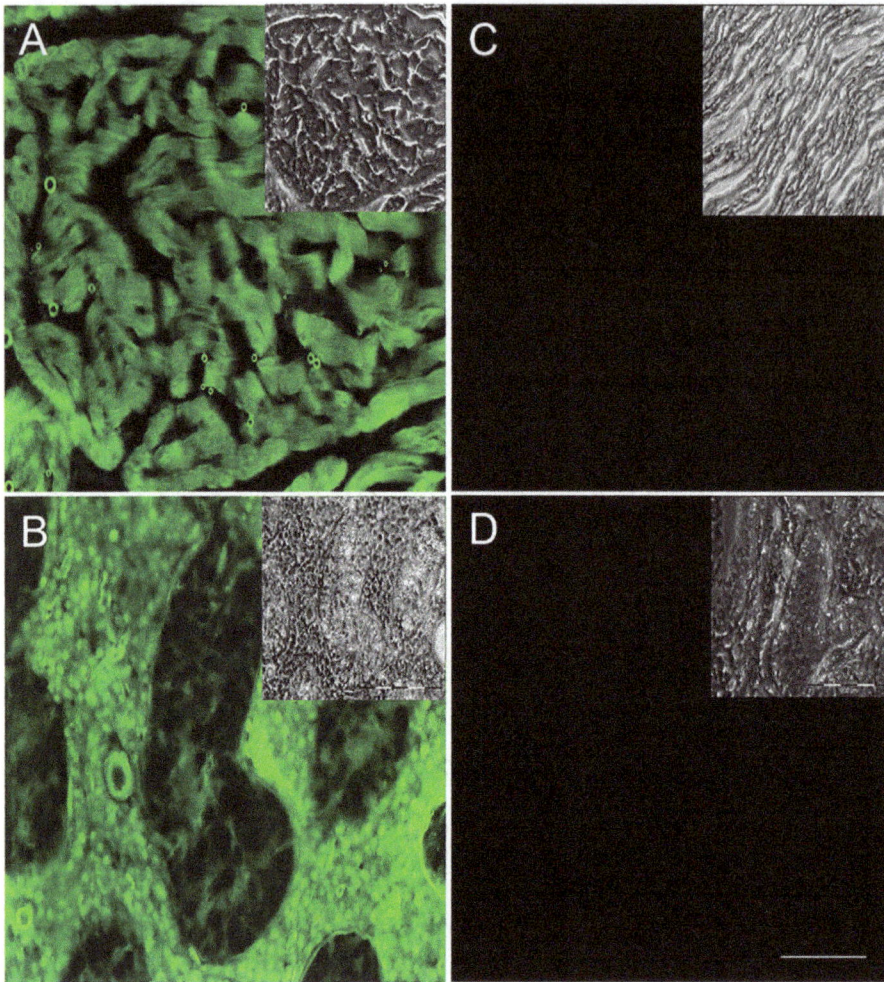

Figure 3. Assessment of reporter-positive cells in solid organs. A, B) Cryosections of heart and testis of a Venus-transposon pig are depicted under specific excitation of Venus and brightfield conditions (insets). C) Heart, and D) testis sections of a non-transgenic littermate were recorded under identical camera settings. White bar = 50 μm.

Venus-positive blood cells (Fig. 2). Even a sow (#404), which had delivered two transgenic litters did not show any Venus positive leukocytes.

To assess the possibility that non-transgenic littermates carry transferred transgenic cells, which do not express the transgene, and consequently would not be detected by flow cytometry, genomic DNA (gDNA) was isolated and analysed by sensitive real time PCR. The highest amount of gDNA per reaction was 25 nanograms (ng), higher amounts of gDNA decreased amplification efficiency. As positive control, *polyadenylate polymerase* (*PAPOLA*) was amplified from all samples. The detection limit of transgenic sequences was determined by assaying a serial dilution of transgenic DNA in wild-type gDNA. It was calculated that the detection limit was 1 transgenic sequence in 10 000 copies of wild-type gDNA (Fig. 2F), corresponding to a threshold cycle of 35.5. Blood was collected from all non-transgenic littermates and wild-type mothers twice at 4 weeks intervals and was used to isolate gDNA. RT-PCR reactions were run in duplicates or triplicates. However, no indication for the presence of *Venus*-specific DNA was found either in samples from non-transgenic littermates or from the sows; all samples produced threshold cycles above 36.

To assess the possibility that solid organs host transferred cells, the seven non-transgenic littermates and three carrier sows were sacrificed and heart, liver, kidney, brain, lung, spleen, skin, muscle and gonade were analysed. In parallel, organ samples from two transgenic piglets were processed. For microscopic analysis, tissues sections (10 μm) were prepared and 20 sections per organ were analysed for Venus-positive cells. Each section covered a minimum of 50 000 cells, thus an estimated number of 1 000 000 cells were screened per organ. Venus-expressing cells were not found in the non-transgenic littermates (Fig. 3).

Discussion

The present findings provide no evidence of fetal cell trafficking in the pig. Depending on the applied technique different detection limits were determined. Real time PCR had a detection limit of 1 target sequence per ~10 000 genome copies, assuming that one diploid cell contains 6 picogram of DNA. The histological examination was estimated to have a detection limit of 1:50 000 cells, and for the flow cytometry a detection limit of 1 in 100 000 cells was determined. The prerequisite for the histological and flow cytometric measurements was the robust and ubiquitous Venus expression in somatic, and germ line cells of the *Venus* transgenic pigs [29,32–34]. The prominent Venus expression allowed the unambiguous detection and identification of transgenic cells.

Whereas flow cytometric and fluorescence microscopic detection depend on the expression of the transgene, real time PCR allows detection of non-expressing cells or free DNA. In contrast to data in cattle, where Y chromosome-specific DNA was detected in up to 73% of blood samples from naturally mated heifers carrying conventional bull calves, and a transgene-specific sequence was detected in up to 50% of recipient cows carrying transgenic fetuses [27], the present study revealed no foreign DNA or cells in either the carrier sows or the non-transgenic piglets. Potentially, the different morphological structures of the porcine placenta diffusa and the bovine placenta cotyledonaria contribute to these divergent results. The premature rupture of fetal membranes and cell exchange shortly before birth might depend on the placenta form [35]. However, whether or not fetal cells exist over long periods of time in bovine foster mothers was not yet investigated [27,28].

Similarly to the present results, no fetal chimerism could be detected between transgenic porcine littermates [36], and in caprine surrogate mothers [37].

In contrast to the absence of fetal microchimerism in the present study, McConico et al. [15] reported fetal chimerism in the pig after xenogenic transplantation. In untreated littermates a frequency of xenogenic cells of 1:10 000–1:100 000 was reported [15]. In this setting, human cord blood cells were injected into the peritoneum of porcine fetuses by surgical intervention around day 40 of gestation. Potentially, the iatrogenic intervention [15,38] itself may have facilitated distribution of human cells to untreated littermates. Consequently, in iatrogenic interventions the carry-over of cells to untreated littermates and mothers should be considered.

Previously, cell tracking experiments with genetically marked cells have been performed in mice to assess fetal chimerism [39–42]. After mating of *CAGGS-EGFP* transgenic males with wild-type females, the highest ratio of fetal cells was 1% in leukocytes prepared from maternal blood during pregnancy and 0.3% in leukocytes after birth [41]. In solid organs of pregnant females Sunami et al. [42] determined 4–191 fetal cells per 100 000 maternal cells, whereas Fujiki et al. [40] found ~1 fetal cell per 1 million maternal cells in leukocytes and 10–40 fetal cells per 1 million maternal cells from lung, liver and spleen at day 18 of

gestation. It was concluded that the availability of transgenic lines with robust EGFP expression in combination with flow cytometric analysis is the most versatile detection method [43]. Most likely, the frequent fetal chimerism in mice is caused by the murine hemichorial structure of the placenta. In species, where the transgenic technology for genetic labelling is not available, PCR approaches for the detection of Y chromosome-specific sequences, or of informative polymorphic sequences are the gold standard for the assessment of fetal chimerism [44,45].

During normal prenatal development the epitheliochorial placenta of the pig seems to effectively prevent cell trafficking between fetuses and mother. Thus the results of the present study substantiate the hypothesis that the epitheliochorial placenta is non-permeable for fetal cells. In addition, no signs for inter-fetal cell transfer were found.

Supporting Information

Figure S1   Microscopic detection of Venus-positive leukocytes in spiking experiments. A) Leukocytes of animal #517 (transgenic littermate) shown under specific fluorescence excitation of Venus, and B) under brightfield illumination. Note that all leukocytes are Venus-positive. A remaining erythrocyte (asterix) expressed no fluorescence [32]. C) Leukocytes of animal #408 (wildtype sow) spiked with leukocytes from #517 (ratio #408 : #517 = 100: 1). A Venus-positive leukocyte (+) is indicated. D) Corresponding brightfield illumination of C). Bar = 25 µm.

Acknowledgments

The expert technical support by Brigitte Barg-Kues, Rolf Poppenga, Edward Kufeld, Rudolf Grossfeld and Mike Diederich is gratefully acknowledged.

Author Contributions

Conceived and designed the experiments: WG WAK. Performed the experiments: WG SH UT DH ZI. Analyzed the data: WG ZI WAK. Wrote the paper: WG HN WAK.

References

1. Clifton VL, Stark MJ, Osei-Kumah A, Hodyl NA (2012) Review: The feto-placental unit, pregnancy pathology and impact on long term maternal health. Placenta 33: Suppl S37–41.
2. Mold JE, McCune JM (2012) Immunological tolerance during fetal development: from mouse to man. Adv Immunol 115: 73–111.
3. Østensen M, Villiger PM, Förger F (2012) Interaction of pregnancy and autoimmune rheumatic disease. Autoimmun Rev 11: A437–446.
4. Schröder J, De la Chapelle A (1972) Fetal lymphocytes in the maternal blood. Blood 39: 153–162.
5. Lapaire O, Hösli I, Zanetti-Daellenbach R, Huang D, Jaeggi C, et al. (2007) Impact of fetal-maternal microchimerism on women's health–a review. J Matern Fetal Neonatal Med 20: 1–5.
6. Reynolds AG (1955) Placental metastasis from malignant melanoma: report of a case. Obstet Gynecol 6: 205–209.
7. Lee ES, Bou-Gharios G, Seppanen E, Khosrotehrani K, Fisk NM (2010) Fetal stem cell microchimerism: natural-born healers or killers? Mol Hum Reprod 16: 869–878.
8. Kallenbach LR, Johnson KL, Bianchi DW (2011) Fetal cell microchimerism and cancer: a nexus of reproduction, immunology, and tumor biology. Cancer Res 71: 8–12.
9. Carter AM, Enders AC (2013) The evolution of epitheliochorial placentation. Annu Rev Anim Biosci 1: 443–467.
10. Engelhardt H, King GJ (1996–1997) Uterine natural killer cells in species with epitheliochorial placentation. Nat Immun 15: 53–69.

11. Anderson D, Billingham RE, Lampkin GH, Medawar PB (1951) The use of skin graftings to distinguish between monozygotic and dizygotic twins in cattle. Heredity 5: 379–397.
12. Billingham RE, Lampkin GH, Medawar PB, Williams HLL (1952) Tolerance to homografts, twin diagnosis, and the feemartin condition in cattle. Heredity 6: 201–212.
13. Niku M, Pessa-Morikawa T, Taponen J, Iivanainen A (2007) Direct observation of hematopoietic progenitor chimerism in fetal freemartin cattle. BMC Vet Res 3: 29.
14. BonDurant RH, McDonald MC, Trommershausen-Bowling A (1980) Probable freemartinism in a goat. J Am Vet Med Assoc 177: 1024–1025.
15. McConico A, Butters K, Lien K, Knudsen B, Wu, et al. (2011) In utero cell transfer between porcine littermates. Reprod Fert Dev 23, 297–302.
16. Kues WA, Niemann H (2004) The contribution of farm animals to human health. Trends Biotechnol 22: 286–294.
17. Aigner B, Renner S, Kessler B, Klymiuk N, Kurome M, et al. (2010) Transgenic pigs as models for translational biomedical research. J Mol Med 88: 653–664.
18. Whyte JJ, Prather RS (2011) Genetic modifications of pigs for medicine and agriculture. Mol Reprod Dev 78: 879–991.
19. Kues WA, Niemann H (2011) Advances in farm animal transgenesis. Prev Vet Med 102: 146–156.
20. Beschorner WE, Sudan DL, Radio SJ, Yang T, Franco KL, et al. (2003) Heart xenograft survival with chimeric pig donors and modest immune suppression. Ann Surg 237: 265–72.

21. Yan Z, Lambert NC, Ostensen M, Adams KM, Guthrie KA, et al. (2006) Prospective study of fetal DNA in serum and disease activtiy during pregnancy in women with inflammatory arthritis. Arthritis Rheum 54: 2069–2073.

22. Gadi VK, Nelson JL (2007) Fetal microchimerism in women with breast cancer. Cancer Res 67: 9035–9038.

23. Gadi VK, Malone KE, Guthrie KA, Porter PL, Nelson JL (2008) Case-control study of fetal microchimerism and breast cancer. PLOS ONE 3: e1706.

24. Cirello V, Recalicati MP, Muszza M, Rossi S, Perriono M, et al. (2008) Fetal cell microchimerism in papillary thryoid cancer: a possible role in tumor damage and tissue repair. Cancer Res 68: 8482–8488.

25. Srivatsa B, Srivatsa S, Johnson KL, Samura O, Lee SL, et al. (2001) Microchimerism of presumed fetal origin in thyroid specimens from women: a case-control study. Lancet 358: 2034–2038.

26. Bogdanova N, Siebers U, Kelsch R, Markoff A, Röpke A, et al. (2010) Blood chimerism in a girl with Down syndrome and possible freemartin effect leading to aplasia of the Müllerian derivatives. Hum Reprod 25: 1339–1343.

27. Turin L, Invernizzi P, Woodcock M, Grati FR, Riva F, et al. (2007) Bovine fetal microchimerism in normal and embryo transfer pregnancies and its implications for biotechnology applications in cattle. Biotechnol J 2: 486–491.

28. Pereira FT, Oliveira LJ, Barreto Rda S, Mess A, Perecin F, et al. (2013) Fetal-maternal interactions in the synepitheliochorial placenta using the eGFP cloned cattle model. PLOS ONE 8: e64399.

29. Garrels W, Mates L, Holler S, Dalda A, Taylor U, et al. (2011). Germline transgenic pigs by Sleeping Beauty transposition in porcine zygotes and targeted integration in the pig genome. PLOS ONE 6, e23573.

30. Garrels W, Holler S, Taylor U, Herrmann D, Struckmann C, et al. (2011) Genotype-independent transmission of transgenic fluorophore protein by boar spermatozoa. PLOS ONE 6: e27563.

31. Garrels W, Cleve N, Niemann H, Kues WA (2012) Rapid non-invasive genotyping of reporter transgenic mammals. BioTechniques 0: 1–4.

32. Garrels W, Holler S, Cleve N, Niemann H, Ivics Z, et al. (2012) Assessment of fecundity and germ line transmission in two transgenic pig lines produced by Sleeping Beauty transposition. Genes 3: 615–633.

33. Ivics Z, Garrels W, Mátés L, Yau TY, Rülicke T, et al. (2014) Germline transgenesis in pigs by cytoplasmic microinjection of *Sleeping Beauty* transposons. Nature Protoc 9: 810–827.

34. Garrels W, Ivics Z, Kues WA (2012) Precision genetic engineering in large mammals. Trends in Biotechnology 30: 386–393.

35. Entrican G (2002) Immune regulation during pregnancy and host-pathogen interactions in infectious abortion. J Comp Pathol 126: 79–94.

36. Tang MX, Zheng XM, Hou J, Qian LL, Jiang SW, et al. (2013) Horizontal gene transfer does not occur between sFat-1 transgenic pigs and nontransgenic pigs. Theriogenology 79: 667–672.

37. Steinkraus HB, Rothfuss H, Jones JA, Dissen E, Shefferly E, et al. (2012) The absence of detectable fetal microchimerism in nontransgenic goats (Capra aegagrus hircus) bearing transgenic offspring. J Anim Sci 90: 481–488.

38. Rubin JP, Cober SR, Butler PE, Randolph MA, Gazelle GS, et al (2001) Injection of allogeneic bone marrow cells into the portal vein of swine in utero. J Surg Res 95: 188–194.

39. Vernochet C, Caucheteux SM, Kanellopoulos-Langevin C (2007) Bi-directional cell trafficking between mother and fetus in mouse placenta. Placenta 28: 639–649.

40. Fujiki Y, Johnson KL, Peter I, Tighiouart H, Bianchi DW (2009) Fetal cells in the pregenant mouse are diverse and express a variety of progenitor and differentiated cell markers. Biol Reprod 81: 26–32.

41. Matsubara K, Uchida N, Matsubara Y, Hyodo S, Ito M (2009) Detection of fetal cells in the maternal kidney during gestation in the mouse. Tohoku J Exp Med 218: 107–113.

42. Sunami R, Komuro M, Yuminamochi T, Hoshi K, Hirata S (2010) Fetal cell microchimerism develops through the migration of fetus-derived cells to the maternal organs early after implantation. J Reprod Immunol 84: 117–23.

43. Fujiki Y, Tao K, Bianchi DW, Giel-Moloney M, Leiter AB, et al. (2008) Quantification of green fluorescent protein by in vivo imaging, PCR, and flow cytometry: comparision of transgenic strains and relevance for fetal cell microchimerism. Cytometry Part A 73A: 111–118.

44. Bakkour S, Baker CA, Tarantal AF, Wen L, Busch MP, et al. (2014) Analysis of maternal microchimerism in rhesus monkeys (*Macaca mulatta*) using real-time quantitative PCR amplification of MHC polymorphisms. Chimerism [Epub ahead of print, 17. Jan 2014].

45. Axiak-Bechtel SM, Kumar SR, Hansen SA, Bryan JN (2013) Y-chromosome DNA is present in the blood of female dogs suggesting the presence of fetal microchimerism. PLOS ONE 8: e68114.

21

The Meganuclease I-SceI Containing Nuclear Localization Signal (NLS-I-SceI) Efficiently Mediated Mammalian Germline Transgenesis via Embryo Cytoplasmic Microinjection

Yong Wang[1]*⁹, **Xiao-Yang Zhou**[1]⁹, **Peng-Ying Xiang**[1], **Lu-Lu Wang**[1], **Huan Tang**[2], **Fei Xie**[1], **Liang Li**[1], **Hong Wei**[1]*

1 Department of Laboratory Animal Science, College of Basic Medical Sciences, Third Military Medical University, Chongqing, China, 2 China Three Gorges Museum, Chongqing, China

Abstract

The meganuclease I-SceI has been effectively used to facilitate transgenesis in fish eggs for nearly a decade. I-SceI-mediated transgenesis is simply via embryo cytoplasmic microinjection and only involves plasmid vectors containing I-SceI recognition sequences, therefore regarding the transgenesis process and application of resulted transgenic organisms, I-SceI-mediated transgenesis is of minimal bio-safety concerns. However, currently no transgenic mammals derived from I-SceI-mediated transgenesis have been reported. In this work, we found that the native I-SceI molecule was not capable of facilitating transgenesis in mammalian embryos via cytoplasmic microinjection as it did in fish eggs. In contrast, the I-SceI molecule containing mammalian nuclear localization signal (NLS-I-SceI) was shown to be capable of transferring DNA fragments from cytoplasm into nuclear in porcine embryos, and cytoplasmic microinjection with NLS-I-SceI mRNA and circular I-SceI recognition sequence-containing transgene plasmids resulted in transgene expression in both mouse and porcine embryos. Besides, transfer of the cytoplasmically microinjected mouse and porcine embryos into synchronized recipient females both efficiently resulted in transgenic founders with germline transmission competence. These results provided a novel method to facilitate mammalian transgenesis using I-SceI, and using the NLS-I-SceI molecule, a simple, efficient and species-neutral transgenesis technology based on embryo cytoplasmic microinjection with minimal bio-safety concerns can be established for mammalian species. As far as we know, this is the first report for transgenic mammals derived from I-SceI-mediated transgenesis via embryo cytoplasmic microinjection.

Editor: Atsushi Asakura, University of Minnesota Medical School, United States of America

Funding: YW was supported by grants from Natural Science Fund of China [31171280, 31271330], National 973 Project of China [2011CB944102], Chongqing Natural Science Fund [cstc2011jjA10049]. HW was supported by grants from Natural Science Fund of China [81173126], National 973 Project of China [2011CBA01006] and National Science and Technology Support Program [2011BAI15BO2]. The funders had no role in study design, data collection and analysis, decision to publish, or preparation of the manuscript.

Competing Interests: The authors have declared that no competing interests exist.

* Email: yongw7528@gmail.com (YW); weihong63528@163.com (HW)

⁹ These authors contributed equally to this work.

Introduction

Genetic modification of mammalian genomes is of great importance for bio-medical researches such as deciphering gene functions, investigating disease mechanisms and searching and validating therapeutic targets, and also a potential method to generate farm animals with improved economic traits for agricultural purposes.

Mammalian genetic modification includes transgenesis, gene disruption and random mutation of genomes. Gene disruption was once a sophisticated and labor-intensive process which was based on DNA homologous recombination (HR) in embryonic stem cells (ESCs). However, this DNA HR-based technology achieved very limited success in mammalian species other than mice due to the lack of ESCs derived from these species. Recently, with the development of powerful site-specific engineered endonucleases(EENs), especially Zinc Finger Nucleases(ZFNs) [1–4], Transcription Activator-like Effector Nucleases (TALENs)[5–10] and Clustered Regularly Interspaced Short Palindromic Repeats/CRISPR-associated system 9 (CRISPR/Cas9) [11–15], which are capable of disrupting genes efficiently by making double strand breaks (DSBs) at target sites, gene disruption has become a much more efficient and convenient process which is independent on ESCs and achieved significant success in mammalian species other than mice. Random mutation of mammalian genomes is regularly efficient using powerful chemical mutagens such as ENU or insertional viral vectors. In contrast, mammalian transgenesis, especially for species other than mice, remains to be further optimized.

Transgenesis is a process of adding exogenous and (or) artificially constructed genes to animal genomes, which is indispensable for generating mammalian models with gain of functions for bio-medical researches or genetically modified farm animals with additional economic traits. Currently, the available technologies for mammalian transgenesis include embryo pronuclear microinjection, somatic cell nuclear transfer (SCNT) using transgenic cells as nuclear donors, sperm-mediated gene transfer (SMGT), lentiviral transgenesis using retro-viral vectors derived from lentiviruses as vehicles to deliver transgenes into animal genomes and transposon-mediated gene transfer. Embryo pronuclear microinjection is a reliable and traditional method to produce transgenic mammals, but the inaccessibility to pronuclear of many mammalian species other than mice and the low efficiency of transgene integration largely limits its effectiveness and utility [16,17]. SCNT is a reproducible method to produce transgenic mammals, but SCNT is a sophisticated and complex procedure with a rather low efficiency [18,19] and a large number of oocytes are needed. Practically, many mammalian species of biological or biomedical importance, such as non-human primates or other none-economic animals, are not able to be cloned due to the lack of regular ovary sources. Besides, the unpredictable abnormalities related to cloned individuals limit the usage of resulted transgenic animals to model human diseases, and the antibiotic resistant genes, the necessary selection markers for transgenic nuclei donor cell culture which are finally added into the genomes of resulted transgenic individuals by SCNT process, brings additional uncertainties for the application of derived transgenic animals. SMGT is reported to be a simple and inexpensive method for transgenic animal production, however extremely variant data has been reported from different labs and the highly unstable outcome of this technology limits its application. Lentiviral transgenesis has been recognized as an extremely efficient method to generate transgenic animals of different species [20–22]. However, the preparation of high titre lentiviral particle suspensions is a complicated procedure and the viral vectors integrated into animal genomes are of bio-safety concerns. Transposon systems have been used for animal transgenesis [23–27], but transposons are mobile genetic elements, and the derived transgenic animals are of similar bio-safety concerns as those derived from lentiviral transgenesis.

On the basis of these points mentioned above, it is valuable to develop an efficient, simple, and species-neutral transgenesis technology for mammals, which is of minimal bio-safety concerns being without the involvement of viral or mobile vectors and independent on SCNT process or the accessibility to embryo pronuclear. The meganuclease I-SceI, which is derived from the mitochondria of *Saccharomyces cerevisiae* and has a long (>18 bp) recognition sequence that does not exist in animal genomes naturally, has been effectively used to facilitate trangenesis in fishes via embryo cytoplasmic microinjection [28–32]. However, in this study we found that the native I-SceI molecule failed to efficiently facilitate transgenesis in mammalian embryos as it did in fish eggs after cytoplasmic microinjection along with the plasmids of transgene vector containing two inversely flanking I-SceI recognition sequences, suggesting that in mammalian embryos, the native I-SceI molecule did not exhibit the efficacy on transgenesis in the same way as that in fish eggs. By adding a mammalian nuclear localization (NLS) signal to the N-terminus of I-SceI molecule, the I-SceI molecule containing NLS (NLS-I-SceI) was found to be capable of translocating DNA fragments from mammalian embryo cytoplasm into nuclear, and the I-SceI recognition sequence-containing transgene vector plasmids, which was injected into cytoplasm along with NLS-I-SceI mRNA,

exhibited expression in both mouse and porcine embryos. By transferring the embryos cytoplasmically co-injected with NLS-I-SceI mRNA and the transgene plasmids into synchronized female recipients, transgenic founder animals were efficiently generated and transgenes were found to be capable of germline transmission. These data suggested that using the NLS-I-SceI molecule, a simple, efficient and species-neutral transgenesis technology, which was based on embryo cytoplasmic microinjection and without the involvement of viral or transposon vectors, can be established for mammals.

Materials and Methods

Animals

Mice of FVBN inbred strain and Bama minipigs, which are of one local minipig strain in China, were used in this study. The mice were purchased from SLAC Laboratory Animal Co., Ltd (Shanghai, China) and maintained under specific pathogen-free conditions in Laboratory Animal Centre of our university. The minipigs used in this study were derived from the closed colony regularly maintained in Laboratory Animal Centre. All the protocols involving the use of animals were approved by the Institutional Animal Care and Use Committee of Third Military Medical University (Approval ID: SYXK-PLA-2007036).

Construction of NLS-I-SceI molecule and transgene vector

The NLS-I-SceI molecule was constructed by adding a modified version of 3×SV40 NLS sequence containing a HA epitope to the N-terminal of the native I-SceI molecule. The coding sequence for NLS-I-SceI, of which the initiation codon was surrounded by a kozak sequence for optimal translation initiation and the codons were optimized for both pigs and mice, was artificially synthesized and subcloned into the mammalian expression vector PCI (Promega) downstream T7 promoter, and the resulted vector was designated as PCI-T7-NLS-I-SceI in this article. For the convenience of transgene vector construction, an intermediate vector designated as p2IS was constructed by subcloning a synthesized DNA fragment containing a long multi cloning sites (MCS) inversely flanked by two I-SceI recognition sequences into pUC18 vector at the two restriction sites BsmBI and SapI to substitute the original MCS region. To construct the transgene vector used in this study, a DNA fragment containing human Ubiquitine C (UBC) promoter, eGFP CDS and a poly (A) signal sequence was cut off from FUGW plasmid (Addgene, #14883) using the two endonucleases PacI and PmeI and then subcloned into p2IS vector at the same two restriction sites, and the resulted transgene vector was designated as p2IS-UBC-eGFP.

Preparation of mRNA

NLS-I-SceI mRNA was prepared by *in vitro* transcription using linearized PCI-T7-NLS-I-SceI plasmid as templates. The plasmid was linearized by restrictive digestion at the ClaI site which was located downstream NLS-I-SceI CDS. After complete digestion, the reaction system were treated with proteinase K (100 μg/mL) and SDS (0.5% (v/v)), and then further treated with one equal volume of phenol:chloroform mixture. After centrifuge at 12000 g, 4°C for 10 min, the supernatant was carefully collected and the DNA was precipitated by adding 2.5 volumes of ice-cold absolute alcohol and one tenth volume of RNase-free 5 M NaAc solution. After washing in 75% alcohol, the DNA precipitate was finally dissolved into RNase-free deionized water after drying. Using the purified linearized plasmids as templates, NLS-I-SceI mRNA was produced by *in vitro* transcription using the mMESSAGE mMACHINE@T7 Ultra Kit (Life Technologies, AM1345) as described in the manual. After transcription was terminated, 1 μL of transcription products was saved prior to poly(A) tailing as a control to assess the tailing quality after poly(A) tailing procedure was completed. To prepare purified mRNA for embryo microinjection, the poly(A)-tailed mRNA products were recovered from reaction system using RNeasy Mini Kit (Qiagen, 74104) and eluted with RNase-free deionized water. The quality of mRNA samples was assessed by agarose gel electrophoresis.

Embryo microinjection, observation and transfer

The circular or linearized transgene vector plasmids p2IS-UBC-eGFP used for embryo microinjection were treated and purified in the same way as that for in vitro transcription templates. For microinjection, the purified p2IS-UBC-eGFP plasmids were mixed with different concentrations of NLS-I-SceI mRNA or included in the digestive reaction system of I-SceI endonuclease (NEB) as the substrate as previously described for fish transgenesis [31]. The I-SceI nuclease was stored at −80°C in 2 μL aliquots and added into the reaction system prior to microinjection as described [31], and its activity was confirmed by digestion of the plasmid p2IS-UBC-eGFP. To observe the localization of the injected DNA, two completely complementary 130 bp-long Cy3-labeled single strand DNA fragments containing two inversely flanking I-SceI recognition sequences at both ends were synthesized, denatured and annealed to be double-stranded, and then used for embryo cytoplasmic microinjection with NLS-I-SceI mRNA in the same way as transgene vector plasmids.

Microinjection was performed as described [33], except that the materials were injected into cytoplasm instead of pronuclear in this study. The mouse or porcine embryos subjected to microinjection were collected from mated female individuals and cultured as described [33,21]. The porcine oocytes were collected from ovaries and subjected to in vitro maturation (IVM) as described [34]. The matured oocytes at metaphase of meiosis II (MII phase) with extruded first polar body were selected and subjected to microinjection post parthenogenetic activation by direct current electrical pulses (1.2 KV/cm, 30 μs, two times, 1 sec interval) as described [34]. The parthenogentically activated porcine oocytes (parthenogenetic embryos) were cultured as that for the collected porcine embryos.

The cultured embryos were observed under fluorescence microscopy or laser scanning confocal microscopy (LSCM, Zeiss LSM 780) to examine transgene expression or the localization of injected Cy3-labeled DNA fragments. To stain chromosomal DNAs, embryos were incubated in culture media containing 15 μg/mL Hoechst 33342(Sigma) for 30 min prior to microinjection and washed thoroughly in fresh media. To obtain transgenic founders, injected embryos were surgically transferred into oviducts of synchronized recipient female mice or sows as described [33,21].

Analysis of the presence of uncut I-SceI recognition site in embryos by polymerase chain reaction (PCR)

Total DNA samples were extracted from individual embryos by incubating each embryo in 10 μL of lysis buffer (KCl: 50 mM; MgCl2:1.5 mM; Tris-Cl (pH8.0): 10 mM; Nonidet P-40:0.5% (w/v); Tween-20:0.5% (v/v); proteinase K: 100 μg/mL) at 65°C for 1 h. After heated at 95°C for 10 min to inactivate proteinase K, the lysate was used as template for PCR. A set of primer pair IS-site-F1/R1 (IS-site-F1:5′-CCACTGACCTTTGGATGGTG-3′; IS-site-R1:5′-TACCGCCTTTGAGTGAGCTG-3′; product size: 518 bp), of which the PCR product covered the I-SceI recognition sequence 3′ to the transgene cassette, was designed to detect the

presence of uncut I-SceI site. Another primer pair set eGFP-F1/R1 (eGFP-F1:5′-ACTGGAGAACTCGGTTTGTCGT-3; eGFP-R1:5′-ACGGCCAGAATTTAGCGGAC -3′; product size: 453 bp) was used to detect the presence of eGFP CDS. The total DNA samples were further subjected to quantitative PCR (qPCR) analysis in a system based on SybrGreen qPCR Master Mix(2×) (ABI). The primer pair set for qPCR analysis of uncut I-SceI sites was IS-site-F2/R2 (IS-site-F2:5′-AACTAGGGAACC-CACTGCTT-3′; IS-site-R2:5′-AACTAGGGAACC-CACTGCTT-3′; product size: 171 bp), and that for qPCR of the eGFP CDS (the internal control) was eGFP-F2/R2 (eGFP-F2:5′-CAGAAGAACGGCATCAAGGT-3′; eGFP-R2:5′-TCTCGTTGGGGTCTTTGCT-3′; product size: 172 bp). Using the p2IS-UBC-eGFP plasmids diluted to different concentrations as standard samples, the qPCR analysis was performed in an absolute quantitation manner.

Transgenic animal screen

Transgenic animals were screened by PCR and Southern blot assay. The primer pair set used for transgenic mouse screen by PCR was eGFP-F3/R3, of which the sequences were 5′-ATGGTGAGCAAGGGCGAGGA-3′ (eGFP-F3) and 5′-TGCCGTCCTCGATGTTGTGG-3′ (eGFP-R3), and the product size was 526 bp. The primer pair used for transgenic pig screen was eGFP-F1/R1 as described above. The probe for Southern blot assay was prepared by PCR using PCR DIG Probe Synthesis Kit (Roche) as described in the kit manual. The primer pair set used for probe preparation was Probe-DIG-F/R, of which the sequences were 5′-GCAGAAGAACGGCATCAAGGT-3′ (Probe-DIG-F) and 5′-TAGGGAGGGGGAAAGCGAA-3′ (Probe-DIG-R), which covered the junction region between eGFP CDS and the poly(A) signal sequence. Southern blot was performed using DIG-High Prime DNA Labeling and Detection Starter Kit II (Roche) as described in manual using genomic DNAs (>10 μg) completely digested by PstI. The in vivo green fluorescence in transgenic animals was detected using a GFP Macroscopy system (BLS, Hungarian) by exposure to blue excitation light with wave length of 460–495 nm and observed through a filter.

Results

The NLS-I-SceI molecule and transgene construct

The NLS-I-SceI molecule consists of 3×SV40 NLS, an HA tag epitope and the native I-SceI molecule as shown in Fig. 1 B. 3×SV40 NLS is highly potent for nuclear localization, and HA tag epitope can be used to detect NLS-I-SceI molecule distribution in cells once it was expressed in cytoplasm. The NLS-I-SceI CDS was optimized for both porcine and murine codon usage preferences (the NLS-I-SceI CDS and amino acid sequence were shown in Fig. S1 and S2). High quality NLS-I-SceI mRNA was produced by T7 promoter-driven in vitro transcription using the linearized PCI-T7-NLS-I-SceI vector as templates (Fig. 1 C). The complete sequence of transgene vector p2IS-UBC-eGFP was shown in Fig. S3. In the transgene vector, a UBC promoter-driven eGFP expression cassette was flanked by two inversed I-SceI recognition sequences at both ends (Fig. 1 A). After cut by NLS-I-SceI molecule, the transgene vector plasmid was linearized, and the NLS-I-SceI protein was expected to be bound to the fragment containing transgene expression cassette, and thereby protect the transgene fragments from degradation and transfer the fragments from cytoplasm into nuclear (Fig. 1 D), for I-SceI protein exhibited high affinity in binding to the downstream cleavage product [36].

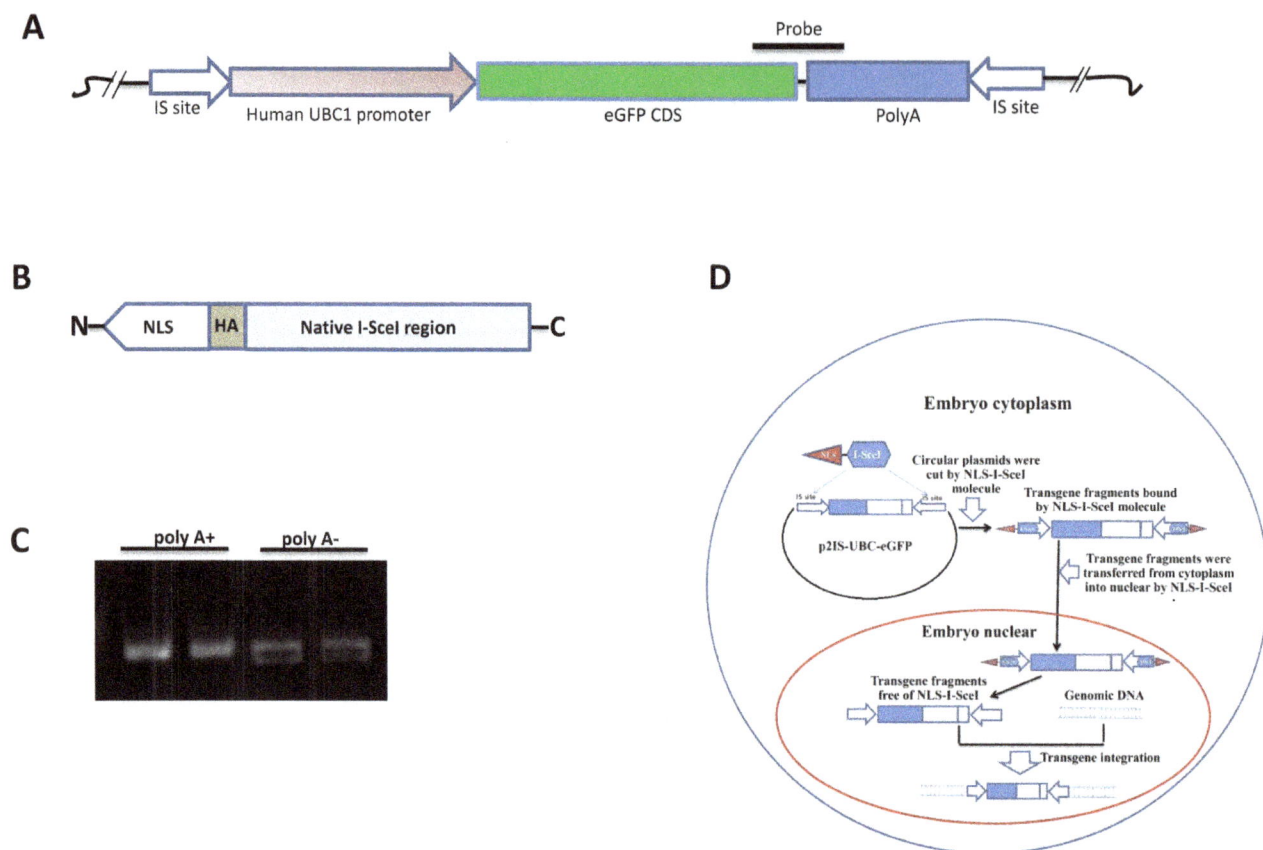

Figure 1. Transgene construct and the NLS-I-SceI molecule. A: The schematic structure of p2IS-UBC-eGFP vector. IS site: the inversely flanking I-SceI recognition sequence; the black bar indicates the position of the probe used for Southern blot assay. B: The schematic structure of NLS-I-SceI molecule. C: The *in vitro* transcribed NLS-I-SceI mRNA. polyA+: the mRNA with polyA tail; polyA-: the mRNA without polyA tail. D: The expected working principle of NLS-I-SceI-mediated transgenesis.

The NLS-I-SceI molecule was capable of cutting circular transgene plasmids and transferring DNA fragments from cytoplasm into nuclear in mammalian embryos

To investigate whether the NLS-I-SceI molecule was capable of cutting the I-SceI recognition sequence-containing circular plasmids in mammalian embryos, total DNAs were extracted from single porcine parthenogenetic blastocysts developed from oocytes co-injected with NLS-I-SceI mRNAs and the circular p2IS-UBC-eGFP plasmids (30 ng/μL each), and subjected to PCR analysis to assess the extent to which the circular plasmids were digested. The uncut I-SceI site was quantitatively detected by qPCR using a primer pair covering the I-SceI recognition sequence 3′ to the transgene expression cassette, and the eGFP CDS detected as internal control. Prior to qPCR, a qualitative PCR was performed to confirm the existence of plasmids in embryos. As shown in Fig. 2 A, in the embryos co-injected with NLS-I-SceI mRNA and the circular transgene plasmids, the band intensities of PCR products covering I-SceI site were remarkably lower than those of eGFP CDS. In contrast, in embryos injected only with circular plasmids, the uncut I-SceI site and eGFP CDS were simultaneously detected or not in these samples (Fig. 2 A), and the band intensities of PCR products covering I-SceI site were comparable to those of eGFP CDS (Fig. 2 A), indicating that in these embryos the levels of uncut I-SceI site and eGFP CDS were comparable and varied proportionally. The samples with expected PCR products were subjected to qPCR analysis, of which the

Amplification Plots, Melt Curves and Standard Curves were shown in Fig. S4. The qPCR data further showed that the levels of uncut I-SceI site relative to eGFP CDS in embryos co-injected with NLS-I-SceI mRNA and circular plasmids were largely lower than those in embryos injected only with circular plasmids (P< 0.001, Fig. 2 B), indicating that the NLS-I-SceI molecule produced from mRNAs in mammalian embryos was bio-active and capable of cutting circular plasmids. Moreover, these data further demonstrated that in the embryos co-injected with NLS-I-SceI mRNA and circular plasmids, although the relative levels of uncut I-SceI site were much lower, the eGFP CDS copy numbers were remarkably higher than those in embryos injected only with circular plasmids (P<0.001, Fig. 2 C), suggesting that the linearized transgene DNA fragments were protected from degradation by NLS-I-SceI molecule after plasmids were cut at I-SceI sites.

To display the localization of injected DNAs in living embryos, Cy3-labeled double-stranded DNA (Cy3-DNA) fragments containing two inversely flanking I-SceI recognition sequences at both ends, of which the schematic structure was shown in Fig. 3 A and sequence Fig. S5, were co-injected with NLS-I-SceI mRNA into the cytoplasm of activated porcine MII oocytes (parthenogenetic embryos) of which the chromosomal DNAs were stained with Hoechst 33342 prior to microinjecction. The injected embryos were cultured and observed under LSCM at 16 and 24 h post activation. In control groups, the embryos were injected with only Cy3-DNA fragments or Cy3-DNA fragments included in the

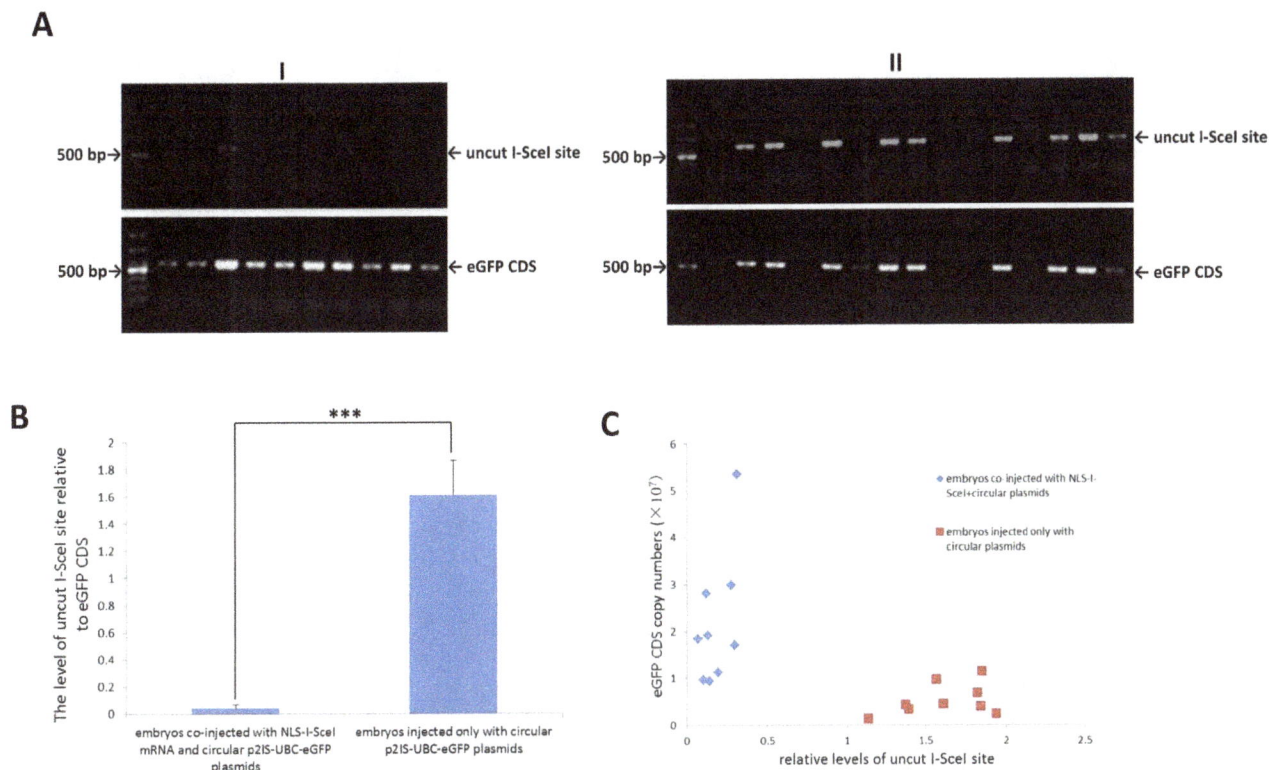

Figure 2. The NLS-I-SceI molecule was capable of cutting circular p2IS-UBC-eGFP plasmids in porcine parthernogenetic embryos. A: Detection of uncut I-SceI site and eGFP CDS by PCR in the embryos cytoplasmically injected with circular p2IS-UBC-eGFP plasmids plus NLS-I-SceI mRNA and only with circular p2IS-UBC-eGFP plasmids. I: embryos cytoplasmically injected with circular p2IS-UBC-eGFP plasmids plus NLS-I-SceI mRNA; II: embryos cytoplasmically injected only with circular p2IS-UBC-eGFP plasmids. B: The levels of uncut I-SceI site relative eGFP CDS detected by qPCR in the injected embryos. C: The eGFP CDS copy numbers in the injected embryos. *: statistical significance.

native I-SceI endonuclease digestive reaction system as that for I-SceI-mediated transgenesis in fish. At 16 h post activation, in the embryos co-injected with NLS-I-SceI mRNAs and Cy3-DNAs, of which the chromosomes were in a relaxed state and loosely assembled suggesting that meiosis was proceeding to the telophase and the nuclear was under construction (Fig. 3 B), the Cy3-DNA fragments (red fluorescence) were found to be clustered and located near to the chromosomes (blue fluorescence) (Fig. 3 B). At 24 h post activation, in the embryos co-injected with NLS-I-SceI mRNAs and Cy3-DNAs, the blue fluorescence was concentrated suggesting that chromosomes were compactly aggregated, meiosis completed and nuclear was constructed, and the clustered Cy3-DNA fragments were observed to be completely co-localized with chromosomes (Fig. 3 B), indicating that the Cy3-DNA fragments were transferred into nuclear. In contrast, in the embryos of control groups, the red fluorescence was scattered and extremely weak, and no Cy3-DNAs were observed to be clustered, located closely to or co-localized with the chromosomes at 16 h or 24 h post activation (Fig. 3 C, D), suggesting that the Cy3-DNA fragments were diffusely distributed in cytoplasm or degraded. These data provided a direct demonstration that the NLS-I-SceI molecule was capable of transferring DNA fragments from cytoplasm into nuclear in mammalian embryos, and this transfer process was co-incident with the process of nuclear formation during meiosis (or mitosis) of embryos, while the native I-SceI molecule was not, being consistent with previously reported data for the native I-SceI-mediated transgenesis in fish embryos [30].

NLS-I-SceI molecule was capable of facilitating transgenesis in mammalian embryos via cytoplasmic microinjection

To investigate whether NLS-I-SceI molecule was capable of mediating transgenesis and resulting in transgene expression in early mammalian embryos, mouse eggs were subjected to cytoplasmic microinjection with the mixture of NLS-I-SceI mRNA and circular transgene plasmid p2IS-UBC-eGFP, for mouse eggs have visible pronuclear and the materials can be confirmed to be injected into cytoplasm. With a given plasmid concentration (30 ng/μL), NLS-I-SceI mRNAs at different concentrations (10, 20 and 30 ng/μL) were co-injected with circular transgene plasmids into cytoplasm of 1-cell mouse eggs, and green fluorescence in the blastocysts developed from injected eggs were observed and counted at 5 d post injection. To avoid cellular lysis after injection, a very small volume (about 5 pL) of solution, which was less than that for pronuclear microinjection, was injected into embryo cytoplasm. Results showed that the NLS-I-SceI molecule mediated transgenesis in a dose-dependent manner (Fig. 4 A). In the group injected with 30 ng/μL of NLS-I-SceI mRNA, the fluorescence intensity was significantly higher than those in groups injected with 10 and 20 ng/μL of NLS-I-SceI mRNA (Fig. 4 A). In contrast, in the group injected with 30 ng/μL of circular p2IS-UBC-eGFP plasmid included in the native I-SceI endonuclease digestive reaction system, no fluorescent blastocysts were observed (Fig. 4 A), indicating that without the added NLS signal, the native I-SceI molecule was not capable of facilitating transgenesis and further resulting in transgene expres-

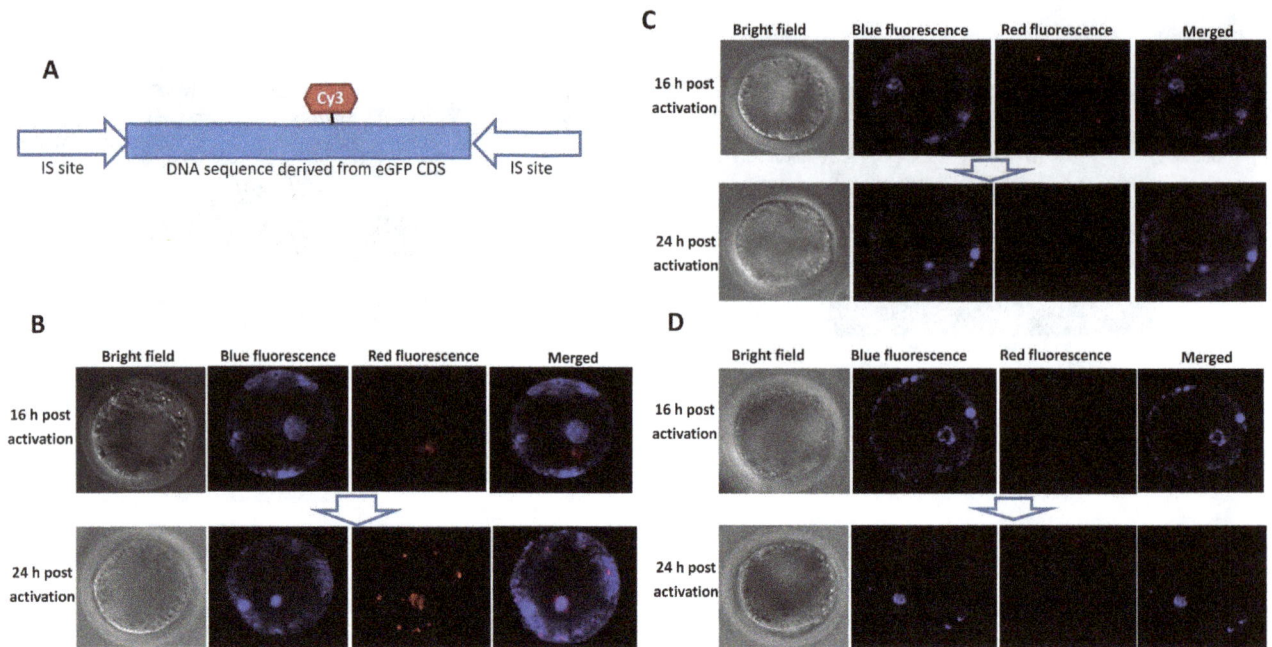

Figure 3. Transfer of DNA fragments from cytoplasm into nuclear by NLS-I-SceI molecule in porcine parthernogenetic embryos. The activated porcine MII oocytes (1-cell parthernogenetic embryos) stained with Hoest33342 were cytoplasmically injected with Cy3-labelled DNA fragments plus NLS-I-SceI mRNA, and the localization of DNA fragments were observed under LSCM at 16 and 24 h post microinjection respectively. In control groups, the embryos were injected with Cy3-DNA fragments included into the native I-SceI endonuclease digestive reaction system or only with Cy3-DNA fragments. A: The structure of Cy3-labeled DNA fragments. B: The localization of Cy3-DNA fragments co-injected with NLS-I-SceI mRNA. C: The localization of Cy3-DNA fragments co-injected with the native I-SceI nuclease. D: The localization of Cy3-DNA fragments injected alone. Red fluorescence: the Cy3-DNA fragments; Blue fluorescence: the chromosomal DNAs.

sion in mouse embryos. In the groups cytoplasmically injected only with circular or linearized plasmids at the same concentration, no fluorescence was observed in the derived blastocysts either, although fluorescence was observed in a few developmentally arrested embryos in the circular plasmid injection group (Fig. 4 A). In all the groups, the blastocyst development rates (blastocysts/ cleaved eggs) were comparable to the untreated group (data not shown), suggesting that the injected materials did not interfere with *in vitro* development once embryos survived the microinjection process. The dynamics of transgene expression in the embryos cytoplasmically co-injected with NLS-I-SceI mRNA and circular transgene plasmid was similar to that in embryos subjected to pronuclear microinjection only with circular transgene plasmid, although the fluorescence intensity in the cytoplasmic injection group was lower (Fig. 4 B), suggesting that the transgene fragments delivered into cytoplasm were transferred into pronuclear by NLS-I-SceI molecule as early as embryo cleavage started, which was consistent with the results of LSCM observation. The lower fluorescence intensity may be due to the less copies of transgene fragment in pronuclear transferred from cytoplasm by NLS-I-SceI molecule compared to those of transgene fragment directly delivered into pronuclear by microinjection.

To address whether NLS-I-SceI molecule was capable of mediating transgenesis in mammalian embryos of species other than mice, 1- or 2-cell porcine eggs surgically collected from mated sows were subjected to cytoplasmic co-injection with NLS-I-SceI mRNAs and circular transgene plasmids (30 ng/μL each), for pig is a typical mammalian species of which the pronuclear is usually invisible and refractory to pronuclear microinjection. Porcine eggs have a relatively larger size and are much more tolerant to cytoplasmic microinjection compared to mouse eggs, and a much

larger volume (40–60 pL) of solution, which contained 1.2–1.8 pg of transgene plasmids and NLS-I-SceI mRNAs respectively, was injected into cytoplasm. As shown in Fig. 5 A, in the porcine embryos derived from eggs co-injected with NLS-I-SceI mRNA and circular p2IS-UBC-eGFP plasmids, strong fluorescence was observed on 3 d post injection, and the majority of derived blastocysts exhibited strong fluorescence on 6 d post injection. In contrast, in the embryos injected with circular p2IS-UBC-eGFP plasmids (30 ng/μL) included into the native I-SceI endonuclease digestive reaction system, only weak fluorescence was observed in a few embryos on 3 d post injection, and on 6 d, no fluorescence was observed in the derived blastocysts, although fluorescence was observed in a few developmentally arrested embryos (Fig. 5 A). This different fluorescence was not because of the difference in eGFP CDS copy numbers, for the eGFP CDS was readily detected in all the injected embryos (Fig. 5 B), and the eGFP CDS copy numbers in the embryos injected with circular plasmids plus NLS-I-SceI mRNA were comparable to those in embryos injected with circular plasmids at the same concentration included into the native I-SceI endonuclease digestive reaction system ($P>0.1$, Fig. 6 A), which were much higher than those in embryos injected only with circular plasmids ($P<0.001$, Fig. 6 A). The uncut I-SceI site was detected in the injected embryos (Fig. 5 B), and its levels relative to eGFP CDS were also comparable between the two groups injected with circular plasmids plus NLS-I-SceI mRNA and native I-SceI nuclease ($P>0.1$, Fig. 6 B), but were significantly lower than those of the group injected only with circular plasmids ($P<0.001$, Fig. 6 B), suggesting that the NLS-I-SceI molecule derived from mRNA cut circular plasmids to a similar degree to the native I-SceI nuclease in porcine embryos. The presence of uncut I-SceI site indicated that there existed residual circular

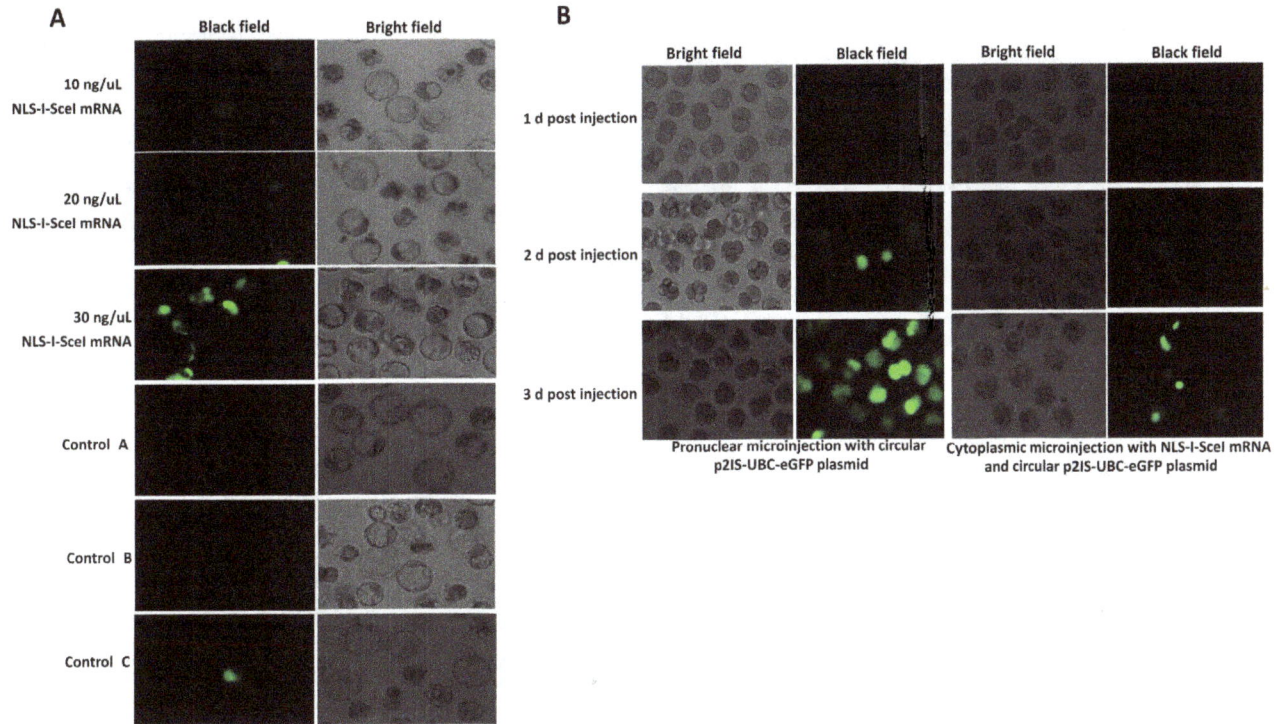

Figure 4. Transgene expression in cytoplasmically injected mouse embryos. A: Mouse eggs cytoplasmically injected with 30 ng/μL of circular p2IS-UBC-eGFP plasmids plus NLS-I-SceI mRNAs at different concentratiions. Controls A-C were the control groups injected with 30 ng/μL circular plasmids included into the native I-SceI endonuclease digestive reaction system (control A), linearized plasmids (control B) or circular plasmids (control C). B: The dynamics of transgene expression in embryos subjected to cytoplasmic microinjection with circular p2IS-UBC-eGFP plasmids plus NLS-I-SceI mRNA or pronuclear microinjection only with circular p2IS-UBC-eGFP plasmids.

plasmids in the injected embryos, which may be a reason for the fluorescence in the few porcine embryos injected with the circular plasmids included into the native I-SceI endonuclease digestive reaction system. The circular plasmids were resistant to endogenous nuclease and could be passively diffused into the nuclear during embryo cleavage as indicated by a previous report [36]. Consistently, in this work, the porcine embryos injected only with circular plasmids also exhibited fluorescence (Fig. 5 A), while those injected with linearized plasmids did not (data not shown). However, the circular plasmid rarely results in transgene integration in mammalian embryos even introduced directly into pronuclear in large amounts [27,36]. Totally, these data indicated that the NLS-I-SceI molecule was capable of efficiently facilitating transgenesis in porcine embryos, while the native I-SceI molecule was not.

The transgenesis mediated by NLS-I-SceI molecule in mammalian embryos efficiently resulted in transgenic animals

To answer whether the NLS-I-SceI-mediated transgenesis in mammalian embryos via cytoplasmic microinjection was able to result in transgenic animals, 411 fertilized mouse eggs were collected from nine super-ovulated and mated female mice, and 330 eggs with visible pronuclear were selected and randomly and equally divided into two groups. One group was subjected to cytoplasmic microinjection with the mixture of NLS-I-SceI mRNA and circular transgene plasmids (30 ng/μL each), and the other group (control) injected with circular transgene plasmid (30 ng/μL) included into the native I-SceI endonuclease digestive reaction system as described above. 116 eggs which survived the

microinjection process and cleaved the next day were transferred into 4 surrogate mice. Totally, 23 founder pups were born, of which 10 pups were derived from eggs of control group, and 13 pups from eggs co-injected with NLS-I-SceI mRNA and circular plasmids. As shown in Fig. 7 A, in the founders derived from eggs co-injected with NLS-I-SceI mRNA and circular plasmids, 6 pups were detected to be transgenic by PCR, while in the control group no transgenic pub was detected. The transgenic rate in founders of NLS-I-SceI-mediated transgenesis group was 46.2% (6/13), and the transgenesis efficiency (transgenic founders/transferred eggs) was 10.7% (6/56), which were both higher than the data for pronuclear microinjection in our lab (unpublished). However, the survival rate of cytoplasmically microinjected mouse eggs (35.2% (116/330)) was remarkably lower than that of eggs subjected to pronuclear microinjection in our lab (usually 50%), indicating that mouse eggs were more vulnerable to cytoplasmic microinjection than to pronuclear microinjection. To test the germline transmission competence of transgene, the transgenic founder mouse with the strongest PCR product band were mated with wild-type mice, and transgenic individuals were detected from the resulted offspring (Fig. 7 B). *In vivo* fluorescence was not observed in the transgenic founder mice or the transgenic individuals of F1 offspring, and transgene integration was detected by Southern blot only in one founder mouse (Fig. 7 C). However, the *in vivo* fluorescence was observed after the transgenes were enriched by mating between transgenic individuals consecutively over at least three generations (Fig. 7 D), indicating that the NLS-I-SceI-mediated transgenesis did resulted in transgene integration in mouse genome although not detected by Southern blot assay in most founders. These results indicated that the NLS-I-SceI-

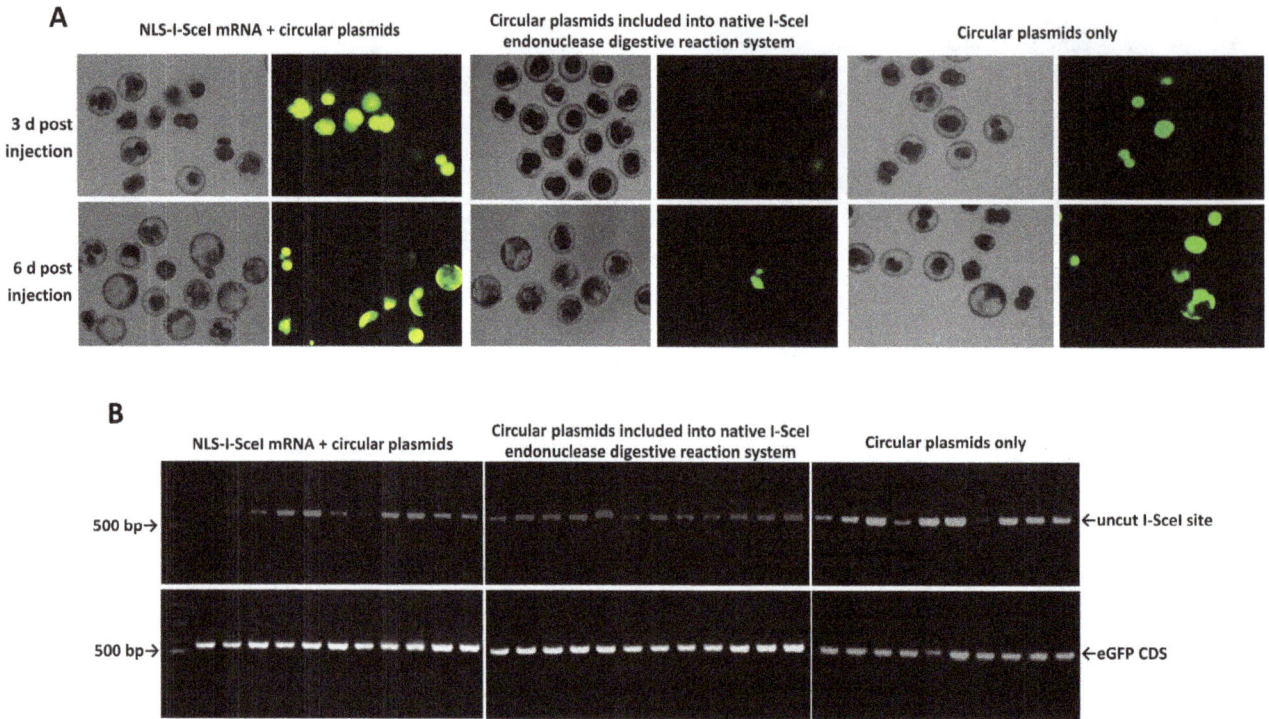

Figure 5. Transgene expression and detection of uncut I-SceI site and eGFP CDS by PCR in cytoplasmically injected porcine embryos. A: Transgene expression in the porcine embryos cytoplasmically injected with circular plasmids (p2IS-UBC-eGFP) plus NLS-I-SceI mRNA, circular plasmids included into the native I-SceI nuclease digestive reaction system and circular plasmids only. B: Detection of uncut I-SceI site and eGFP CDS by PCR in the cytoplasmically injected porcine embryos as described in A.

mediated transgenesis was capable of resulting in transgenic mice, while the native I-SceI nuclease was not.

To further test whether the NLS-I-SceI-mediated transgenesis would result in transgenic animals of species other than mice, 36 porcine eggs at 1- or 2-cell stage surgically collected from mated sows were subjected to cytoplasmic co-injection with NLS-I-SceI mRNA and the circular p2IS-UBC-eGFP plasmids, and then transferred into two synchronized surrogate sows. One recipient

was pregnant and four piglets were born. The *in vivo* fluorescence was observed in three of the four founder pigs (Fig. 8 A). However, all of the founder pigs were detected to be transgenic by PCR screen and transgene integration was confirmed by Southern blot assay (Fig. 8 B, C). The lack of *in vivo* fluorescence in one transgenic founder pig (4#) may be due to the low copy number of integrated transgenes as indicated by the Southern blot data (Fig. 8 C). The founder pig with the strongest fluorescence (1#)

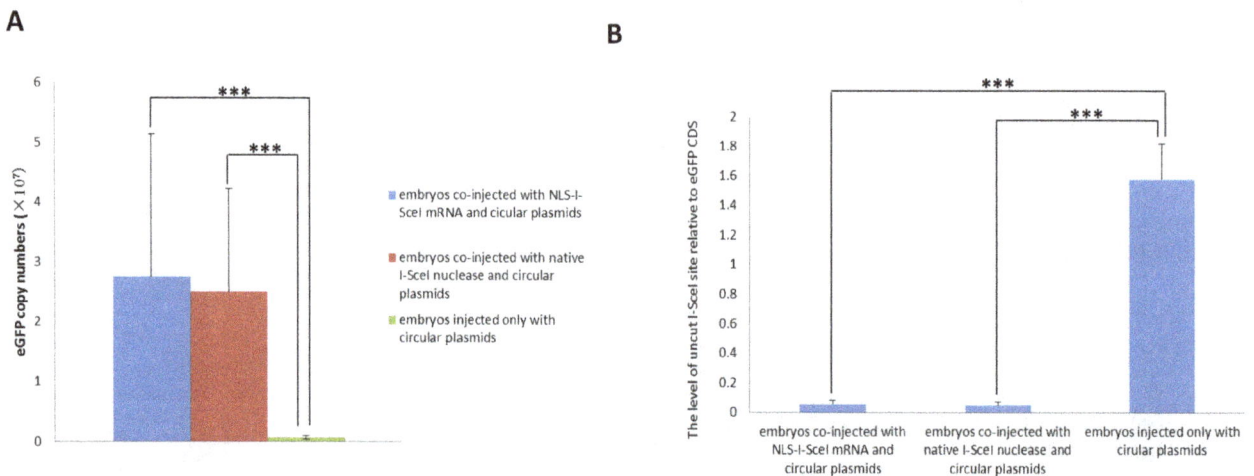

Figure 6. Quantitative analysis of uncut I-SceI site and eGFP CDS by qPCR in cytoplasmically injected porcine embryos. A: The eGFP CDS copy numbers in the cytoplasmically injected porcine embryos as described in Fig. 5. B: The uncut I-SceI site levels relative to eGFP CDS in the cytoplasmically injected porcine embryos as described in Fig. 5. *: statistical significance.

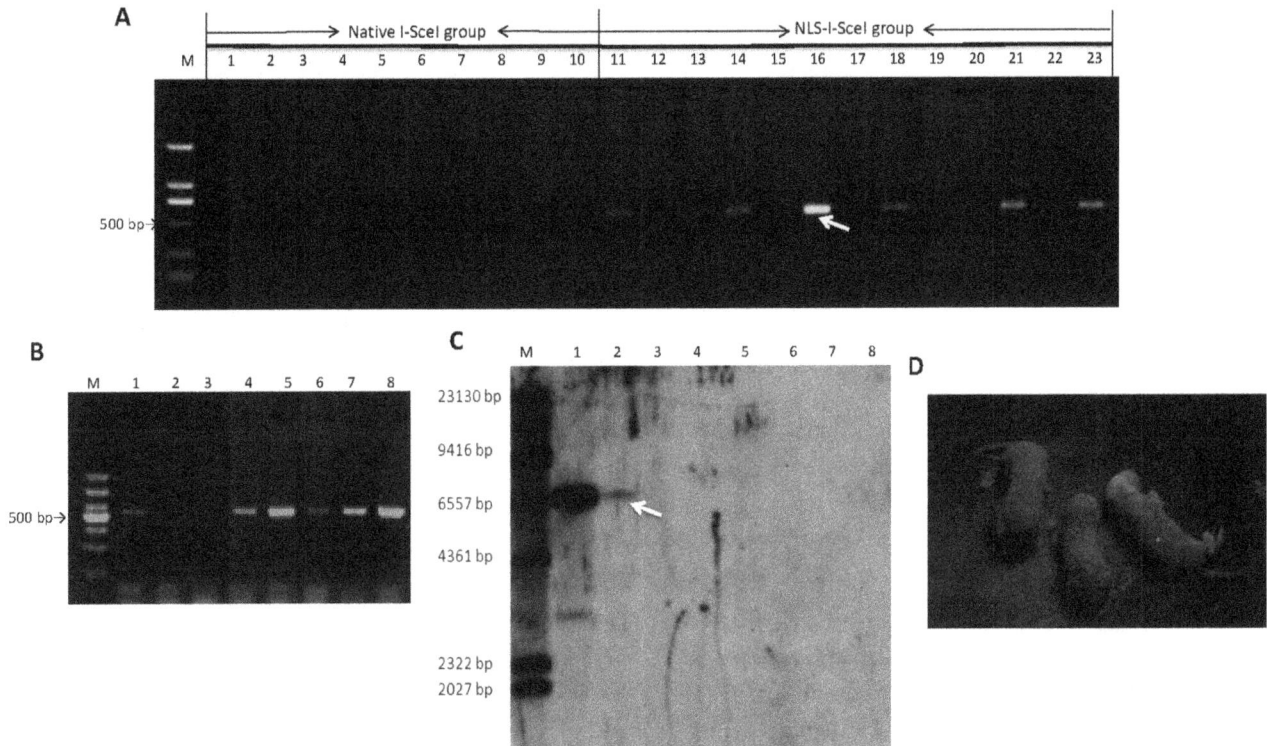

Figure 7. Genetic screen of transgenic mice derived from cytoplasmically microinjected eggs. A: Screen of transgenic founder mice by PCR. M: DL2000 DNA marker; 1–10: the founder mice derived from cytoplasmic microinjection with circular p2IS-UBC-eGFP plasmids (30 ng/μL) included into the native I-SceI nuclease digestive reaction system; 11–23: The founder mice derived from cytoplasmic microinjection with circular p2IS-UBC-eGFP plasmids plus NLS-I-SceI mRNA (30 ng/μL each). B: Screen of transgenic individuals of F1 offspring derived from transgenic founder mice by PCR. M: DNA marker; 1–8: Genomic DNA samples of F1 individuals. C: Genetic screen of transgenic founder mice by Southern blot assay. M: DNA molecular weight marker II; 1: plasmids; 2–7: genomic DNA samples of founder mice; 8: negative control (wild-type mouse genomic DNA). D: The transgenic mice exhibiting *in vivo* fluorescence derived from breeding between transgenic individuals over three consecutive generations. The arrow indicates the founder mouse detected to be transgenic by both Southern blot and PCR screen.

was mated with wild-type pig to test the germline transmission competence of transgenes. As shown in Fig. 8 D, in the seven individuals of F1 offspring, four were detected to be transgenic by Southern blot, indicating that the transgenes were capable of germline transmission. After gemline transmission was confirmed, the founder pig (1#) was sacrificed due to disease related to respiratory system infection, and genomic DNA samples of different organs were subjected to Southern blot assay. As shown in Fig. 8 E, transgene was detected in all the organs except skin and lung in a similar band distribution pattern. However, the failure to detect transgene in these two organs was due to the experimental procedure but not to the lack of transgene integration, for the genomic DNAs of the two organs were not thoroughly digested and separated in gel electrophoresis as a result before DNA was transferred to membrane (Fig. S6 A), and transgene was finally detected in these two organs with a similar band distribution pattern by a repeated Southern blot assay after the genomic DNAs were completely digested (Fig. S6 B, C), suggesting that this founder pig was not transgenically mosaic and transgene integration occurred at a very early stage of embryo development. The death of the founder pig was not due to transgenesis, for some wild-type pigs in the farm also died of the same disease at that time. The rest transgenic pigs, including the offspring of the dead founder pig, kept healthy. These results demonstrated that the NLS-I-SceI-mediated transgenesis in mammalian embryos was capable of efficiently resulting in transgenic animals with germline transmission competence,

especially in species other than mice which was refractory to embryo pronuclear microinjection but exhibited higher tolerance to embryo cytoplasmic microinjection.

Discussion

Embryo microinjection is a simple and reproducible method for mammalian transgenesis, however the dependence on visible pronuclear largely limits its application to mammalian species other than mice, especially those large animal species of which the pronuclear is usually invisible. Currently, transgenisis via embryo cytoplasmic microinjection has achieved limited success in mammalian species. Page et al (2005) produced transgenic mice using Polylysine/DNA mixture by cytoplasmic microinjection of eggs, however the transgenic rate (born transgenic pups/ transferred embryos) was much lower than that of pronuclear microinjection (12.8% vs 21.7%) [37]. Garrels et al (2011) efficiently produced transgenic pigs by Sleeping Beauty (SB) transposon-mediated transgenesis via embryo cytoplasmic micro-injection with circular plasmids of SB transposon-based transgene vector and SB tranposase expression vector, and the transgenic rate of founder pigs was as high as 47.3% [23]. Nonetheless, transposons are mobile genetic elements and transgenic organisms derived from transposon-mediated transgenesis would be of bio-safety concerns. Recently, Wilson et al (2013) has described a sophisticated system termed intracellular electroporetic nanoinjec-tion (IEN) to propel transgene fragments from cytoplasm into

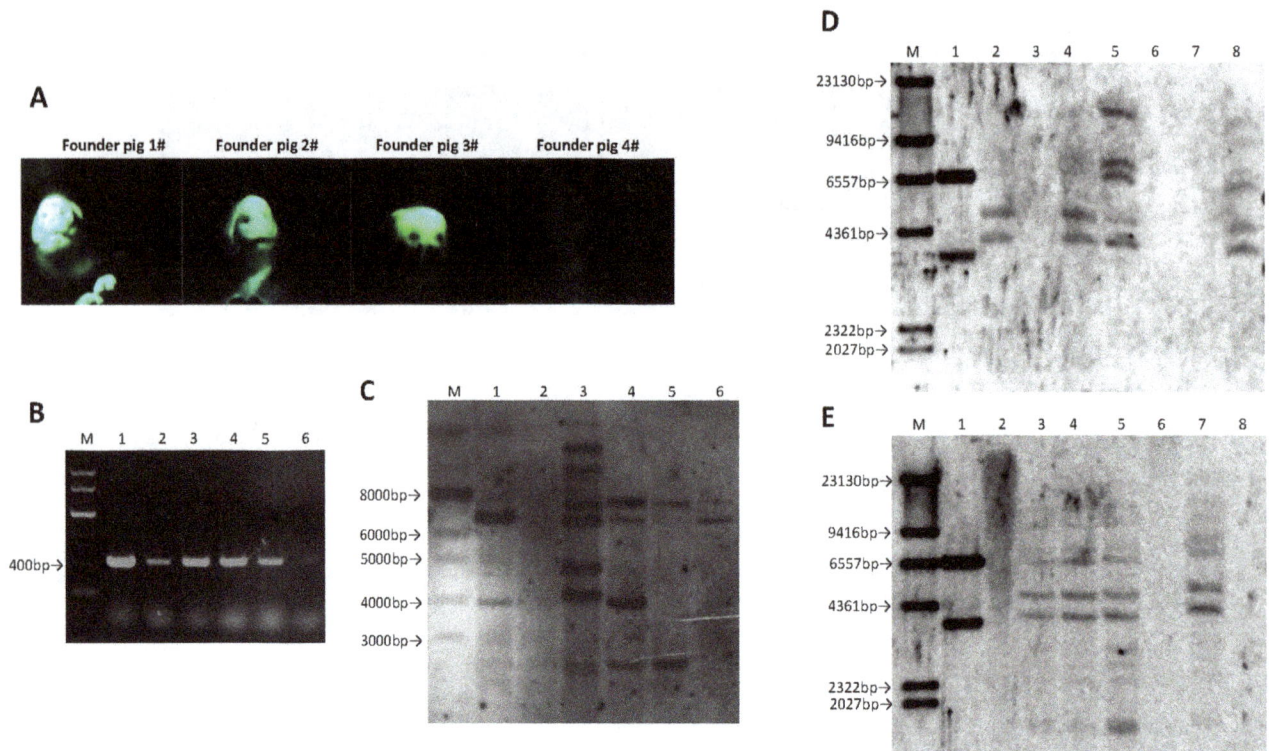

Figure 8. Genetic screen of transgenic pigs derived from embryos cytoplasmically microinjected with circular p2IS-UBC-eGFP plasmids plus NLS-I-SceI mRNA. A: *In vivo* fluorescence in founder pigs. B: Screen of transgenic founder pigs by PCR. M: DNA marker; 1–4: genomic DNA samples of 1–4# founder pigs; 5: positive controls (wild-type pig genomic DNAs containing p2IS-UBC-eGFP plasmids); 6: negative control (wild-type pig genomic DNA). C: Southern blot assay of transgenic founder pigs. M: DNA marker (1KB DNA Ladder); 1: positive control (plasmids); 2: wild-type pig genomic DNA as negative control; 3–6:1–4# founder pigs. D: Southern blot analysis of F1 offspring individuals derived from founder pig 1#. M: DNA molecular weight marker II; 1: plasmid as positive control; 2–8: the F1 offspring individuals. E: Southern blot analysis of genomic DNAs extracted from different organs of founder pig 1#. M: DNA molecular weight marker II; 1: positive control (plasmids); 2: skin; 3: heart; 4: liver; 5: spleen; 6: lung; 7: kidney; 8: wild-type pig genomic DNA as negative control.

pronuclear, however this method required additional complicated equipment and experimental skills besides conventional microinjection, and more importantly, the transgenesis efficiency of IEN system was not higher (actually slightly lower) than that of pronuclear microinjection [38].

I-SceI has been effectively used to facilitate transgenesis in fish eggs for several years. Because only plasmid vectors containing I-SceI recognition sequences are involved in the I-SceI-mediated transgenesis, regarding the transgenesis process and the application of the resulted transgenic organisms, the I-SceI-mediated transgenesis is of minimal bio-safety concerns. In this work, we efficiently generated transgenic mammals (pigs and mice) simply by co-injecting circular transgene vector plasmids containing I-SceI recognition sequences and the mRNAs coding NLS-I-SceI molecule into embryo cytoplasm. As far as we know, this is the first report for efficient generation of transgenic mammals via embryo cytoplasmic microinjection using the I-SceI molecule.

Our work demonstrated that the native I-SceI molecule was not capable of efficiently facilitating transgenesis in mammalian embryos as it did in fish eggs, which may be due to the much smaller size of mammalian embryos compared to that of fish eggs and much less plasmid copies that can be delivered into mammalian embryos as a result. In contrast, the NLS-I-SceI molecule, which contains mammalian NLS sequence at its N-terminal, was shown to be capable of cutting transgene fragments off from circular plasmids, protecting transgene fragments from degradation and efficiently facilitating transgenesis in both mouse

and porcine embryos, indicating that the artificially added mammalian NLS signal largely promoted the efficacy of I-SceI-mediated transgenesis. The ability of NLS-I-SceI molecule to facilitate transgenesis in mammalian embryos was directly demonstrated by the localization of Cy3-labeled DNA fragments containing inversely flanking I-SceI cutting sites at both ends which were co-injected with NLS-I-SceI mRNA into the cytoplasm of porcine pathenogenically activated oocytes at MII stage (parthenogenetic embryos). The reason for the use of porcine MII oocytes was that the nuclear was breakdown at this stage and to be constructed upon activation, providing a time window to observe the localization of DNA fragments during the process of mammalian pronuclear construction, and that in addition, the lack of nuclear excluded the probability that materials happened to be injected into nuclear by chance due to the invisibility of pronuclear. Data showed that only the DNA fragments co-injected with NLS-I-SceI molecule were clustered and co-localized with chromosomes in parthenogenetic porcine embryos, while those co-injected with the native I-SceI molecule were diffusely distributed in the cytoplasm and not clustered or co-localized with chromosomes, indicating that the NLS-I-SceI was capable of transferring DNA fragments from cytoplasm into nuclear, while the native I-SceI molecule was not, and the transferring process was co-incident with the procedure of nuclear formation. These results were consistent with the observation that the porcine blastocysts developed from eggs co-injected with NLS-I-SceI mRNA and circular transgene plasmids p2IS-UBC-eGFP exhibited strong fluorescence, while those co-injected with the native I-SceI nuclease and circular transgene plasmids at the same concentration did not, although the eGFP CDS was detected at similar levels in these embryos, suggesting that although the transgene fragments were efficiently cut off from circular plasmids and protected from degradation by the native I-SceI nuclease in porcine embryos, the transgene fragments were not efficiently translocated from cytoplasm into nuclear by this molecule to result in expression. The efficient production of transgenic mice and pigs and the germline transmission competence of the resulted transgenic animals further confirmed that the NLS-I-SceI molecule can be used as a potent tool to facilitate mammalian transgenesis.

The NLS-I-SceI-mediated transgenesis resulted in random integration in mammalian genome. With the advent of powerful EENs such as ZFN, TALEN and CRIPR/Cas9 system, the NLS-I-SceI molecule can be used in combination with EENs to facilitate targeted transgene integration into mammalian embryo genomes. Recently, it has been reported that *in vivo* cleavage of circular plasmids by EENs effectively facilitated targeted integration of transgenes into the DSBs created by the same or another EEN molecule through none-homologous end joining (NHEJ) mechanism in the genomes of mammalian somatic cells [39,40]. However, the ability of the EENs to bind to the cleaved DNAs and further transfer DNA fragments from cytoplasm into nuclear of mammalian embryos remains to be investigated, although these molecules have NLS signal. More recently, Cas9-sgRNA complex was shown to be capable of binding cleaved DNA with high affinity, however the both ends of cleaved DNA were tightly bound to Cas9-sgRNA complex and the cleaved circular plasmids were still in circular form [41], which would hinder transgene integration. On this basis, considering the confirmed ability of NLS-I-SceI molecule to cut transgene fragments off from circular transgene plasmids, protect transgene fragments from degradation and transfer transgene fragments from cytoplasm into nuclear in mammalian embryos, NLS-I-SceI molecule can be used in combination with EENs to facilitate targeted transgene integration into mammalian embryo genomes, and thereby a simple, efficient and species-neutral technology for targeted transgenesis in mammalian animals can be established, especially for large mammalian species such as pig, cattle and none-human primates.

In this work, we found that the native I-SceI molecule without mammalian NLS signal did not efficiently facilitate transgenesis in mouse or porcine embryos. However, Bevacqua et al (2013) recently reported that the native I-SceI-mediated transgenesis resulted in transgenic eGFP expression in bovine blastocysts derived from *in vitro* fertilization [42]. This inconsistency may be partly due to the much higher concentration (50 ng/μL) of circular transgene plasmids used in this study compared to that in our work (30 ng/μL). Such a high concentration may result in the presence of more uncut circular plasmids in embryos, which were resistant to degradation in cells and can be passively diffused into nuclear during embryo cleavage as suggested by a previous report [36] and our data in this work. However, circular plasmids rarely integrated into genome even directly delivered into pronuclear in a large amount, and no transgenic cattle was produced in this report either. Besides, the fluorescence in bovine blastocysts resulting from the native I-SceI-mediated transgenesis with plasmids of a natural promoter (Pax6)-driven eGFP expression vector was rather weak, which was comparable to that in the few fluorescent porcine embryos co-injected with the native I-SceI nuclease and circular p2IS-UBC-eGFP plasmids in this study. The moderately stronger fluorescence, of which the intensity was remarkably lower compared to that in the porcine embryos co-injected with NLS-

I-SceI mRNAs and p2IS-UBC-eGFP plasmids in our work, resulted from transgenesis with another artificially synthesized strong promoter(CAG)-driven eGFP expression vector plasmids, suggesting that the relatively stronger fluorescence was due to the much higher activity of the CAG promoter, but not to the more transgene copies in nuclear, and the native I-SceI molecule did not actively or efficiently transfer transgene fragments from cytoplasm into nuclear in the *in vitro* fertilized bovine embryos either.

Because the circular DNA plasmids can be passively diffused into nuclear during embryo cleavage and the NLS-I-SceI molecule is nuclear-localized, we can't exclude the possibility that the efficient NLS-I-SceI-mediated transgenesis in mammalian embryos was partly derived from *in situ* cleavage of circular plasmids by NLS-I-SceI molecule in nuclear. *In situ* cleavage of circular transgene plasmids in cells was shown to protect transgene fragments from degradation and facilitate transgene integration as a result [39,40]. On this basis, considering that cytoplasmic microinjection with circular bacterial artificial vector (BAC) plasmids also resulted in transgene expression in mammalian embryos, suggesting that circular BAC plasmids can be passively diffused into nuclear once introduced into cytoplasm of embryos [35], the NLS-I-SceI molecule can be used to facilitate BAC transgenesis in mammalian embryos only if the I-SceI recognition sequences were included in BAC vectors.

In summary, this work demonstrated that the NLS-I-SceI molecule was capable of efficiently facilitating mammalian transgenesis, and using this molecule, a simple and efficient general transgenesis technology with minimal bio-safety concerns can be established for mammals. For fully validating this method, a transgenic animal model with exclusive characteristics can be generated via NLS-I-SceI-mediated transgenesis as a quality control, such as the transgenic pig model for human Huntington's disease exhibiting apoptosis in brain neurons similar to human that is not observed in murine models harboring the same transgene [43]. In addition, to fully characterize this technology, the variation of transgene integration sites can be investigated in the future when more transgenic individuals were derived from NLS-I-SceI-mediated transgenesis.

Supporting Information

Figure S1　The coding sequence of NLS-I-SceI molecule. The codon usage was optimized for both mice and pigs on the basis that possible splice sites were excluded.

Figure S2　The amino acid sequence of NLS-I-SceI molecule.

Figure S3　The sequence of p2IS-UBC-eGFP vector. The bold and underlined sequences are inversely flanking I-SceI recognition sequences, and the bold sequence in green is the eGFP CDS.

Figure S4　The Amplification Plots, Melt Curves and Standard Curves for qPCR of the uncut I-SceI site and eGFP CDS.

Figure S5　The sequence of the Cy3-labeled DNA fragment. The underlined sequences are the inversely flanking I-SceI recognition sequences, and the bold base in red is the one where the Cy3 fluorophore is linked.

Figure S6 Repeated Southern blot analysis of transgene integration in the skin and lung of transgenic founder pig 1#. The genomic DNA samples of the skin and lung of founder pig1# were not thoroughly digested with PstI endonuclease in the first Southern blot assay (A). In the repeated Southern blot analysis, the same genomic DNA samples were completely digested as indicated by gel electrophoresis (B), and then transgene integration was detected by Southern blot in the two organs (C). M: DNA marker (1 Kb ladder in gel electrophoresis, and DNA molecular weight marker II in Southern blot assay).

Acknowledgments

We'd like to thank Dr. Baltimore of California Institute of Technology (USA) for his kind offer of the vector FUGW.

Author Contributions

Conceived and designed the experiments: YW HW. Performed the experiments: XYZ PYX LLW HT FX LL. Analyzed the data: YW XYZ. Contributed reagents/materials/analysis tools: LLW. Contributed to the writing of the manuscript: YW.

References

1. Yang D, Yang H, Li W, Zhao B, Ouyang Z, et al. (2011) Generation of PPARγ mono-allelic knockout pigs via zinc-finger nucleases and nuclear transfer cloning. Cell Research, 21: 979–982.

2. Orlando SJ, Santiago Y, DeKelver RC, Freyvert Y, Boydston EA, et al. (2010) Zinc-finger nuclease-driven targeted integration into mammalian genomes using donors with limited chromosomal homology. Nuclear Acids Research, 38 : e152.

3. Mashimo T, Takizawa A, Voigt B, Yoshimi K, Hiai H, et al. (2010) Generation of knockout rats with X-linked severe combined immunodeficiency (X-SCID) using Zinc-Finger Nucleases. PLoS ONE, 2010, 5: e8870.

4. Geurts AM, Cost GJ, Freyvert Y, Zeitler B, Miller JC, et al. (2009) Knockout Rats via Embryo Microinjection of Zinc-Finger Nucleases. Science, 325: 433–434.

5. Boch J, Scholze H, Schornack S, Landgraf A, Hahn S, et al. (2009) Breaking the Code of DNA Binding Specificity of TAL-Type III Effectors. Science, 326: 1509–1512.

6. Liu H, Chen Y, Niu Y, Zhang K, Kang Y, et al. (2014) TALEN-mediated gene mutagenesis in Rhesus and Cynomolgus monkeys. Cell Stem Cell, 14: 323–328.

7. Carlson DF, Tan W, Lillico SG, Stverakova D, Proudfoot C, et al. (2012) Efficient TALEN-mediated gene knockout in livestock. PNAS, 43: 17382–17387.

8. Mussolino C, Morbitzer R, Lutge F, Dannemann N, Lahaye T, et al. (2011) A novel TALE nuclease scaffold enables high genome editing activity in combination with low toxicity. Nuclear Acids Research, 39: 9283–9293.

9. Miller CJ, Tan S, Qiao G, Barlow KA, Wang J, et al. (2011) A TALE nuclease architecture for efficient genome editing. Nature Biotechnology, 29: 143–149.

10. Cermak T, Doyle EL, Christian M, Wang L, Zhang Y, et al. (2011) Efficient design and assembly of custom TALEN and other TAL effector-based constructs for DNA targeting. Nuclear Acids Research, 39: e82.

11. Niu Y, Shen B, Cui Y, Chen Y, Wang J, et al. (2014) Generation of gene-modified Cynomolgus monkey via Cas9/RNA-mediated gene targeting in one-cell embryos. Cell, 156: 1–8.

12. Hai T, Teng F, Guo R, Li W and Zhou Q (2014) One-step generation of knockout pigs by zygote injection of CRISPR/Cas9 system. Cell Research, doi:10.1038/cr.2014.11.

13. Wang H, Yang H, Shivalila CS, Dawlaty MM, Cheng AW, et al. (2013) One-step generation of mice carrying mutation in multiple genes by CRISPR/Cas9-mediated genome engineering. Cell, 153: 1–9.

14. Cong L, Ran FA, Cox D, Lin S, Barretto R, et al. (2013) Multiplex genome engineering using CRISPR/Cas9 system. Science, 339: 819–823.

15. Mali P, Yang L, Esvelt KM, Aach J, Guell M, et al. (2013) RNA-guided genome engineering via Cas9. Science, 339: 823–826.

16. Wall RJ (1996) Transgenic livestock: Progress and prospects for the future. Therionology, 45: 57–68.

17. Hammer RE, Pursel VG, Rexroad CE Jr, Wall RJ, Bolt DJ, et al. (1985) Production of transgenic rabbits, sheep and pigs by microinjection. Nature, 315: 680–683.

18. Zhao J, Ross JW, Hao Y, Spate LD, Walters EM, et al. (2009) Significant Improvement in Cloning Efficiency of an Inbred Miniature Pig by Histone Deacetylase Inhibitor Treatment after Somatic Cell Nuclear Transfer. Biology of Reproduction, 81: 525–530.

19. Huang Y, Tang X, Xie W, Zhou Y, Li D, et al. (2011) Histone Deacetylase Inhibitor Significantly Improved the Cloning Efficiency of Porcine Somatic Cell Nuclear Transfer Embryos. Cellular Reprogramming, 13: 513–520.

20. Lois C, Hong EJ, Pease S, Brown EJ, Baltimore D (2002) Germline Transmission and Tissue-Specific Expression of Transgenes Delivered by Lentiviral Vectors. Science, 295: 868–872.

21. Whitelaw CB, Radcliffe PA, Ritchie WA, Carlisle A, Ellard FM, et al. (2004) Efficient generation of transgenic pigs using equine infectious anaemia virus (EIAV) derived vector. FEBS Letter, 571: 233–236.

22. Brem G, Wolf E, Pfeifer A (2003) Efficient transgenesis in farm animals using lentiviral vectors. EMBO Reports, 4: 1054–1060.

23. Garrels W, Mátés L, Holler S, Dalda A, Taylor U, et al. (2011) Germline transgenic pigs by Sleeping beauty transposition in porcine zygote and targeted integration in pig genome. PLoS ONE, 6: e23573.

24. Ivics Z, Garrels W, Mátés L, Yau TY, Bashir S, et al. (2014) Germline transgenesis in pigs by cytoplasmic microinjection of Sleeping Beauty transposons. Nature Protocol, 2014, 9: 810–827.

25. Ivics Z, Hiripi L, Hoffmann OI, Mátés L, Yau TY, et al. (2014) Germline transgenesis in rabbits by pronuclear microinjection of Sleeping Beauty transposons. Nature Protocol, 2014, 9: 794–809.

26. Rostovskaya M, Naumann R, Fu J, Obst M, Mueller D, et al. (2013) Transposon mediated BAC transgenesis via pronuclear injection of mouse zygotes. Genesis, 51: 135–141.

27. Ding S, Wu X, Li G, Han M, Zhuang Y, et al. (2005) Efficient Transposition of the piggyBac (PB) Transposon in Mammalian Cells and Mice. Cell, 122: 473–483.

28. Ogino H, McConnell WB, Grainger RM (2006) Highly efficient transgenesis in Xenopus tropicalis using I-SceI endonuclease. Mechanism of Development, 123: 103–113.

29. Thermes V, Grabher C, Ristoratore F, Bourrat F, Choulika A, et al. (2002) I-SceI mediated highly efficient transgenesis in fish. Mechanism of Development, 118: 91–98.

30. Pan FC, Chen Y, Loeber J, Henningfeld K, Pieler T (2006) I-SceI meganuclease-mediated transgenesis in Xenopus. Developmental Dynamics, 235: 247–252.

31. Rembold M, Lahiri K, Foulkes NS, Wittbrodt J (2006) Transgenesis in fish: efficient selection of transgenic fish by co-injection with a fluorescent report construct. Nature Protocols, 1: 1133–1139.

32. Grabher C, Wittbrodt J (2007) Meganuclease and transposon mediated transgenesis in medaka. Genome Biology, 8: S10.

33. Brigid H, Rosa B, Frank C (1996) Mouse embryo manipulation (2nd edition). Cold Spring Habour Press.

34. Betthauser J, Forsberg E, Augenstein M, Childs L, Eilertsen K, et al. (2000) Production of cloned pigs from in vitro system. Nature Biotechnology, 18: 1055–1059.

35. Perrin A, Buckle M, Dujon B (1993) Asymmetrical recognition and activity of the I-SceI endonuclease on its site and on intron-exon junctions. The EMBO Journal, 12: 2939–2947.

36. Iqbal K, Barg-Kues B, Broll S, Bode J, Niemann H, et al. (2009) Cytoplasmic injection of circular plasmids allows targeted expression in mammalian embryos. Biotechniques, 47: 959–68.

37. Page RL, Butler SP, Subramanian A, Gwazdauskas FC, Johson JL, et al. (1995) Transgenesis in mice by cytoplasmic injection of plylysine/DNA mixtures. Transgenic Research, 4: 353–360.

38. Wilson AM, Aten QT, Toone NC, Black JL, Jensen BD, et al. (2013) Transgene delivery via intracellular electrophoretic nanoinjection. Transgenic Research, 22: 993–1002.

39. Maresca M, Lin VG, Guo N, Yang Y (2013) Obligate ligation-gated recombination (ObLigaRe): custom-designed nuclease-mediated targeted integration through nonhomologous end joining. Genome Research, 23: 539–546.

40. Cristea S, Freyvert Y, Santiago Y, Holmes MC, Urnov FD, et al. (2013) In vivo Cleavage of Transgene Donors Promotes Nuclease-Mediated Targeted Integration. Biotechnology and Bioengineering, 110: 871–880.

41. Sternberg SH, Redding S, Jinek M, Greene EC, Doudna JA (2014) DNA interrogation by the CRISPR RNA-guided endonuclease Cas9. Nature, 507: 62–67.

42. Bevacqua RJ, Canel NG, Hiriart MI, Sipowicz P, Rozenblum GT, et al. (2013) Simple gene transfer technique based on I-SceI meganuclease and cytoplasmic injection in IVF bovine embryos. Theriogenology, 80: 104–113.

43. Yang DS, Wang CE, Zhao BT, Li W, Ouyang Z, et al. (2010) Expression of Huntington's disease protein results in apoptotic neurons in the brains of cloned transgenic pigs. Human Molecular Genetics, doi:10.1093/hmg/ddq313.

22

Reduction of T Cell Receptor Diversity in NOD Mice Prevents Development of Type 1 Diabetes but Not Sjögren's Syndrome

Joanna Kern, Robert Drutel[¤], Silvia Leanhart, Marek Bogacz, Rafal Pacholczyk*

Center for Biotechnology and Genomic Medicine, Georgia Regents University, Augusta, Georgia, United States of America

Abstract

Non-obese diabetic (NOD) mice are well-established models of independently developing spontaneous autoimmune diseases, Sjögren's syndrome (SS) and type 1 diabetes (T1D). The key determining factor for T1D is the strong association with particular MHCII molecule and recognition by diabetogenic T cell receptor (TCR) of an insulin peptide presented in the context of I-A^{g7} molecule. For SS the association with MHCII polymorphism is weaker and TCR diversity involved in the onset of the autoimmune phase of SS remains poorly understood. To compare the impact of TCR diversity reduction on the development of both diseases we generated two lines of TCR transgenic NOD mice. One line expresses transgenic TCRβ chain originated from a pathogenically irrelevant TCR, and the second line additionally expresses transgenic TCRα^{mini} locus. Analysis of TCR sequences on NOD background reveals lower TCR diversity on Treg cells not only in the thymus, but also in the periphery. This reduction in diversity does not affect conventional CD4$^+$ T cells, as compared to the TCR^{mini} repertoire on B6 background. Interestingly, neither transgenic TCRβ nor TCR^{mini} mice develop diabetes, which we show is due to lack of insulin B:9–23 specific T cells in the periphery. Conversely SS develops in both lines, with full glandular infiltration, production of autoantibodies and hyposalivation. It shows that SS development is not as sensitive to limited availability of TCR specificities as T1D, which suggests wider range of possible TCR/peptide/MHC interactions driving autoimmunity in SS.

Editor: John A. Chiorini, National Institute of Dental and Craniofacial Research, United States of America

Funding: This research was supported by grants from National Institute of Allergy and Infectious Diseases of the National Institutes of Health (R01AI081798) and Juvenile Diabetes Research Foundation (RP). The funders had no role in study design, data collection and analysis, decision to publish, or preparation of the manuscript.

Competing Interests: The authors have declared that no competing interests exist.

* Email: rpacholczyk@gru.du

¤ Current address: Medical University of South Carolina, College of Medicine, Charleston, South Carolina, United States of America

Introduction

NOD mice serve as well-established models of independently developing autoimmune diseases, Type 1 Diabetes (T1D) and Sjögren's syndrome (SS) [1,2]. T1D is characterized by autoimmune attacks against the pancreatic beta-cells with T cells playing an essential role in the initiation and progression of the disease, leading to hyperglycemia and vascular complications [3,4]. SS is an autoimmune disease with local and systemic manifestations, characterized by mononuclear infiltrates into salivary and lacrimal glands leading to clinical symptoms of dry mouth and dry eyes [5,6]. Glandular infiltrates consist mostly of CD4$^+$ T cells with lesser amounts of CD8$^+$ T cells and B cells. Although factors like viral or bacterial infections, aberrant glandular development or cytokine production are important in the initial phase of the pathogenesis of SS, CD4$^+$ T cells are important players in the onset of autoimmunity and disease progression.

Autoimmunity in NOD mice is attributed to several different events occurring in the thymus and in the periphery. Studies in these mice showed a defect in negative selection [7], perturbed αβ/γδ lineage decision leading to a shift in selection niches [8], reduced relative diversity of thymic Treg cells [9], peripheral hyper-responsiveness of effector CD4$^+$ T cells [10], multiple binding registers of insulin B:9–23 peptide resulting in poor negative selection in the thymus [11,12], or peripheral post-translational modification of self-peptides/neo-antigens [13]. Despite genetic predispositions, the key component in the development of autoimmune diseases is the recognition of a particular antigen in the context of MHC Class II molecule by CD4$^+$ T cells. The development of diabetes in NOD mice is associated with the key I-A^{g7} molecule (HLA-DQ8 in humans) in the absence of a functional I-E molecule [14,15]. Co-expression of other MHC molecules with I-A^{g7} can prevent development of diabetes in a dominant fashion [14,15]. Replacement of I-A^{g7} with other MHC molecules, like I-A^b, I-A^p or I-A^q, does not promote the development of diabetes yet mice continue to develop autoimmune exocrinopathy and the severity of the SS and the profile of antibodies' specificities vary between congenic mice [16]. In large-scale association study of SS in humans, HLA was found to have the strongest linkage to the disease [17].

The strict dependence of T1D on the particular MHC allele correlates with its primary antigen requirement where insulin B:9–23 peptide has been identified as the epitope necessary for onset of the disease in NOD mice [18]. In SS, no key epitope(s) are identified, although several proteins have been implicated as a source of antigens: Ro/SSA 52 kDa, αFodrin, Muscarinic Acetylcholine 3 Receptor (M3R), α-amylase, islet cell autoantigen-69, kallikrein-13 [19–24]. Recently it has been shown that the transfer of T cells from M3R-immunized M3R$^{-/-}$ mice into Rag$^{-/-}$ mice leads to development of sialadenitis, showing pathogenic potential of M3R specific T cells [25].

Despite the strict requirement of the presence of the insulin B:9–23/I-A^{g7} combination, the development of T1D in NOD mice proceeds even when total TCR diversity and precursor frequency of diabetogenic TCRs is limited. The reduction of TCR diversity by use of TCRβ transgenic mice [26], or great reduction of precursor frequency relying on allelic exclusion escapees on NOD background does not prevent development of T1D [27], although not all endogenous TCRβ chains are permissive for the development of insulin B:9–23 specific TCRs [28]. In SS it is not clear as to what role diversity of interactions between TCRs and different peptide/MHCII complexes play in the onset and development of the disease. Previous studies in patients with SS found that TCR repertoire of infiltrating T cells is to some extent restricted with different dominant clonotypes of Vβ families [29–33]. Despite the lack of dominant Vα/Vβ families or dominant specificity in different patients these studies show clonal expansion of infiltrating T cells, which suggests that the number of epitopes participating in the autoimmunity of the disease is limited [34,35]. However, the weaker dependence of SS on MHC polymorphism suggests broad diversity of possible TCR/peptide/MHCII interactions participating in the pathogenesis of the disease. As the diversity of antigenic specificities and TCR repertoire on T cells involved in SS development is not well understood, we wanted to compare sensitivity of development of SS versus T1D to the diminishing diversities of TCRs in the presence of the same I-A^{g7} molecule in NOD mice. To reduce the diversity of TCRs in NOD mice we generated two types of transgenic strains, in which all T cells express either one transgenic TCRβ chain or additionally co-express TCRα chains from TCRα^{mini} locus [36]. Interestingly, these mice do not develop T1D but still develop SS with glandular infiltrations, autoantibody presence and hyposalivation. We investigated the reasons for the lack of T1D development and the role of reduced diversity on generated repertoires of TCRs on conventional and regulatory T cells in the thymus and periphery of NOD mice.

Materials and Methods

Ethics statement

All mice used in this study were housed in the animal care facility at the Georgia Regents University (GRU). All work involving animals was conducted under protocols approved by the Animal Care and Use Committee at the GRU (#2008−0231). All efforts were made to minimize suffering. Mice were euthanized by CO2 followed by cervical dislocation.

Mice

Production of TCRβ^{Tg} and TCRα^{mini} constructs and generation of transgenic mice on C57BL/6 (B6) background was described previously (Pacholczyk 2006). A similar strategy was used to microinject one or both DNA constructs into zygotes of NOD mice (Transgenic Mice Core Facility, GRU). To eliminate expression of endogenous TCRα chains, NOD.TCRβ^{Tg}.TCRα^{mini} transgenic mice were crossed with NOD.TCRα$^{-/-}$ (NOD.129P2(C)-Tcratm1Mjo/DoiJ) mice purchased from The Jackson Laboratory (Bar Harbor, ME). To facilitate identification

of Treg cells both transgenic lines then were crossed with NOD.FoxP3GFP/cre mice (NOD/ShiLt-Tg(Foxp3-EGFP/cre)1Jbs/J) purchased from The Jackson Laboratory. The B6.Aec1Aec2 (B6.DC) mice were kindly provided by Dr. Ammon Peck [37].

Histology

Organs were removed from each mouse at the time of euthanasia, placed in 10% phosphate-buffered formalin for at least 24 h and then embedded in paraffin. Sections were taken at 5 μm of thickness 200 μm apart. The tissue sections were stained with hematoxylin and eosin (H&E) at the Histology Core Laboratory, GRU. One infiltrate was defined as a cluster of at least 50 nucleated cells, scoring described in figures.

Measurement of saliva flow

Mice were given an i.p. injection of 100 μl of pilocarpine (0.05 mg/ml in PBS) per 20 g of body weight. Saliva was collected for 10 min., starting 1 min. after injection of pilocarpine. The volume of saliva was measured and normalized to the mouse body weight.

Detection of auto-antibodies

Auto-antibodies and total IgG1 were measured using mouse serum with the following ELISA kits: αFodrin (American Research Products), ANA, ssDNA, dsDNA (Immuno-Biological Laboratories) and IgG1 (Immunology Consultants Laboratory). Assays were performed according to manufacturer's protocols with serum dilution 1:100 and 1:50,000 for IgG1. OD$_{450}$ values of negative samples were subtracted from the OD$_{450}$ of experimental samples.

Cell preparation, flow cytometry and cell sorting

Cells were isolated from peripheral lymph nodes (axillary, brachial and inguinal) and thymii by mechanical disruption through nylon mesh. Salivary mandibular and extraorbital lacrimal glands were first cut and digested using collagenase (1 ug/ml) for 30 min in 37°C. Cells were washed and counted (Countess, Invitrogen) and used for staining with monoclonal antibodies: CD4 (clone RM4-5), CD8α (53-6.7), B220 (RA3-6B2), TCRVα2 (B20.1), TCRVβ14 (14-2), TCRβ (H57-597), CD25 (PC61), CD45RB (16A), CD62L (MEL-14), all from BD Biosciences. Stained cells were either analyzed using FACS Canto (BD Biosciences) or sorted on MoFlo Sorter (Cytomation). Dead cells were excluded using forward vs side scatter dot plots and doublets discrimination was accomplished using forward scatter height vs width dot plots. Purities of all sorted populations were above 98%.

Immunization and generation of T cell hybridomas

Mice were immunized at the base of the tail with 50 μg of insulin B:9–23 peptide emulsified in Complete Freund's Adjuvant (Thermo Scientific). One week later lymphocytes were isolated from draining lymph nodes and cultured in vitro for 3–4 days with insulin B:9–23 peptide (50 μg/ml), followed by 3–4 days of expansion with murine recombinant IL-2 (20 U/ml, Peprotech). For generation of allo-specific T cell hybridomas, CD4$^+$ T cells sorted from un-immunized experimental mice were stimulated in vitro by co-culture with splenocytes from B6.TCRα$^{-/-}$ mice, followed by expansion with IL-2. Activated T cells were fused to BW5147 TCRα-β- NFAT-EGFP cells as previously described [38]. BW NFAT-EGFP fusion partner expresses GFP protein under the minimal human IL-2 promoter, which contains NFAT-binding sites [39]. T cell hybridomas were selected using HAT

(Cellgro) selection media by limiting dilution method. For stimulation, cloned T cell hybridomas (10^5 cells) were co-cultured overnight with 5×10^5 splenocytes (from NOD.TCRα−/− or B6.TCRα−/− mice) with or without insulin B:9–23 peptide (50 μg/ml) or anti-CD3 antibody (1 μg/ml). Specific activation of the hybridomas was measured by detection of GFP-positive cells or by detection of IL-2 in culture supernatant using CBA Mouse IL-2 Flex Set (BD Biosciences).

MTT assay

The proliferation of the cells was measured using 3-(4,5-dimethythiazol-2-yl)-2,5-diphenyltetrazolium bromide (MTT; Sigma) assay. On the third day of *in vitro* stimulation 10 μl of MTT (5 mg/ml) was added to each well and incubated for 4 h. After discarding supernatant, the remaining formazan precipitates were dissolved in 150 ul of 70% isopropanol solution (70% isopropanol, 30% water, 0.02N hydrochloric acid) overnight. Absorbance was measured at 570 nm using Microplate Reader (Biotek Synergy HT).

Sequencing

Single cell sorting and sequencing was done as previously described [36]. High-throughput sequencing was done using Ion Torrent platform (Life Technologies) by Genomic Core Facility at GRU. Libraries of the TCRs were prepared according to protocol. Shortly, CDR3α regions were amplified using primers specific for Vα2 and Cα segments with integrated adapters and barcodes, provided by manufacturer. Before sequencing, consistency of samples was checked by 2D-F-SSCP analysis of PCR products from three aliquots of cDNA per each sample [36]. Only samples with three similar/identical profiles were considered without PCR bias and were used for further purification using Agencourt AMPure XP reagent (Beckman Coulter) and used for sequencing. FASTQ files with sequences were processed and analyzed using custom-written program in Pearl (ActivePearl, ActiveState Software Inc.). Sequences with quality score of CDR3α region above 27 were used for analysis. The length of CDR3α region was defined by counting from the third amino acid after the invariant C residue in all Vα regions (Y-L/F-C-A-X-first) to the amino acid immediately preceding common Jα motif (last-F-G-X-F-G-T).

Statistical analysis

The similarity, diversity and richness estimators were calculated using programs designed to measure biodiversity: EstimateS8.2 (Colwell, R. EstimateS: Biodiversity Estimation Software. Program and User's Guide at http://viceroy.eeb.uconn.edu/estimates) and SPADE (Chao, A. and Shen, T.-J. (2010) Program SPADE (Species Prediction And Diversity Estimation). Program and User's Guide published at http://chao.stat.nthu.edu.tw).

Results

Development of T cells in NOD^{βTg} and NOD^{mini} mice

We generated TCRβ transgenic mice, in which all T cells use the same TCRβ^{Tg} chain (Vβ14Dβ2Jβ2.6) and one of the endogenous TCRα chains. The TCRβ transgenic chain originated from I-A^b restricted TCR specific to Ep63 K peptide an analog of Eα 52–68 peptide in which the residue at position 63(I) was substituted with lysine [40]. To further reduce diversity of TCRs we used TCRα^{mini} transgenic construct, used previously to create TCR^{mini} mouse on B6 background (B6^{mini}) [36]. The TCRα^{mini} transgene allows a single Vα2.9 segment to rearrange to one of the two Jα (Jα26 and Jα2) segments. The generated transgenic mice were further crossed with TCRα$^{−/−}$ mice to ensure that all

developing T cells use transgenic TCRα^{mini} locus. To track CD4$^+$ Foxp3$^+$ regulatory T (Treg) cells we crossed transgenic mice with NOD.Foxp3.EGFP/cre mice as described in methods. Final characteristics of mice used in this paper were NOD^{mini} (NOD.TCRα^{mini}.TCRβ^{Tg}.TCRα$^{−/−}$.Foxp3EGFP/cre), NOD^{βTg} (TCRβ^{Tg}.TCRα$^{+/−}$.Foxp3EGFP/cre) and NOD (NOD.Foxp3EGFP/cre).

In NOD^{mini} and NOD^{βTg} transgenic mice thymic development of T cells proceeds normally, and selection of single positive (SP) thymocytes is very efficient with bias toward CD4$^+$ T cells in the thymus and in the periphery, similarly to a previously reported bias on B6 background (Fig. 1A, 1B) [36]. TCRβ transgenic mice are characterized by allelic exclusion, resulting in all T cells to express only one transgenic TCRβ chain. It has been proposed that allelic exclusion in NOD mice is less efficient [27]. In our mice all SP thymocytes and peripheral T cells have exclusive expression of transgenic TCRβ^{Tg} chain (TCRVβ14) in both types of transgenic mice and exclusive expression of TCRα^{mini} transgene (TCRVα2) in NOD^{mini} mice (Fig. 1A, 1B). Also we did not observe emergence of the cells bearing TCRs with Vβ segments other than Vβ14 even in older mice. Furthermore, we observed lower percentages of Treg cells in thymii and periphery of transgenic mice as compared to NOD mice and the percentages correlated with diversities of TCRs in analyzed mice (Fig. 1C, 1E). Nevertheless, the total numbers of peripheral Treg cells were similar in all of analyzed mice and reduced percentages of Treg cells were due to more efficient selection of CD4$^+$ T cells, rather than diversity of TCRs, as we observed this correlation on B6 background and in other types of mice that differ by efficiency of CD4$^+$ T cells selection (Fig. 1C, 1E) [36,41]. Finally, to check whether this particular TCRVβ14 transgenic chain can impose an unusual restriction on the TCRα repertoire, we evaluated the frequency of individual Vα families. We found no bias in TCR repertoire as the frequency of Vα families usage by CD4$^+$Vβ14$^+$ T cells was similar between NOD and NOD^{βTg} mice (Fig. 1D).

Lack of T cells specific to insulin B:9–23 in NOD transgenic mice

To test development of diabetes we measured blood glucose levels in experimental mice. Surprisingly, neither NOD^{βTg} nor NOD^{mini} transgenic mice develop diabetes (Fig. 2A). Evaluation of H&E stained pancreatic sections showed lack of lymphocytic infiltrates in 25 week old transgenic mice, with only insignificant infiltrates found in a few sections of NOD^{βTg} mice (Fig. 2B). As a control we used non-transgenic littermates of NOD^{βTg}, NOD.T-CRα$^{+/−}$ mice. These mice developed diabetes and by 30 weeks of age more than 75% of females had high levels of blood glucose (Fig. 2A). One of the possibilities was that the lack of development of diabetes in transgenic mice is due to perturbed proportions of regulatory to effector T cell ratios, which could result in more efficient suppression of potentially diabetogenic T cells. We took advantage of cyclophosphamide treatment, which selectively affects numbers and function of Treg cells and accelerates development of diabetes in NOD mice [42,43]. This treatment however did not cause insulitis nor diabetes in transgenic mice indicating inability of effector T cells to initiate autoimmunity (unpublished data).

The onset of spontaneous diabetes in NOD mice is dependent upon presence of effector T cells specific to insulin B:9–23 antigen and lack of such specificity results in lack of early infiltrates into pancreatic islets [18]. As our mice did not develop islet-infiltrates, we tested their ability to respond to stimulation by insulin B:9–23 peptide (Fig. 3). CD4$^+$ T cells isolated from NOD^{mini}, NOD^{βTg} and NOD mice responded to stimulation by allogeneic spleno-

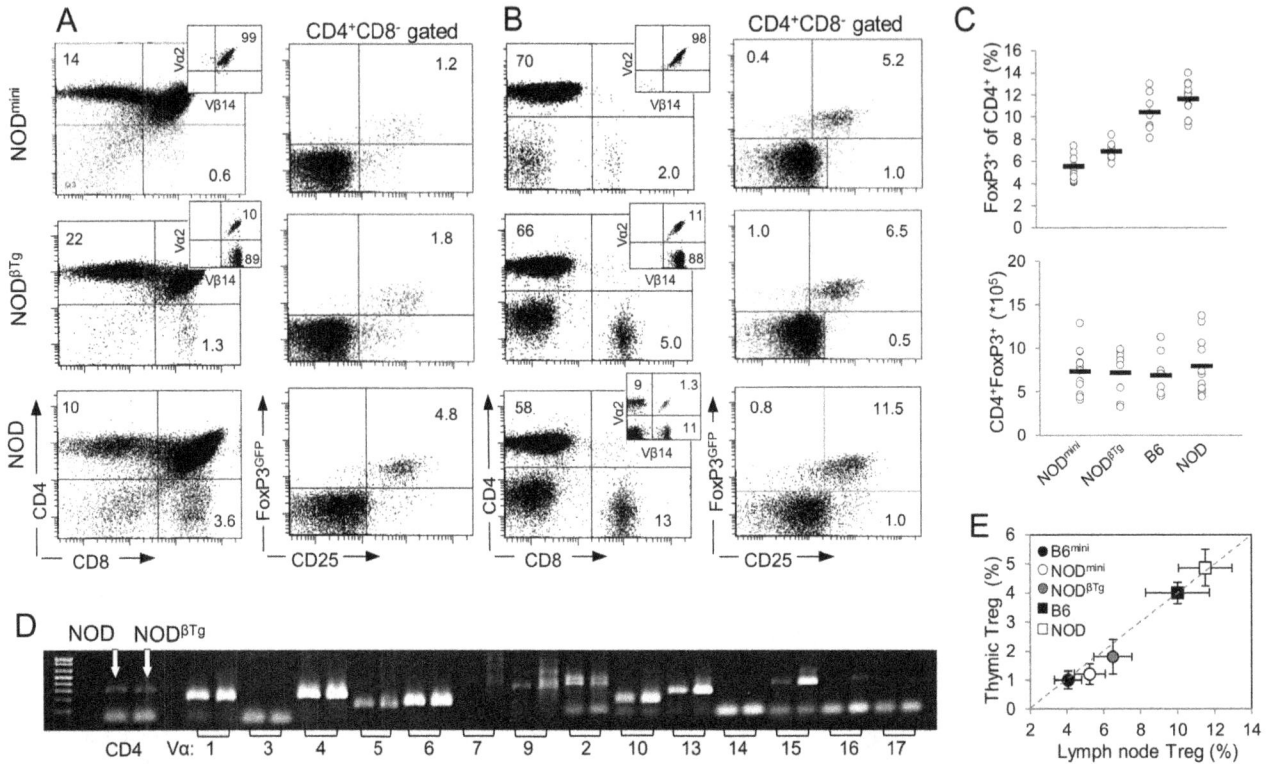

Figure 1. Efficient selection of CD4[+] T lymphocytes in NOD[mini] and NOD[βTg] mice. Lymphocytes isolated from thymii (A) and lymph nodes (B) of indicated mice were stained with monoclonal antibodies and analyzed by flow cytometry. Numbers in quadrants are representative percentages of at least six mice (6 week old) per group. The numbers of thymocytes and lymphocytes ± SD recovered from NOD[mini], NOD[βTg] and NOD mice were: from thymii $78.5 \pm 16.7 \times 10^6$, $72.05 \pm 13.8 \times 10^6$ and $70.1 \pm 14.0 \times 10^6$, and from lymph nodes (axillary, brachial and inguinal) $17.2 \pm 5.8 \times 10^6$, $14.1 \pm 2.5 \times 10^6$, and $13.2 \pm 3.0 \times 10^6$, respectively. (C) Percentages (top) and total numbers (bottom) of CD4[+]Foxp3[+] T cells in peripheral lymph nodes of 6 week old mice; each circle represents individual mouse. (D) Expression of mRNA of TCRVα genes in sorted CD4[+] T cells isolated from NOD and NOD[βTg] mice. Analysis was done by RT-PCR using primers specific to indicated Vα segments and Cα region. (E) Comparison of CD4[+]Foxp3[+] T (Treg) cells from thymii and peripheral lymph nodes of transgenic and wild type B6 and NOD mice. Mean percentage and SD of six young mice per group are shown.

cytes, however only CD4[+] T cells from immunized NOD mice were able to respond to B:9–23 peptide in the presence of syngeneic splenocytes (Fig. 3). As of note, B:9–23 peptide does not bind to I-A[b] molecule and in the presence of B6 splenocytes does not lead to CD4 T cell response [44]. We were also unable to generate B:9–23 specific T cell hybridomas from immunized NOD[mini] and NOD[βTg] mice, however we had no problem generating allo-specific T cell hybridomas from transgenic mice. These results showed that changes introduced by transgenic chains to the TCR repertoire in NOD mice resulted in lack of peripheral specificity to the key antigen required for onset of T1D.

Development of SS in NOD[βTg] and NOD[mini] mice

The lack of T1D in transgenic mice prompted us to test whether reduced TCR diversity will affect development of Sjögren's syndrome – secondary autoimmune disease in NOD mice. It is characterized by lymphocytic infiltrates into salivary and lacrimal glands, production of auto-antibodies and the loss of saliva and tear production by 20 weeks of age [5,45]. Histological evaluation of tissue sections of submandibular salivary and extraorbital lacrimal glands showed focal lymphocytic infiltrates, in both transgenic NOD[mini] and NOD[βTg] and parental NOD mice (Fig. 4). Similarly to the NOD mice, males of both transgenic mice had milder infiltration of salivary glands than females, while the infiltration of lacrimal glands was more prominent in male

NOD[βTg] and NOD[mini] mice. This infiltration is organ specific, as we did not detect infiltrates in lungs, kidney, liver, as well as sublingual and parotid salivary glands and harderian lacrimal glands of 20 week old transgenic mice, similarly to NOD mice (Fig. 4B). Of note, in some of 30 week old NOD[mini] females, but not NOD[βTg] mice, we noticed development of lymphoprolifera-tion, manifested by enlarged lymph nodes and spleen and infiltration into multiple organs (unpublished data). FACS analysis of lymphocytic infiltrates in affected exocrine glands showed that all of the infiltrating CD4[+] T cells expressed transgenic Vβ14 chain in NOD[βTg] mice and Vβ14/Vα2 transgenic chains in NOD[mini] mice (Fig. 4C). This also showed the aforementioned stability of expression of transgenic TCRs in experimental mice.

Development of SS in humans is often correlated with presence of anti-nuclear antibodies (ANA), anti-Ro/SSA, anti-La/SSB, anti-dsDNA, anti-αFodrin and anti-M3R [5,22,46]. In the NOD mouse model of SS, anti-SSA and anti-SSB auto-antibodies are rarely present and they are found at the very low levels [47]. To determine the presence of auto-antibodies in experimental mice, we used ELISA-based assay. For negative control, we used sera from NOD.TCRα[−/−] mice, that do not have T cells. As shown in Fig. 5, transgenic and parental NOD mice from 14–17 week old group had elevated levels of antibodies against αFodrin, ssDNA, dsDNA and ANA. Additionally, we determined the staining pattern of ANA auto-antibodies using immunofluorescent staining of HEp-2 cells (Fig. 5B). The majority (>90%) of NOD[mini] and

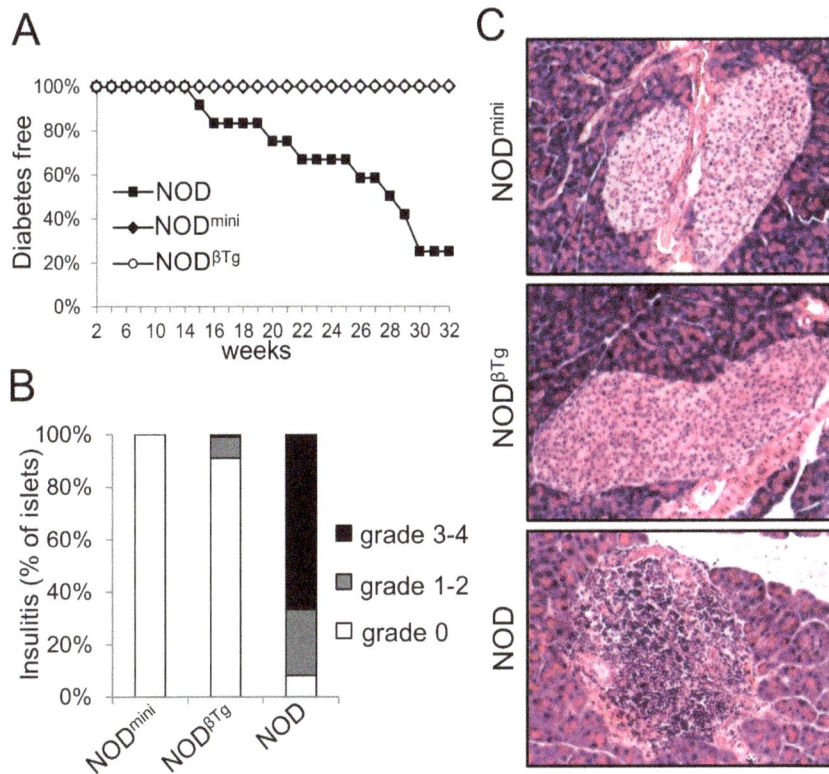

Figure 2. NOD$^{\beta Tg}$ and NOD^{mini} mice do not develop T1D. (A) Incidence of diabetes in mice shown as Kaplan–Meier survival curve. Mice with three consecutive measurements of blood glucose level above 250 mg/dL were considered diabetic. At least twelve mice per group were analyzed; p<0.05. (B) H&E staining of pancreatic tissue sections were analyzed for lymphocytic infiltrates and percentages of islets with grade of insulitis were counted. Insulitis was graded based on following criteria: no infiltrates – grade 0; peri-insulitis - grade 1; insulitis <25% - grade 2; insulitis <50% - grade 3; insulitis >50% - grade 4. At least six mice at 20 weeks of age were analyzed with the total number of 140 (NOD^{mini}), 155 (NOD$^{\beta Tg}$) and 160 (NOD) islets. (C) Example of H&E staining of pancreatic tissue sections of indicated mice at 20 weeks of age.

NOD$^{\beta Tg}$ mice tested had a speckled nuclear staining pattern, characteristic for SS development and found in parental NOD mice.

The onset of salivary gland dysfunction and presence of autoantibodies relies on B cells involvement and is dependent on IL-4 mediated IgM to IgG1 class switching [48–50]. Analysis of IgG1 concentration in sera of experimental mice revealed increased levels in older mice with the highest level in NOD^{mini} mice (Fig. 5C). This increase in IgG1 titer correlates with detection of auto-antibodies. Finally, to evaluate glandular dysfunction, we quantified secretion of saliva in 20 wks old mice. Both NOD$^{\beta Tg}$ and NOD^{mini} mice had reduced saliva flow as compared to healthy B6 mice and at the level of the reference parental NOD mice (Fig. 5D). Taken together, glandular infiltration, autoantibody production, and defective salivary secretion is diagnostic of Sjögren's syndrome in transgenic NOD$^{\beta Tg}$ and NOD^{mini} mice, similarly to the parental strain of NOD mice. Interestingly, the timing of infiltrates and levels of autoantibodies and total IgG1 production vary between analyzed mice, which differ only by TCR repertoire diversity.

Diversity of TCR repertoire on CD4$^+$ T cells in NOD^{mini} mice

Lack of certain specificities in NOD^{mini} TCR repertoire and the previously reported lower TCR diversity of thymic Treg cells on NOD background prompted us to take a closer look at the similarity and diversity of TCRα^{mini} repertoires. To determine the influence of TCR diversity reduction on the selection efficiency of TCRs in NOD mice we started with a single cell analysis to compare to a previously analyzed similar model of TCR^{mini} transgenic mice on "healthy" C57BL/6 background (B6^{mini}) [36]. We compared the similarity between TCR sequences from single cell sorted T$_N$ (CD4$^+$Foxp3$^-$CD45RB$^+$CD62L$^+$) and Treg (CD4$^+$Foxp3$^+$) cells from thymii and peripheral lymph nodes of NOD^{mini} mice. As expected, based on the Morisita-Horn index, the highest similarity was observed between thymic and peripheral subsets of T$_N$ or Treg cells, whereas comparison between populations of T$_N$ and Treg cells showed mostly non-overlapping repertoires with values similar to those observed on B6 background (Fig. 6A and [36]). Previously we've shown that based on abundance coverage estimator (ACE), estimated richness (total unique CDR3α clonotypes in the population) of Treg cells in B6^{mini} mice significantly exceeded estimated richness of T$_N$ cells [36]. Although ACE underestimates true richness at low sample size, it accounts for "unseen sequences" based on low abundance data and is suitable for comparative analyses. We combined thymic and peripheral sequences for each population and calculated the ACE index, based on 578 DNA sequences per subset. The ACE values for T$_N$ and Treg cells were respectively 1187 and 994 for NOD^{mini} mice, and 1184 and 1815 for B6^{mini} mice. Interestingly, estimated ACE value for TCRs on T$_N$ cells from NOD^{mini} mice was comparable to the ACE value for TCRs on T$_N$ cells from B6^{mini} mice. However when we compared the ratio of ACE values for Treg cells between NOD^{mini} and B6^{mini}

A

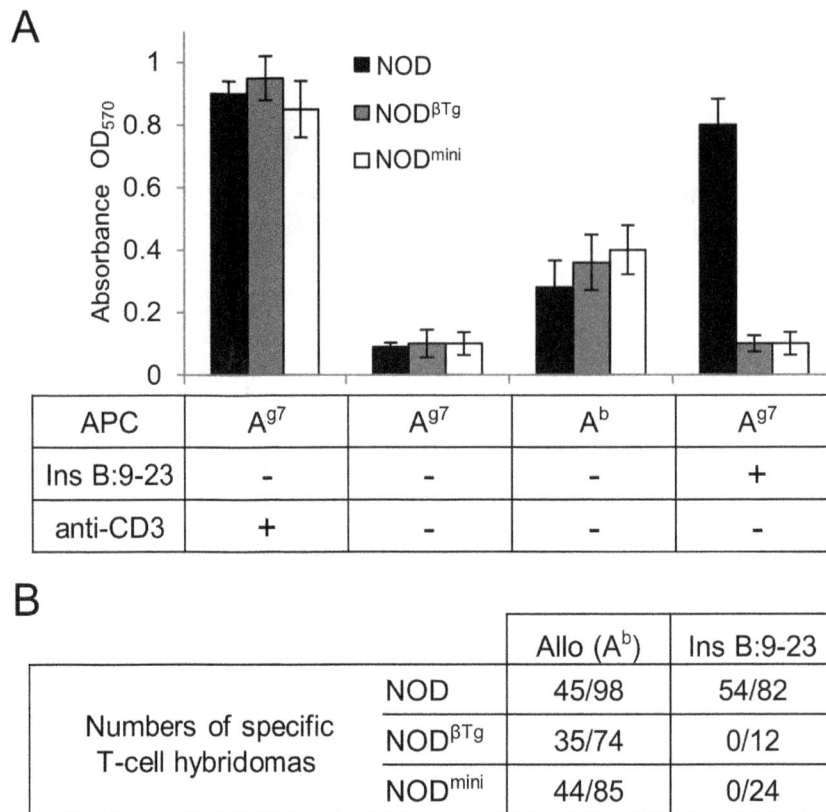

APC	A^{g7}	A^{g7}	A^b	A^{g7}
Ins B:9-23	-	-	-	+
anti-CD3	+	-	-	-

B

		Allo (A^b)	Ins B:9-23
Numbers of specific T-cell hybridomas	NOD	45/98	54/82
	NOD$^{\beta Tg}$	35/74	0/12
	NOD^{mini}	44/85	0/24

Figure 3. Lack of response to insulin B:9–23 peptide in transgenic NOD mice. (A) 5×10^4 of CD4$^+$ T cells sorted from lymph nodes of NOD, NOD$^{\beta Tg}$ and NOD^{mini} mice were cultured in the presence of 5×10^5 splenocytes from NOD.TCR$\alpha^{-/-}$ (A^{g7}) or B6.TCR$\alpha^{-/-}$ (A^b) mice and soluble anti-CD3 (1 μg/ml) or insulin B:9–23 peptide (50 μg/ml), as indicated. Proliferation of cells was measured after 3 days by MTT assay [38]. Experiments were done twice with 3 mice per group. (B) T-cell hybridomas specific to allo-antigens or insulin B:9–23 peptide were generated from indicated mice. For generation of B:9–23 specific hybridomas, mice were immunized with the peptide 7 days prior to isolation of lymph nodes for *in vitro* blasts generation [38]. Generated hybridomas were tested for their ability to respond to syngeneic (NOD.TCR$\alpha^{-/-}$) or allogeneic (B6.TCR$\alpha^{-/-}$) splenocytes with or without B:9–23 peptide or anti-CD3. Table shows numbers of identified hybridomas specific to indicated antigens and numbers of hybridomas responding to anti-CD3 stimulation. Table is representative of 3 independent experiments with two mice per group.

mice the number of possible unique TCRs on NOD background was reduced by almost 50% (Fig. 6B).

Previously it has been reported that based on analysis of two selected VJ (TCRα) or one VDJ (TVRβ) rearrangements in wild type mice, TCR repertoire of Treg cells in NOD mice was less diverse as compared to conventional T cells in the thymus, but also less diverse in the thymus of B6 mice [9]. These differences were observed based on calculation of Shannon entropy and normalization to the logarithm of unique sequences [9]. Such transformation is a measure of distribution of frequency of individual species and is a good measure of relative evenness of assemblage [51,52]. Together with richness, evenness is a descriptive measure of diversity not the diversity *per se* [51]. Therefore, as explained by Jost, to put the estimates in perspective, we converted Shannon entropy (diversity index) to "true diversity" by calculating "numbers equivalent" also called "effective number of species" (ENS), to preserve linear scale of comparison [51,53]. The ENS measure represents diversity of a particular sample, and the numeric value represents the theoretical number of equally common unique sequences in the assemblage. Comparison of "true diversity" between B6^{mini} and NOD^{mini} mice showed almost reversal of ratios of diversities between TCRs on T$_N$ and Treg cells, with the differences more profound in the thymus than in the periphery (Fig. 6C). These differences in NOD^{mini} mice were confirmed by high throughput sequencing, and were consistent

regardless of total numbers of sequences analyzed (Fig. 6C, D). Lower diversity of Treg TCR repertoire was visualized empirically by plotting accumulation curves of observed sequences from peripheral T$_N$ and Treg cells (Fig. 6E). These curves show that accumulation of unique CDR3 regions from the first 120 thousands of sequences for each population gives twice as many unique DNA clonotypes in T$_N$ (9757) as compared to Treg cells (4968). This ratio is reversed in comparison to accumulation curves observed in B6^{mini} mice [36]. Collectively our data show that although TCR diversity on Treg cells in NOD^{mini} mice is reduced, conventional T cells retain diverse TCR repertoire at least at the levels found on B6 background.

Discussion

In this study we investigated the impact of the reduction of TCR diversity on the development of two autoimmune diseases in NOD mice; T1D and SS. Previously it has been shown that T1D can develop despite use of transgenic TCRβ chains or reduction of precursor frequency of potentially diabetogenic T cell clones. In our model overall diversity of TCRs was reduced by allelic exclusion caused by use of transgenic TCRβ chain that was not only pathogenically irrelevant, but also was originally selected in B6 mice on I-A^b molecule. Despite normal distribution of Vα families in NOD$^{\beta Tg}$ mice, neither insulitis nor diabetes developed.

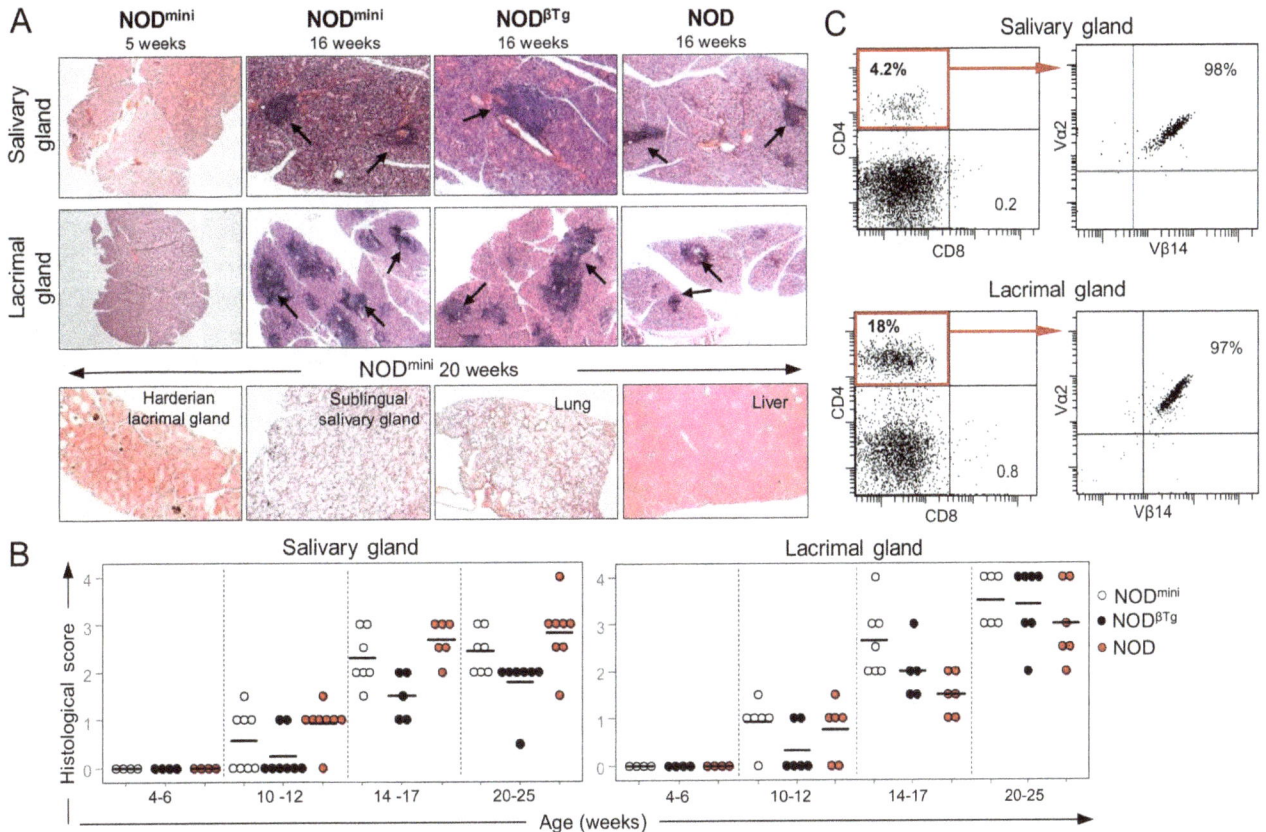

Figure 4. Lymphocytic infiltrates in mandibular salivary and extraorbital lacrimal glands. (A) H&E staining of tissue sections from indicated organs of analyzed mice, showing lymphocytic infiltrates indicated by arrows. (B) Histological score of infiltrated glands in indicated age groups. Scoring criteria: score 0, no infiltrates; score 1–1.5, 1–2 foci per section; score 2–2.5, 3–5 foci per section; score 3, 6–10 foci per section; score 4, more than 10 foci per section. Infiltrate is considered as focus when number of infiltrating cells in continuous space is greater than 50. Three sections at different anatomical locations per organ were analyzed with at least 5 mice per age group. (C) FACS analysis of CD4 T cells infiltrating into salivary and lacrimal glands in 16 week old NOD[mini] mice. Dot plots on the right show expression of transgenic TCR on CD4+ gated cells.

Conversely, these mice developed infiltrates in salivary and lacrimal glands, leading to autoantibody production and exocrine gland dysfunction. Further reduction of TCR diversity by generation of transgenic mice with TCR[mini] repertoire, where one Vα segment is allowed to rearrange to only two Jα segments, did not prevent development of SS. Our results indicate that the difference between T1D and SS regarding the dependence on MHC polymorphism is directly correlated to the magnitude of possible TCR/peptide/MHCII interactions participating in the autoimmune phase of the disease.

We show that the lack of development of diabetes or even insulitis in our NOD[βTg] or NOD[mini] transgenic mice is due to lack of specificity to the key immunodominant insulin B:9–23 peptide, which is known to be instrumental for the onset of the T1D in NOD mice. The use of transgenic TCRVβ14 chain in our mice did not dramatically influence the ability of its binding to different TCRα chains, as T cells from NOD[βTg] mice use all Vα families with frequencies found in NOD mice (Fig. 1). This includes efficient amplification of the Vα13 family which contains TRAV5D-4 chain (Vα13s1) that was shown to be sufficient to elicit anti-insulin autoimmunity without bias toward particular Vβ family of TCRβ chain partners [28,54]. Moreover, previous studies show that T cells using Vβ14 family were found on T cells specific to insulin antigen, T cells expanding in pancreatic lymph nodes or in T cells infiltrating pancreatic islets, showing that the Vβ14 family is not negatively influencing the development of

diabetogenic TCRs [55–58]. One cannot exclude the possibility that this particular transgenic TCRVβ14 chain may be unable to pair with appropriate TCRα chain, preventing the ability of the expressed αβTCR to recognize the B:9–23 peptide. This selective requirement for a TCRβ chain would reinforce our observation that development of SS is less dependent than development of T1D on overall TCR diversity and a particular peptide/MHC combination.

Development of T1D relies on different insulin B:9–23 register recognition, allowing escape of specific T cells due to register shifting [11–13]. The lack of peripheral recognition of insulin B:9–23 in our transgenic mice can also be due to the impact of the limited TCR repertoire. Reduction of overall TCR diversity can influence (reduce) the precursor frequency of DP thymocytes bearing potentially autoreactive TCRs, resulting in more efficient negative selection in the thymus of NOD mice. It has been shown that early expression of transgenic αβTCR, due to ERK1/2 defect on NOD background results in greater commitment of the DN thymocytes to αβ lineage "overcrowding" DP compartment [8]. Our TCRα[mini] transgene has natural timing of expression, similar to B6[mini] and polyclonal NOD mice, where pre-TCR signaling is not perturbed by early expression of the transgene [36]. As suggested by Mingueneau et. al., in the polyclonal repertoire on the NOD background, the ERK1/2 defect increases the affinity threshold of positive selection, shifting the selection window of thymocytes toward self-reactivity, however not impacting the

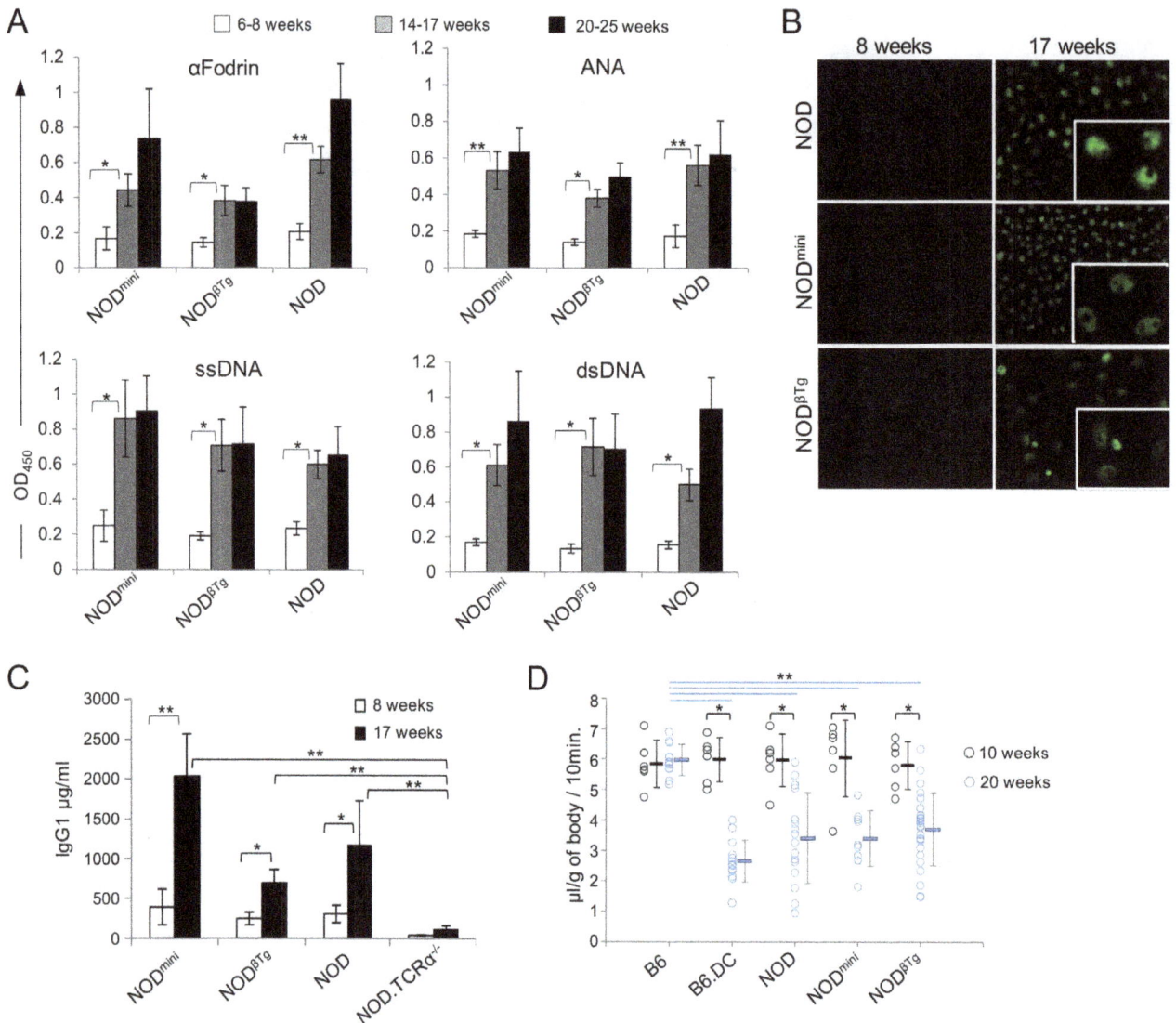

Figure 5. Detection of autoantibodies and hyposalivation in NOD^{mini} and NOD$^{\beta Tg}$ mice. (A) ELISA assay was performed using mouse serum (1:100) from indicated age groups. Each graph represents mean value of OD$_{450}$ and standard deviation for indicated antigens. At least six mice were used per each group. *p<0.05, **p<0.005. (B) Detection of ANAs pattern using Hep-2 cell line. Sera from mice were diluted 1:40, incubated with HEp-2-fixed slides and evaluated under fluorescent microscope at x20 magnifications. Representative images are shown. (C) Quantitative ELISA analysis of IgG1 levels in sera (1:50,000) from indicated mice. Each bar represents mean value and standard deviation from six mice per experimental groups. (D) Salivary flow rates after pilocarpine injection in indicated mice at 10 and 20 wks of age. Double congenic B6.NODIdd3.NODIdd5 (B6.DC) mice, that develop SS on B6 genetic background, were used as a control. Saliva volume was measured and calculated in mg per mouse body mass. Each circle represents one mouse and horizontal lines indicate mean values of the experimental groups. T-test was used to calculate differences between groups. *p<0.002, **p<0.0002.

efficiency of negative selection. This results in higher overall self-reactivity of peripheral effector T cells and possibly explains lower diversity of thymic Treg cells on NOD background [8,9]. Considering partial overlap of specificities between Treg and autoreactive T cells, one could suggest similar effect of lower diversity on autoreactive population. However weak and unstable peptide-binding property of I-A^{g7} molecule does not favor the elimination or inactivation of autoreactive T cells [59]. This instability may have additional influence on "a leak" of autoreactive T cells, but also on inefficient generation of Treg population, which may require longer or stronger interactions with self MHC/peptide complexes [60,61]. Therefore the shift in selection window will not impact autoreactive T cells as much as it

will impact Treg cells, after all, Treg development relies on recognition of self-peptide/MHCII complexes in the thymus, whereas thymic escape of autoreactive T cells relies on avoidance of such complexes during negative selection. It is possible that in our model, we reached the threshold of diversity required to generate autoreactive TCR repertoire without "holes" in specificities. Therefore despite reduced TCR diversity of Treg cells mice do not develop diabetes and we were unable to detect insulin B:9–23 specific T cells in the periphery. Similarly, the comparison of TCR^{mini} repertoires between B6 and NOD backgrounds shows minimal impact of the NOD genotype on TCR diversity of conventional CD4^{+} T cells however it substantially reduces the TCR diversity on Treg cells in NOD mice.

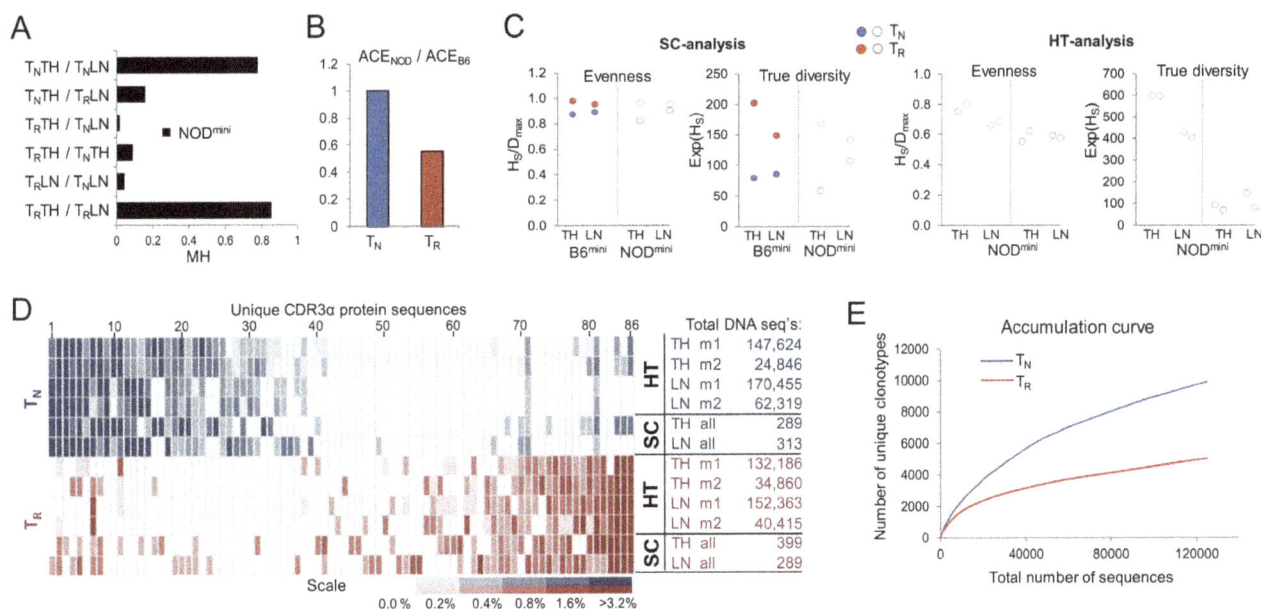

Figure 6. TCR repertoire of naïve and regulatory T cells in NOD^mini mice. (A) Similarity between indicated populations (first 289 single cell sequences per each population) was estimated based on Morisita-Horn index (MH). T_N - naïve CD4+Foxp3−CD45RB+CD62L+ T cells, T_R - regulatory CD4+Foxp3+ T cells, TH - thymus, LN - lymph nodes. (B) Ratio of richness of TCRs on T_N and T_R cells between NOD^mini and B6^mini mice. Abundance coverage estimator (ACE) was calculated based on 578 sequences for each population combined from thymus and peripheral lymph nodes. (C) Evenness and effective number of species (ENS) for analyzed TCR repertoires. Shannon evenness index was calculated as Shannon entropy (H_s) divided by maximum diversity D_{max}, where D_{max} equals natural logarithm of number of unique sequences in analyzed population. ENS (true diversity) was calculated as exponential of Shannon entropy. (D) Comparison of frequency of 20 most dominant unique protein CDR3 clonotypes found in each population indicated on the right of the heat map. Table indicates analyzed populations from lymph nodes and thymii by single cell analysis (SC) and high throughput sequencing (HT). Experimental mouse 1 and 2 are marked as m1 and m2. Numbers next to each population indicate total numbers of DNA sequences analyzed. All 86 unique CDR3α protein sequences in the heat map are shown in Table S1. (E) Accumulation curve of unique DNA clonotypes observed after accumulation of 124,730 sequences for T_N cells and 124,696 for T_R cells. (A–C) All indices were computed based on DNA sequences for each population using software SPADE and EstimateS8.2.

Development of SS in NOD^mini mice is especially interesting, since CD4+ T cells are instrumental in immunopathogenesis and their recognition of self antigens is essential for the onset and progression of the disease [34,62]. It shows that the TCR^mini repertoire is diverse enough not only to drive glandular infiltration and activation of the CD4+ T cells but also the repertoire is still diverse enough to support the full development of the disease with production of Th2-dependent IgG1 pathogenic autoantibodies (Fig. 5C) [50]. Moreover, we noticed differences in timing of infiltrates, levels of autoantibodies and total IgG1 production between transgenic mice and parental NOD mice, which indicates different frequencies of certain TCR specificities between mice. In congenic strains of NOD mice models of SS (NOD.B10-H2^b, NOD.H2^p, NOD.H2^q, NOD.H2^h4) replacement of I-A^g7 with other MHC molecules does not prevent salivary and lacrimal gland infiltration and decreased saliva and tear production [16,48,63]. Interestingly, in NOD.H2^h4, contrary to the parental NOD strain, there is a high frequency of ANA with a high proportion of SSA/Ro and SSB/La observed [63]. Also, NOD.H2^q mice exhibit increased production of lupus-like types of autoantibodies and develop nephritis, as compared to NOD and NOD.H2^p mice [16]. This weak dependence of SS on a particular MHC haplotype in NOD mice correlates with our data that show development of SS despite limited TCR diversity. In human studies it was suggested that production profiles of certain autoantibodies were associated with HLA-DR haplotypes rather than with clinical manifestations [64], however studies of familial inheritance in patients with SS showed a linkage between particular HLA and disease susceptibility [65,66]. The most

recent comprehensive analysis by Sjögren's Genetics Network showed that HLA has the strongest linkage to the SS, although it is not on the level of T1D [17]. Certain HLA haplotypes will influence binding diversity of self or environmental peptides and the nature of antigen presentation to T cells during thymic development or during immune responses in the periphery. Our results from the mouse model emphasize that SS may be less affected by requirement of a unique key antigen/MHCII combination but rather may be more influenced by a wider range of overall TCR/peptide/MHC interactions involved in the onset/progression of the disease. It can be due to a combination of cross-reactivity of the TCR repertoire on SS-specific T cells, wider range of antigens presented by MHCII molecules, higher peripheral self-reactivity of effector T cells, increased tissue expression of MHCII complexes on salivary epithelial cells and de novo expression or post-translational modification of self-antigens [67–69].

Acknowledgments

Cell sorting was performed by Jeanene Pihkala in the GRU FACS Core, the sequencing was performed by John Nechtman in the GRU Genomic Core, and transgenic mice were generated by Gabriela Pacholczyk in the GRU Transgenic Core.

Author Contributions

Conceived and designed the experiments: JK RP. Performed the experiments: JK RD SL MB. Analyzed the data: JK RD SL MB RP. Contributed to the writing of the manuscript: JK RP.

References

1. Chaparro RJ, Dilorenzo TP (2010) An update on the use of NOD mice to study autoimmune (Type 1) diabetes. Expert Rev Clin Immunol 6: 939–955.
2. Lavoie TN, Lee BH, Nguyen CQ (2011) Current concepts: mouse models of Sjogren's syndrome. J Biomed Biotechnol 2011: 549107.
3. Anderson MS, Bluestone JA (2005) The NOD Mouse: A Model of Immune Dysregulation. Annu Rev Immunol 23: 447–485.
4. Mathis D, Vence L, Benoist C (2001) beta-Cell death during progression to diabetes. Nature 414: 792–798.
5. Nguyen CQ, Peck AB (2009) Unraveling the pathophysiology of Sjogren syndrome-associated dry eye disease. Ocul Surf 7: 11–27.
6. Fox RI (2005) Sjogren's syndrome. Lancet 366: 321–331.
7. Kishimoto H, Sprent J (2001) A defect in central tolerance in NOD mice. Nat Immunol 2: 1025–1031.
8. Mingueneau M, Jiang W, Feuerer M, Mathis D, Benoist C (2012) Thymic negative selection is functional in NOD mice. J Exp Med 209: 623–637.
9. Ferreira C, Singh Y, Furmanski AL, Wong FS, Garden OA, et al. (2009) Non-obese diabetic mice select a low-diversity repertoire of natural regulatory T cells. Proc Natl Acad Sci U S A 106: 8320–8325.
10. D'Alise AM, Auyeung V, Feuerer M, Nishio J, Fontenot J, et al. (2008) The defect in T-cell regulation in NOD mice is an effect on the T-cell effectors. Proceedings of the National Academy of Sciences 105: 19857–19862.
11. Stadinski BD, Zhang L, Crawford F, Marrack P, Eisenbarth GS, et al. (2010) Diabetogenic T cells recognize insulin bound to IAg7 in an unexpected, weakly binding register. Proceedings of the National Academy of Sciences 107: 10978–10983.
12. Mohan JF, Petzold SJ, Unanue ER (2011) Register shifting of an insulin peptide–MHC complex allows diabetogenic T cells to escape thymic deletion. J Exp Med 208: 2375–2383.
13. Marrack P, Kappler JW (2012) Do MHCII-Presented Neoantigens Drive Type 1 Diabetes and Other Autoimmune Diseases? Cold Spring Harbor Perspectives in Medicine 2.
14. Wicker LS, Appel MC, Dotta F, Pressey A, Miller BJ, et al. (1992) Autoimmune syndromes in major histocompatibility complex (MHC) congenic strains of nonobese diabetic (NOD) mice. The NOD MHC is dominant for insulitis and cyclophosphamide-induced diabetes. J Exp Med 176: 67–77.
15. Li X, Golden J, Faustman DL (1993) Faulty major histocompatibility complex class II I-E expression is associated with autoimmunity in diverse strains of mice. Autoantibodies, insulitis, and sialadenitis. Diabetes 42: 1166–1172.
16. Lindqvist AKB, Nakken B, Sundler M, Kjellen P, Jonsson R, et al. (2005) Influence on Spontaneous Tissue Inflammation by the Major Histocompatibility Complex Region in the Nonobese Diabetic Mouse. Scand J Immunol 61: 119–127.
17. Lessard CJ, Li H, Adrianto I, Ice JA, Rasmussen A, et al. (2013) Variants at multiple loci implicated in both innate and adaptive immune responses are associated with Sjogren's syndrome. Nat Genet 45: 1284–1292.
18. Nakayama M, Abiru N, Moriyama H, Babaya N, Liu E, et al. (2005) Prime role for an insulin epitope in the development of type[thinsp]1 diabetes in NOD mice. Nature 435: 220–223.
19. Arakaki R, Ishimaru N, Saito I, Kobayashi M, Yasui N, et al. (2003) Development of autoimmune exocrinopathy resembling Sjögren's syndrome in adoptively transferred mice with autoreactive CD4+ T cells. Arthritis Rheum 48: 3603–3609.
20. Takada K, Takiguchi M, Konno A, Inaba M (2005) Autoimmunity against a tissue kallikrein in IQI/Jic Mice: a model for Sjogren's syndrome. J Biol Chem 280: 3982–3988.
21. Winer S, Astsaturov I, Cheung R, Tsui H, Song A, et al. (2002) Primary Sjögren's syndrome and deficiency of ICA69. Lancet 360: 1063–1069.
22. Haneji N, Nakamura T, Takio K, Yanagi K, Higashiyama H, et al. (1997) Identification of alpha-fodrin as a candidate autoantigen in primary Sjogren's syndrome. Science 276: 604–607.
23. Naito Y, Matsumoto I, Wakamatsu E, Goto D, Ito S, et al. (2006) Altered peptide ligands regulate muscarinic acetylcholine receptor reactive T cells of patients with Sjogren's syndrome. Ann Rheum Dis 65: 269–271.
24. Matsumoto I, Maeda T, Takemoto Y, Hashimoto Y, Kimura F, et al. (1999) Alpha-amylase functions as a salivary gland-specific self T cell epitope in patients with Sjogren's syndrome. Int J Mol Med 3: 485–490.
25. Iizuka M, Wakamatsu E, Tsuboi H, Nakamura Y, Hayashi T, et al. (2010) Pathogenic role of immune response to M3 muscarinic acetylcholine receptor in Sjogren's syndrome-like sialoadenitis. J Autoimmun 35: 383–389.
26. Lipes MA, Rosenzweig A, Tan KN, Tanigawa G, Ladd D, et al. (1993) Progression to diabetes in nonobese diabetic (NOD) mice with transgenic T cell receptors. Science 259: 1165–1169.
27. Serreze DV, Johnson EA, Chapman HD, Graser RT, Marron MP, et al. (2001) Autoreactive diabetogenic T-cells in NOD mice can efficiently expand from a greatly reduced precursor pool. Diabetes 50: 1992–2000.
28. Zhang L, Jasinski JM, Kobayashi M, Davenport B, Johnson K, et al. (2009) Analysis of T cell receptor beta chains that combine with dominant conserved TRAV5D-4*04 anti-insulin B: 9–23 alpha chains. J Autoimmun 33: 42–49.
29. Dwyer E, Itescu S, Winchester R (1993) Characterization of the primary structure of T cell receptor beta chains in cells infiltrating the salivary gland in the sicca syndrome of HIV-1 infection. Evidence of antigen-driven clonal selection suggested by restricted combinations of V beta J beta gene segment usage and shared somatically encoded amino acid residues. J Clin Invest 92: 495–502.
30. Matsumoto I, Okada S, Kuroda K, Iwamoto I, Saito Y, et al. (1999) Single cell analysis of T cells infiltrating labial salivary glands from patients with Sjogren's syndrome. Int J Mol Med 4: 519–527.
31. Pivetta B, De Vita S, Ferraccioli G, De RV, Gloghini A, et al. (1999) T cell receptor repertoire in B cell lymphoproliferative lesions in primary Sjogren's syndrome. J Rheumatol 26: 1101–1109.
32. Sumida T, Yonaha F, Maeda T, Tanabe E, Koike T, et al. (1992) T cell receptor repertoire of infiltrating T cells in lips of Sjogren's syndrome patients. J Clin Invest 89: 681–685.
33. Yonaha F, Sumida T, Maeda T, Tomioka H, Koike T, et al. (1992) Restricted junctional usage of T cell receptor V beta 2 and V beta 13 genes, which are overrepresented on infiltrating T cells in the lips of patients with Sjogren's syndrome. Arthritis Rheum 35: 1362–1367.
34. Singh N, Cohen PL (2012) The T cell in Sjogren's syndrome: Force majeure, not spectateur. J Autoimmun 39: 229–233.
35. Sumida T, Tsuboi H, Iizuka M, Hirota T, Asashima H, et al. (2014) The role of M3 muscarinic acetylcholine receptor reactive T cells in Sjögren's syndrome: A critical review. J Autoimmun.
36. Pacholczyk R, Ignatowicz H, Kraj P, Ignatowicz L (2006) Origin and T cell receptor diversity of Foxp3+CD4+CD25+ T cells. Immunity 25: 249–259.
37. Cha S, Nagashima H, Brown VB, Peck AB, Humphreys-Beher MG (2002) Two NOD Idd-associated intervals contribute synergistically to the development of autoimmune exocrinopathy (Sjogren's syndrome) on a healthy murine background. Arthritis Rheum 46: 1390–1398.
38. Pacholczyk R, Kern J, Singh N, Iwashima M, Kraj P, et al. (2007) Nonself-antigens are the cognate specificities of Foxp3+ regulatory T cells. Immunity 27: 493–504.
39. Kisielow P, Tortola L, Weber J, Karjalainen K, Kopf M (2011) Evidence for the divergence of innate and adaptive T-cell precursors before commitment to the alphabeta and gammadelta lineages. Blood 118: 6591–6600.
40. Kraj P, Pacholczyk R, Ignatowicz L (2001) Alpha beta TCRs differ in the degree of their specificity for the positively selecting MHC/peptide ligand. J Immunol 166: 2251–2259.
41. Pacholczyk R, Kraj P, Ignatowicz L (2002) Peptide specificity of thymic selection of CD4+CD25+ T cells. J Immunol 168: 613–620.
42. Lutsiak MEC, Semnani RT, De Pascalis R, Kashmiri SVS, Schlom J, et al. (2005) Inhibition of CD4+25+ T regulatory cell function implicated in enhanced immune response by low-dose cyclophosphamide. Blood 105: 2862–2868.
43. Harada M, Makino S (1984) Promotion of spontaneous diabetes in non-obese diabetes-prone mice by cyclophosphamide. Diabetologia 27: 604–606.
44. Michels AW, Ostrov DA, Zhang L, Nakayama M, Fuse M, et al. (2011) Structure-Based Selection of Small Molecules To Alter Allele-Specific MHC Class II Antigen Presentation. The Journal of Immunology 187: 5921–5930.
45. Cha S, Peck AB, Humphreys-Beher MG (2002) Progress in understanding autoimmune exocrinopathy using the non-obese diabetic mouse: an update. Crit Rev Oral Biol Med 13: 5–16.
46. Atkinson JC, Travis WD, Slocum L, Ebbs WL, Fox PC (1992) Serum anti-SS-B/La and IgA rheumatoid factor are markers of salivary gland disease activity in primary Sjogren's syndrome. Arthritis Rheum 35: 1368–1372.
47. Skarstein K, Wahren M, Zaura E, Hattori M, Jonsson R (1995) Characteriza-tion of T cell receptor repertoire and anti-Ro/SSA autoantibodies in relation to sialadenitis of NOD mice. Autoimmunity 22: 9–16.
48. Robinson CP, Yamachika S, Bounous DI, Brayer J, Jonsson R, et al. (1998) A novel NOD-derived murine model of primary Sjogren's syndrome. Arthritis Rheum 41: 150–156.
49. Brayer JB, Cha S, Nagashima H, Yasunari U, Lindberg A, et al. (2001) IL-4-dependent effector phase in autoimmune exocrinopathy as defined by the NOD.IL-4-gene knockout mouse model of Sjogren's syndrome. Scand J Immunol 54: 133–140.
50. Gao J, Killedar S, Cornelius JG, Nguyen C, Cha S, et al. (2006) Sjogren's syndrome in the NOD mouse model is an interleukin-4 time-dependent, antibody isotype-specific autoimmune disease. J Autoimmun 26: 90–103.
51. Jost L (2010) The Relation between Evenness and Diversity. Diversity 2: 207–232.
52. Pielou EC (1966) The measurement of diversity in different types of biological collections. J Theor Biol 13: 131–144.

53. Adelman MA (1969) Comment on the H Concentration Measure as a Numbers-Equivalent. The Review of Economics and Statistics 51: 99–101.

54. Nakayama M, Castoe T, Sosinowski T, He X, Johnson K, et al. (2012) Germline TRAV5D-4 T-Cell Receptor Sequence Targets a Primary Insulin Peptide of NOD Mice. Diabetes 61: 857–865.

55. Simone E, Daniel D, Schloot N, Gottlieb P, Babu S, et al. (1997) T cell receptor restriction of diabetogenic autoimmune NOD T cells. Proceedings of the National Academy of Sciences 94: 2518–2521.

56. Baker FJ, Lee M, Chien Y-h, Davis MM (2002) Restricted islet-cell reactive T cell repertoire of early pancreatic islet infiltrates in NOD mice. Proceedings of the National Academy of Sciences 99: 9374–9379.

57. Marrero I, Hamm DE, Davies JD (2013) High-Throughput Sequencing of Islet-Infiltrating Memory CD4$^+$ T Cells Reveals a Similar Pattern of TCR Vβ Usage in Prediabetic and Diabetic NOD Mice. PLoS ONE 8: e76546.

58. Petrovc Berglund J, Mariotti-Ferrandiz E, Rosmaraki E, Hall H, Cazenave P-A, et al. (2008) TCR repertoire dynamics in the pancreatic lymph nodes of non-obese diabetic (NOD) mice at the time of disease initiation. Mol Immunol 45: 3059–3064.

59. Carrasco-Marin E, Shimizu J, Kanagawa O, Unanue ER (1996) The class II MHC I-Ag7 molecules from non-obese diabetic mice are poor peptide binders. J Immunol 156: 450–458.

60. Aschenbrenner K, D'Cruz LM, Vollmann EH, Hinterberger M, Emmerich J, et al. (2007) Selection of Foxp3+ regulatory T cells specific for self antigen expressed and presented by Aire+ medullary thymic epithelial cells. Nat Immunol 8: 351–358.

61. Fontenot JD, Rasmussen JP, Williams LM, Dooley JL, Farr AG, et al. (2005) Regulatory T cell lineage specification by the forkhead transcription factor foxp3. Immunity 22: 329–341.

62. Sumida T, Tsuboi H, Iizuka M, Nakamura Y, Matsumoto I (2010) Functional role of M3 muscarinic acetylcholine receptor (M3R) reactive T cells and anti-M3R autoantibodies in patients with Sjögren's syndrome. Autoimmunity Reviews 9: 615–617.

63. Burek CL, Talor MV, Sharma RB, Rose NR (2007) The NOD.H2h4 mouse shows characteristics of human Sjögren's Syndrome. J Immunol 178: S232–S223d.

64. Gottenberg JE, Busson M, Loiseau P, Cohen-Solal J, Lepage V, et al. (2003) In primary Sjögren's syndrome, HLA class II is associated exclusively with autoantibody production and spreading of the autoimmune response. Arthritis Rheum 48: 2240–2245.

65. Fox RI, Kang HI (1992) Pathogenesis of Sjögren's syndrome. RheumDisClinNorth Am 18: 517–538.

66. Manoussakis MN, Georgopoulou C, Zintzaras E, Spyropoulou M, Stavropoulou A, et al. (2004) Sjögren's syndrome associated with systemic lupus erythematosus: clinical and laboratory profiles and comparison with primary Sjögren's syndrome. Arthritis Rheum 50: 882–891.

67. Anderton SM (2004) Post-translational modifications of self antigens: implications for autoimmunity. Curr Opin Immunol 16: 753–758.

68. Engelhard VH, Altrich-Vanlith M, Ostankovitch M, Zarling AL (2006) Post-translational modifications of naturally processed MHC-binding epitopes. Curr Opin Immunol 18: 92–97.

69. Moutsopoulos HM, Hooks JJ, Chan CC, Dalavanga YA, Skopouli FN, et al. (1986) HLA-DR expression by labial minor salivary gland tissues in Sjögren's syndrome. Ann Rheum Dis 45: 677–683.

23

Transgenic Mice Expressing Yeast CUP1 Exhibit Increased Copper Utilization from Feeds

Xiaoxian Xie[9], Yufang Ma[9], Zhenliang Chen, Rongrong Liao, Xiangzhe Zhang, Qishan Wang, Yuchun Pan*

School of Agriculture and Biology, Department of Animal Sciences, Shanghai Jiao Tong University, Shanghai, PR China, Shanghai Key Laboratory of Veterinary Biotechnology, Shanghai, PR China

Abstract

Copper is required for structural and catalytic properties of a variety of enzymes participating in many vital biological processes for growth and development. Feeds provide most of the copper as an essential micronutrient consumed by animals, but inorganic copper could not be utilized effectively. In the present study, we aimed to develop transgenic mouse models to test if copper utilization will be increased by providing the animals with an exogenous gene for generation of copper chelatin in saliva. Considering that the *S. cerevisiae CUP1* gene encodes a Cys-rich protein that can bind copper as specifically as copper chelatin in yeast, we therefore constructed a transgene plasmid containing the *CUP1* gene regulated for specific expression in the salivary glands by a promoter of gene coding pig parotid secretory protein. Transgenic CUP1 was highly expressed in the parotid and submandibular salivary glands and secreted in saliva as a 9-kDa copper-chelating protein. Expression of salivary copper-chelating proteins reduced fecal copper contents by 21.61% and increased body-weight by 12.97%, suggesting that chelating proteins improve the utilization and absorbed efficacy of copper. No negative effects on the health of the transgenic mice were found by blood biochemistry and histology analysis. These results demonstrate that the introduction of the salivary *CUP1* transgene into animals offers a possible approach to increase the utilization efficiency of copper and decrease the fecal copper contents.

Editor: Vladimir V. Kalinichenko, Cincinnati Children's Hospital Medical Center, United States of America

Funding: This work was supported by the National Transgenic Breeding Program (grant no.: 2014ZX08006-004; 2014ZX08009-003-006). The funders had no role in study design, data collection and analysis, decision to publish, or preparation of the manuscript.

Competing Interests: The authors have declared that no competing interests exist.

* Email: panyuchun1963@aliyun.com

[9] These authors contributed equally to this work.

Introduction

Copper is an essential trace element and required for survival by a wide range of species, from yeast to mammals [1]. It functions as a cofactor and is required for structural and catalytic properties of a variety of enzymes because of its capacity to act as an intermediary in the transfer of electrons that makes it central to the catalytic activity of the enzymes [2,3], which are involved in a number of vital biological processes, such as cellular respiration and iron transport, required for growth and development [4]. Thus, Cu supplements were used to treat anemia in animals in the 1920s, and later in chicks [5], pigs [6], infants [7], and adult humans with good success [8].

Feeds provide most of the copper as an essential micronutrient consumed by animals, and drinking water contributes about 6–13% of average daily intake of copper [9,10,11,12]. Most of ingested copper is absorbed in the small intestine, and very small amounts in the stomach [10]. The absorption of copper in the body depends on a variety of factors including its chemical form [10]. Chelated copper has been proven to improve the utilization of copper, which is absorbed more efficiently through an amino acid transport system [13] by increasing intestinal absorption and renal tubular reabsorption of copper, and the chelated form displays increased retention in the body compared with its inorganic form [14,15,16], as has been demonstrated in many compounds, such as copper-lysine [17], organic copper chelates [18], copper carbonate [19] and copper-metallothionein (copper-MT) complex [15].

If inorganic coppers are transformed to copper-MT through binding by the MT produced endogenously by animals, such organic copper could be utilized effectively, and then lower doses of inorganic copper could be added in feed, which would, in turn, lead to reduced fecal copper contents. To investigate the feasibility of this hypothesis, we developed transgenic mice that secrete MT proteins in their saliva. The transgenes used in this study contain the *S. cerevisiae CUP1* gene. It encodes a Cys-rich protein with a low molecular weight that can bind copper as specifically as copper chelatin in yeast. The protein is characterized principally by its high copper-binding capacity and unusual amino acid composition, and it contains 20% cysteine residues [20,21]. Mammalian MT is a metal-binding protein that is present in most tissues. The protein was first found to bind cadmium and zinc [22], but it also binds copper and is a major copper-binding protein in the liver, in addition to binding other metals [16]. Yeast CUP1 has highly divergent primary sequences compared to mammalian MT by reconstructing the phylogenetic tree [23]. However, these proteins all possess identical functional sequence motifs, Cys-X-Cys or Cys-X-X-Cys, and binding of copper to these motifs occurs through the Cys residues [21,24]. In this study, we took advantage of the yeast *CUP1* gene to establish a transgenic mouse model to determine whether endogenous expression of CUP1 can increase copper utilization by mice.

Materials and Methods

Ethics Statement

The FVB and ICR mice varieties were used in this research. All animal procedures received approval from the Institutional Animal Care and Use Committee (IACUC) of Shanghai city, China. The mice were housed in the Animal Care Facility at Shanghai Jiao Tong University (IACUC permit numbers: SYXK (Shanghai) 2013-0052).

Construction of the recombinant plasmids expressing the *CUP1* gene

A fragment, which contained the complete open reading frame (ORF) of the *CUP1* gene, was synthesized based on the published sequence in GenBank (NM_001179185). The vector pPSP (pig parotid secretory protein) was a gift from Ning Li (College of Biological Sciences, China Agricultural University, Beijing, China). The recombinant plasmid pPSP-CUP1 was constructed by insertion of the fragment containing the ORF of the *CUP1* gene, which was digested with *Asc* I (TaKaRa, Japan), into the same endonuclease-digested pPSP vector. The recombinant plasmid was confirmed by restriction analysis and DNA sequencing.

Transgene purification, quantification and pronuclear microinjection

The linear DNA fragment containing the pPSP promoter, signal peptide and *CUP1* gene was obtained by digestion with *Xho* I and *Not* I and subsequently purified by agarose gel electrophoresis as described by Yin et al. [25]. DNA was resuspended in microinjection buffer, which consisted of 0.1 mmol/L ethylenedi-aminetetraacetic acid and 10 mmol/L Tris Cl (pH 7.4) at a concentration of 20 ng/μL, and stored at $-20°C$.

The injection of transgene DNA was performed according to Hogan et al. [26].

Transgenic examination by PCR and southern blot

The presence of the *CUP1* transgene in the transgenic founders and offspring was confirmed by PCR analysis of genomic DNA derived from tail biopsies and DNA sequencing. PCR was performed with specific primers (forward primer 5′TGTGTAAGCGTGGTAGGTGCTCATC 3′, reverse primer 5′GACACCTACTCAGACAATGCGATGC 3′), and the transgene length was 337 bp. The transgenic founders (G0) were confirmed by Southern blot analysis. Genomic DNA was isolated from the mouse tails using a ZR Genomic DNA-Tissue MiniPrep Kit (Zymo Research, USA). Twenty micrograms of DNA was restriction digested with *Bgl* II and *Ssp* I, fractionated in a 0.8% agarose gel electrophoresis, and transferred to a nylon membrane (Millipore, UK). The fragment containing the complete ORF of yeast *CUP1* was amplified by PCR using specific primers (forward primer 5′TGGGGAATCAGTAGGAAGTCTTGGC 3′, reverse primer 5′CCCCAGAATAGAATGACACCTACTC 3′), and the fragment length was 832 bp. The fragment was then purified with Qiagen PCR purification kits before its use as a probe. The membrane was hybridized with a DIG-labeled CUP1 DNA probe (20 ng). Pre-hybridization and hybridization were performed according to the procedures described by Van Rijs et al. [27].

Pre-hybridization was performed for 1 h at 45°C, hybridization for 6 h at 45°C, and then the membrane was washed twice in 2×standard saline citrate (SSC), 0.1% SDS at 65°C for 20 min, and 0.5×SSC with 0.1% SDS once. Hybridization signals were examined using a Roche DIG DNA Labeling and Detection Kit according to the manufacturer's instructions.

Western blot analysis

Polyclonal antibodies (Santa Cruz Biotechnology, USA) were raised against amino acids 1–61 taken from the CUP1 and represented the full-length CUP1 sequence of *S. cerevisiae*. The β-actin antibodies were purchased from Sigma-Aldrich (St Louis, MO). Western blot analysis was performed according to the methods of Spencer et al. [28].

Approximately 200 μL of saliva per mouse was collected from 6-wk-old mice as described by Hu et al. [29] and stored at -80°C. The proteins were extracted from the tissues and saliva by homogenization in lysis buffer, separated by SDS-PAGE electrophoresis, and transferred to nitrocellulose. Immunoblotting was performed with antibodies against CUP1 (1:400) and β-actin (1:3000), which served as loading controls. As the secondary antibody, a goat anti-rabbit horseradish peroxidase-conjugated antibody was diluted to 1:5000.

Detection of the copper content in mouse manure and changes of body weight

Prepared feed was purchased from Shanghai SLAC Laboratory Animal Co., Ltd., China. The feed contained 10 mg/kg content of copper, which is in concordance with the national nutrition standard GB 14924.3–2010 [30].

The G1 offspring were weaned at 4 wk of age, and the transgenic mice were confirmed as described above. Then, the transgenic and control mice were individually caged under controlled temperature (22±2°C), humidity (40–60%) and lighting (12 h light; 12 h darkness) and fed the prepared feed and water ad libitum. Body weight was recorded once daily for 2 wk. Fecal samples were collected once daily for 2 wk and then placed into 2 mL sterile tubes and dried immediately. The samples were dried (130°C) for 48 h and ashed at 600°C for 4 h in a muffle furnace. Next, they were cooled, weighed, and digested in nitric acid (Merck, Germany) at 95°C for at least 2 h. After filtration, the contents of copper in mouse manure were measured by ICP-MS (7500 Series ICP-MS system; USA). Each digested sample volume was standardized to 5 mL.

Blood biochemistry and histology analysis

Blood samples of approximately 1 mL per mouse were obtained from the retro-orbital venous plexus of the transgenic and control mice using heparinized capillary tubes. Five mice at 6 wk of age and ten mice at 1 yr of age in each group were used for the blood biochemistry analysis.

The blood samples were centrifuged at 3000 rpm for 10 min for the sera. The sera were stored at $-80°C$ prior to blood biochemistry analysis. Nineteen blood biochemical parameters, including Ca (Calcium ion), Fe (ferrum ion), GLU (glucose), CRE (creatinine), CHO (cholesterol), BUN (blood urea nitrogen), AMY (amylase), ALT (alanine aminotransferase), AST (aspartate aminotrasferase), and ALP (alkaline phosphatase), were detected using an auto-analyzer (Hitachi 7180, Hitachi, Japan).

After blood drawing, the mice were sacrificed for histopathology analysis. Tissues (heart, liver, spleen, stomach, kidney, intestine, brain, parotid gland and submandibular gland) were collected and fixed in PBS buffered 10% formalin. The specimens, after paraffin embedding, were sectioned horizontally at 5 µm thickness, stained with hematoxylin and eosinaccording to standard protocol, and observed using a microscope (Nikon, Japan) at an excitation wavelength of 559 nm.

Statistical analysis

The phenotypic data (the fecal ash copper contents and the body-weight increases of transgenic and control mice) were analyzed separately based on a general linear model (SAS 9.3): $y_{ijk} = \mu + s_i + d_j + g_k + e_{ijk}$ Where

y_{ijk}: the phenotypic value

μ: an overall mean

s_i: a fixed paternal effect

d_j: a fixed maternal effect

g_k: a fixed CUP1 gene effect

e_{ijk}: a residual error effect with a normal distribution N $(0, \sigma^2)$

Results

Generation of transgenic mice

The 12.5-kb linear transgene pPSP-CUP1 (construction shown in Fig. S1) was generated by digestion with *Xho* I and *Not* I and introduced into fertilized mouse oocytes through pronuclear injection.

Four male (No. 5, 6: FVB mice; No. 20, 22: ICR mice) and two female (No. 15: FVB mouse; No. 26: ICR mouse) transgenic founder (G0) mice obtained from 29 mice were confirmed by PCR screening and DNA sequencing. Southern blotting was further used to identify the transgene integrated into the genome of the transgenic mice. The transgenes shares the same 832-bp sequences containing the complete ORF of yeast *CUP1* as the probe. The results indicated that the transgene was integrated into the genome (Fig. 1A). Six transgenic founders were mated twice with wild-type mice, of which 4 males transmitted the transgene to their offspring, and 2 females did not pass the transgene to their progeny. A total of 46 G1 transgenic mice were confirmed by PCR amplification from genomic DNA and sequencing among 77 offspring (Table S1).

Expression of yeast *CUP1* transgene in the salivary glands

The *CUP1* transgene mRNA expression in the salivary glands of transgenic founder was analyzed by reverse transcription PCR. The results revealed that *CUP1* was expressed in the parotid and submandibular glands and was barely expressed in the heart, liver, spleen, stomach, kidney, intestine, and brain tissue of the transgenic founders. However, the CUP1 gene was not expressed in all tissues of the control mice (Fig. S2).

The CUP1 protein was detected in the parotid and submandibular glands using anti-CUP1 antibodies to probe western blot analysis in transgenic founders. The level of β-actin in each sample was determined as the control for protein loading. The results indicated the presence of CUP1 in both detected tissues and indicated relatively high expression in the submandibular glands after normalization against β-actin. In contrast, relatively low expression was observed in the parotid glands (Fig. 1B). CUP1 protein was also detected in the salivary fluid of the transgenic mice by western blot analysis (Fig. 1B), and no CUP1 protein was detected in the saliva of the control mice. The molecular mass of the protein containing CUP1 was 9 kDa as identified in the salivary glands and the secreted saliva.

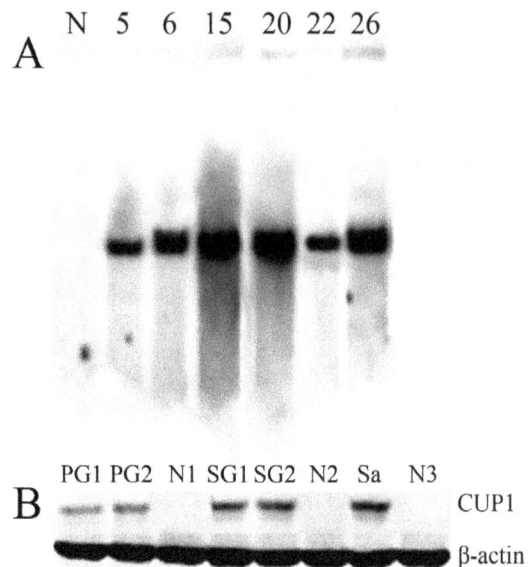

Figure 1. Identification of yeast *CUP1* transgene by southern blot and western blot analysis. (A) Southern blot analysis of the transgenes. Purified genomic DNA from each transgenic founder was digested with *Bgl* II and *Ssp* I, and analyzed by southern blotting using probes specific for *CUP1*. Transgenic founders numbered 5, 6, 15, 20, 22, and 26; N: genomic DNA of control mice as a negative control. (B) Western blot was performed to confirm expression of CUP1 in the parotid and submandibular glands and in the saliva of transgenic lines. The expected protein size is 9 kDa. The lower band is β-actin (42 kDa, used as an internal control). The results demonstrated high expression of CUP1 in the parotid and submandibular glands after normalization against β-actin, as well as that in the saliva of transgenic mice. PG1 and PG2: the parotid glands of transgenic mice; SG1 and SG2: the submandibular glands of transgenic mice; Sa: the saliva of transgenic mice; N1, N2, and N3: the parotid gland, the submandibular gland and the saliva of control mice as negative controls, respectively.

The contents of copper in mouse manure ash and changes of body weight

Forty-four G1 offspring were selected from the total G1 mice considering similar weights and used for further experiments, of which 28 were transgenic mice, and 16 were control mice. To determine the effect of the expressed CUP1 on the transgenic mice, the usage efficiency of copper in the prepared feed was investigated by detecting the contents of copper in mouse manure from the transgenic and control mice at a dietary level of 10 mg/kg copper. For the first week, the transgenic mice exhibited manure ash copper contents of 168.285 ± 18.849 mg/kg, a reduction of 18.41% $(P = 0.0022 < 0.01)$ compared with the control mice $(206.263 \pm 42.307$ mg/kg) raised under the same conditions. At the second week, the manure ash copper contents of transgenic mice $(171.449 \pm 10.767$ mg/kg) were significantly lower $(P = 0.0003 < 0.01)$ compared with that $(218.713 \pm 49.831$ mg/kg) of the control mice. This represents a reduction of 21.61% in the ash copper contents under the same conditions (Fig. 2A). The effect of the expressed CUP1 on body weight of the transgenic mice was also analyzed. At day 0, the body weight of the control group was 20.531 ± 1.099 g and that of the transgenic group was 20.835 ± 1.214 g. After 1 wk, the body-weight increases of the transgenic mice $(6.906 \pm 0.998$ g) were significantly greater $(P = 0.025 < 0.05)$ compared with those $(5.063 \pm 1.214$ g) of the control mice (Fig. 2B). On average, the transgenic mice were 7.2% heavier than the control mice raised under the same conditions.

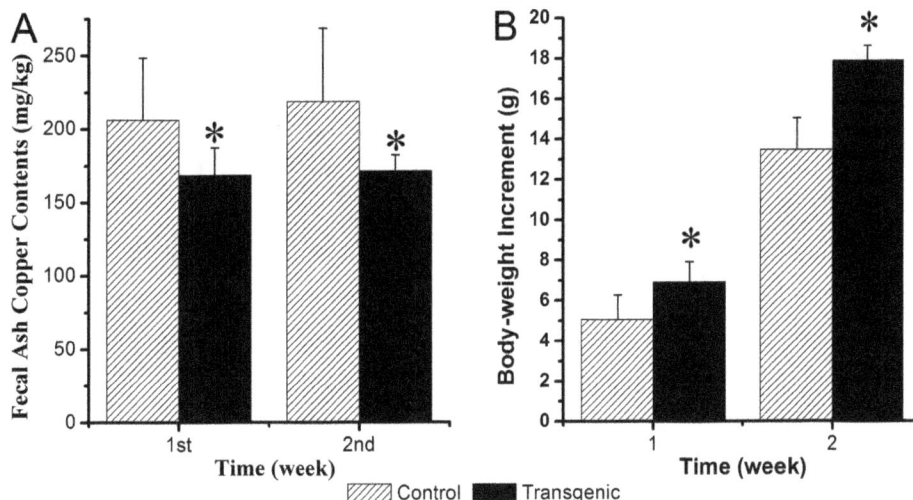

Figure 2. Fecal ash copper contents and changes of body weight of the transgenic and control mice. At 4 wk of age, all G1 offspring were weaned and fed with the prepared feed (copper content, 10 mg/kg). Subsequently, the fecal samples were collected and the body weights of the mice were recorded once daily for 2 wk. (A) The fecal copper contents of the mice were analyzed. The fecal ash copper contents of transgenic mice (n = 28) were significantly lower than the control mice (n = 16), presenting a reduction of 18.41% (*$P = 0.0022 < 0.01$) and 21.61% (*$P = 0.0003 < 0.01$) for the first week and the second week, respectively. (B) The changes of body weight of the mice were analyzed. After 1 and 2 wk, the body-weight of transgenic mice (n = 28) increased significantly (*$P = 0.025 < 0.05$; *$P = 0.019 < 0.05$, respectively) compared with those of the control mice (n = 16) at 7.2% and 12.97%, respectively. 1st: the first; 2nd: the second.

After 2 wk, the body-weight increases of the transgenic mice (17.884 ± 0.728 g) were significantly greater ($P = 0.019 < 0.05$) compared with those (13.475 ± 1.556 g) of the control mice (Fig. 2B). On average, the transgenic mice were 12.97% heavier than the control mice raised under the same conditions.

Analysis of blood biochemistry and histology

The blood biochemistry results revealed that the levels of GLU, BUN, AMY, ALT, AST, and ALP in serum were slightly elevated in the transgenic mice at 6 wk of age compared with the control mice. In contrast, the blood concentration of CRE was slightly decreased, and all differences were not significant ($P > 0.05$; Table 1). The similar results were observed in the mice at 1 yr of age, but the level of ALP was slightly decreased in the transgenic mice (Table S2). Differences between the transgenic and control

mice were hardly observed for other serum biochemical parameters.

At the ages of 6 wk and 1 yr, the transgenic mice were in good health and did not exhibit any gross pathological abnormalities or illness. Histological analysis was performed in the tissues (heart, liver, spleen, stomach, kidney, intestine, brain, parotid gland, and submandibular gland) of the mice, and no obvious changes were observed in the tissues of the transgenic mice compared with those of the control mice, and the results were shown in Fig. 3 and Fig. S3, respectively.

Discussion

Metallothionein is a highly conserved family of closely related proteins.

Table 1. Blood biochemistry results in the transgenic and control mice at 6 wk of age.

	ALB (g/L)	GLOB (g/L)	A/G	TP (g/L)	GLU (mmol/L)	CHO (mmol/L)	TG (mmol/L)
Transgenic	18.167±1.002	24.333±0.577	0.757±0.038	42.2±1.323	5.165±0.332	2.133±0.153	1.3±0.386
Control	18.400±1.365	24.5±2.082	0.7±0.026	42.967±3.362	4.335±0.458	1.717±0.097	1.347±0.375
	Ca (mmol/L)	**Fe (mmol/L)**	**HDL (mmol/L)**	**LDL (mmol/L)**	**UA (μmol/L)**	**BUN (mmol/L)**	**CRE (μmol/L)**
Transgenic	1.58±0.115	36.833±1.501	1.513±0.101	0.463±0.148	187.5±25.03	6.6±0.917	3.145±0.518
Control	1.433±0.076	33.767±3.465	1.533±0.163	0.427±0.154	176.867±15.689	5.567±0.586	6.76±0.679
	LDH (U/L)	**AMY (U/L)**	**ALT (U/L)**	**AST (U/L)**	**ALP (U/L)**		
Transgenic	1165.867±161.645	1801.43±195.493	35.433±3.047	130.633±15.314	64.5±8.839		
Control	1173.167±97.779	1627.097±94.849	29.6±0.566	105.467±10.919	51.7±2.83		

The differences between the transgenic (n = 5) and control mice (n = 5) were not significant (P > 0.05) for all examined serum biochemical parameters.
ALB: albumin; GLOB: globulin; A/G: ALB/GLOB; TP: total protein; GLU: glucose; CHO: cholesterol; TG: triglyceride; Ca: calcium ion; Fe: ferrum ion; HDL: high density lipoprotein; LDL: low-lipid lipoprotein; UA: uric acid; BUN: blood urea nitrogen; CRE: creatinine; LDH: lactate dehydrogenase; AMY: amylase; ALT: alanine aminotransferase; AST: aspartate aminotrasferase; ALP: alkaline phosphatase.

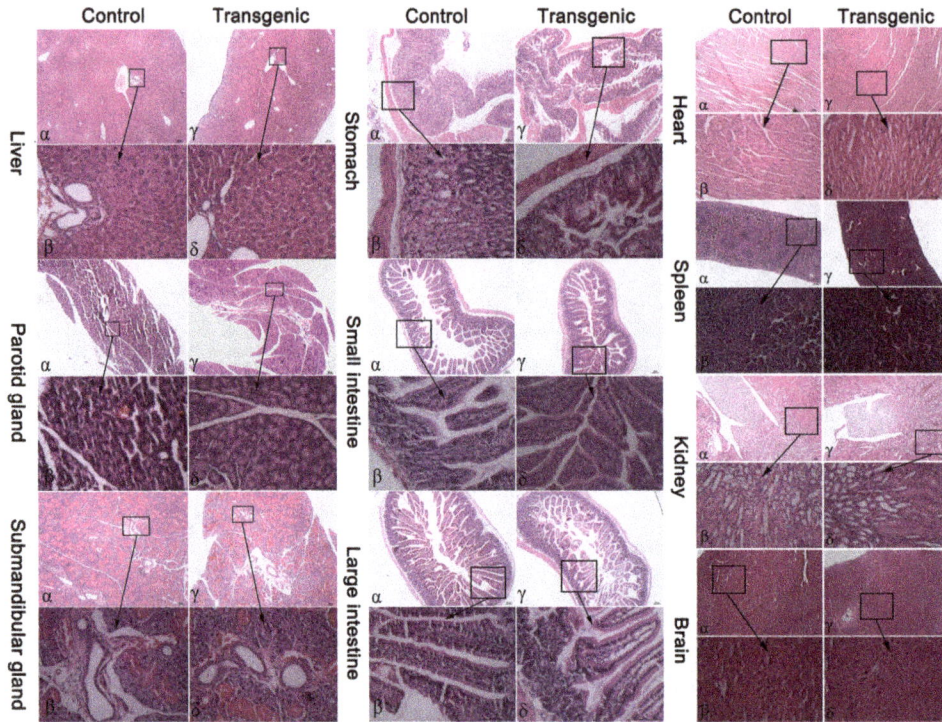

Figure 3. Histological analysis of the tissues of the transgenic and control mice. The heart, liver, spleen, stomach, kidney, small intestine, large intestine, brain, parotid gland, and submandibular gland tissue samples from the transgenic mice (transgenic; n = 5) and control mice (n = 5) at 6 wk of age were analyzed by histology observation. In the above pictures, α and γ are whole tissues, and β and δ are amplified regions of the tissues. The length of the scale bar is 100 μm in all micrographs. The profiles of the tissues of the transgenic and control mice were determined. No obvious changes were observed in the tissues of the transgenic mice compared with those of the control mice.

Yeast CUP1 is a member of the MT family and accounts for copper-binding in *S. cerevisiae* [31]. Mammals have the *MT* gene, and it possesses multiple isoforms [32]. In this study, we took advantage of yeast *CUP1* for transgene as it shares functional sequence identity to mammalian *MT*. Thus, yeast CUP1 binds to copper through Cys residues by the formation of metal-thiolate linkages, as well as the mammalian MT proteins [33], suggesting that this gene may be effective in copper-binding in mammals. This gene has been used for transgene in some organisms. Yeast *CUP1* was introduced to tobacco plants, and its expression contributed to copper content because of its role in copper-binding [32]. In *Drosophila*, this gene was selected as the transgene instead of the endogenous *Mtn* gene, which has a similar structure and function with yeast *CUP1*, to determine its role in binding copper [34].

Many researchers have used the promoter of the mouse salivary gland-specific *PSP* gene to express the exogenous genes such as *phytase* gene in the saliva of transgenic mice, which has been confirmed to be feasible [25,35,36]. A similar phenomenon was examined in our results, and constitutive expression in the PSP/CUP1 mice was notably specific for the parotid and submandibular glands. A three-fold higher expression was detected in submandibular glands compared with the parotid glands, and a similar phenomenon was detected by Mikkelsen et al. (1992) [37], and the opposite observation was reported by Golavan et al. (2001) [36].

Copper as a feed additive is effective in growth enhancement and disease prevention in weanling pigs, and it is widely used in pork production around the world, especially in China [38,39]. However, copper in pig diets heavily exceeds the minimum requirements for normal performance (5–25 mg/kg copper for different classes of pigs) [40], and most of the indigested copper, acting as promotants, by pigs is excreted in the manure (>90%) [41]. The concentrations of copper in pig manure are 5–12 times those in pig feeds with additives [42], which are higher than for other agricultural animals, such as cattle and sheep [43]. The application of pig manure directly onto agricultural land as fertilizer is common practice in China [39]. Because of copper's low mobility and non-degradation, copper can accumulate in soils [44], which leads to environmental consequences. When manure is repeatedly applied as fertilizer, copper can cause surface pollution with severe biological consequences, e.g., causing toxicity to plants, elevating bacterial resistance to toxic metals and increasing human exposure to copper via the food chain [39,40,45]. Still, the present inputs of copper are too high and reducing the contents of copper in the diet should reduce concentrations in the pig manure [46]. In the present study, we validated a method to increase copper utilization efficiency and to decrease the fecal copper content by providing the animals with an exogenous gene for generation of copper chelatin in the saliva. We determined that this approach can reduce mouse fecal copper content by 21.61%. Therefore, this might provide an important clue for preventing the pollution caused by the fecal copper in the pig production.

In addition, we reduced the dietary supply of copper with a 10 mg/kg concentration and demonstrated its feasibility for decreasing fecal copper content. Similar results were observed by Jondreville et al. (2003) [47]. The *CUP1* transgene mice, at a dietary level of 10 mg/kg copper, displayed a body weight-increase response, with the transgene mice 12.97% heavier compared with the control mice after 2 wk. These results are similar to those obtained when animals were fed 250 mg/kg of dietary copper [48]. The decreased fecal copper contents and increased body-weight increases caused by the transgene yeast CUP1 suggest that the *CUP1* transgene most likely enhanced the usage of copper in the diet. Copper is able to stimulate the secretion of several neuropeptides and growth hormones [49], in addition to being a component of the growth factor Iamin [48]. Therefore, copper could influence the growth regulatory system in many ways and might be the main reason for the growth stimulation.

No significant difference was found in a range of markers and the histology of tissues of the transgenic mice compared with the controls. These results suggest that the *CUP1* transgene did not affect the blood composition and histology of the mice. The transgenic mice were confirmed to be in good health and did not exhibit any gross pathological abnormalities or illness.

In summary, we have demonstrated that the repertoire of copper-chelating proteins produced by a model animal can be modified by introduction of *CUP1* transgene into its genome. The salivary copper-chelating proteins in these mice lead to a significant reduction of fecal copper levels and a significant increase of body weight, suggesting the enhancement of the utilization efficiency of the dietary copper by transgenic mice. Our findings provide the essential data toward elucidating the physiological functions of *MT* gene on copper metabolism.

Supporting Information

Figure S1   Construction of the recombinant plasmids expressing *CUP1* gene and confirmation by PCR and restriction. (A) The recombinant plasmid pPSP-CUP1 was constructed by insertion of the fragment containing the ORF of the *CUP1* gene into the same endonuclease-digested pPSP vector.

(B) The recombinant plasmid was confirmed by restriction analysis, DNA sequencing, and by PCR.

Figure S2   RT-PCR analysis of yeast *CUP1* transgene expression. (A, B) The *CUP1* transgene mRNA expression was analyzed by RT-PCR in the salivary glands of the transgenic founders and the control mice, respectively.

Figure S3   Histological analysis of the tissues of the transgenic and control mice at 1 yr of age. The heart, liver, spleen, stomach, kidney, small intestine, large intestine, brain, parotid gland, and submandibular gland tissue samples from the transgenic mice (transgenic; n = 10) and control mice (n = 10) at 1 yr of age were analyzed by histology observation. In the above pictures, α and γ are whole tissues, and β and δ are amplified regions of the tissues. The length of the scale bar is 100 μm in all micrographs. The profiles of the tissues of the transgenic and control mice were determined. No obvious changes were observed in the tissues of the transgenic mice compared with those of the control mice.

Table S1   Generation of G1 transgenic mice.

Table S2   Blood biochemistry results in the transgenic and control mice at 1 yr of age.

Author Contributions

Conceived and designed the experiments: YP XX. Performed the experiments: XX YM ZC RL. Analyzed the data: QW XX XZ. Contributed reagents/materials/analysis tools: QW XX. Wrote the paper: XX.

References

1. Pena MM, Lee J, Thiele DJ (1999) A delicate balance: homeostatic control of copper uptake and distribution. J Nutr 129: 1251–1260.
2. Shils ME, Shike M (2006) Modern nutrition in health and disease: Lippincott Williams & Wilkins.
3. Gambling L, Kennedy C, McArdle HJ (2011) Iron and copper in fetal development. Semin Cell Dev Biol 22: 637–644.
4. Gaetke LM, Chow CK (2003) Copper toxicity, oxidative stress, and antioxidant nutrients. Toxicology 189: 147–163.
5. Elvehjem C, Hart E (1929) The relation of iron and copper to hemoglobin synthesis in the chick. J Biol Chem 84: 131–141.
6. Elvehjem C, Hart E (1932) The necessity of copper as a supplement to iron for hemoglobin formation in the pig. J Biol Chem 95: 363–370.
7. Elvehjem C, Duckles D, Mendenhall DR (1937) Iron versus iron and copper in the treatment of anemia in infants. Am J Dis Child 53: 785–793.
8. Harris ED (2003) Basic and clinical aspects of copper. Crit Rev Clin Lab Sci 40: 547–586.
9. Sandstead HH (1995) Requirements and toxicity of essential trace elements, illustrated by zinc and copper. Am J Clin Nutr 61: 621S–624S.
10. Turnlund JR, Scott KC, Peiffer GL, Jang AM, Keyes WR, et al. (1997) Copper status of young men consuming a low-copper diet. Am J Clin Nutr 65: 72–78.
11. Fitzgerald DJ (1998) Safety guidelines for copper in water. Am J Clin Nutr 67: 1098S–1102S.
12. Potrykus J, Ballou ER, Childers DS, Brown AJ (2014) Conflicting Interests in the Pathogen–Host Tug of War: Fungal Micronutrient Scavenging Versus Mammalian Nutritional Immunity. PLoS Pathog 10: e1003910.
13. Jacob RA, Skala JH, Omaye ST, Turnlund JR (1987) Effect of varying ascorbic acid intakes on copper absorption and ceruloplasmin levels of young men. J Nutr 117: 2109–2115.
14. Coffey R, Cromwell G, Monegue H (1994) Efficacy of a copper-lysine complex as a growth promotant for weanling pigs. J Anim Sci 72: 2880–2886.
15. Bunch R, McCall J, Speer V, Hays V (1965) Copper supplementation for weanling pigs. J Anim Sci 24: 995–1000.
16. Bremner I (1987) Involvement of metallothionein in the hepatic metabolism of copper. J Anim Sci 117: 19–29.
17. Apgar G, Kornegay E, Lindemann M, Notter D (1995) Evaluation of copper sulfate and a copper lysine complex as growth promoters for weanling swine. J Anim Sci 73: 2640–2646.
18. Stansbury W, Tribble L, Orr D (1990) Effect of chelated copper sources on performance of nursery and growing pigs. J Anim Sci 68: 1318–1322.
19. Armstrong T, Cook D, Ward M, Williams C, Spears J (2004) Effect of dietary copper source (cupric citrate and cupric sulfate) and concentration on growth performance and fecal copper excretion in weanling pigs. J Anim Sci 82: 1234–1240.
20. Fogel S, Welch JW, Cathala G, Karin M (1983) Gene amplification in yeast: CUP1 copy number regulates copper resistance. Curr Genet 7: 347–355.
21. Karin M, Najarian R, Haslinger A, Valenzuela P, Welch J, et al. (1984) Primary structure and transcription of an amplified genetic locus: the CUP1 locus of yeast. Proc Natl Acad Sci U S A 81: 337–341.
22. Kägi JH, Vallee BL (1960) Metallothionein: a cadmium-and zinc-containing protein from equine renal cortex. J Biol Chem 235: 3460–3465.
23. Ecker DJ, Butt T, Sternberg E, Neeper M, Debouck C, et al. (1986) Yeast metallothionein function in metal ion detoxification. J Biol Chem 261: 16895–16900.
24. Jensen LT, Howard WR, Strain JJ, Winge DR, Culotta VC (1996) Enhanced effectiveness of copper ion buffering by CUP1 metallothionein compared with CRS5 metallothionein in Saccharomyces cerevisiae. J Biol Chem 271: 18514–18519.
25. Yin H, Fan B, Yang B, Liu Y, Luo J, et al. (2006) Cloning of pig parotid secretory protein gene upstream promoter and the establishment of a transgenic mouse model expressing bacterial phytase for agricultural phosphorus pollution control. J Anim Sci 84: 513–519.
26. Hogan B, Costantini F, Lacy E (1986) Manipulating the mouse embryo: a laboratory manual: Cold spring harbor laboratory Cold Spring Harbor, NY.
27. Van Rijs J, Giguère V, Hurst J, Van Agthoven T, van Kessel AG, et al. (1985) Chromosomal localization of the human Thy-1 gene. Proc Natl Acad Sci U S A 82: 5832–5835.
28. Spencer RL, Kalman BA, Cotter CS, Deak T (2000) Discrimination between changes in glucocorticoid receptor expression and activation in rat brain using western blot analysis. Brain Res 868: 275–286.

29. Hu Y, Nakagawa Y, Purushotham KR, Humphreys-Beher MG (1992) Functional changes in salivary glands of autoimmune disease-prone NOD mice. Am J Physiol- Endocrinol Metab 263: E607–E614.

30. Lv J, Nie Z-K, Zhang J-L, Liu F-Y, Wang Z-Z, et al. (2013) Corn Peptides Protect Against Thioacetamide-Induced Hepatic Fibrosis in Rats. J Med Food 16: 912–919.

31. Richards MP (1989) Recent developments in trace element metabolism and function: role of metallothionein in copper and zinc metabolism. J Nutr 119: 1062–1070.

32. Thomas JC, Davies EC, Malick FK, Endreszl C, Williams CR, et al. (2003) Yeast metallothionein in transgenic tobacco promotes copper uptake from contaminated soils. Biotechnol Prog 19: 273–280.

33. Kaegi JH, Schaeffer A (1988) Biochemistry of metallothionein. Biochemistry 27: 8509–8515.

34. Meyer JL, Hoy MA, Jeyaprakash A (2006) Insertion of a yeast metallothionein gene into the model insect Drosophila melanogaster (Diptera: Drosophilidae) to assess the potential for its use in genetic improvement programs with natural enemies. Biol Control 36: 129–138.

35. Madsen HO, Hjorth JP (1985) Molecular cloning of mouse PSP mRNA. Nucleic Acids Res 13: 1–13.

36. Golovan SP, Hayes MA, Phillips JP, Forsberg CW (2001) Transgenic mice expressing bacterial phytase as a model for phosphorus pollution control. Nat Biotechnol 19: 429–433.

37. Mikkelsen TR, Brandt J, Larsen HJ, Larsen BB, Poulsen K, et al. (1992) Tissue-specific expression in the salivary glands of transgenic mice. Nucleic Acids Res 20: 2249–2255.

38. Cromwell GL, Stahly TS, Monegue HJ (1989) Effects of source and level of copper on performance and liver copper stores in weanling pigs. J Anim Sci 67: 2996–3002.

39. Xiong X, Yanxia L, Wei L, Chunye L, Wei H, et al. (2010) Copper content in animal manures and potential risk of soil copper pollution with animal manure use in agriculture. Resour Conserv Recy 54: 985–990.

40. de Lange K, Nyachoti M, Birkett S (1999) Manipulation of diets to minimize the contribution to environmental pollution. Adv Pork Prod 10: 173–186.

41. Delahaye R, Fong P, Van Eerdt M, Van der Hoek K, Olsthoorn C (2003) Emissie van zeven zware metalen naar landbouwgrond. CBS, Voorburg/Heerlen.

42. Isobe H, Sekimoto H (1999) A survey of the contents of heavy metals in blended feeds, feces and composts of swine. Jpn J Soil Sci Plant Nutr 70: 39–44.

43. Ogiyama S, Sakamoto K, Suzuki H, Ushio S, Anzai T, et al. (2005) Accumulation of zinc and copper in an arable field after animal manure application. Soil Sci Plant Nutr 51: 801–808.

44. Graber I, Hansen JF, Olesen SE, Petersen J, Ostergaard H, et al. (2005) Accumulation of copper and zinc in Danish agricultural soils in intensive pig production areas. Geografisk Tidsskrift-Danish J 105: 15.

45. Poulsen HD (1998) Zinc and copper as feed additives, growth factors or unwanted environmental factors. J Anim Feed Sci 7: 135–142.

46. Aarnink A, Verstegen M (2007) Nutrition, key factor to reduce environmental load from pig production. Livest Sci 109: 194–203.

47. Jondreville C, Revy P, Dourmad J (2003) Dietary means to better control the environmental impact of copper and zinc by pigs from weaning to slaughter. Livest Prod Sci 84: 147–156.

48. Zhou W, Kornegay ET, Lindemann MD, Swinkels JW, Welten MK, et al. (1994) Stimulation of growth by intravenous injection of copper in weanling pigs. J Anim Sci 72: 2395–2403.

49. Tsou R, Dailey R, McLanahan C, Parent A, Tindall G, et al. (1977) Luteinizing hormone releasing hormone (LHRH) levels in pituitary stalk plasma during the preovulatory gonadotropin surge of rabbits. Endocrinology 101: 534–539.

Permissions

The contributors of this book come from diverse backgrounds, making this book a truly international effort. This book will bring forth new frontiers with its revolutionizing research information and detailed analysis of the nascent developments around the world.

We would like to thank all the contributing authors for lending their expertise to make the book truly unique. They have played a crucial role in the development of this book. Without their invaluable contributions this book wouldn't have been possible. They have made vital efforts to compile up to date information on the varied aspects of this subject to make this book a valuable addition to the collection of many professionals and students.

This book was conceptualized with the vision of imparting up-to-date information and advanced data in this field. To ensure the same, a matchless editorial board was set up. Every individual on the board went through rigorous rounds of assessment to prove their worth. After which they invested a large part of their time researching and compiling the most relevant data for our readers.

The editorial board has been involved in producing this book since its inception. They have spent rigorous hours researching and exploring the diverse topics which have resulted in the successful publishing of this book. They have passed on their knowledge of decades through this book. To expedite this challenging task, the publisher supported the team at every step. A small team of assistant editors was also appointed to further simplify the editing procedure and attain best results for the readers.

Apart from the editorial board, the designing team has also invested a significant amount of their time in understanding the subject and creating the most relevant covers. They scrutinized every image to scout for the most suitable representation of the subject and create an appropriate cover for the book.

The publishing team has been an ardent support to the editorial, designing and production team. Their endless efforts to recruit the best for this project, has resulted in the accomplishment of this book. They are a veteran in the field of academics and their pool of knowledge is as vast as their experience in printing. Their expertise and guidance has proved useful at every step. Their uncompromising quality standards have made this book an exceptional effort. Their encouragement from time to time has been an inspiration for everyone.

The publisher and the editorial board hope that this book will prove to be a valuable piece of knowledge for researchers, students, practitioners and scholars across the globe.

List of Contributors

Yuriko Matsuzaki and Yukiyo Mizuguchi
Division of Gene Regulation, Institute for Advanced Medical Research, School of Medicine, Keio University, Tokyo, Japan

Haru Hosokai
Division of Material and Biological Sciences, Japan Women's University, Tokyo, Japan

Shoji Fukamachi
Laboratory of Evolutionary Genetics, Department of Chemical and Biological Sciences, Japan Women's University, Tokyo, Japan

Atsushi Shimizu
Department of Molecular Biology, School of Medicine, Keio University, Tokyo, Japan

Hideyuki Saya
Division of Gene Regulation, Institute for Advanced Medical Research, School of Medicine, Keio University, Tokyo, Japan
Japan Science and Technology Agency, Core Research for Evolutional Science and Technology, Tokyo, Japan

Shengwei Hu, Wei Ni and Chuangfu Chen
College of Animal Science and Technology, Shihezi University, Shihezi, China Key Laboratory of Agrobiotechnology, Shihezi University, Shihezi, China

Wujiafu Sai, Ha Zi, Jun Qiao, Pengyang Wang and Jinliang Sheng
College of Animal Science and Technology, Shihezi University, Shihezi, China

Yuan Yu, Yongsheng Wang, Qi Tong, Xu Liu, Feng Su, Fusheng Quan, Zekun Guo and Yong Zhang
Key Laboratory of Animal Biotechnology of the Ministry of Agriculture, College of Veterinary Medicine, Northwest A&F University, Yangling, Shaanxi, People's Republic of China

Li Jiang, Tansy Z. Li and Jeanne M. Frederick
Department of Ophthalmology and Visual Sciences, University of Utah Health Science Center, Salt Lake City, Utah, United States of America

Shannon E. Boye and William W. Hauswirth
Department of Ophthalmology, University of Florida College of Medicine, Gainesville, Florida, United States of America

Wolfgang Baehr
Department of Ophthalmology and Visual Sciences, University of Utah Health Science Center, Salt Lake City, Utah, United States of America
Department of Biology, University of Utah, Salt Lake City, Utah, United States of America
Department of Neurobiology and Anatomy, University of Utah Health Science Center, Salt Lake City Utah, United States of America

Paméla Thébault
Hospital Maisonneuve-Rosemont Research Center, Montreal, Quebec, Canada

Héloïse Frison, Gloria Giono and Marilaine Fournier
Hospital Maisonneuve-Rosemont Research Center, Montreal, Quebec, Canada
Department of Microbiology, Infectiology and Immunology, University of Montreal, Montreal, Quebec, Canada

Janet J. Bijl
Hospital Maisonneuve-Rosemont Research Center, Montreal, Quebec, Canada
Department of Medicine, University of Montreal, Montreal, Quebec, Canada

Nathalie Labrecque
Hospital Maisonneuve-Rosemont Research Center, Montreal, Quebec, Canada
Department of Microbiology, Infectiology and Immunology, University of Montreal, Montreal, Quebec, Canada
Department of Medicine, University of Montreal, Montreal, Quebec, Canada

Inja Radman, Sebastian Greiss and Jason W. Chin
Medical Research Council Laboratory of Molecular Biology, Cambridge, United Kingdom

Xianshu Bai, Aiman S. Saab, Wenhui Huang, Isolde K. Hoberg, Frank Kirchhoff and Anja Scheller
Department of Molecular Physiology, University of Saarland, Homburg, Germany

Shouqi Xie and Wei Hu
State Key Laboratory of Freshwater Ecology and Biotechnology, Institute of Hydrobiology, Chinese Academy of Sciences, Wuhan, China

Yuhe Yu
Key Laboratory of Aquatic Biodiversity and Conservation of Chinese Academy of Sciences, Institute of Hydrobiology, Chinese Academy of Sciences, Wuhan, China

Qingyun Yan
Key Laboratory of Aquatic Biodiversity and Conservation of Chinese Academy of Sciences, Institute of Hydrobiology, Chinese Academy of Sciences, Wuhan, China
State Key Laboratory of Freshwater Ecology and Biotechnology, Institute of Hydrobiology, Chinese Academy of Sciences, Wuhan, China

Zihua Hu
Center for Computational Research, New York State Center of Excellence in Bioinformatics and Life Sciences, Department of Ophthalmology, Department of Biostatistics, Department of Medicine, State University of New York at Buffalo, Buffalo, New York, United States of America

Xuemei Li
Key Laboratory of Aquatic Biodiversity and Conservation of Chinese Academy of Sciences, Institute of Hydrobiology, Chinese Academy of Sciences, Wuhan, China
Key Laboratory of Freshwater Biodiversity Conservation, Ministry of Agriculture of China, Yangtze River Fisheries Research Institute, Chinese Academy of Fishery Sciences, Wuhan, China

Shinji Mii, Takuya Kato, Masato Asai, Akiyoshi Hoshino, Masahide Takahashi
Department of Pathology, Nagoya University Graduate School of Medicine, Nagoya, Aichi, Japan

Hiroki Sakakura
Department of Pathology, Nagoya University Graduate School of Medicine, Nagoya, Aichi, Japan
Department of Oral and Maxillofacial Surgery, Nagoya University Graduate School of Medicine, Nagoya, Aichi, Japan

Yoshiki Murakumo
Department of Pathology, Nagoya University Graduate School of Medicine, Nagoya, Aichi, Japan
Department of Pathology, Kitasato University School of Medicine, Sagamihara, Kanagawa, Japan

Noriyuki Yamamoto, Minoru Ueda and Sumitaka Hagiwara
Department of Oral and Maxillofacial Surgery, Nagoya University Graduate School of Medicine, Nagoya, Aichi, Japan
Department of Pathology, Kitasato University School of Medicine, Sagamihara, Kanagawa, Japan

Sayaka Sobue and Masatoshi Ichihara
Department of Biomedical Sciences, College of Life and Health Sciences, Chubu University, Kasugai, Aichi, Japan

Huanming Yang
BGI-Shenzhen, Shenzhen, Guangdong, China

Huan Liu, Yong Li, Qiang Wei and Chunxin Liu
BGI-Shenzhen, Shenzhen, Guangdong, China
ShenZhen Engineering Laboratory for Genomics-Assisted Animal Breeding, BGI-Shenzhen, Shenzhen, Guangdong, China

Lars Bolund
BGI-Shenzhen, Shenzhen, Guangdong, China
Department of Biomedicine, University of Aarhus, Aarhus C, Denmark

Gábor Vajta
BGI-Shenzhen, Shenzhen, Guangdong, China
Central Queensland University, Rockhampton, Queensland, Australia

Hongwei Dou, Wenxian Yang, Ying Xu and Jing Luan
BGI Ark Biotechnology, BGI-Shenzhen, Shenzhen, Guangdong, China
ShenZhen Engineering Laboratory for Genomics-Assisted Animal Breeding, BGI-Shenzhen, Shenzhen, Guangdong, China

Nicklas Heine Staunstrup
BGI Ark Biotechnology, BGI-Shenzhen, Shenzhen, Guangdong, China
Department of Biomedicine, University of Aarhus, Aarhus C, Denmark

Yutao Du
BGI-Shenzhen, Shenzhen, Guangdong, China

BGI Ark Biotechnology, BGI-Shenzhen, Shenzhen, Guangdong, China
ShenZhen Engineering Laboratory for Genomics-Assisted Animal Breeding, BGI-Shenzhen, Shenzhen, Guangdong, China

Jun Wang
BGI-Shenzhen, Shenzhen, Guangdong, China
Department of Biology, University of Copenhagen, Copenhagen, Denmark King Abdulaziz University, Jeddah, Saudi Arabia

Huanxing Sun, Rayman Choo-Wing, Angara Sureshbabu and Vineet Bhandari
Department of Pediatrics, Yale University, New Haven, Connecticut, United States of America

Juan Fan, Lin Leng, Shuang Yu and Richard Bucala
Department of Medicine, Yale University, New Haven, Connecticut, United States of America

Dianhua Jiang and Paul Noble
Department of Medicine, Duke University School of Medicine, Durham, North Carolina, United States of America

Robert J. Homer
Department of Pathology, Yale University, New Haven, Connecticut, United States of America

Nathaniel Safren and Mervyn J. Monteiro
Neuroscience Graduate Program, School of Medicine, University of Maryland, Baltimore, Maryland, United States of America
Center for Biomedical Engineering and Technology, School of Medicine, University of Maryland, Baltimore, Maryland, United States of America
Department of Anatomy and Neurobiology, School of Medicine, University of Maryland, Baltimore, Maryland, United States of America

Lydia Chang
Center for Biomedical Engineering and Technology, School of Medicine, University of Maryland, Baltimore, Maryland, United States of America
Department of Anatomy and Neurobiology, School of Medicine, University of Maryland, Baltimore, Maryland, United States of America

Todd D. Gould
Neuroscience Graduate Program, School of Medicine, University of Maryland, Baltimore, Maryland, United States of America

Department of Anatomy and Neurobiology, School of Medicine, University of Maryland, Baltimore, Maryland, United States of America
Department of Psychiatry, School of Medicine University of Maryland, Baltimore, Maryland, United States of America
Department of Pharmacology, School of Medicine University of Maryland, Baltimore, Maryland, United States of America

Chantelle E. Terrillion
Neuroscience Graduate Program, School of Medicine, University of Maryland, Baltimore, Maryland, United States of America
Department of Psychiatry, School of Medicine University of Maryland, Baltimore, Maryland, United States of America

Amina El Ayadi and Darren F. Boehning
Department of Neuroscience and Cell Biology, University of Texas Medical Branch, Galveston, Texas, United States of America

Hongjie Wang, Ruizhi Wang and Madepalli K. Lakshmana
Section of Neurobiology, Torrey Pines Institute for Molecular Studies, Port Saint Lucie, Florida, United States of America

Shaohua Xu
Department of Biological Sciences, Florida Institute of Technology, Melbourne, Florida, United States of America

Daniel Harari, Renne Abramovich and Gideon Schreiber
Department of Biological Chemistry, The Weizmann Institute of Science, Rehovot, Israel

Alla Zozulya, Paul Smith and Sandrine Pouly
MS Platform, Merck-Serono, (a division of Merck KGaA), Geneva, Switzerland

Mario Köster and Hansjörg Hauser
Helmholtz Centre for Infection Research, Dept. Gene Regulation and Differentiation, Braunschweig, Germany

Saurabh Chattopadhyay, Sean P. Kessler, Juliana Almada Colucci, Michifumi Yamashita and Preenie Ganes C. Sen
Department of Molecular Genetics, Lerner Research Institute, Cleveland Clinic, Cleveland, Ohio, United States of America

deS Senanayake
Department of Ophthalmic Research, Cole Eye Institute, Cleveland Clinic, Cleveland, Ohio, United States of America

Lei Chen, Longxin Chen and Pingping Tan
State Key Laboratory of Molecular and Developmental Biology, Institute of Genetics and Developmental Biology, Chinese Academy of Sciences, Beijing, China

Peng Liu, Hongwei Dou and Lin Lin
BGI ARK Biotechnology Co., Ltd, Shenzhen, China

Gabor Vajta
BGI ARK Biotechnology Co., Ltd, Shenzhen, China
IRIS, Central Queensland University, Rockhampton, Australia

Jianfeng Gao
School of Life Sciences, Shihezi University, Shihezi, China

Yutao Du
BGI ARK Biotechnology Co., Ltd, Shenzhen, China
BGIShenzhen, Shenzhen, China

Runlin Z. Ma and Peng Zhang
State Key Laboratory of Molecular and Developmental Biology, Institute of Genetics and Developmental Biology, Chinese Academy of Sciences, Beijing, China
Graduate University of the Chinese Academy of Sciences, Beijing, China

Brett K. Baillie and Caryl E. Hill
Blood Vessel Laboratory, The John Curtin School of Medical Research, The Australian National University, Canberra, Australia

Susan K. Morton and Daniel J. Chaston
Blood Vessel Laboratory, The John Curtin School of Medical Research, The Australian National University, Canberra, Australia
Stem Cell & Gene Targeting Laboratory, The John Curtin School of Medical Research, The Australian National University, Canberra, Australia

Klaus I. Matthaei
Stem Cell & Gene Targeting Laboratory, The John Curtin School of Medical Research, The Australian National University, Canberra, Australia

Ilya Mertsalov
Institute of Gene Biology, Russian Academy of Sciences, Moscow, Russia

Nadezda Kust
Institute of Gene Biology, Russian Academy of Sciences, Moscow, Russia
Ltd Apto-pharm, Moscow, Russia

Ekaterina Savchenko
Institute of Medicine and Cell Transplantation, Moscow, Russia
Ltd Apto-pharm, Moscow, Russia

Ekaterina Rybalkina
Ltd Apto-pharm, Moscow, Russia

Alexander Revishchin and Gali Pavlova
Institute of Gene Biology, Russian Academy of Sciences, Moscow, Russia
Institute of Medicine and Cell Transplantation, Moscow, Russia
Ltd Apto-pharm, Moscow, Russia

Nathalie Daniel-Carlier, Louis-Marie Houdebine, Bruno DaSilva, Geneviève Jolivet, Erwana Harscoët and Céline Viglietta
INRA UMR1198, Biologie du Développement et Reproduction, Jouy en Josas, France

Itzik Harosh and Sandrine Braud
ObeTherapy Biotechnology, Evry, France

Thomas Gautier and Laurent Lagrost
INSERM UMR866, Université de Bourgogne, Dijon, France

Bruno Passet
INRA UMR1313, Génétique Animale et Biologie Intégrative, Jouy-en-Josas, France

Wiebke Garrels, Stephanie Holler, Ulrike Taylor, Doris Herrmann, Heiner Niemann and Wilfried A. Kues
Institut für Nutztiergenetik, Friedrich-Loeffler-Institut, Mariensee, Germany

Zoltan Ivics
Paul-Ehrlich-Institute, Langen, Germany

Yong Wang, Xiao-Yang Zhou, Peng-Ying Xiang, Lu-Lu Wang, Fei Xie, Liang Li and Hong Wei
Department of Laboratory Animal Science, College of Basic Medical Sciences, Third Military Medical University, Chongqing, China

Huan Tang
China Three Gorges Museum, Chongqing, China

Joanna Kern, Robert Drutelᴼ, Silvia Leanhart, Marek Bogacz and Rafal Pacholczyk
Center for Biotechnology and Genomic Medicine, Georgia Regents University, Augusta, Georgia, United States of America

Xiaoxian Xie, Yufang Ma, Zhenliang Chen, Rongrong Liao, Xiangzhe Zhang, Qishan Wang and Yuchun Pan
School of Agriculture and Biology, Department of Animal Sciences, Shanghai Jiao Tong University, Shanghai, PR China, Shanghai Key Laboratory of Veterinary Biotechnology, Shanghai, PR China

Index